International Advances in the Ecology, Zoogeography, and Systematics of Mayflies and Stoneflies

Edited by

F. R. Hauer, J. A. Stanford and, R. L. Newell

International Advances in the Ecology, Zoogeography, and Systematics of Mayflies and Stoneflies

Edited by

F. R. Hauer, J. A. Stanford, and R. L. Newell

University of California Press

Berkeley • Los Angeles • London

University of California Press, one of the most distinguished university presses in the United States, enriches lives around the world by advancing scholarship in the humanities, social sciences, and natural sciences. Its activities are supported by the UC Press Foundation and by philanthropic contributions from individuals and institutions. For more information, visit www.ucpress.edu.

University of California Publications in Entomology, Volume 128

University of California Press
Berkeley and Los Angeles, California

University of California Press, Ltd.
London, England

Library of Congress Cataloging-in-Publication Data

International Conference on Ephemeroptera (11th : 2004 : Flathead Lake Biological Station, The University of Montana)
 International advances in the ecology, zoogeography, and systematics of mayflies and stoneflies / edited by F.R. Hauer, J.A. Stanford, and R.L. Newell.
     p. cm. – (University of California publications in entomology ; 128)
 "Triennial Joint Meeting of the XI International Conference on Ephemeroptera and XV International Symposium on Plecoptera held August 22-29, 2004 at Flathead Lake Biological Station, The University of Montana, USA." – Pref.
 Includes bibliographical references and index.
 ISBN 978-0-520-09868-8 (pbk. : alk. paper)
 1. Mayflies–Congresses. 2. Stoneflies–Congresses. I. Hauer, F. Richard. II. Stanford, Jack Arthur, 1947– III. Newell, Robert L. IV. International Symposium on Plecoptera (15th : 2004 : Flathead Lake Biological Station, The University of Montana) V. Title.
 QL505.I56 2004
 595.7'34–dc22
                         2008017087

Contents

Preface

We had the honor of hosting the triennial Joint Meeting of the XI International Conference on Ephemeroptera and XV International Symposium on Plecoptera held August 22–29, 2004 at Flathead Lake Biological Station, The University of Montana, USA.

The purpose of this congress was to encourage and facilitate focused research and provide a forum for scholarly exchange about the status of Mayfly and Stonefly science. A secondary, but important objective was engendering camaraderie amongst investigators with mutual interests and in the course of informal activities surrounding the formal agenda, discuss research directions—evaluating and outlining fundamental research problems and sharing information on methodology and analysis.

Scientists and scholars from around the world gathered and engaged in an exchange of information on their latest research and investigations on Ephemeroptera (Mayfly) and Plecoptera (Stonefly). At this congress, attendees came from 18 countries. The host country had a good turnout and despite the remote location within the Rocky Mountains, we were joined by scientists and students from Argentina, Australia, Austria, Brazil (2), Czech Republic, France, Germany (3), Italy (2), Japan (5), Korea (2), Mongolia (2), New Zealand, Norway (2), P. R. China, Russia, Slovenia, South Africa, Switzerland and USA (26). Drs. J. A. Stanford and F. R. Hauer convened the meetings and also participated along with other members of the Flathead Lake Biological Station faculty and staff.

Professor John Brittain, whose research is focused on freshwater entomology, especially egg development and life cycle strategies of Ephemeroptera and Plecoptera, presented *Mayflies in a Changing World,* reflecting on the quality of mayflies as good indicators of global warming and the quality of streams and lakes. Professor Emeritus Andrew Sheldon, whose interests have encompassed community and population ecology of aquatic animals over a span of more than 40 years, especially insects and fishes, gave a talk about *Scale, Hierarchy and Perspectives in the Ecology of Plecoptera*, discussing how studies emphasizing scale and perspective reveal importance of stoneflies to ecosystems. We sincerely thank Drs. Brittain and Sheldon for their plenary participation.

Other abstracts covered a broad base of disciplines including morphology, physiology, phylogeny, taxonomy, ecology and conservation. Manuscripts submitted post meeting were peer-reviewed and compiled into three sections for this volume: Ecology, Zoogeography and Systematics.

We wish to thank all those that made the Congress possible including Sponsors, fellowship providers, and Flathead Lake Biological Station faculty and staff. Meeting sponsors were Flathead Lake Biological Station, The University of Montana, Jessie M. Bierman Professorship of Ecology and the FLBS Limnology Professorship. The Permanent Committee on Ephemeroptera Conferences and Flathead Lake Biological Station provided partial financial support for students.

We greatly appreciate the efforts of Flathead Lake Biological staff especially Sue Gillespie and Mark Potter for administrative and facilities operations support and Jeremy Nigon for the registration website and technical support. A very special thanks goes to Marie Kohler for her tireless effort and skills in the publishing layout and proofreading of these proceedings for publication.

The conveners also thank the participants who journeyed to the Biological Station and joined us for a short while in northwest Montana in the pursuit of Mayfly-Stonefly science. We also thank Joe Giersch for his design of the meeting logo.

F. Richard Hauer
Jack A. Stanford
Robert L. Newell

2004 Joint Meeting

Flathead Lake
Biological **Station**
The University of Montana

Proceeding of the 2004 International Joint Meeting
XI International Conference on Ephemeroptera
XV International Symposium on Plecoptera
22–29 August 2004

MAYFLIES, BIODIVERSITY AND CLIMATE CHANGE

John E. Brittain

*Natural History Museum, University of Oslo, P.O. Box 1172,
Blindern, NO-0318 Oslo, Norway
Norwegian Water Resources and Energy Directorate, P.O. Box 5091,
Majorstua, NO-0301 Oslo, Norway*

Abstract

Mayflies (Ephemeroptera) are an ancient order of insects that are globally distributed in both northern and southern hemispheres and have survived major environmental shifts. Despite the problems associated with selection processes operating in both terrestrial and aquatic environments, mayflies have successfully colonized a wide range of freshwater habitats from the tropics to the arctic, a somewhat greater range than other hemimetabolic aquatic insects such as the Plecoptera and Odonata. While many species of Ephemeroptera require specific environmental cues, others display considerable flexibility in life cycle length and timing in relation to environmental changes. This is particularly apparent in arctic and alpine species. Climate change scenarios predict rapid shifts across many environmental gradients, including temperature and the frequency and magnitude of floods and droughts. Changes in the mayfly fauna are hypothesized in the light of the environmental tolerances, life cycle plasticity and the dispersal mechanisms of present day mayflies. During periods of rapid environmental transition certain species traits will be beneficial. Generalists will do better; specialists with strict environmental limits and poor powers of dispersal may become extinct.

Key words: mayflies; Ephemeroptera; climate change; life cycles; alpine; streams; lakes; temperature; ice cover; floods; dispersal; species traits.

Introduction

Mayflies (Ephemeroptera) are an ancient order of insects, first appearing in the fossil record in the Upper Carboniferous in excess of 250 million years ago (Brittain 1980). Mayflies have a complex life cycle, involving both aquatic and terrestrial phases. Such life cycles create evolutionary dichotomy with selection pressures operating in two, more or less, independent environments (Wilbur 1980). In theory, this dichotomy will lead to the reduction of one of these phases. This is clearly seen in the extremely short-lived adult stages of the Ephemeroptera, whose sole, but crucial roles are reproduction and dispersal.

The Ephemeroptera are a small order of insects, numbering about 3000 described species within 37 families (Brittain and Sartori 2003). Despite the problems associated with selection processes operating in both terrestrial and aquatic environments, mayflies have survived many climatic shifts and have successfully colonized a wide range of freshwater habitats from the tropics to the Arctic and from large rivers to small ponds. In comparison with the stoneflies (Plecoptera), they have made a greater intrusion into the tropics, both in terms of diversity and abundance, at the same time as they are more abundant and diverse than the dragonflies (Odonata) in Arctic and alpine regions (Brittain 1990, Fig. 1). In this paper, future trends in the mayfly fauna as a result of climate change are hypothesized in the light of the environmental tolerances, life cycle plasticity and the dispersal mechanisms of present day mayflies.

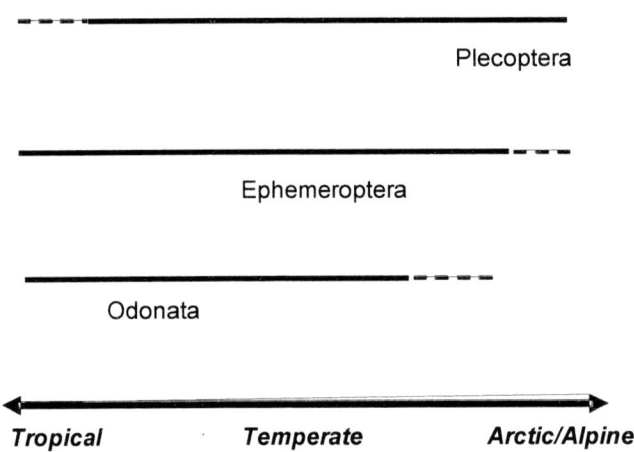

Figure 1. Optimal and suboptimal geographic and climatic regions for the aquatic insects orders, Plecoptera, Ephemeroptera and Odonata.

Climate Change

Climate has changed many times over the course of the Earth's history. Over the past two million years there have been numerous Ice Ages, but at present we are in a warm interglacial period. The geological record reveals that past climate change has not always been gradual, rather there have been many changes that were very rapid; over periods as short as centuries or decades (Bonan 2002). Beginning around 1550 A.D. and lasting until 1850 was a period of cold temperatures, known as the "Little Ice Age" (Lamb 1995). Temperatures warmed by over 0.5°C between the mid 1800s and the 1940s, but cooled again over the next 25 years. Since then temperatures have been rising and current predictions indicate increases of annual mean air temperatures as much as 8°C by the end of the century in some regions (IPCC 2001).

These changes are predicted to be greater in arctic and alpine regions and thus emphasis in this paper is given to aquatic ecosystems in these areas.

Response to Temperate Changes

Aquatic organisms and ecosystems respond to climate change in a variety of different ways. There are four main consequences of climate change at the species or population level: simple adaptation, demographic change, emigration/immigration and extinction. Adaptation can involve inherent plasticity in response to environmental change or be accomplished by longer-term evolutionary change. Demographic effects include changes in recruitment, mortality and growth, while emigration and immigration will produce changes in species' distributions. These processes, including extinction, can give rise to changes, both in ecosystem structure and function. At the ecosystem level, the primary effects of temperature and runoff changes will give rise to a number of secondary effects, affecting physico-chemical aspects such as length of the ice-free season, decomposition, weathering and water residence time, as well as affecting levels of primary and secondary production (Hauer et al. 1997, Rouse et al. 1997).

Water temperature steers many of the physiological processes in aquatic insects and all life cycle stages respond to changes in temperature (e.g., Brittain 1980, Ward and Stanford 1982, Sweeney 1984). This is particularly apparent in the egg stage, and in many mayflies there is a clear relationship between water temperature and the length of egg development, as well as distinct temperature limits for successful development (Elliott and Humpesch 1980). The effect of such limitations on distribution can be illustrated in the Australian mayfly genus, *Coloburiscoides*. This genus is widespread in the higher regions of mainland southeastern Australia, but is absent from apparently suitable habitats in Tasmania. Laboratory studies demonstrated that despite its assumed adaptation to cooler environments, the eggs only hatch successfully between 15 and 25°C (Brittain and Campbell 1991). It is, therefore, likely that this genus became extinct in Tasmania during a recent glacial period as water temperatures were too low for successful egg development.

Adult body size in many hemimetabolous insects depends largely on thermal conditions during development, while adult size in mayflies is closely correlated with fecundity (Brittain 1980, Elliott and Humpesch 1980). It therefore follows that smaller adults and lowered fecundity result when temperatures are suboptimal (Sweeney and Vannote 1978). Thus, a species' distribution both locally and regionally is limited, at least in part, by lowered fecundity as adult size gradually diminishes in streams that are either too warm or too cold. Such relationships will lead to changed mayfly distributions under global warming.

Many mayfly species, especially among the Baetidae, display considerable life cycle plasticity, being able to change the number of generations per year in response to changes in temperature. For example, the widespread western Palaearctic species, *Baetis rhodani*, that has several generations a year in warmer lowland habitats, is univoltine in cooler streams and even displays a two-year semivoltine life cycle in alpine areas (Sand 1997). In a warmer climate, such species will be able to shorten generation time and increase the number of generations per year.

The timing of mayfly emergence also frequently depends on temperature (Brittain 1980) and earlier emergence is likely in many mayfly species in a warmer climate. Relatively small shifts in temperature in warmer years or at lower altitudes have been shown to cause significant changes in the timing of the onset of mayfly emergence (Brittain 1976, 1978, 1980), although not always (Langford 1975). In laboratory studies, emergence has been advanced by several months by artificially increasing temperatures (Nebeker 1971). A shift in emergence as a result of rising temperatures has already been reported by fly fishermen who are especially focused on the "mayfly hatch" in an English river (The European Times 1992).

In arctic and alpine areas, the balance of water sources between glacial or kryal, spring-fed or krenal and snowmelt/precipitation or rhithral is crucial in determining biological communities (Ward 1994, Füreder 1999). This interplay of these different water sources operates via two main pathways, fluvial processes and water quality (Fig. 2). The important role of water source was clearly demonstrated in a study of the benthic macroinvertebrate fauna of two contrasting alpine catchments, the one glacier-fed and the other dominated by groundwater inputs (Füreder et al. 2003). Mayflies were much more abundant in the nonglacial system, although the wider food niche of species inhabiting glacial rivers was much broader than in the spring-fed system, demonstrating the adaptive nature of certain mayfly species faced by paucity of food resources.

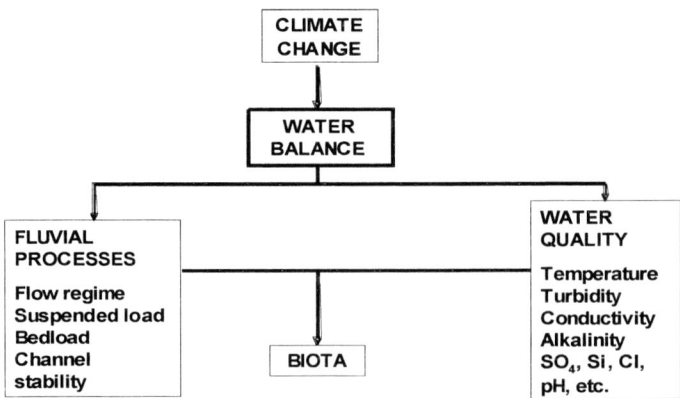

Figure 2. The conceptual relationship between stream biota and climate change in Arctic and Alpine ecosystems.

Climate change, by altering the water balance in arctic and alpine regions, will cause changes in mayfly species' distribution and abundance. In general, an increase in diversity of mayflies in temperate and arctic regions, as well as expansion further up into alpine areas is to be expected as a result of global warming (Lillehammer and Brittain 1978, Hauer et al. 1997). However, this could be confounded by the upward and northward movement of the tree line and a concomitant decrease in primary production as a result of increased shading. This may negatively affect the mayflies, which are predominantly grazers. In addition, there will be the increase, at least in the short-term, of cold adapted species in watercourses affected by glacial runoff (Melack et al. 1997). Increased discharge in glacier-fed watercourses as a result of warmer summers will cause reduction in water temperatures, increased turbidity and greater channel instability, which in turn will result in changes in benthic composition and abundance (McGregor et al. 1995, Milner et al. 2001a). Among the changes predicted downstream is an expansion of cold-adapted species such as the Diamesinae at the expense of other invertebrate taxa such as the mayflies. Primary production would also be reduced, again making the environment less favourable for most mayflies. However, in the southern hemisphere leptophlebiid mayflies occur close to glacial margins and such species are unlikely to be adversely affected by increased glacial runoff (Milner et al. 2001a). These include the mayfly genera *Deleatidium* in New Zealand (Milner et al. 2001b) and *Meridialaris* in Patagonia (Wais and Bonetto 1988).

There may be other unexpected effects of global warming on the balance of water sources. In the cold regions of the world, permafrost limits the availability of groundwater to lakes and rivers. However, under polythermal glaciers, temperatures at parts of the glacier base along with a steep hydraulic gradient may permit groundwater intrusion. However, the retreat of the front position of many glaciers as a result of global warming can expose new areas to permafrost, thereby cutting off groundwater sources (Haldorsen et al. 2002). In fact, under a rapidly changing climate regime, such as predicted for future anthropogenic global warming, groundwater influxes will be crucial in maintaining spatial and temporal heterogeneity in freshwater systems (Malard et al. 1999, Ward and Uehlinger 2003) and will provide refugia for mayfly species affected by climate change and other human impacts. Groundwater influenced systems are characterised by stability in environmental characteristics and will in general be less affected by changing air temperatures, although changes in runoff will undoubtedly affect water table levels.

Response to Other Effects of Climate Change

Climate change scenarios predict rapid shifts in environmental conditions, not only air temperatures, but also in precipitation and runoff. This will change the frequency and magnitude of floods and droughts. Precipitation will also affect the extent and duration of ice cover in cold regions.

Air pressures in the Atlantic Ocean affect ocean currents, temperature and precipitation in the North Atlantic, northern Europe and even further a field (Straile et al. 2003). A high winter North Atlantic Oscillation (NAO) index, based on the difference in sea-level air pressure measured close to the centre of the Azores High and that measured at a station in Iceland, results in mild, moist air currents over northern Europe. An increase in the NAO index is predicted in climate change scenarios and is already taking place, giving a deeper snow pack in the mountains of western Norway. A deeper snow pack delays ice break resulting in a shorter ice-free season in mountain lakes (Borgstrøm 2001). In the lakes of southern Norway, the number of mayfly species has been shown to be highly correlated with the length of the ice-free season (Brittain 1974). A shorter ice-free season is likely to give rise to lower summer water temperatures and lower primary production, both disadvantageous for mayfly diversity, while a longer ice-free season will have the opposite effect. It is predicted that in mountains areas, especially in the high precipitation areas of western Norway, the winter snowpack and the length of the period of ice cover will increase. However, in lowland areas the extent of ice cover will decrease and winter floods will become more frequent. The absence of ice cover during winter permits some degree of primary production and it has been shown that mayfly densities and growth rates can be significantly higher in ice-free sections of alpine rivers (Schütz et al. 2001).

Floods have a major structuring effect on aquatic communities, triggering species replacement and succession. Certain climate scenarios predict an increase in the frequency and magnitude of floods. Floods, especially those with a long return time, can have a catastrophic effect on mayfly communities (Pupilli and Puig 2003). In less catastrophic floods, there may be a shift in dominance in benthic communities as a result of high discharge (e.g., Anderson and Lehmkuhl 1968, Fjellheim and Raddum 1993). In alpine streams in central southern Norway, an unusually high spring flood significantly decreased densities of the mayfly, *Baetis rhodani*, which was present as nymphs at that time. In contrast, densities of the summer species, *Acentrella lapponica*, present in the egg stage down in the substrate, increased in the two years after the flood (Sand and Brittain 1997). Floods are also important in terms of dispersal, opening up waterways and establishing connectivity between watercourse elements and permitting the establishment of new populations.

While runoff will increase in some regions as a result of climate change, it will decrease in others. Droughts and low flows will also impact mayfly communities. In rivers regulated for hydropower, flows are frequently reduced for all or part of the year. Such flow reductions favor genera characteristic of lentic habitats, such as *Cloeon, Paraleptophlebia* and *Siphlonurus* at the expense of the typical lotic genera such as *Baetis, Rhithrogena* and *Epeorus* (Brittain and Saltveit 1989). A similar shift may take place as a result of climate change.

Dispersal and Gene Flow

In a changing environment brought on by climate change, dispersal ability will be crucial to species survival. If a species is unable to adapt physiologically to a changed climate regime in its present habitat, then it must be able to reach a suitable environment by dispersal. If not, extinction will result and it has recently been predicted that on the basis of climate scenarios for 2050, that between 18% and 35% of all plant and animal species on the Earth are committed to extinction (Thomas et al. 2004).

Mayflies, particularly Baetidae, often dominate stream drift (Brittain and Eikeland 1988), providing an efficient mechanism for downstream dispersal of lotic species. However, upstream movement and colonization of adjacent watercourses and lakes occurs primarily through adult flight (e.g., Peckarsky et al. 2000).

Mayflies have winged adults, but they are small, fragile and short-lived (Brittain 1980). This limits their powers of dispersal. Mountain ranges, deserts and oceans present major barriers to dispersal. Climate change can also have impacts on the mayfly fauna of islands, especially those far away from continental landmasses. In this respect, the time frame and the rapidity of climatic changes are critical. The mayfly fauna of Madagascar displays a high degree of endemism, although showing strong affinities with the African fauna, while Oriental and Oceanian elements are negligible (Gattolliat and Sartori 2003). This contradicts with the separation sequence of Gondwanaland and suggests the dispersal powers of the Baetidae are greatly underestimated. Three mayfly families, Leptophlebiidae, Baetidae and Prosopistomatidae are present on the Comoros Islands, young volcanic islands at least 300 km from Madagascar and Africa (Starmühlner 1979). In the Arctic, there is already evidence of the migration of insects over large distances by the help of favorable winds generated by a changing climate (Coulson et al. 2002).

Colonization of new habitats by mayflies can take place fairly rapidly, although this depends on a supply of colonizers and environmental conditions being suitable. After the commencement of liming of a Norwegian river affected by acidification, the mayfly *Baetis rhodani* returned within two years (Raddum and Fjellheim 2003). In Canada, the mayflies *Stenonema femoratum* and *Stenacron interpuctatum* recolonized acid-damaged lakes from nearby refugia less than 4–8 years after pH reached threshold levels (Snucins 2003). However, initial small populations are very vulnerable to stochastic and Alle effects, such as low encounter rates for fertile adults (Frank and Brickman 2000, Yan et al. 2003). In an Alaskan glacial river, created in the 1960s as a result of glacial retreat, it was not until the mid 1980s that Baetidae were first recorded (Milner 1994). However, this delay was more a result of low water temperatures than the species' inability to colonize from nearby streams. Apart from the abiotic factors such as temperature affecting the ability of a species to colonize new habitats, species interactions may also be important, and the debate on the relative importance of biotic versus abiotic factors in determining the distribution and abundance of animals continues in the context of global warming (Hodkinson 1999).

Although it is possible to mark insects using stable isotopes for example (Hershey et al. 1993), the degree of dispersal can also be estimated indirectly by studying the genetic structure of a population. If dispersal is high, little genetic variation is to be expected, and vice versa (Slatkin 1985). However, Bunn and Hughes (1997) in a study of a subtropical stream in northern Australia found that genetic variation in *Baetis* was low at larger spatial scales indicating that there was considerable dispersal by winged adults. However, at the reach scale genetic variation was high, suggesting that recruitment was the result of oviposition by only a few females, a conclusion supported by work on populations of *Baetis alpinus* (Monaghan et al. 2001, 2002). The presence of greater genetic diversity at large spatial scales may be evidence of nonequilibrium between gene flow and genetic drift, resulting from historical gene flow that continues to mask reduced dispersal (Monaghan et al. 2002). In contrast, *Rhithrogena loyolaea* exhibited little genetic differentiation within and among streams but significant differentiation among drainages, suggesting that dispersal occurs readily among stream reaches and between adjacent valleys and that equilibrium has been reached between gene flow and genetic drift (Monaghan et al. 2002). Clearly there are differences in the dispersal ability of mayfly taxa and under climate change those species with greater powers of dispersal will have a better chance of survival.

As a result of their manipulation study of the effects of global warming on a first-order stream, Hogg and Williams (1996) concluded that the level of gene flow among habitats may be critical to the degree of impact as a result of global warming. Fragmentation, both natural and anthropogenic, represents a potential barrier to gene flow and dispersal (Zwick 1992). However, Sweeney et al. (1986) observed no genetic differentiation between populations of two mayflies, *Ephemerella subvaria* and *Euryophella verisimilis*, above and below reservoirs of the Delaware River, USA. In a study of populations of *Baetis alpinus* in alpine streams, where populations were fragmented either by natural lakes or by reservoirs, Monaghan et al. (2001) concluded that lentic water bodies act as barriers to gene flow in *B. alpinus*, but that the low divergence between fragments separated by reservoirs did not indicate high levels of gene flow, but rather showed that genetic differentiation is not detectable within the first 100–1000 years of habitat fragmentation.

Species Traits

During periods of rapid environmental transition certain species traits will be beneficial (Brittain 1991, Table 1). Mayfly eggs are frequently located down in the substrate where they are protected at least in part from environmental vicissitudes. A relatively long egg development period will increase the chance of the species' survival. This was clearly seen in a Norwegian alpine area in connection with an extreme spring flood. The summer species that were still in the egg stage survived and in fact increased in numbers at the expense of species that were present as nymphs during the spring (Sand and Brittain 1997). The "hedging of bets" associated with asynchronous egg hatching as well as variation in nymphal growth rates is also clearly advantageous in a changing environment (e.g., Bretschko 1990). A smaller size and a more cylindrical body shape enable nymphs to move down into the substrate during periods of high discharge and other unfavourable environmental conditions.

Table 1. Aquatic insect life history attributes generally advantageous or disadvantageous in disturbed habitats or in rapidly changing environments (modified from Brittain 1991).

Species Trait	Advantageous	Disadvantageous
Egg development	Long	Short
Egg hatching	Asynchronous	Synchronous
Nymphal development	Asynchronous	Synchronous
Nymphal size and shape	Small and cylindrical	Large
Temperature relationships	Eurytherm; temperature independent	Stenotherm; temperature dependent
Life cycle	Flexible; multivoltine	Fixed; univoltine

Strict temperature limits or specific temperature cues will be disadvantageous, as will growth relationships that show a strong dependence on temperature. The ability to shift life cycle duration, for example from univoltine to multivoltine in a warmer climatic regime will increase a species' annual production, while species with a fixed voltinism are likely to be disadvantaged or even become extinct. Although hatching success may be lower, parthenogenesis, both facultative and obligatory, is advantageous for dispersal of mayflies to new and more suitable habitats as it does not require mating and fertilization before oviposition (Sweeney and Vannote 1987). As a rule, it is envisaged that generalists will do better in a rapidly changing environment, while specialists with strict environmental limits and poor powers of dispersal may face extinction.

Temporal and Spatial Differences

There will be both temporal and spatial differences in the effects of climate change. Short-term and long-term changes are likely to differ as secondary and tertiary effects come into play. There will also be major regional differences, depending on the extent and nature of climatic changes (see Cushing 1997). For example, temperature increases are predicted to be greater in northern and Arctic regions and less in the southern hemisphere. In some areas precipitation will increase while in others it will remain stable or decrease. The effects of increased precipitation will also differ in mountain areas compared to lowlands. However, the accuracy of climate change predictions becomes lower when downscaled to smaller regional and national scales (Hauer et al. 1997). Nevertheless, the small order of aquatic insects, the mayflies, by virtue of their worldwide distribution and response to environmental cues, have the potential to function as sensitive indicators of present and future climate change.

Acknowledgments

Svein Jakob Saltveit, University of Oslo, as well as the participants of the Mayfly-Stonefly Conference, provided constructive ideas and comments.

Literature Cited

Anderson, N. H., and D. M. Lehmkuhl 1968. Catastrophic drift of insects in a woodland stream. Ecology **49**:198–206.

Bonan, G. 2002. Ecological Climatology: Concepts and Applications. Cambridge University Press, New York, USA.

Borgstrøm, R. 2001. Relationship between spring snow depth and growth of brown trout, *Salmo trutta*, in an alpine lake: predicting consequences of climate change. Arctic, Antarctic and Alpine Research **33**:476–480.

Bretschko, G. 1990. A flexible larval development strategy in *Siphlonurus aestivalis* Eaton exploiting an unstable biotope. Pages 17–25 *in* I. C. Campbell, editor. Mayflies and Stoneflies: Life History and Biology. Kluwer Academic Publishers, Dordrecht, The Netherlands.

Brittain, J. E. 1974. Studies on the lentic Ephemeroptera and Plecoptera of southern Norway. Norsk entomologisk tidsskrift **21**:135–154.

Brittain, J. E. 1975. The temperature of two Welsh lakes and its effect on the distribution of two freshwater insects. Hydrobiologia **48**:37–49.

Brittain, J. E. 1979. Emergence of Ephemeroptera from Øvre Heimdalsvatn, a Norwegian subalpine lake. Pages 115–123 *in* Proceedings 2nd International Conference on Ephemeroptera 1975, Krakow, Poland.

Brittain, J. E. 1980. The biology of mayflies. Annual Review of Entomology **27**:119–147.

Brittain, J. E. 1990. Life history strategies in Ephemeroptera and Plecoptera. Pages 1–12 *in* I. C. Campbell, editor. Mayflies and Stoneflies: Life History and Biology. Kluwer Academic Publishers, Dordrecht, The Netherlands.

Brittain, J. E. 1991. Life history characteristics as a determinant of the response of mayflies and stoneflies to man-made environmental disturbance. Pages 539–545 *in* J. Alba-Tercedo, editor. Overview and Strategies of Ephemeroptera and Plecoptera. Sandhill Press, Gainesville, USA.

Brittain, J. E., and I. C. Campbell. 1991. The effect of temperature on egg development in the Australian mayfly genus *Coloburiscoides* (Ephemeroptera:

Coloburiscoidae) and its relationship to distribution and life history. Journal of Biogeography **18**:231–235.

Brittain, J. E., and T. J. Eikeland. 1988. Invertebrate drift - a review. Hydrobiologia **166**:77–93.

Brittain, J. E., and S. J. Saltveit. 1989. A review of the effect of river regulation on mayflies (Ephemeroptera). Regulated Rivers: Research and Management **3**:191–204.

Brittain, J. E., and M. Sartori. 2003. Ephemeroptera (Mayflies). Pages 373–380 *in* V. H. Resh and R. Cardé, editors. Encyclopedia of Insects. Academic Press, Boston, USA.

Bunn, S. E., and J. M. Hughes. 1997. Dispersal and recruitment in streams: evidence from genetic studies. Journal of the North American Benthological Society **16**:338–346.

Coulson, S. J., I. D. Hodkinson, N. R. Webb, K. Mikkola, J. A. Harrison, and D. E. Pedgley. 2002. Aerial colonization of high Arctic islands by invertebrates: the diamondback moth *Plutella xylostella* (Lepidoptera: Yponomeutidae) as a potential indicator species. Diversity and Distributions **8**:327–334.

Cushing, C. E., editor. 1997. Freshwater Ecosystems and Climate Change in North America. Advances in Hydrological Processes, Wiley, Chichester, UK.

Elliott, J. M., and U. H. Humpesch. 1980. Eggs of Ephemeroptera. Annual Report of the Freshwater Biological Association, UK **48**:41–52.

Fjellheim, A., and G. G. Raddum. 1993. Effects of increased discharge on benthic invertebrates in a regulated river. Regulated Rivers: Research and Management **8**:179–187.

Frank, K. T., and D. Brickman. 2000. Allee effects and compensatory population dynamics within a stock complex. Canadian Journal of Fisheries and Aquatic Sciences **57**:513–517.

Füreder, L., C. Welter, and J. Jackson. 2003. Dietary and stable isotope (δ^{13}C, δ^{15}N) analyses in alpine Ephemeroptera and Plecoptera. Pages 39–46 *in* E. Gaino, editor. Research Update on Ephemeroptera and Plecoptera. University of Perugia, Italy.

Füreder L. 1999. High alpine streams: cold habitats for insect larvae. Pages 181–196 *in* R. Margesin, and F. Schinner, editors. Cold Adapted Organisms. Ecology, Physiology, Enzymology and Molecular Biology. Springer-Verlag, Berlin, Germany.

Gattolliat, J-L., and M. Sartori. 2003. An overview of the Baetidae of Madagascar. Pages 135–144 *in* E. Gaino, editor. Research Update on Ephemeroptera and Plecoptera. University of Perugia, Italy.

Haldorsen, S., M. Heim, B. Lefauconnier, L.-E. Pettersson, M. Røros, and K. Sandsbråten. 2002. The water balance of an arctic lake and its dependence on climate change: Tvillingvatnet in Ny-Ålesund, Svalbard. Norwegian Journal of Geography **56**:146–151.

Hauer, F. R., J. S.Baron, D. H. Campbell, K. D. Fausch, S. W. Hostetler, G. H. Leavesley, P. R. Leavitt, D. M. McKnight, and J. A. Stanford. 1997. Assessment of climate change and freshwater ecosystems of the Rocky Mountains, USA and Canada. Hydrological Processes **11**:903–924.

Hershey, A. E., J. Pastor, B. J. Peterson, and G. W. Kling. 1993. Stable isotopes resolve the drift paradox for Baetis mayflies in an Arctic river. Ecology **74**:2315–2325.

Hodkinson, I. 1999. Species response to global change or why ecophysiological models *are* important: a reply to Davis et al. Journal of Animal Ecology **68**:1259–1262.

Hogg, I. D., and D. D. Williams. 1996. Response of stream invertebrates to a global-warming thermal regime: an ecosystem level manipulation. Ecology **77**:395–407.

IPCC. 2001. Climate Change 2001: Synthesis Report. Third Assessment Report. Cambridge University Press, Cambridge, UK.

Lamb, H. H. 1995. Climate, History and the Modern World. Routledge, London, UK.

Langford, T. E. 1975. The emergence of insects from a British river, warmed by power station cooling water. Part 2: The emergence patterns of some Ephemeroptera and Trichoptera and Megaloptera in relation to water temperature and river flow, upstream and downstream of cooling-water outfalls. Hydrobiologia **47**:91–133.

Lillehammer, A., and J. E. Brittain. 1978. The invertebrate fauna of the streams in Øvre Heimdalen. Holarctic Ecology. **1**:271–276.

Malard F., K. Tockner, and J. V. Ward. 1999. Shifting dominance of subcatchment water sources and flow paths in a glacial Floodplain, Val Roseg. Switzerland. Arctic, Antarctic, and Alpine Research **31**:135–150.

McGregor, G., G. E. Petts, A. M. Gurnell, and A. M. Milner. 1995. Sensitivity of alpine stream ecosystems to climate change and human impacts. Aquatic Conservation **5**:233–247.

Melack, J. M., J. Dozier, C. R. Goldman, D. Greenland, A. M. Milner, and R. J. Naiman. 1997. Effects of climate change on inland waters of the Pacific Coastal Mountains and Western Great Basin of North America. Hydrological Processes **11**:971–99.

Milner A. M. 1994. Colonization and succession of invertebrate communities in a new stream in Glacier Bay National Park, Alaska. Freshwater Biology **32**:387–400.

Milner, A. M., J. E. Brittain, E. Castella, and G. E. Petts. 2001a. Trends of macroinvertebrate community structure in glacier-fed rivers in relation to environmental conditions: a synthesis. Freshwater Biology **46**:1833–1847.

Milner, A. M., R. C. Taylor, and M. J. Winterbourn. 2001b. Longitudinal distribution of macroinvertebrates in two glacier-fed New Zealand rivers. Freshwater Biology **46**:1765–1775.

Monaghan, M. T., P. Spaak, C. T. Robinson, and J. V. Ward. 2001. Genetic differentiation of *Baetis alpinus* Pictet (Ephemeroptera: Baetidae) in fragmented alpine streams. Heredity **86**:395–403.

Monaghan, M. T., P. Spaak, C. T. Robinson, and J. V. Ward. 2002. Population genetic structure of 3 alpine stream insects: influences of gene flow, demographics, and habitat fragmentation. Journal of the North American Benthological Society **21**:114–131.

Nebeker, A. V. 1971. Effect of high winter water temperatures on adult emergence of aquatic insects. Water Research **5**:777–783.

Peckarsky, B. L., B. W. Taylor, and C. C. Caudill. 2000. Hydrologic and behavioral constraints on oviposition of stream insects: implications for adult dispersal. Oecologia **125**:186–200.

Pupilli, E., and M. A. Puig. 2003. Effects of a major flood on the mayfly and stonefly populations in a Mediterranean stream (Matarranya Stream, Ebro River basin, North East of Spain). Pages 381–389 *in* E. Gaino, editor. Research Update on Ephemeroptera and Plecoptera. University of Perugia, Italy.

Raddum, G. G. and A. Fjellheim. 2003. Liming of River Audna, Southern Norway: a large-scale experiment of benthic invertebrate recovery. Ambio **32**:230–234.

Rouse, W. R., M. S. V. Douglas, R. E. Hecky, A. E. Hershey, G. W. Kling, L. Lesack, P. Marsh, M. McDonald, B. J. Nicholson, N. T. Roulet, and J. P. Smol. 1997. Effects of climate change on the freshwaters of arctic and subarctic North America. Hydrological Processes **11**:873–902.

Sand, K. 1997. Life cycle studies of Ephemeroptera and Plecoptera in a Norwegian mountain area, Øvre Heimdalen. Cand. scient. thesis, University of Oslo, Norway.

Sand, K., and J. E. Brittain. 1997. [Benthos in alpine streams.] Pages 22–29 *in* Å. Brabrand, editor. Effects of floods on aquatic organisms. HYDRA report Mi02. Norwegian Water Resources and Energy Directorate, Oslo, Norway. In Norwegian.

Schütz, C., M. Wallinger, R. Burger, and L. Füreder. 2001. Effects of snow cover on the benthic fauna in a glacier-fed stream. Freshwater Biology **46**:1691–1704.

Slatkin, M. 1985. Gene flow in natural populations. Annual Review of Ecology and Systematics **16**:393–430.

Snucins, E. 2003. Recolonisation of acid-damaged lakes by the benthic invertebrates *Stenacron interpunctatum, Stenonema femoratum* and *Hyalella azteca*. Ambio **32**:225–229.

Starmühlner, F. 1979. Results of the Austrian Hydrobiological Mission, 1974, to the Seychelles-, Comores-, and Macarene Archipelagos: Part 1 Preliminary report. Annales Naturhistorische Museum Wien **82**:621–742.

Straile, D., D. M. Livingstone, G. A. Weyhenmeyer, and D. G. George. 2003. The response of freshwater ecosystems to climate variability associated with the North Atlantic Oscillation. Geophysical Monograph **134**:263–279.

Sweeney, B. W. 1984. Factors influencing life-history patterns of aquatic insects. Pages 56–100 *in* V. H. Resh, and D. M. Rosenberg, editors. The Ecology of Aquatic Insects. Praegar Publishers, New York, USA.

Sweeney, B. W., and R. L. Vannote. 1987. Geographic parthenogenesis in the stream mayfly *Eurylophella funeralis* in eastern North America. Holarctic Ecology **10**:52–59.

Sweeney, B. W., D. H. Funk, and R. L. Vannote. 1986. Population genetic structure of two mayflies (*Ephemerella subvaria, Eurylophella versimilis*) in the Delaware River drainage basin (USA). Journal of the North American Benthological Society **5**:253–262.

Sweeney, B. W., and R. L. Vannote. 1978. Size variation and the distribution of hemimetabolous aquatic insects: two thermal equilibrium hypotheses. Science **200**:444–446.

The European Times. 1992. Mayfly are reacting to global warming. Thursday, 18 June 1992.

Thomas, C. D., A. Cameron, R. E. Green, M. Bakkenes, and L. J. Beaqumont. 2004. Extinction risk from climate change. Nature **427**:145–148.

Wais, I. R., and A. A. Bonetto. 1988. Analysis of allochthonous organic matter and associated macroinvertebrates in some streams of Patagonia (Argentina). Verhandlungen der Internationale Vereinigung für Theoretische und Angewante Limnologie **23**:1455–1459.

Ward, J. V. 1994. Ecology of alpine streams. Freshwater Biology **32**:277–294.

Ward, J. V., and J. A. Stanford. 1982 Thermal responses in the evolutionary ecology of freshwater insects. Annual Review of Entomology **27**:97–117.

Ward, J. V., and U. Uehlinger, editors. 2003. Ecology of a Glacial Flood Plain. Kluwer Academic Publishers, Dordrecht, The Netherlands.

Wilbur, H. M. 1980. Complex life cycles. Annual Review of Ecology and Systematics **11**:67–93.

Yan, N. D., B. Leung, W. Keller, S. E. Arnott, J. M. Gunn, and G. G. Raddum. 2003. Developing conceptual frameworks for the recovery of aquatic biota from acidification. Ambio **32**:165–169.

Zwick, P. 1992 Stream habitat fragmentation – a threat to biodiversity. Biodiversity and Conservation **1**:80–97.

SCALE AND HIERARCHY IN THE ECOLOGY OF STONEFLIES

Andrew L. Sheldon

*Division of Biological Sciences, The University of Montana,
Missoula, Montana 59812 USA*

Abstract

Stoneflies, a small portion of the global insect fauna, are abundant and play important ecological roles in streams. As shredders, stoneflies are best understood at the ecosystem level but are poorly resolved at smaller scales and lower hierarchical levels. In contrast, predatory stoneflies are best understood at smaller scales; community and system effects are less explored. Population biology is an underdeveloped perspective in stonefly ecology. New developments in egg and adult biology, descriptive life histories and dispersal measurement (genetics, stable isotopes) open doorways to demographic analysis and metapopulation dynamics. Existing, but scattered, faunistic research is an unexploited resource for investigations of global change and macroecology. I use this literature to explore large-scale patterns of relative diversity across families. Major structural differences exist within North America and over the Holarctic. I sketch a research agenda in stonefly ecology emphasizing collaboration and a geographical perspective.

Key words: biogeography; family richness; hierarchy; macroecology; Plecoptera; population; predation; scale; shredder.

Introduction

In the global scheme of things, Plecoptera is a minor order of insects; a counterpoint is that stoneflies are important components in running waters, the most pervasive of continental ecosystems. This shift in viewpoint illustrates the importance of perspective and scale in the ways we study and interpret the natural world. In this very selective review, I consider what we know and what we might know about the ecology of stoneflies against a logical template of scale and inclusiveness.

Table 1 is one possible view of the world and the ways ecologists and evolutionary biologists see it. The spatial and temporal columns are scalars with direction and magnitude and all columns are hierarchies with higher levels inclusive of lower ones. The columns are not independent: flood disturbances of large magnitude originate from large areas at long intervals, major evolutionary change requires time and typically involves large areas, and ecological and evolutionary responses are linked. Stream networks inhabited by most stoneflies are hierarchical (Frissell et al.1986, Hildrew 1996) and hierarchy theory (Allen and Starr 1982) is

Table 1. Scalars and hierarchies relevant to the ecology of stoneflies.

Space	Time	Ecology	Evolution
Planet	Deep Time	Biosphere	Life
Continent	Cenozoic	Ecosystem	Order
Region	Quaternary	Community	Family
Catchment	Holocene	Assemblage	Genus
Reach	Century	taxocene	Species
Macrohabitat	Decade	guild	Metapopulation
Microhabitat	Year	functional group	Population
	Day	Metapopulation	Individual
	Hour	Population	Chromosome
	Second	Generation	Molecule
		Life History	
		Stage	
		egg	
		nymph	
		adult	
		Individual	

generally useful in ecology. Wiegert's (1988) perspective is especially relevant at this meeting where attendees' interests and specializations span much of Table 1. He concluded patterns observed at one scale or hierarchical level have explanations at lower or higher levels than the one under study, thus ecosystem behavior influences population biology but so does individual behavior. In a previous meeting of this group, Statzner (1997) advocated simple, general models in stream ecology which amounts to emphasis of the mid scale portions of Table 1. However, observed truth is not necessarily the same at all levels (Wiley et al. 1997) and real understanding requires research at multiple scales. In the following sections, I examine three aspects of stonefly ecology (shredding, predation, population biology) with special attention to scale; present preliminary original work on stonefly macroecology; and end with some personal suggestions concerning the way forward.

Shredders

The study of shredding stoneflies began at the ecosystem level with the recognition that many streams are driven by terrestrial inputs (Hynes 1975). Initial efforts were more concerned with the processed material, including the microbial component, than the insect processors. The small size, diversity and difficult taxonomy of the shredding assemblage, including stoneflies, encouraged the view that shredders are a functional group of interchangeable individuals and taxa.

Moving downscale (Table 1) has revealed important details. Differences in life history timing (Grubbs and Cummins 1996, Richardson 2001) produce waves of shredders with poorly resolved interactions with the timing of terrestrial inputs. Both flexibility and unexpected specializations appear in some nominal shredder taxa. Some Nemouridae are important scraper-grazers in acid streams (Ledger and Hildrew 2000). The widespread western North American *Zapada cinctipes* (Nemouridae) manifests a preference or tolerance of cedar needles which distinguishes it from coexisting shredders (Richardson et al. 2004). Members of the genus are significant consumers of salmon carcasses in Alaskan streams (Chaloner et al. 2002) and nemourids and leuctrids are implicated in the rapid disappearance of dead trout fry (Elliott 1997). Leaf packs and dead fish are patchy resources; over short times and distances the drift and settlement behaviors by which stoneflies occupy these patches are barely explored (Ledger et al. 2002).

Investigations at small and large scales are complementary; experimental evidence that increased species richness of shredding stoneflies increases leaf processing rates (Jonsson and Malmqvist 2000, Dangles et al. 2002) is supported by landscape-scale field studies (Jonsson et al. 2001, Huryn et al. 2002). Species and assemblage attributes (Grubbs and Cummins 1996, Richardson 2001, Richardson et al. 2004) suggest mechanisms for diversity effects. Landscape and ecosystem research is enhanced by autecological and population details. Sweeney (1993) and Friberg et al. (2002) observed effects on stonefly growth, survivorship, condition and production in streams affected by conifer plantations and invading exotic riparian plants. Food web analyses of streams in postfire landscapes (Mihuc and Minshall 1995) and streams in which leaf inputs were manipulated (Hall et al. 2000) capture the functional roles of shredding stoneflies.

Predators

Our knowledge of predation ecology of stoneflies began with analysis of gut contents, i.e., the last meals of individual nymphs. The large nymphs of Perlodidae and Perlidae are good subjects for observation and experimentation in laboratory systems so the next advances focused on individual behavior over short intervals; home range and territorial behaviors (Sjostrom 1985a, Feltmate and Williams 1991a), interactions on and dispersal from patchy habitats (Rader and McArthur 1995), hunting behavior (Sjostrom 1985b, Williams et al. 1993, Tikkanen et al. 1997) and habitat modification (Statzner et al. 1996) brought predaceous stoneflies to life. Elliott (2000, 2003a, 2003b, 2004) used similar techniques but designed a series of experiments at different levels (Table 1). He described activity patterns and hunting behaviors of several species, moved up the ecological hierarchy to questions of prey choice and the functional response (prey consumption as a function of prey density) and finally to competitive effects among predaceous stoneflies. Williams et al. (1993) described complicated interactions of stonefly and prey densities with habitat complexity and growth rate.

Stoneflies produce significant effects on their prey, often through behavioral responses rather than direct mortality (Peckarsky et al. 1993). Comparable effects of fish on stoneflies occur at different scales. Feltmate et al. (1992) showed, in a laboratory system, that two perlids differing in color pattern were active at different times of day and on backgrounds which minimized their risk from rainbow trout, a visual predator. Enclosure experiments in a natural stream documented negative effects of trout on numbers and size or condition of stoneflies (Feltmate and Williams 1991b). At a regional scale, Harvey (1993) found higher densities of Perlidae in streams without trout and of Perlodidae in trout streams. It is difficult to generalize about predation effects because nonlinear and nonadditive effects of multiple predators are commonly reported (Soluk 1993, Huhta et al. 1999). Duvall and Williams (1995) observed considerable variation in metabolic and behavioral responses of individual conspecific stoneflies to trout and Kratz (1996) described density-dependent changes in responses of both stoneflies and their mayfly prey to mayfly density.

The role of stoneflies as predators in intact, diverse stream communities is problematic. In a detailed stream food web (Tavares-Cromar and Williams 1996), two of the three top carnivores were perlid stoneflies, suggesting a considerable potential for top-down regulation; stoneflies are prominent in other stream food webs (Mihuc and Minshall 1995, Hall et al. 2000, Benke et al. 2001) although they often share the higher trophic positions with other taxa (odonates, megalopterans, caddises). Dudgeon (2000) considered stoneflies to be generalized predators taking prey in proportion to their abundances and thus unlikely to affect community composition. (This argument overlooks the varied demographies of prey and their sensitivity to predation rates.) Experiments cited earlier suggest differences in preferences and vulnerability and availability in different habitats. Wipfli and Merritt (1994) manipulated prey availability with a bacterial larvacide; one perlid predator switched to alternative prey but another was unable to compensate for the loss of favored prey and simply ate less. Another compensatory mechanism is the surprising amount of algal consumption (omnivory) by some predaceous stoneflies (Lancaster et al. 2005), which may support large populations capable of switching to periodically abundant prey. Direct measurements of predation effects include Cooper et al. (1998) who manipulated perlid densities in enclosures; predator effects were greater in smaller enclosures. How do we extrapolate to entire streams? Benke et al. (2001) estimated that all invertebrate predators consumed 50% of annual prey production. Investigations at smaller scales suggest a rich variety of predator actions so responses at larger scales may vary with diversity and composition of the predator assemblage and with the physical background including hydrological disturbance (Thomson et al. 2002). Designing observational and experimental programs of realistic duration and spatial extent will be a challenge.

Population Biology

Good life history descriptions are the foundation of population biology but ~45% of North American genera (Stewart and Stark 2002) lack all but minimal life history information. Furthermore, life histories are not invariant. *Pteronarcys* (DeWalt 1995, Townsend and Pritchard 1998) and *Dinocras* (Frutiger 1987, Sanchez-Ortega and Alba-Tercedor 1991, Zwick 1996, 2002, Sand and Brittain 2001, Erkinaro and Erkinaro 2003) extend life histories one to several years at high latitudes or elevations. European *Nemurella pictetii* provides the only known example of bivoltinism in stoneflies (Wolf and Zwick 1989) but other populations are univoltine or semivoltine (Brittain 1978). North American *Nemoura trispinosa* nymphs develop in one year in intermittent streams and two years in permanent spring brooks (Williams et al. 1995).

Past life history studies emphasized nymphal development and the egg and adult components of the complete life cycle (Table 1) were inaccessible. Rearing experiments have opened new dimensions in stonefly ecology. Delayed egg development including diapause (Schwarz 1970, Taylor et al. 1999) adds a year to the life cycle and uncouples nymphal recruitment from adult emergence in that year. Rates of population increase are decreased by delays in age at maturity so egg diapause shifts stonefly demography even further into "life in the slow lane." Variable diapause durations coupled with varying nymphal growth rates can lead to cohort splitting for which there is developmental (Moreira and Peckarsky 1994, Townsend and Pritchard 1998) and genetic (White 1989, Schultheis et al. 2002a) evidence. An egg bank (Zwick 1996) with variable developmental durations (Sandberg and Stewart 2005) results in spreading of risk and stabilization of populations in cases of catastrophic mortality. Although at the assemblage level of organization rather than the population, the possibility that close synchronization of hatching dates within species and differences among species would permit division of resources and coexistence of related species (Elliott 1995) is another example of the importance of the egg stage.

Adults are somewhat better known although most observations concern reproductive behavior (Hanada et al. 1994, Alexander and Stewart 1996, Taylor et al. 1998). Adult food (Tierno de Figueroa and Sanchez-Ortega 1999, Smith and Collier 2000) and shelter (Collier and Smith 2000) requirements are poorly understood. Myers and Resh (2000) observed many adult aquatic insects sheltering, perhaps from nightly frosts or predators, beneath overhanging banks in high elevation meadows; grazing cattle, common in the western USA, break down such overhangs and may impact insect populations by this mechanism as well as the changes in sediment, channel morphology and water quality often associated with grazing. More generally, we know very little about adult movements or habitat use. Holognath stoneflies walk away from streams into the forest (Thomas 1966, Kuusela and Huusko 1996) but some perlodids (Jop and Szczytko 1984) depart for destinations unknown within hours of emergence. Quantitative estimates of mortality and longevity are lacking. Emergence sampling (Haro et al. 1994, Sheldon 1999, Tierno de Figueroa et al. 2003) provides important life history data but leaves quantitative adult biology unresolved, although Zwick (1990) identified the adult as the likely controlling stage in stonefly population dynamics. (Basic demographic theory implies the same thing since adults are the few survivors of a nymphal cohort and will reproduce very soon; hence their reproductive value is nearly maximal.)

Production estimates (Benke 1984) incorporate growth and mortality rates but the results are usually given as annual production and turnover (P/B) for use in ecosystem research (Oertli and Dall 1993). Only rarely is any attempt made to quantify growth and mortality separately or the season- and size/age-specific rates of production (Sheldon 1972, Derka et al. 2004). Especially good examples of stream insect demography involve taxa other than stoneflies. Enders and Wagner (1996) quantified the complete life history of a caddis; adult mortality was extremely important as it was for two mayflies in the same stream (Werneke and Zwick 1992). McPeek and Peckarsky (1998) explored the fitness consequences of predator effects on growth, mortality and fecundity of a damselfly and a mayfly. For the mayfly, sublethal effects of a predaceous stonefly were strong. Adequate demographic data to conduct similar analyses for stoneflies are not available. However, the perspectives provided by these few papers are very different from the production emphasis of much of stream ecology and suggest new ways to approach stonefly ecology.

In an exceptional data set from Broadstone Stream in southern England over three decades, relative stability of stonefly numbers for periods of a few years was coupled with long-term abundance trends driven by decreasing acidity (Woodward et al. 2002). A model of this system (Speirs et al. 2000, Hildrew 2004) implies stable dynamics of two stoneflies important both as consumers and as prey.

Data at extended spatial scales are equally scarce. Lavandier (1982) demonstrated a shift from two year nymphal development at lower elevations to three years at higher sites. The additional year reduced survival to emergence from ~6% to ~1%; Lavandier suggested that populations at high elevations could not replace themselves without immigration of adult females from downstream populations with shorter life cycles and lower mortality. Such source-sink dynamics (Pulliam 1988) may be common in stream insects.

Dispersal by adults across the landscape may have important consequences; relative wing length of stoneflies is positively correlated with geographic range (Malmqvist 2000). Direct measurement of stonefly dispersal (Griffith et al. 1998, Petersen et al. 1999) by trapping has been supplemented by mass marking with a stable isotope (Brier et al. 2004, Macneale et al. 2004) and molecular techniques (Hughes et al. 1999, Ketmaier et al. 2001, Schultheis et al. 2002b). Collective evidence suggests low dispersal among streams. Hughes et al. (1999) working with *Yoraperla brevis* in mountainous terrain observed limited genetic differentiation within canyon streams (evidence of high dispersal), low genetically effective dispersal among streams and additional genetic differentiation between two adjacent streams and the remaining five.   The latter observation suggests genetic differentiation at still larger scales. Support for this idea comes from another taxon; the grazing tadpoles of the frog *Ascaphus montanus* are widespread in the northern Rocky Mountains and commonly co-occur with nymphs of *Y. brevis*. Nielson et al. (2001) observed haplotype differences over distances of 100–300 km; these patterns, and possibly some of the structure in *Y. brevis*, probably involve evolutionary changes at longer time scales. Returning to scales of ecological time and distance, the evidence for limited dispersal is cause for concern.   Local extirpation may represent the loss of unique genotypes (Wishart and Davies 2003), may not be followed by recolonization and gene flow may be necessary to allow adaptation to impacts such as global warming (Hogg and Williams 1996).   The concept of stoneflies as metapopulations subject to fragmentation (Zwick 1992) and extinction in dendritic stream networks (Fagan 2002) supports the need for research at larger spatial scales. More needs to be done, both conceptual and empirical. Most of the dispersal research involves small streams and few stonefly taxa; large Perlidae enter light traps kilometers away from likely streams of origin (Poulton and Stewart 1991) so existing data may not be general.

Faunistics and Phenology

A significant part of the stonefly literature consists of regional lists and local studies of spatial distribution and emergence phenology. This scattered literature is a valuable but unsynthesized resource that provides a baseline against which local anthropogenic impacts and remediation can be measured. More generally, widely distributed sites could be used to measure latitudinal and elevational changes induced by global climate change. Repeating thorough surveys, e.g., Ravizza and Ravizza Dematteis (1986), at decadal intervals would vastly increase the power and sensitivity of such an enterprise.

A simple example of data mining is Richardson's (2001) use of published emergence data to develop a regression model predicting emergence dates of the widespread nemourid *Zapada cinctipe* from latitude and elevation.   In the next section, I use published data to explore macroecological patterns.

Family Profiles

Macroecology focuses on statistical patterns of abundance, distribution and diversity (Brown 1995). In this venture into stonefly macroecology, I examine patterns higher in the evolutionary scale (families) at continental and Holarctic scales (Table 1). (Southern hemisphere stoneflies are not included.) The strikingly different profiles of Fig. 1 introduce the problem. In these graphs, the numbers of species in each of the indicated families are plotted. The four histograms differ in total species richness and, of more interest, in the relative diversity of the families or shapes of the profiles. The profile for the Great Smoky Mountains (B. Kondratieff, *unpublished data*) in the

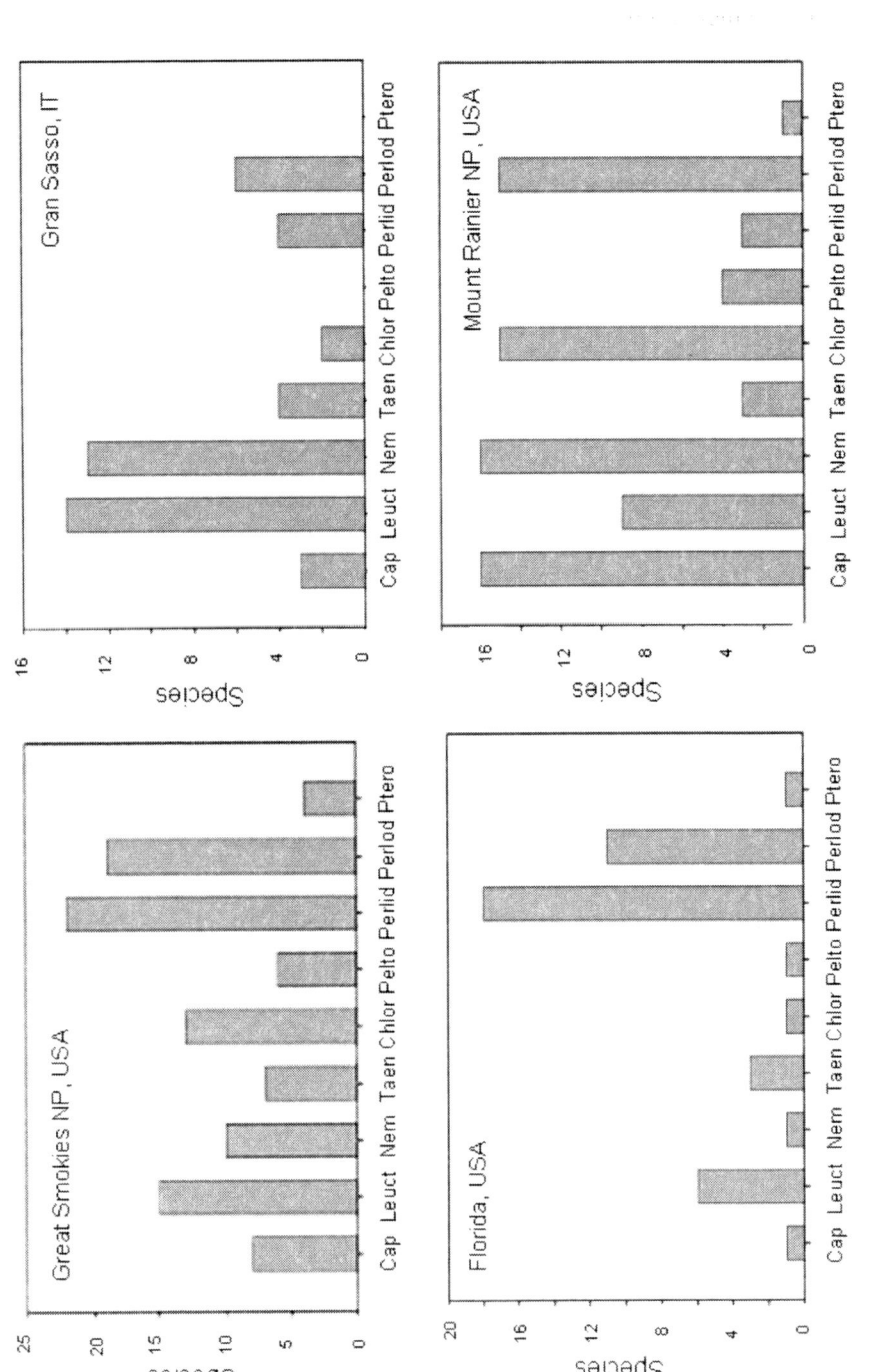

Figure 1. Family profiles of species richness by family in four regions.

southeastern USA is relatively even or equitable when contrasted with the Italian profile(http://utenti.lycos.it/ecologia_aquila/sasso.htm?), which lacks two families (Peltoperlidae, Pteronarcyidae) confined to North America and Asia; more notable is the dominance of Leuctridae and Nemouridae. The Florida profile (Pescador et al 2000) is still different; it resembles the Great Smokies in diversity of Perlodidae and Perlidae but the remaining families are represented by few species. Comparison of a species-rich site (Mount Rainier NP) in western North America (Kondratieff and Lechleitner 2002) with the Great Smokies illustrates more subtle differences; the western site is impoverished in Perlidae and relatively rich in Capniidae, Nemouridae and Chloroperlidae.

High diversity of Leuctridae in Europe and increased richness, both relative and absolute, of Perlidae in warmer regions are well known; the visual differences in the selected cases in Fig. 1 suggest more inclusive distinctions. Are the patterns general? I used Stark's (2002) list (http://www.mc.edu/campus/users/stark/Sfly0102.htm) of North American stoneflies to construct a species richness x family table for states and provinces. (These are political units varying substantially in area and also in collecting effort. Small adjacent states, e.g., CT, MA, RI were combined to reduce these problems.) Species counts were log transformed to emphasize proportional differences in a principal components (PCA) ordination (McCune and Mefford 1999) based on the correlation matrix. The first two components (Fig. 2), both with eigenvalues >1) account for 64% and 16% of total variance. PC1 is dominated by richness; vectors for all families are directed leftward. (The same feature in morphometric analyses is termed a size component with shape information in higher components.) PC 2 distinguishes eastern versus western units; vectors are positive (eastern) for Perlidae and less so for Taeniopterygidae whereas Nemouridae and Chloroperlidae vectors are negative (western). Mapping the PC scores (Fig. 3) shows the geographical coherence of the patterns summarized by the ordination scores. High diversity (negative scores on PC1) political units, e.g., VA, CA occur in the southern Appalachians and western coastal mountains. Diversity is lower (positive scores) in higher latitudes and regions of low relief or precipitation. PC2 demonstrates the structural distinction of eastern and western stonefly faunas. Sites in the Great Plains are seen as impoverished outliers of the eastern fauna. Northern units, including the northeastern USA and eastern Canada are more similar to western ones than to the core area of the eastern USA. Worth emphasizing at this point is the fact that species identities were not included in the analysis; the maps and ordinations summarize patterns in the absolute and relative diversities of families (Fig. 1). East and west share only a few stonefly species so the pattern on PC2 in Fig. 3 resembles that expected on compositional grounds; however, the northeastern USA and eastern Canada are compositionally similar to the rest of the eastern USA, but are structurally similar to the west.

To extend this analysis across the northern hemisphere, I assembled data from locations (single sites to small regions) throughout the Holarctic (sources available from the author). Only the seven families shared across the entire region were

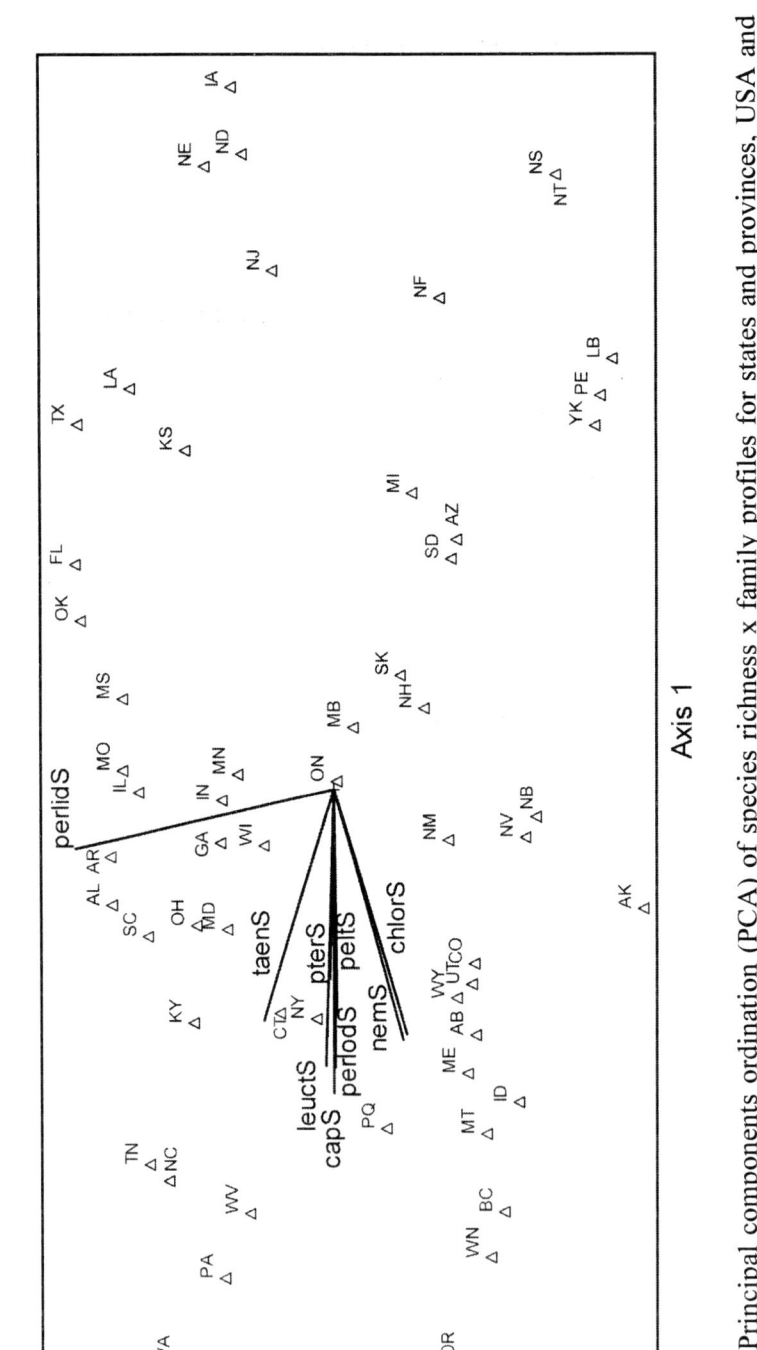

Figure 2. Principal components ordination (PCA) of species richness x family profiles for states and provinces, USA and Canada. Vectors indicate the weighted contributions of stonefly families.

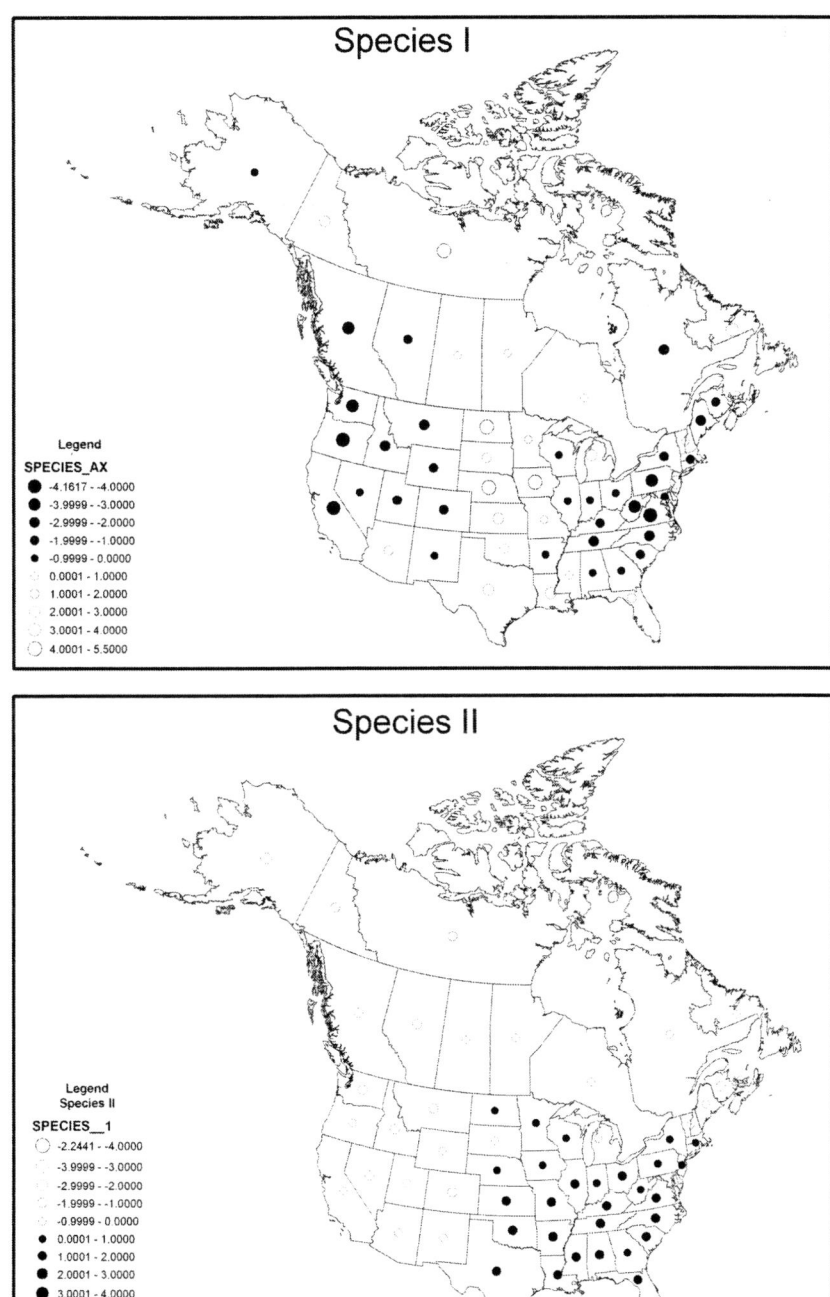

Figure 3. Maps of ordination scores (Axis 1 and Axis 2) of family richness profiles from Fig. 2.

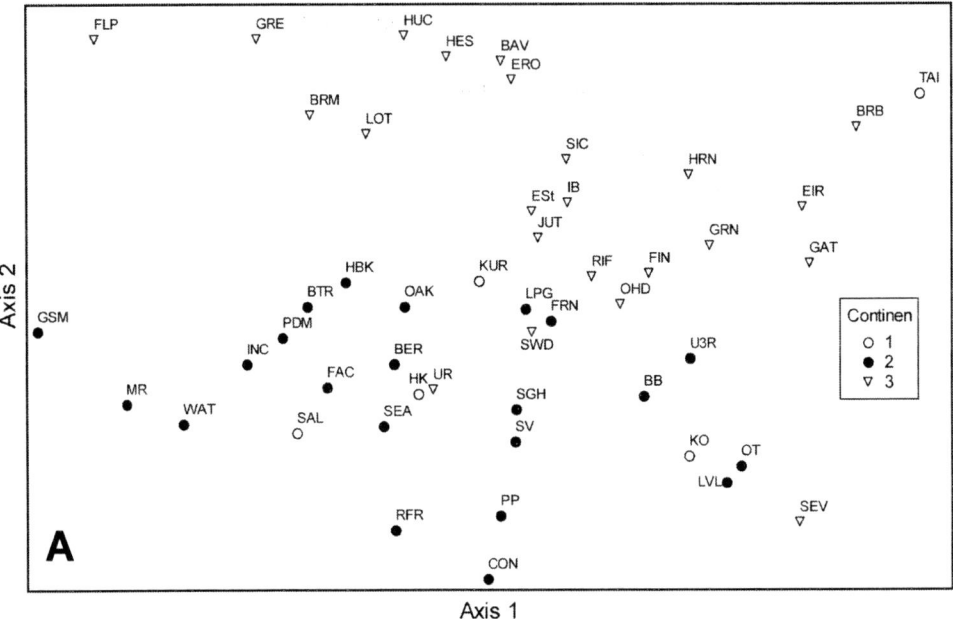

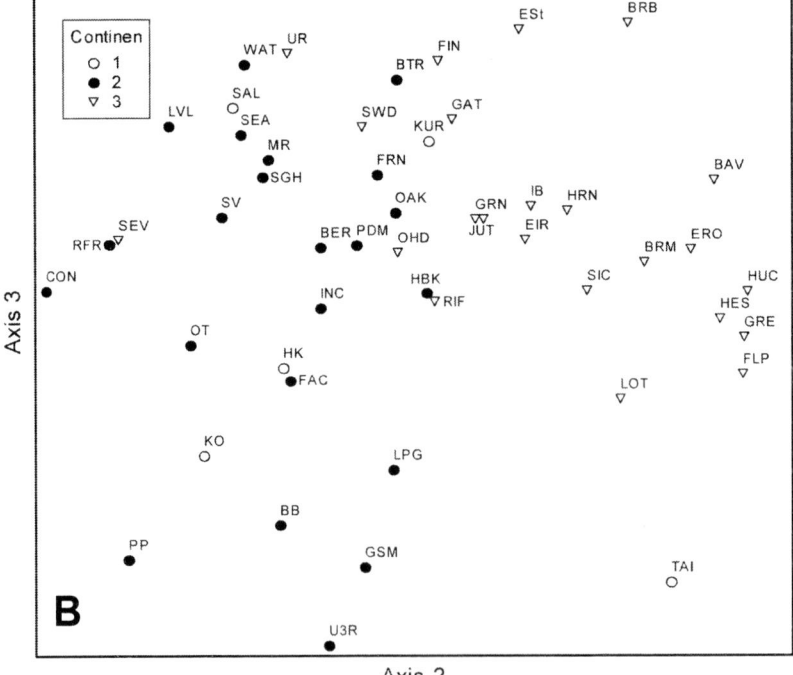

Figure 4. Ordinations (PCA) of family richness profiles from Asia (1), North America (2) and Europe (3).

included; Asian-North American Peltoperlidae and Pteronarcyidae and Asian Scopuridae and Styloperlidae were excluded. Three principal components with eigenvalues >1 accounted for 43%, 21% and 14% of total variance (Fig. 4). As in the North American analysis, PC1 is heavily weighted by richness but PC2 separates European sites from North American ones and most of the small set of Asian localities; Taiwan (TAI), which lacks four of the seven families considered, is an extreme outlier. PC2 and PC3, which contain most of the information on proportional representation of families, retain geographic distinctions but with considerable fuzziness. Extremes on PC2 are European (positive) and western North American (negative). Positive scores on PC3 include western North American and European sites at high latitudes (Urals, Finland) or elevation (Estaragne R. in the French Pyrenees); negative scores are sites in the eastern USA. The distribution of the small sample of Asian sites is especially interesting. Northern sites (Kurile Islands, Sihote Alin in the Russian Far East) group with western North America and other northern locales; the Japanese island of Hokkaido and a Korean site align with temperate localities in the eastern USA. Impoverished Taiwan groups weakly with some Perlidae-dominated sites in the southern USA.

The influence of total species richness is very strong in these ordinations. Larger regions or species-rich locations are seen as more distinctive than single localities or impoverished sites which tend to converge in the ordinations. I evaluated this effect by ordinating nested data from regional or catchment scales to single sites. A Slovakian (Krno 2000) and an Italian (Ravizza 1975) stream formed the European set. North American data are for the Ozark-Ouachita region of the central USA and its Ouachita Mountains subregion (Poulton and Stewart 1991). Within the Ouachitas, 38 sites were collected in two catchments, one with three subcatchments (A. Sheldon and M. Warren, *unpublished data*). Ordination of these nested data (Fig. 5) shows the strength of the richness effect. The clouds of points do converge as smaller units with shorter species lists are compared. However the continental signal persists; a single Slovakian site, polluted and with only three species, joins the Ouachita sites but all others remain distinct.

The profiles (Fig. 1) suggest substantial differences in stonefly assemblages and the various ordinations demonstrate geographical coherence of these patterns over large regions. If, as seems reasonable, family membership entails shared ecological attributes, then stonefly assemblages in different parts of the world may be functionally quite different. Investigations of community-level patterns in Europe and eastern or western North America may reveal common elements but also major departures from accepted generalizations. At the very least, family profiles (Fig. 1) should be a useful comparative tool and the ordinations and maps suggest a framework for comparative research; the similarity of some Asian sites to North American ones is striking as are the differences between European and North American assemblages (Fig. 4).

Causation of these patterns must have roots in deep time. Patterns of vicariance and dispersal followed by the interplay of speciation and extinction must determine modern diversity and the shapes of the family profiles. Phylogenies and estimates of speciation rates (Xiang et al. 2004) are needed to understand regional diversities within and among families. (Illies [1969] proposed that diversity of European *Leuctra* was driven by high speciation rates.)

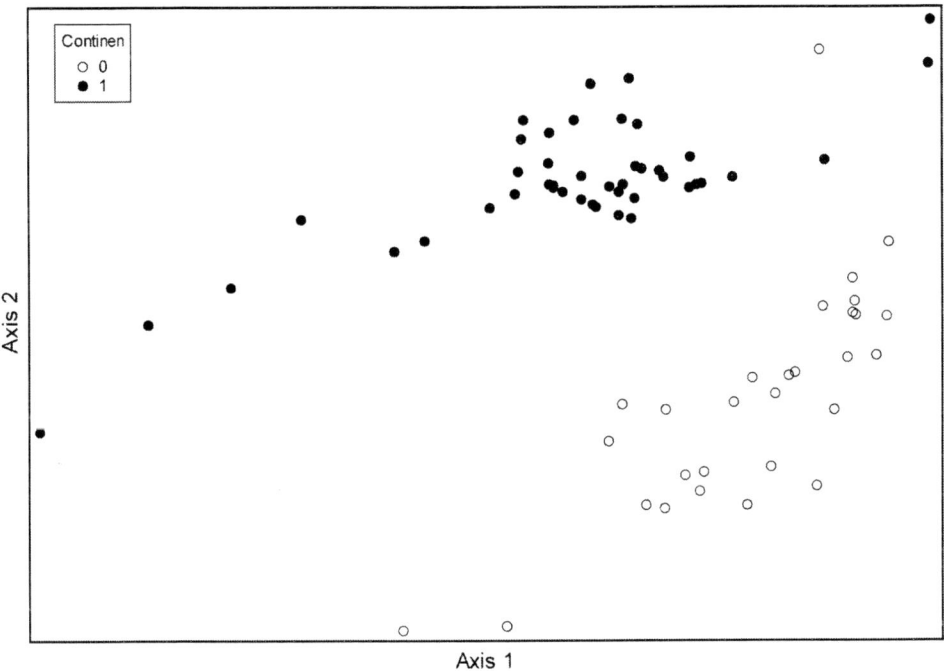

Figure 5. Ordination (PCA) of family profiles nested from regional to local scale in Europe (0) and the Ozark-Ouachita region of North America (1). Larger regions with more species are toward the left.

All the ordinations, especially Fig. 5, imply that family profiles of local sites are more similar than the regional profiles of which they are a part. One explanation is that local lists are simply too small a sample to provide statistical power to detect differences. Another, and ecologically more interesting, possibility is that local assemblages are structured by interactions and follow assembly rules (Illies 1952, Belyea and Lancaster 1999, Weiher and Keddy 1999). If regional diversity of a family is primarily among sites (β) diversity and only a few confamilial species coexist, the modes in Fig. 1 would flatten to produce a more uniform family profile. Resolution of this issue will require ecological/faunistic research including synthesis of published faunistic studies, careful use of null models (Gotelli and Graves 1996), and research at the interface of ecology and phylogeny (Webb et al. 2002). McPeek and Brown (2000) concluded that nonadaptive speciation events played a large role in structuring damselfly assemblages. Work on other taxa (Cavender-Bares et al. 2004, Graham et al. 2004) suggests a greater role for species interactions and habitat specificity and Silvertown et al. (2001) concluded that niche differences have accumulated throughout the evolutionary history of species. Considering stoneflies, what are the patterns of coexistence and assemblage structure from a single riffle to catchment, regional and continental scales? This essay in macroecology identifies questions but does not answer them. However, stoneflies distributed in globally replicated stream networks may hold answers to very general ecological and evolutionary questions.

Complementarity, Collaboration and Conservation

The previous sections make it clear that the ecological and evolutionary columns (Table 1) are complementary and that understanding any level of a hierarchy complements work done at other levels. Since few researchers have the enthusiasm, knowledge and technical skills to work effectively at all levels, collaboration is essential. Some collaboration is unplanned in any formal sense. Accessible taxonomy is essential for ecological research so the production of regional monographs or the nymphal treatise of Stewart and Stark (2002) advance the ecological enterprise. Ecologists should return the favor by depositing material, especially associated nymphs and adults, in major collections. Ross and Ricker (1971) developed a loose network of volunteers, The Winter Stonefly Club, whose widespread collections formed the basis for definitive taxonomic treatments of the winter-emerging stonefly families of eastern North America. Brittain et al. (2003) organized a formal comparative study of glacial streams, which could serve as a model for ecological investigations at continental or global scales.

Opportunities for concept-based collaboration are numerous. Molecular biologists should be included in groups investigating problems in systematics, biogeography, ecology and conservation. A coordinated geographically extensive network, including previously studied sites, of investigations of distribution, diversity and phenology would contribute to systematics, community ecology, global change and conservation issues. Note that such research could profitably involve amateurs. Other possibilities come to mind but problem definition should be done by those doing the work, so I will not attempt to chart the full course.

Conservation of stonefly diversity requires solid taxonomy and good distributional data supported by environmental data and management plans. Stoneflies have not played a major role in conservation planning, partly because stoneflies are not seen as important components of ecosystems, but also because distributional data are seldom good enough to define conservation status. Lists such as Morse et al. (1997) are a substantial step forward. However, most of the data are based on intuitive and opportunistic choices of collecting locales rather than an objective sampling design. (White [2004] nicely contrasted the Traditional and the

Structured approaches to biodiversity sampling.)   The collaborative enterprise I
sketched above could provide a structural framework.   Most stonefly conservation
efforts will involve habitat protection but opportunities for restoration, especially in
large rivers, exist.   We know that stonefly species have vanished from parts of the
landscape (Kury 1997, Landa et al. 1997, Favret and DeWalt 2002, Webb 2002).
Improvements in land and water management may bring them back through natural
recolonization, but reintroduction, from populations genetically matched with
museum specimens, may be highly desirable (R. E. DeWalt, *pers. comm.*).   My
personal favorites for monitoring and, perhaps, active intervention, are large,
regulated rivers such as the French Broad in eastern Tennessee, USA, where
effective management of discharge and oxygen concentrations plus introductions
have re-established populations of endangered fishes and mollusks.   Will we see the
return of large river assemblages of stoneflies?   And do we know what they were?

Taxon-focused ecology is a questionable enterprise since biota and ecosystems
are far more inclusive than a single order of insects.   Alternatively, stoneflies warrant
attention as important actors in running waters, the most pervasive of continental
ecosystems.   Stoneflies also provide different environmental information than other
aquatic insect orders (Heino et al. 2003, Park et al. 2003).   Hawkins (1996)
questioned the value of the mixed bag of papers resulting from meetings like this
one.   However, ecologists should "know their organisms" and there is real value in
the integration of topics from ultrastructure to palaeobiogeography.   The challenge
will be melding broad concepts of ecology, behavior and evolution with our detailed
knowledge of this fascinating group of insects.   Plecoptera, with their ancient roots
and predilection for globally distributed streams and rivers, can contribute
fundamental ecological understanding.

Acknowledgments

I thank the conference organizers for the invitation and support.   T. Losano prepared
the maps.   I thank L. Sheldon for encouragement and technical support.

Literature Cited

Alexander, K. D., and K. W. Stewart. 1996. Description and theoretical
    considerations of mate finding and other adult behaviors in a Colorado
    population of *Claassenia sabulosa* (Plecoptera: Perlidae). Annals of the
    Entomological Society of America **89**:290–296.
Allen, T. F. H., and T. B. Starr. 1982. Hierarchy: perspectives for ecological
    complexity. University of Chicago Press, Chicago, USA.
Belyea, L. R., and J. Lancaster. 1999. Assembly rules within a contingent ecology.
    Oikos **86**:402–416.

Benke, A. C. 1984. Secondary production of aquatic insects. Pages 289–322 *in* V. H. Resh, and D. M. Rosenberg, editors. The Ecology of Aquatic Insects. Praeger Publishers, New York, USA.

Benke, A. C., J. B. Wallace, J. W. Harrison, and J.W. Koebel. 2001. Food web quantification using secondary production analysis: predaceous invertebrates of the snag habitat in a subtropical river. Freshwater Biology **46**:329–346.

Briers, R. A., J. H. R. Gee, H. M. Cariss, and R. Geoghegan. 2004. Inter-population dispersal by adult stoneflies detected by stable isotope enrichment. Freshwater Biology **49**:425–431.

Brittain, J. E., E. Castella, S. Knispel, V. Lencioni, B. Lods-Crozet, B. Maiolini, A. M. Milner, S. J. Saltveit, and D. L. Snook. 2003. Ephemeroptera and Plecoptera communities in glacial rivers. Pages 271–277 *in* E. Gaino, editor. Research Update on Ephemeroptera and Plecoptera. University of Perugia, Perugia, Italy.

Brown, J. H. 1995. Macroecology. University of Chicago Press, Chicago, USA.

Cavender-Bares, J., D. D. Ackerly, D. A. Baum, and F. A. Bazzaz. 2004. Phylogenetic overdispersion in Floridian oak communities. American Naturalist **163**:823–843.

Chaloner, D. T., and M. S. Wipfli, and J. P. Caquette. 2002. Mass loss and macroinvertebrate colonization of Pacific salmon carcasses in south-eastern Alaskan streams. Freshwater Biology **47**:263–273.

Collier, K. J., and B. J. Smith. 2000. Interactions of adult stoneflies (Plecoptera) with riparian zones I. Effects of air temperature and humidity on longevity. Aquatic Insects **22**:275–284.

Cooper, S. D., S. Diehl, K. Kratz, and O. Sarnelle. 1998. Implications of scale for patterns and processes in stream ecology. Australian Journal of Ecology **23**:27–40.

Dangles, O., M. Jonsson, and B. Malmqvist. 2002. The importance of detritivore species diversity for maintaining stream ecosystem functioning following invasion of a riparian plant. Biological Invasions **4**:441–446.

Derka, T., J. M.Tierno de Figueroa, and I. Krno. 2004. Life cycle, feeding and production of *Isoptena serricornis* (Pictet, 1841) (Plecoptera, Chloroperlidae). International Review of Hydrobiology **89**:165–174.

DeWalt, R. E., and K. W. Stewart. 1995. Life histories of stoneflies (Plecoptera) in the Rio Conejos of southern Colorado. Great Basin Naturalist **55**:1–18.

Dudgeon, D. 2000. Indiscriminate feeding by a predatory stonefly (Plecoptera: Perlidae) in a tropical Asian stream. Aquatic Insects **22**:39–47.

Duvall, C. J., and D. D. Williams. 1995. Individuality in the growth of stonefly nymphs in response to stress from a predator. Archiv für Hydrobiologie **133**:273–286.

Elliot, J. M. 1995. Egg hatching and ecological partitioning in carnivorous stoneflies (Plecoptera). Comptes Rendus de L'Academie des Sciences, Serie III, Sciences de al Vie **318**:237–243.

Elliott, J. M. 1997. An experimental study on the natural removal of dead trout fry in a Lake District stream. Journal of Fish Biology **50**:870–877.

Elliott, J. M. 2000. Contrasting diel activity and feeding patterns of four species of carnivorous stoneflies. Ecological Entomology **25**:26–34.

Elliott, J. M. 2003a. A comparative study of the functional response of four species of carnivorous stoneflies. Freshwater Biology **48**:191–202.

Elliott, J. M. 2003b. Interspecific interference and the functional response of four species of carnivorous stoneflies. Freshwater Biology **48**:1527–1539.

Elliott, J. M. 2004. Prey switching in four species of carnivorous stoneflies. Freshwater Biology **49**:709–720.

Enders, G., and R. Wagner. 1996. Mortality of *Apatania fimbriata* (Insecta: Trichoptera) during embryonic, larval and adult life stages. Freshwater Biology **36**:93–104.

Erkinaro, H., and J. Erkinaro. 2003. Distribution and nymphal age of *Dinocras cephalotes* (Curtis) (Plecoptera: Perlidae) in northern Finland. Entomologica Fennica **14**:125–128.

Fagan, W. F. 2002. Connectivity, fragmentation, and extinction risk in dendritic metapopulations. Ecology **83**:3243–3249.

Favret, C., and R. E. DeWalt. 2002. Comparing the Ephemeroptera and Plecoptera specimen databases at the Illinois Natural History Survey and using them to document changes in the Illinois fauna. Annals of the Entomological Society of America **95**:35–40.

Feltmate, B. W., and D. D. Williams. 1991a. Path and spatial learning in a stonefly nymph. Oikos **60**:64–68.

Feltmate, B. W., and D. D. Williams. 1991b. Evaluation of predator-induced stress on field populations of stoneflies (Plecoptera). Ecology **72**:1800–1806.

Feltmate, B. W., D. D. Williams, and A. Montgomery. 1992. Relationship between diurnal activity patterns, cryptic coloration, and subsequent avoidance of predaceous fish by perlid stoneflies. Canadian Journal of Fisheries and Aquatic Sciences **49**:2630–2634.

Friberg, N., and A. D. Larsen, A. Rodkjaer, and A. D. Thomsen. 2002. Shredder guilds in three Danish forest streams contrasting in forest type. Archiv für Hydrobiologie **153**:197–215.

Frissell, C. A., W. J. Liss, C. E. Warren, and M. D. Hurley. 1986. A hierarchical framework for stream habitat classification: viewing streams in a watershed context. Environmental Management **10**:199–214.

Frutiger, A. 1987. Investigations on the life-history of the stonefly *Dinocras cephalotes* Curt. (Plecoptera: Perlidae). Aquatic Insects **9**:51–63.

Gotelli, N. J., and G. R. Graves. 1996. Null models in ecology. Smithsonian Institution Press, Washington, District of Columbia, USA.

Graham, C. H., S. R. Ron, J. C. Santos, C. J. Schneider, and C. Moritz. 2004. Integrating phylogenetics and environmental niche models to explore speciation mechanisms in dendrobatid frogs. Evolution **58**:1781–1793.

Griffith, M. B., E. M. Barrows, and S. A. Perry. 1998. Lateral dispersal of adult aquatic insects (Plecoptera, Trichoptera) following emergence from headwater streams in forested Appalachian catchments. Annals of the Entomological Society of America **91**:195–201.

Grubbs, S. A., and K. W. Cummins. 1996. Linkages between riparian forest composition and shredder voltinism. Archiv für Hydrobiologie **137**:39–58.

Hall, R. O., J. B. Wallace, and S. L. Eggert. 2000. Organic matter flow in stream food webs with reduced detrital resource base. Ecology **81**:3445–3463.

Hanada, S., Y. Isobe, K. Wada, and M. Nagoshi. 1994. Drumming behavior of two stonefly species, *Microperla brevicauda* Kawai (Peltoperlidae) and *Kamimuria tibialis* (Pictet) (Perlidae), in relation to other behaviors. Aquatic Insects **16**:75–89.

Haro, R. J., K. Edley, and M. J. Wiley. 1994. Body size and sex ratio in emergent stonefly nymphs (*Isogenoides olivaceus*: Perlodidae): variation between cohorts and populations. Canadian Journal of Zoology **72**:1371–1375.

Harvey, B. C. 1993. Benthic assemblages in Utah headwater streams with and without trout. Canadian Journal of Zoology **71**:896–900.

Hawkins, C. P. 1996. Review of Current Ddirections in Rresearch on Ephemeroptera. Journal of the North American Benthological Society **15**:136–138.

Heino, J., T. Muotka, R. Paavola, and L. Paasivirta. 2003. Among-taxon congruence in biodiversity patterns: can stream insect diversity be predicted using single taxonomic groups? Canadian Journal of Fisheries and Aquatic Sciences **60**:1039–1049.

Hildrew, A. G. 1996. Whole river ecology: spatial scale and heterogeneity in the ecology of running waters. Archiv für Hydrobiologie, Supplement **113**:25–43.

Hildrew, A. G., G. Woodward, J. H. Winterbottom, and S. Orton. 2004. Strong density dependence in a predatory insect: large-scale experiments in a stream. Freshwater Biology **73**:448–458.

Hogg, I. D., and D. D. Williams. 1996. Response of stream invertebrates to a global-warming thermal regime: an ecosystem-level manipulation. Ecology **77**:395–407.

Huryn, A. D., V. M. Butz Huryn, C. J. Arbuckle, and L. Tsomides. 2002. Catchment land-use, macroinvertebrates and detritus processing in headwater streams: taxonomic richness versus function. Freshwater Biology **47**:401–415.

Huhta, A., T. Muotka, A. Juntunen, and M. Yrjonen. 1999. Behavioural interactions in stream food webs: the case of drift-feeding fish, predatory invertebrates and grazing mayflies. Journal of Animal Ecology **68**:917–921.

Illies, J., 1952. Die Plecopteren und das Monardsche prinzip. Berichte Limnologischen Flussstation Freudenthal **3**:53–69.

Illies, J., 1969. Biogeography and ecology of neotropical freshwater insects especially those from running waters. Pages 685–708 *in* E. J. Fittkau, J. Illies, H. Kling, G. H. Schwab, and H. Sioli, editors. Biogeography and ecology in South America. Vol. 2. Dr. W. Junk, The Hague, Netherlands.

Küry, D. 1997. Changes in the Ephemeroptera and Plecoptera populations of a Swiss Jura stream (Röserenbach) between 1935 and 1990. Pages 296–301 *in* P. Landolt, and M. Sartori (editors). Ephemeroptera and Plecoptera. Biology-Ecology-Systematics. MTL—Mauron + Tinguely and Lachat SA, Fribourg, Switzerland.

Landa, V., J. Helesic, T. Soldan, and S. Zahradkova. 1997. Stoneflies (Plecoptera) of the River Vlatava, Czech Republic: a century of extinction. Pages 288–295 *in* P. Landolt, and M. Sartori (editors). Ephemeroptera and Plecoptera. Biology-Ecology-Systematics. MTL—Mauron+Tinguely and Lachat SA, Fribourg, Switzerland.

Jonsson, M., and B. Malmqvist. 2000. Ecosystem process rate increases with animal species richness: evidence from leaf-eating, aquatic insects. Oikos **89**:519–523.

Jonsson, M., B. Malmqvist, and P.-O. Hoffsten. 2001. Leaf litter breakdown rates in boreal streams: does shredder richness matter? Freshwater Biology **46**:161–171.

Jop, K., and S. W. Szczytko. 1984. Life cycle and production of *Isoperla signata* (Banks) in a central Wisconsin trout stream. Aquatic Insects **6**:81–100.

Ketmaier, V., R. Fochetti,V. Iannilli, and E. de Mattheis. 2001. Patterns of genetic differentiation and gene flow in central Italian populations of *Dinocras cephalotes* (Curtis, 1827) (Insecta, Plecoptera). Archiv für Hydrobiologie **150**:457–472.

Kondratieff, B. C., and R. A. Lechleitner. 2002. Stoneflies (Plecoptera) of Mount Rainier National Park, Washington. Western North American Naturalist **62**:385–404.

Kratz, K. W. 1996. Effects of stoneflies on local prey populations: mechanisms of impact across prey density. Ecology **77**:1573–1585.

Krno, I. 2000. Stoneflies (Plecoptera) in some volcanic ranges of the West Carpathians (Slovakia) and the impact of human activities. Limnologica **30**:341–350.

Kuusela, K., and A. Huusko. 1996. Post-emergence migration of stoneflies (Plecoptera) into the nearby forest. Ecological Entomology **21**:171–177.

Lancaster, J., D. C. Bradley, A. Hogan, and S. Waldron. 2005. Intraguild omnivory in predatory stream insects and the ecological consequences. Journal of Animal Ecology **74**:619–629.

Lavandier, P. 1982. Developpement larvaire—regime alimentaire, production d'*Isoperla viridinervis* Pictet (Plecoptera, Perlodidae) dans un torrent froid de haute montagne. Annales de Limnologie **18**:301–318.

Ledger, M. E., and A. G. Hildrew. 2000. Herbivory in an acid stream. Freshwater Biology **43**:545–556.

Ledger, M. E., A. M. Crowe, G. Woodward, and M. J. Winterbourne. 2002. Is the mobility of stream insects related to their diet? Archiv für Hydrobiologie **154**:41–59.

Macneale, K. H., B. L. Peckarsky, and G. E. Likens. 2004. Contradictory results from different methods for measuring direction of insect flight. Freshwater Biology **49**:1260–1268.

Malmqvist, B. 2000. How does wing length relate to distribution patterns of stoneflies (Plecoptera) and mayflies (Ephemeroptera)? Biological Conservation **93**:271–276.

McCune, B., and M. J. Mefford. 1999. PC-ORD. Multivariate analysis of ecological data, version 4. MjM Software Design, Gleneden Beach, Oregon, USA.

McPeek, M. A., and J. M. Brown. 2000. Building a regional species pool: diversification of the *Enallagma* damselflies in eastern North America. Ecology **81**:904–920.

McPeek, M. A., and B.L. Peckarsky. 1998. Life histories and the strengths of species interactions: combining mortality, growth, and fecundity effects. Ecology **79**:867–879.

Mihuc, T. B., and G. W. Minshall. 1995. Trophic generalists vs. trophic specialists: implications for food web dynamics in post-fire streams. Ecology 76:2361–2372.

Moreira, G. R. P., and B. L. Peckarsky. 1994. Multiple developmental pathways of *Agnetina capitata* (Plecoptera: Perlidae) in a temperate forest stream. Journal of the North American Benthological Society 13:19–29.

Morse, J. C., B. P. Stark, W. P. McCafferty, and K. J. Tennessen. 1997. Southern Appalachian and other southeastern streams at risk: implications for mayflies, dragonflies and damselflies, stoneflies, and caddisflies. Pages 17–42 *in* G. W. Benz, and D. E. Collins (editors). Aquatic Fauna in Peril: the Southeastern Perspective. Lenz Design and Communications, Decatur, Georgia, USA.

Myers, M. J., and V. H. Resh. 2000. Undercut banks: a habitat for more than just fish. Transactions of the American Fisheries Society 129:594–597.

Nielson, M., K. Lohman, and J. Sullivan. 2001. Phylogeography of the tailed frog (*Ascaphus truei*): implications for the biogeography of the Pacific Northwest. Evolution 55:147–160.

Oertli, B., and P. C. Dall. 1993. Population dynamics and energy budget of *Nemoura avicularis* (Plecoptera) in Lake Esrom, Denmark. Limnologica 23:115–122.

Park, Y.-S., R. Cereghino, A. Compin, and S. Lek. 2003. Applications of artificial neural networks for patterning and predicting aquatic insect species richness in running waters. Ecological Modelling 160:265–280.

Peckarsky, B. L., C. A. Cowan, M. A. Penton, and C. Anderson. 1993. Sublethal consequences of stream-dwelling predatory stoneflies on mayfly growth and fecundity. Ecology 74:1836–1846.

Pescador, M. L., A. K. Rasmussen, and B. A. Richard. 2000. A Guide to the Stoneflies (Plecoptera) of Florida. Florida Dept. of Envir. Protection, Tallahassee, USA.

Petersen, I., J. H. Winterbottom, S. Orton, N. Friberg, A. G. Hildrew, D. C. Speirs, and W. S. C. Gurney. 1999. Emergence and lateral dispersal of adult Plecoptera and Trichoptera from Broadstone Stream, U.K. Freshwater Biology 42:401–416.

Poulton, B. C., and K. W. Stewart. 1991. The stoneflies of the Ozark and Ouachita Mountains (Plecoptera). Memoirs of the American Entomological Society 38:1–116.

Pulliam, H. R. 1988. Sources, sinks, and population regulation. American Naturalist 132:652–661.

Rader, R. B., and J. V. McArthur. 1995. The relative importance of refugia in determining the drift and habitat selection of predaceous stoneflies in a sandy-bottomed stream. Oecologia 103:1–9.

Ravizza, C. 1975. Faunistica, ecologia e fenologia immaginale dei plecotteri reofili nella Val Brembana (Lombardia) con descrizione di una specie nuova (Plecoptera). Redia 56:271–373.

Ravizza, C., and E. Ravizza Dematteis. 1986. Les Plecopteres du Grana (Alpes Cottiennes meridionales) (Plecoptera). Bolletino del Museo Regionale di Scienze Naturali Torino 4(2):311–339.

Richardson, J. S. 2001. Life cycle phenology of common detritivores from a temperate rainforest stream. Hydrobiologia 455:87–95.

Richardson, J. S., C. R. Shaughnessy, and P. D. Harrison. 2004. Litter breakdown and invertebrate association with three types of leaves in a temperate rainforest stream. Archiv für Hydrobiologie **159**:309–325.

Ross, H., and W. E. Ricker. 1971. The classification, evolution and dispersal of the winter stonefly genus *Allocapnia*. Illinois Biological Monographs **45**:1–166.

Sanchez-Ortega, A., and J. Alba-Tercedor. 1991. The life cycle of *Perla marginata* and *Dinocras cephalotes* in Serria (sic!) Nevada (Granada, Spain) (Plecoptera:Perlidae). Pages 493–501 *in* J. Alba-Tercedor and A. Sanchez-Ortega, editors. Overviews and Strategies of Ephemeroptera and Plecoptera. Sandhill Crane Press, Gainesville, USA.

Sand, K., and J. E. Brittain. 2001. Egg development in *Dinocras cephalotes* (Plecoptera, Perlidae) at its altitudinal limit in Norway. Pages 209–216 *in* E. Dominguez , editor. Trends in Research in Ephemeroptera and Plecoptera. Kluwer Academic/ Plenum Publishers, New York, USA

Sandberg, J. B., and K. W. Stewart. 2004. Capacity for extended diapause in six *Isogenoides* Klapalek species (Plecoptera:Perlodidae). Transactions of the American Entomological Society **130**:411–423.

Schultheis, A. S., A. C. Hendricks, and L. A. Weigt. 2002a. Genetic evidence for 'leaky' cohorts in the semivoltine stonefly *Peltoperla tarteri* (Plecoptera: Peltoperlidae). Freshwater Biology **47**:367–376.

Schultheis, A. S., L. A. Weigt, and A. C. Hendricks. 2002b. Gene flow, dispersal, and nested clade analysis among populations of the stonefly *Peltoperla tarteri* in the southern Appalachians. Molecular Ecology **11**:317–327.

Sheldon, A. L. 1972. Comparative ecology of *Arcynopteryx* and *Diura* (Plecoptera) in a California stream. Archiv für Hydrobiologie **69**:521–546.

Sheldon, A. L. 1999. Emergence patterns of large stoneflies (Plecoptera: *Pteronarcys Calineuria, Hesperoperla*) in a Montana river. Great Basin Naturalist **59**:169–174.

Silvertown, J., M. Dodd, and D. Gowing. 2001. Phylogeny and the niche structure of meadow plant communities. Journal of Ecology **89**:428–435.

Sjostrom, P. 1985a. Territoriality in nymphs of *Dinocras cephalotes*. Oikos 45:353–357.

Sjostrom, P. 1985b. Hunting behaviour of the perlid stonefly nymph *Dinocras cephalotes* (Plecoptera) under different light conditions. Animal Behaviour **33**:534–540.

Smith, B. J., and K. J. Collier. 2000. Interactions of adult stoneflies (Plecoptera) with riparian zones II. Diet. Aquatic Insects **22**:285–296.

Soluk, D. A. 1993. Multiple predator effects: predicting combined functional response of stream fish and invertebrate predators. Ecology **74**:219–225.

Speirs, D. C., W. S. C. Gurney, A. G. Hildrew, and J. H. Winterbottom. 2000. Long-term demographic balance in the Broadstone stream insect community. Journal of Animal Ecology **69**:45–58.

Statzner, B. 1997. Complexity of theoretical concepts in ecology and predictive power: patterns observed in stream organisms. Pages 211–218 *in* P. Landolt and M. Sartori , editors. Ephemeroptera and Plecoptera. Biology-Ecology-Systematics. MTL—Mauron+Tinguely & Lachat SA. Fribourg, Switzerland.

Statzner, B., U. Fuchs, and L. W. G. Higler. 1996. Sand erosion by mobile predaceous stream insects: implications for ecology and hydrology. Water Resources Research **32**:2279–2287.

Stewart, K. W., and B. P. Stark. 2002. Nymphs of the North American stonefly genera. Caddis Press, Columbus, Ohio, USA.

Sweeney, B. W. 1993. Effects of streamside vegetation on macroinvertebrate communities of White Clay Creek in eastern North America. Proceedings of The Academy of Natural Sciences of Philadelphia **144**:291–340.

Tavares-Cromar, A. F., and D. D. Williams. 1996. The importance of temporal resolution in food web analysis: evidence from a detritus-based stream. Ecological Monographs **66**:91–113.

Taylor, B. W., C. R. Anderson, and B. L. Peckarsky. 1998. Effects of size at metamorphosis on stonefly fecundity, longevity and reproductive success. Oecologia **114**:494–502.

Taylor, B. W., C. R. Anderson, and B. L. Peckarsky. 1999. Delayed egg hatching and semivoltinism in Nearctic stonefly *Megarcys signata* (Plecoptera: Perlodidae). Aquatic Insects **21**:179–185.

Tierno de Figueroa, J. M., and A. Sanchez-Ortega. 1999. Imaginal feeding of certain systellognathan stonefly species (Insecta: Plecoptera). Annals of the Entomological Society of America **92**:218–221.

Tierno de Figueroa, J. M., J. M. Luzon-Ortega, and A. Sanchez-Ortega. 2003. Protandry and its relationship with adult size in some Spanish stoneflies species (Plecoptera). Annals of the Entomological Society of America **96**:560–562.

Thomas, E. 1966. Orientierung der imagines von *Capnia atra* Morton (Plecoptera). Oikos **17**:278–280.

Thomson, J. R., P. S. Lake, and B. J. Downes. 2002. The effect of hydrological disturbance on the impact of a benthic invertebrate predator. Ecology **83**:628–642.

Tikkanen, P., T. Muotka, A. Huhta, and A. Juntunen. 1997. The role of active predator choice and prey vulnerability in determining the diet of predatory stonefly (Plecoptera) nymphs. Journal of Animal Ecology **66**:36–48

Townsend, G. D., and G. Pritchard. 1998. Larval growth and development of the stonefly *Pteronarcys californica* (Insecta: Plecoptera) in the Crowsnest River, Alberta. Canadian Journal of Zoology **76**:2274–2280.

Webb, C. O., D. D. Ackerly, M.A. McPeek, and M. J. Donoghue. 2002. Phylogenies and community ecology. Annual Review of Ecology and Systematics **33**:475–505.

Webb, D. W. 2002. The winter stoneflies of Illinois (Insecta: Plecoptera): 100 years of change. Illinois Natural History Survey Bulletin **36**:195–274.

Weiher, E., and P. Keddy. 1999. Ecological assembly rules: perspectives, advances, retreats. Cambridge University Press, Cambridge, UK.

Werneke, U., and P. Zwick. 1992. Mortality of the terrestrial adult and aquatic nymphal life stages of *Baetis vernus* and *Baetis rhodani* in the Breitenbach, Germany (Insecta: Ephemeroptera). Freshwater Biology **28**:249–255.

White, M. M. 1989. Age class and population genic differentiation in *Pteronarcys proteus* (Plecoptera: Pteronarcyidae). American Midland Naturalist **122**:242–248.

White, P. 2004. Exploring and sampling for biodiversity: another way we can lead! ATBI (All Taxa Biodiversity Inventory) Quarterly **5**:2.

Wiley, M. J., S. L. Kohler, and P. W. Seelbach. 1997. Reconciling landscape and local views of aquatic communities: lessons from Michigan trout streams. Freshwater Biology **37**:133–148.

Wishart, M.J., and B. R. Davies. 2003. Beyond catchment considerations in the conservation of lotic biodiversity. Aquatic Conservation: Marine and Freshwater Ecoystems **13**:429–437.

Williams, D. D., J.-A. Barnes, and P. C. Beach. 1993. The effects of prey profitability and habitat complexity on the foraging success and growth of stonefly (Plecoptera) nymphs. Freshwater Biology **29**:107–117.

Williams, D. D., N. E. Williams, and I. D. Hogg. 1995. Life history plasticity of *Nemoura trispinosa* (Plecoptera: Nemouridae) along a permanent-temporary water habitat gradient. Freshwater Biology **34**:155–163.

Wipfli, M. S., and R. W. Merritt. 1994. Disturbance to a stream food web by a bacterial larvacide specific to black flies: feeding responses of predatory macroinvertebrates. Freshwater Biology **32**:91–103.

Wiegert, R. C. 1988. Holism and reductionism in ecology: hypotheses, scale and systems models. Oikos **53**:267–269.

Wolf, B., and P. Zwick. 1989. Plurimodal emergence and plurivoltinism of Central European populations of *Nemurella pictetii* (Plecoptera: Nemouridae). Oecologia **79**:431–438.

Woodward, G., J. I. Jones, and A.G. Hildrew. 2002. Community persistence in Broadstone Stream (U.K.) over three decades. Freshwater Biology **47**:1419–1435.

Xiang, Q.-Y., W. H.Zhang, R. E. Ricklefs, H. Qian, Z. D. Chen, J. Wen, and J. H. Li. 2004. Regional differences in rates of plant speciation and molecular evolution: a comparison between eastern Asia and eastern North America. Evolution **58**:2175–2184.

Zwick, P. 1990. Emergence, maturation and upstream oviposition flights of Plecoptera from the Breitenbach, with notes on the adult phase as a possible control of stream insect populations. Hydrobiologia **194**:207–223.

Zwick, P. Stream habitat fragmentation—a threat to biodiversity. Biodiversity and Conservation **1**:80–97.

Zwick, P. 1996. Variable egg development of *Dinocras* spp. (Plecoptera: Perlidae) and the stonefly seed bank theory. Freshwater Biology **35**:81–100.

Zwick, P. 2002. The stonefly (Insecta: Plecoptera) seed bank theory: new experimental data. Verhandlungen der Internationale Vereinigung für Theoretische und Angewante Limnologie **28**:1–7.

MAYFLY POPULATION DENSITY, PERSISTENCE AND GENETIC STRUCTURE IN FRAGMENTED HEADWATER HABITATS

L. C. Alexander and W. O. Lamp

*Department of Entomology, University of Maryland,
College Park, Maryland 20742 USA*

Abstract

We assessed the effects of stream habitat loss and fragmentation on the density, genetic diversity and persistence of a mayfly (*Ephemerella inconstans* Traver, 1932) in 24 first-order streams across nine headwater stream networks in Maryland and Virginia. We present differences in population density and local extinction in forested versus deforested headwater streams, as well as a preliminary analysis of genetic diversity in populations of *E. inconstans* and three closely related species. Because the sampling period spanned two years of drought (2001–2002) followed by two years of recovery (2003–2004), we predicted that mayfly density would be higher and population extinction rates lower at forested sites compared with deforested (agricultural and residential) sites. We found no difference in initial density at forested and deforested sites and no difference in the level of population decline across all sites by the end of the drought. However, one year after the drought had ended, population density was significantly higher in forested streams compared with streams flowing through agricultural and residential areas. Further, while only 1 of 11 populations at forested sites remained extirpated in 2004, populations in 4 of the 13 deforested streams were extirpated at the end of the study. These results suggest that recovery and recolonization following a major regional disturbance was more successful in the intact, forested stream networks than in the altered networks. To examine the population genetic effects of the demographic decline, extinction events and post-drought recovery, we sequenced a region of mitochondrial DNA in 10 populations. However, we found very low polymorphism across the entire 200 km range of the study, suggesting that a prehistoric bottleneck occurred in this species.

Key words: Ephemeroptera; *Ephemerella*; dispersal; population extinction; first-order streams; population density; genetic diversity.

Introduction

Extensive deforestation of small watersheds in the Mid-Atlantic Piedmont region of North America has altered the structure and function of headwater streams by reducing their number, disrupting critical ecosystem processes, and fragmenting surviving headwaters into isolated or semi-isolated habitat patches. Predicting the impact of habitat loss on aquatic insect species depends, in part, on understanding how individuals move among resource patches and population units (Turner et al. 2001, Goodwin and Fahrig 2002). Some movements, such as annual migration, are predictable; others, such as stream drift by insects, are responses to random events or local conditions (Humphries 2002, Ledger et al. 2002, Anholt 1995). Whatever the reason and mechanism, the movements of individuals shape the spatial structure of populations and the species, and play an important role in their persistence (Hanski and Ovaskeinen 2002, Lowe 2002).

Headwaters are naturally patchy habitats that support diverse communities of aquatic insects. Individual headwater streams may be only a few hundred meters in length and flow through watersheds less than one square kilometer in area. However, in their natural state in the Mid-Atlantic Piedmont region of North America, headwaters rarely exist in isolation; rather, they form complex networks covering large areas over which flow is diffused through a dendritic network of many small channels. Preserving natural connections for movement of individuals can significantly affect the population extinction probabilities in habitats with this particular geographic structure (Fagan 2002).

Small as they are, headwater streams and their associated wetlands perform multiple ecological functions of critical importance to the larger ecosystem. Headwaters retain sediment and slow runoff; recharge groundwater sources; take up chemicals and excess nutrients that would otherwise be transported to bays, lakes and oceans; process and transport beneficial organic matter to downstream ecosystems; and provide refuge and habitat for mayflies and other aquatic organisms (Peterson et al. 2001, Wallace et al. 1997). Recent surveys estimate that headwater streams comprise at least 80% of total stream miles in the United States (Meyer et al. 2003) and at least 66% of stream miles in Maryland (MDNR 1997, 2001). However, because of their small size and ubiquitous presence in areas with high dollar-value real estate, headwater networks in the Central Piedmont are highly susceptible to development. Viewed as nuisances to some property owners, the wet lowlands and small channels associated with headwaters are filled, diverted or piped underground to prevent flooding of roads, fields, lawns and buildings. In recent years, simplification and degradation of small stream networks in Maryland and Virginia has greatly reduced the quantity and quality of habitats available to headwater-specific organisms (MDNR 1997, 2001).

We tracked 24 populations of the mayfly *Ephemerella inconstans* Traver, 1932, over two to four generations to ask how recent changes in the landscape structure of headwater steam networks have affected the interaction and persistence of mayfly populations living in them. Here we present results of a study of population density and patch extinction, an initial study of genetic diversity among the sampled populations of the species *E. inconstans* and two other species recently synonymized with *Ephemerella invaria* (Walker), 1853 (Jacobus and McCafferty 2003), and plans for the continuing the population genetic and phylogenetic analyses in the coming year.

Methods and Materials

Study Organism. The genus *Ephemerella* is Holarctic in distribution, encompassing a range of Europe and nontropical Asia, Africa north of the Sahara and North America south to the Mexican desert region. Until recently it was thought that the species *E. inconstans* was found only in the southeast United States, in the Central Piedmont and Southern Appalachian regions (Allen and Edmunds 1965). A revision by Jacobus and McCafferty (2003) recognizes eight previously distinct species, including *E. inconstans*, as morphological synonyms of the more widely distributed species *E. invaria*. For clarity in the presentation of molecular data for *E. invaria* synonyms, and because the project described herein was well underway before the 2003 revision, thus we use the historical species names.

The naturally patchy distribution of headwater stream habitat is reflected in the population distribution of *E. inconstans* in the states of Maryland and Virginia, USA. In a preliminary survey of first-to-third order streams in the Maryland Piedmont in 2001, *E. inconstans* was common throughout the area but was found only in small, relatively undisturbed streams (unpublished data). Based on the 2001 survey, the preference of *E. inconstans* for headwater habitat over larger streams appears to depend more on flow regime and substrate than on nutrient levels or chemical composition of the stream water (unpublished data). This species was selected for the project because its distribution, habitat preferences, univoltine life cycle, synchronized emergence, equal sex-ratio (determined through emergence trapping) and limited flight period make it a good model for examining the role of dispersal in the patch dynamics of insects in headwater stream networks.

Study Sites. A total of 24 headwater streams in 9 headwater stream networks were sampled. Ten streams were sampled in 2001, 20 streams in 2002 and 24 streams in both 2003 and 2004. The names and locations of the study sites, which fall within four major river watersheds in Maryland and Virginia, are provided in Table 1. Each stream network consists of one to four adjacent headwater streams containing one or more populations of *E. inconstans*. All study sites are located within the region of Central Piedmont between 37Y20' and 39Y20' latitude, bounded to the west by the Appalachian mountains and to the east by the Coastal Plain. Although lengths and flow regimes vary among the streams due to differences in local topography, groundwater sources and land use, a typical stream in this study drains an area < 1 km^2 with baseflow discharge < 0.03 m^3/s.

Sampling. In 2001, 2002 and 2004, nymph samples were collected using moss-packs (colonizing samplers) consisting of a fixed amount of dried moss enclosed in plastic mesh bags and tied with string to roots or stakes along the stream margin for a period of three weeks in March and April, when late instar *E. inconstans* nymphs are present in the stream margins. Moss-packs are designed to move freely with stream flow to imitate natural moss or root-wad habitats. They are readily colonized by *E. inconstans* and other aquatic invertebrate taxa. Eight moss-packs were placed in each stream, positioned in pairs along a 75 m reach so that a total of four subsamples were taken in each stream.

Table 1. Sample sites and years sampled.

County and State (USA)	8-Digit Watershed (HUC#)	Headwater stream network identifier	No. of streams sampled			
			2001	2002	2003	2004
Baltimore MD	Patapsco River Lower North Branch (02130906)	Daniels Creek	-	3	3	3
		MPEA	3	3	3	3
Howard MD	Middle Patuxent River (02131106)	Homewood	-	1	2	2
		UMD Dairy Farm	1	3	3	3
	Rocky Gorge Dam (02131107)	Rocky Gorge	1	1	1	1
	Brighton Dam (02131108)	Cattail Creek	2	3	3	3
Montgomery MD	Seneca Creek (02140208)	Little Seneca	1	3	3	3
Appomattox VA	Appomattox River (02080207)	Saunders Creek	2	3	3	3
Buckingham VA	Slate River (0208020)	Jamison Creek	-	-	3	3
		Total per year	10	20	24	24

Samples were bagged in stream water and sorted while specimens were still alive. Ephemerellid mayflies were identified and sorted by species using Allen and Edmunds' key (1965), counted and stored in 100% ethyl alcohol at $-20\Upsilon C$. When a stream sample contained fewer than 16 individuals of *E. inconstans*, that stream was resampled with a D-frame net to increase the size of the sample available for population genetic analysis. The extra samples were labeled appropriately, stored separately from the moss-pack samples and excluded from the population density counts.

In 2003, nymph samples were collected with a D-frame net. From a comparison of samples taken using both methods in one stream, active search with a D-frame net produced larger sample counts and thus would overestimate the density relative to the mosspack samples. To make the D-frame samples comparable to the moss-pack samples for categorical estimates of population density, D-frame sampling for 2003 season was constrained to three 25-m sections selected at random from a 150-m stream reach. All suitable habitats in the substrate and stream margins within the three randomly selected sections were sampled extensively. Processing of samples in 2003 was done as in 2001, 2002 and 2004, described above.

Population Density and Persistence. Nymph sample counts were converted to a categorical variable with four levels: none (sample count=0), rare (0 < sample count < 10), common (10 ≤ sample count < 20) and abundant (sample count ≥ 20). The counts in each density category were plotted to visually check for trends in the density distribution in forested versus deforested streams within each year. The density categories were then combined to create two broader density categories: low (sample count < 10) and high (sample count ≥ 10), to compare the densities in forested and deforested streams using Fisher's Exact Test. A separate statistical test was conducted for each of the last three years (2002, 2003, 2004). The results of 2002 reflect density during the drought; the results of 2003 reflect the population response to the final year of drought (summer 2002); and the results of 2004 reflect the population recovery one year after the end of the drought.

Genetic Analysis: E. inconstans. We sequenced a portion of the mitochondrial DNA (mtDNA) cytochrome oxidase (CO) I gene in 10 populations in along a north-south gradient in Maryland and Virginia to determine the scale at which regional genetic variation occurs in this species. Mitochondrial DNA markers were selected because a high mutation rate and maternal inheritance result in a smaller effective population size, so that mtDNA can accumulate population genetic variation at a faster rate than nuclear DNA. This process, which occurs over long periods of time, provides the background of genetic variation against which reductions in diversity that occur in more recent time-frames may become visible. Also, mtDNA haplotypes make it possible to separate population history from current population structure through the use of gene genealogies. This is a powerful factor in separating effects from historical events, such as geographic isolation due to formation of major land barriers, from recent events, such as habitat fragmentation. Mitochondrial DNA is also relatively easy to use, providing results more rapidly than methods that require development of new markers (e.g., microsatellites), and has been used successfully in other studies of small scale aquatic insect population genetics (e.g., Hughes et al. 2003, Galacatos et al. 2002, Myers et al. 2001)

Genetic Analysis: E. invaria. To estimate genetic similarity among the historical species recently synonymized with *E. invaria*, we also sequenced samples of *Ephemerella rotunda* Morgan, 1911, *Ephemerella floripara* McCafferty, 1985 and *E. invaria*. Tissue specimens for this analysis were obtained from the primary author of the 2003 revision.

Results

Population Density and Persistence. The population density distribution for each year in (a) all streams; (b) forested streams only; and (c) deforested streams only, is described and plotted in Fig. 1. The plot of combined sites (Fig. 1a) shows the general trend of population decrease during the drought, followed by population increase during recovery. The apparent symmetry of the combined response is a composition of inverse patterns of response by forested streams and deforested stream populations (Figs. 1b and 1c). One year after the drought (2004), 91% of forested streams were classified as having high (= abundant + common) mayfly density, none were classified rare and the population in one of the 11 streams (9%) was extinct. In the deforested streams, 46% of streams were classified as high density, 23% were rare and populations in four of 13 streams (31%) were extinct. Fisher's Exact Tests of categorical density in forested and deforested sites within each year (Table 2) show that the population density does not differ between forested and deforested sites in years 2002 or 2003 (p=0.4 and p=0.6), indicating that streams could not be distinguished by site (forested or unforested) at the start of the study or at end of the drought. However, one year after the drought had ended (spring 2004) the proportion of streams with high population density was significantly greater in forested than in deforested streams (p<0.04).

Table 2. Fisher's Exact Test of density across sites, within years. Test of Site (Forested, Unforested) by Density (Low=None + Rare, High=Common +Abundant).

a. 2002	Low	High	b. 2003	Low	High	c. 2004	Low	High
Forested	1	7	Forested	5	6	Forested	1	10
Deforested	4	8	Deforested	9	4	Deforested	7	6

a) 2002: No difference in density by site during drought (p=0.4).

b) 2003: No difference in density by site at the end of the drought (p=0.6).

c) 2004: Significant difference in density one year after drought (p<0.04).

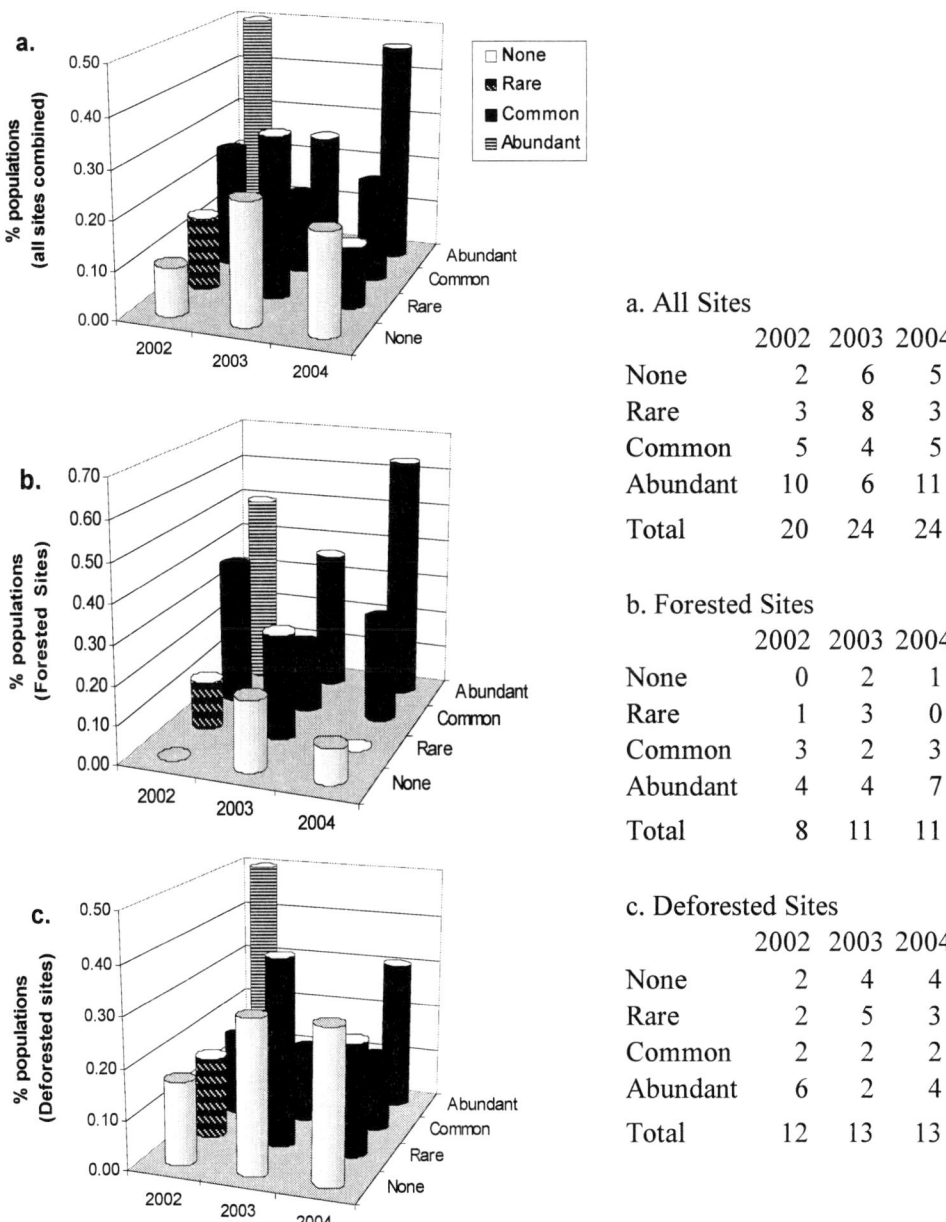

a. All Sites

	2002	2003	2004
None	2	6	5
Rare	3	8	3
Common	5	4	5
Abundant	10	6	11
Total	20	24	24

b. Forested Sites

	2002	2003	2004
None	0	2	1
Rare	1	3	0
Common	3	2	3
Abundant	4	4	7
Total	8	11	11

c. Deforested Sites

	2002	2003	2004
None	2	4	4
Rare	2	5	3
Common	2	2	2
Abundant	6	2	4
Total	12	13	13

Figure 1. Density Distribution by Year. Counts represent the density distribution for each year in (a) all streams, (b) forested streams and (c) deforested streams.

Genetic Analysis: E. inconstans. We found extremely low variation among the 10 study populations of *E. inconstans* across the ~200 km range of sites in Maryland and Virginia. There were two haplotypes, a "southern" haplotype and a "northern" haplotype, which differed at only two loci (2 alleles per locus) in the sequenced region of mtDNA (~450 bp). This lack of polymorphism in mtDNA is probably the result of a bottleneck following recolonization of the Central Piedmont at the end of the last Ice Age. Thus, isolation that might be occurring now from current ecological processes (e.g., habitat fragmentation) is not detectable in these populations using mtDNA.

Genetic Analysis: E. invaria. The estimated genetic distances among five populations of *E. invaria* synonyms from Maryland (MD), Virginia (VA), Tennesee (TN) and North Carolina (NC) are given in Table 3. *E. inconstans* and *E. invaria* specimens had the lowest among-population genetic distances (0.053 to 0.059), whereas populations of *E. rotunda* (VA) and *E. floripara* (NC) showed the equally high genetic divergence in all pairwise comparisons (mean distance = 0.12 for both species). Within-population distance ranged from .005 to .022 in all samples, with mean within-population genetic distance < 0.01. The genetic divergences among these populations could be the result of speciation, geographic distance, founder effect, or local adaptation. Since the amount of genetic distance among species varies considerably among taxonomic orders and families, interpretation of the observed differences among these populations will require the addition of molecular characters from other, morphologically-distinct *Ephemerella* species.

Table 3. Mean genetic distance among 5 populations of *Ephemerella invaria* synonyms.

spp	A	B	C	D	E	Species Key		
A	-	0.059	0.121	0.058	0.105	A	*E. inconstans,* TN	n=2
B		-	0.111	0.053	0.116	B	*E. inconstans,* D+VA	n >100
C			-	0.123	0.124	C	*E. rotunda,* VA	n=2
D				-	0.137	D	*E. invaria,* NC	n=6
E					-	E	*E. floripara,* NC	n=1

Discussion

The results of this study show that population recovery and habitat recolonization following a major regional disturbance was more successful in the intact, forested headwater stream networks than in the altered, deforested networks. This could be because dispersal from surviving populations to uninhabited patches was more effective in forested stream networks, resulting in a higher probability of recolonization as well as a larger founding population in these streams. Or, it could be that refugia in the forested sites (e.g., in the hyporheic zone) provided protection to a small number of individuals who were able to regenerate large population sizes in a single generation. These findings are consistent with studies that have found habitat type to be a significant predictor of local extinction, even after the effect of regional distribution has been removed (e.g., Korkeamaeki and Suhonen 2002).

We would like to know what role mayfly dispersal plays in maintaining population abundances and population genetic diversity in the face of local and regional disturbances, and how changes to the structure of headwater stream networks may affect demographic processes. To that end, we are now using Amplified Fragment Length Polymorphism (AFLP) to evaluate nuclear DNA diversity in *E. inconstans*. This technique samples the entire nuclear genome for polymorphisms at restriction enzyme sites, and thus provides high temporal and spatial sensitivity for detecting changes in genetic composition of populations at small scales (Mueller and Wolfenbarger 1999). If fragment length polymorphism exists in this region and species, then our samples of four generations of mayflies (spanning a period of drought, local extinction and recovery) will provide a good opportunity to relate observed events in the demographic history of these populations to changes in their genetic composition through space and time (Barrett et al. 2005).

We are also developing a spatially explicit model of population growth with migration to model different theoretical scenarios of genetic variation in subdivided mayfly populations (as in Johst et al. 2002) and, when AFLP data become available, to test alternative hypotheses of dispersal against the observed changes in allele frequencies (Felsenstein 1982, Takami et al. 2004). The theoretical implications of alternative dispersal mechanisms in a spatially structured insect habitat are interesting and may have applications to conservation of headwater species. For example, Fagan (2002) showed that the geometry of dendritic stream systems affects the persistence of interacting populations, especially in landscapes subjected to fragmentation or natural disturbance. Thus, a specific landscape structure, interacting with mayfly dispersal behavior, could increase or decrease the environmental pressures faced by small populations of these insects.

Lastly, we are working in collaboration with Luke Jacobus at Purdue University and David H. Funk at the Stroud Water Research Center in a continuing molecular analysis of species in the genus *Ephemerella*. Samples of 9 additional species and subspecies: *Ephemerella excrucians* Walsh, 1862; *Ephemerella aurivillii* (Bengtsson), 1908; *Ephemerella alleni* Jensen & Edmunds, 1966; *Ephemerella hispida* Allen & Edmunds, 1965; *Ephemerella catawba* Traver, 1932; *Ephemerella subvaria* McDunnough, 1931; *Ephemerella dorothea dorothea* (Needham), 1908; *Ephemerella dorothea infrequens* (McDunnough), 1924; *Ephemerella rossi* Allen and Edmunds, 1965; from populations broadly distributed across the United States and Canada have been added to *E. inconstans*, *E. floripara*, *E. rotunda* and *E. invaria* Walker, for analysis. This work will provide molecular characters for constructing a partial phylogeny of the genus based on both morphological and molecular data. In combination with the study of population diversity and dispersal, it may also provide a small link from local genetic processes by which populations become structured and differentiated over relatively short periods of time, to the long-term, large-scale genetic processes underlying the evolution of new characters and species.

Acknowledgments

We thank Melanie Delion, Lauren Moffatt and Joshua Han for their assistance in the lab; David Hawthorne for advice about population genetics techniques; Steve Burian (Southern Connecticut State University) for help with taxonomic identification; Luke Jacobus (Purdue University) for mayfly tissue samples; Jeff Schwierjohann and Cheryl Farfaras (Middle Patuxent Environmental Area, MD) for access to field sites. This work was funded in part by grants from the Environmental Protection Agency and the Middle Patuxent Valley Association.

Literature Cited

Allen, R. K., and G. F. Edmunds, Jr. 1965. A revision of the genus *Ephemerella* (Ephemeroptera, Ephemerellidae). VIII. The subgenus *Ephemerella* in North America. Miscellaneous Publications of the Entomological Society of America **4**:244–282.

Anholt, B. R. 1995. Density-dependence resolves the stream drift paradox. Ecology **76**:2235–2239.

Barrett, L. G., T. He, B. B. Lamont, and S. L. Krauss. 2005. Temporal patterns of genetic variation across a 9-year-old aerial seed bank of the shrub *Banksia hookeriana* (Proteaceae). Molecular Ecology **14**:4169–4179

Fagan, W. F. 2002. Connectivity, fragmentation, and extinction risk in dendritic metapopulations. Ecology **83**:3243–3249.

Felsenstein, J. 1982. How can we infer geography and history from gene frequencies? Journal of Theoretical Biology **96**:9–20.

Galacatos, K., A. I. Cognato, and F. A. H. Sperling. 2002. Population genetic structure of two water strider species in the Ecuadorian Amazon. Freshwater Biology **47**:391–399.

Goodwin, B. J., and L. Fahrig. 2002. How does landscape structure influence landscape connectivity? Oikos **99**:552–570.

Hanski, I., and O. Ovaskeinen. 2002. Extinction debt at extinction threshold. Conservation Biology **16**:666–673.

Hughes, J. M., P. B. Mather, M. J. Hillyer, C. Cleary, and B. Peckarsky. 2003. Genetic structure in a montane mayfly *Baetis bicaudatus* (Ephemeroptera: Baetidae), from the Rocky Mountains, Colorado. Freshwater Biology **48**:2149–2162.

Humphries, S. 2002. Dispersal in drift-prone macroinvertebrates: a case for density independence. Freshwater Biology **47**:921–929.

Jacobus, L. M., and W. P. McCafferty. 2003. Revisionary contributions to North American *Ephemerella* and *Serratella* (Ephemeroptera: Ephemerellidae). Journal of the New York Entomological Society **111**:174–193.

Johst, K., R. Brandl, and S. Eber. 2002. Metapopulation persistence in dynamic landscapes: the role of dispersal distance. Oikos **98**:263–270.

Korkeamaeki, E., and J. Suhonen. 2002. Distribution and habitat specialization of species affect local extinction in dragonfly (Odonata) populations. Ecography **25**:459–465.

Ledger, M. E., A. L. M. Crowe, G. Woodward, and M. J. Winterbourn. 2002. Is the mobility of stream insects related to their diet? Archiv für Hydrobiologie **154**:41–59.

Lowe, W. 2002. Landscape-scale spatial population dynamics in human-impacted stream systems. Environmental Management **30**:225–233.

Maryland Department of Natural Resources. 1997. Maryland Biological Stream Survey Results 1995–1997 (EA-99-6). http://www.dnr.state.md.us/streams /mbss/mbss_pubs.html.

Maryland Department of Natural Resources. 2001. Maryland Biological Stream Survey 2000–2004, Vol.1: Watersheds sampled in 2000 (EA-01-5). http://www.dnr.state.md.us/streams/mbss/mbss_pubs.html.

Meyer, J. L., L. A. Kaplan, D. Newbold, D. L. Strayer, C. J. Woltemade, J. B. Zedler, R. Beilfuss, Q. Carpenter, R. Semlitsch, M. C. Watzin, and P. H. Zedler. 2003. Where rivers are born: the scientific imperative for defending small streams and wetlands. American Rivers and the Sierra Club. http://www.amrivers.org /whereriversareborn.html.

Mueller, U. G., and L. L. Wolfenbarger. 1999. AFLP genotyping and fingerprinting. Trends in Ecology and Evolution **14**:389–394.

Myers, M. J., F. A. H. Sperling, and V. H. Resh. 2001. Dispersal of two species of Trichoptera from desert springs: Conservation Implications for isolated vs. connected populations. Journal of Insect Conservation **5**:207–215.

Pannell, J. R., and B. Charlesworth. 1999. Neutral genetic diversity in a metapopulation with recurrent local extinction and recolonization. Evolution **53**:664–676.

Pannell, J. R. 2003. Coalescence in a metapopulation with recurrent local extinction and recolonization. Evolution **57**:949–961.

Peterson, B. J., W. M. Wolheim, P. J. Mulholland, J. R. Webster, J. L. Meyer, J. L. Tank, E. Marti, W. B. Bowden, H. M. Valett, A. E. Hershey, W. H. McDowell, W. K. Dodds, S. K. Hamilton, S. Gregory, and D. D. Morrall. 2001. Control of nitrogen export from watersheds by headwater streams. Science **292**:86–90.

Petersen, I., Z. Masters, A. G. Hildrew, and S. J. Ormerod. 2004. Dispersal of adult aquatic insects in catchments of differing land use. Journal of Applied Ecology **41**:934–950.

Takami, Y., C. Koshio, M. Ishii, H. Fujii, T. Hidaka, and I. Shimizu. 2004. Genetic diversity and structure of urban populations of *Pieris* butterflies assessed using amplified fragment length polymorphism. Molecular Ecology **13**:245–258.

Turner, M. G., R. H. Gardner, and R. V. O'Neill. 2001. Landscape ecology in theory and practice: pattern and process. Springer, New York, USA.

Wallace, J. B., S. L. Eggert, J. L. Meyer, and J. R. Webster. 1997. Multiple trophic levels of a stream linked to terrestrial litter inputs. Science **277**:102–104.

STONEFLY EGGS IN THE SANDY SHORE AND THE ECOLOGICAL IMPORTANCE OF AN OVOID EGG IN A MOUNTAIN STREAM

Yu Isobe[1], Mayumi Yoshimura[2] and Tadashi Oishi[3]

[1] Faculty of Eco Liberal-arts, Nara Bunka Women's College, Higashinaka, Yamato-Taka, Nara Pref. 635-8530, Japan
[2] Kansai Research Center, Forestry and Forest Products Research Institute, Fushimi, Momoyama, Kyoto Pref. 612-0855, Japan
[3] KYOUSEI Science Center for Life and Nature and Graduate School of Humanities and Sciences, Nara Women's University, Nara 630-8506, Japan

Abstract

Most of the eggs of Perloidea (Plecoptera) have highly developed structures for attachment to the substrate in streams, but there are ovoid eggs without these attachment structures among some species. Thus, ovoid eggs must have another mechanism of spatial stability. We investigated the oviposition site, egg settling sites and the behavior and timing of female oviposition in the family group Perloidea (Plecoptera) in the field.

The eggs of *Sweltsa* (Chloroperlidae) without attachment structures are found mostly in sand at the water edge of current near the riffle habitats. We also found the eggs of Perlodidae and Perlidae with attachment structures, but rarely. Those of Perlidae were found in deeper waters. The females of these three families oviposit on the same riffle surfaces and at similar times. These factors showed that *Sweltsa* eggs are caught in the sands at water edge after floating from the riffle surface where they are oviposited. This is a unique tactic of eggs in Perloidea to settle in the running water.

Key Words: egg distribution; oviposition timing; egg morphology; Plecoptera.

Introduction

Many studies on egg morphology have been conducted in Plecoptera (Hynes 1974, Isobe 1988, 1997), and phylogenetic relations among species were discussed regarding egg morphology (Stark and Szczytko 1982, 1984, 1988). In the family group Perloidea (Plecoptera), it is especially characteristic that the eggs have a collar and an anchor that function as attachment structures. Some species, however, lay ovoid eggs without collars or anchors. Stark and Szczytko (1988) proposed in Archynopterygini (Perlodidae) such eggs of *Setvena* and *Pseudomegarcys* might have lost collar independently. The features on the surfaces of the ovoid eggs likely have an ecological function, however we know of no studies specifically investigating this topic.

We observed reactions of the living eggs to water in the laboratory and searched for the presence of eggs in the field. We found that ovoid eggs of *Sweltsa* sp. (Chloroperlidae) floated on the water in the laboratory dish, while perlid eggs with attachment structures sank into the water and stayed at the bottom of the dish (Yoshimura et al. 2004). The observation and a preliminary egg-searching trial suggested that the ovoid eggs might float into the sands on the shore after oviposition (Yoshimura et al. 2004). Since we could find only small number of eggs at the trial, it was necessary to confirm the observation described above. In order to clarify the ecological significance of the egg shape in Perloidea, we further studied the oviposition site and the behaviors of several species and the locations of egg settling on the streambed, with special reference to sandy shores.

Study Site

We sampled both egg-laying females and eggs under water in a second order mountain stream of the Yoshino-river system, Nara Pref., Japan (Fig. 1A). The stream at the site is covered with broad-leaved trees and surrounded with planted Japanese cedars (Fig. 1B). The stream is approximately 5-m wide with three riffle-pool sets along the 25-m length of the study reach (Fig. 2). Many females were observed to lay eggs along the stream at the site.

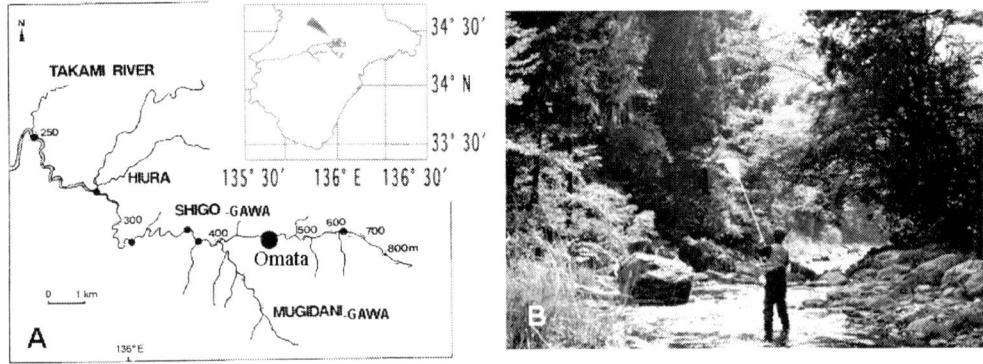

Figure 1. A map of Shigo-gawa (A) and the stream (B). The altitude at the sampling site, Omata, is 450 m above sea level.

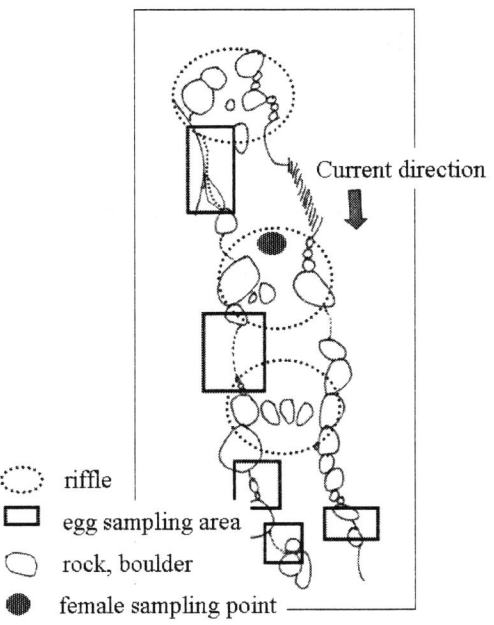

Figure 2. Sampling areas.

Methods

Samples of the egg laying females were collected in the evening on 23 May and 5 June 1997. We caught the females flying on a riffle by using an aerial net (Fig. 1B and Fig. 2). Five min.-sampling and -interval periods were repeated from the first female appearance till dark and air temperature and light intensity were measured during the sampling period.

The egg-samples were collected at several sandy shores (Fig. 2) from May to July 1999. A grab-sample of the sands at each sampling point was screened through a series of sieves (φ8 cm) with five mesh sizes. On and after 1 June, we added ten cups (10 cc/cup) of welled water from the hollow formed by sand scraping to the sands for screening. The samples left in the 500 and 125 μm sieves were taken to the laboratory and eggs were searched with a light microscope.

Results

Egg-Laying Site, Behavior and Timing. It was sunny on both sampling days and it was fairly warm on 5 June 1997. For oviposition, females of several species appeared from the riparian trees and all of them gradually flew down to the middle of riffles with broken waves. Females released their eggs in a mass. We observed two types of the egg-laying behavior, dropping down from the air (*Oyamia lugubris* and *Sweltsa* sp.) and releasing on the water surface (*Kamimuria uenoi, Stavsolus* sp.).

Females of two species were collected on 23 May and five species were captured on 5 June. *Sweltsa* sp. and *K. uenoi,* collected on 23 May, flew onto the riffle earlier than those on 5 June (Fig. 3). The other species, *Oyamia lugubris, Isoperla nipponica* and *Stavsolus* sp. came together with *Sweltsa* sp. and *K. uenoi* on 5 June.

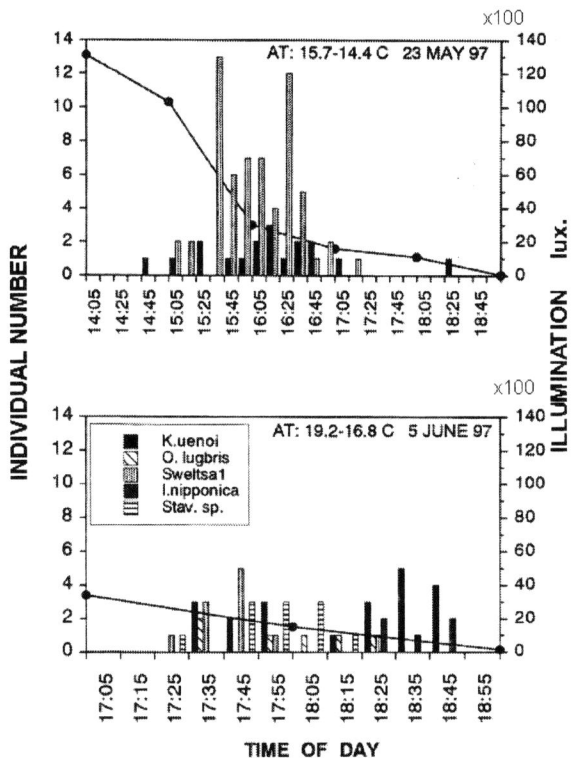

Figure 3. Number of egg laying females caught by a net and illuminations. The range of air temperature (AT) is shown during the sampling period on each day. Bar: females. Solid line: illumination

Egg-Settling Area. We could find the eggs of ten taxa in three families, Chloroperlidae, Perlodidae and Perlidae, in the 12-μm mesh samples (Table 1). Four species, *Isoperla nipponica, Oyamia lugubris, Oyamia* sp. and *Kamimuria quadrata* were identified to species level and others were identified to the genus level. The egg features have been shown in Isobe (1988) and Yoshimura et al. (2004). Only *Sweltsa* has the egg without a collar or an anchor among those taxa.

Table 1. Egg taxa found in the sandy bottoms.

	Chloroperlidae	Perlodidae	Perlidae
18-May	*Sweltsa*	*Stavsolus*	*Kamimuria**
1-Jun	*Sweltsa*	*Stavsolus*	*Kamimuria**
		Ostrovus	*Oyamia* sp.
		I. nipponica	
23-Jun	*Sweltsa*	*Ostrovus*	*Kamimuria**
			Oyamia sp.
			Oyamia lugubris
31-Jul			*Kamimuria quadrata*
			Oyamia lugubris
			Neoperla, Kiotina

*Kamimuria** : *K. uenoi* and/or *K. tibialis.*

Sweltsa eggs were dominated in our samples (184/256 =71.9%, Table 2). Most of them were found on 1 June while the eggs of Perlidae were found later in the season than those of *Sweltsa*. The number of egg-found points was approximately a half of the total sampling points (Table 2).

Table 2. Egg numbers found in the sandy bottoms.

Date	Total sampling points	Egg number (egg-found point)							
		Sweltsa		Perlodidae		Perlidae		Total	
18-May	25	13	(5)	6	(3)	2	(2)	21	(7)
1-Jun	18	162	(7)	14	(14)	6	(4)	182	(9)
23-Jun	21	9	(4)	2	(2)	25	(6)	36	(10)
31-Jul	17	0	(0)	0	(0)	17	(8)	17	(8)
Total	81	184	(16)	22	(11)	50	(20)	256	(34)

Sweltsa eggs were found near the water edge (Fig. 4). We could not find any eggs in the upper and deeper areas from the water edge on 18 May 1999.

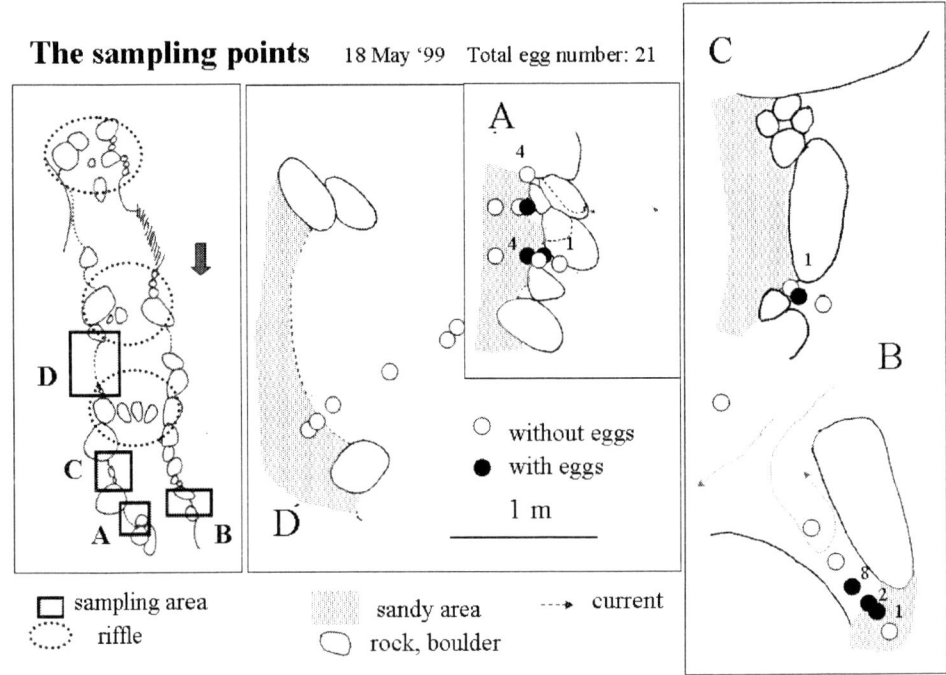

Figure 4. The sampling and egg-found points on 18 May 1999. The eggs were found at the points near the edge of the water. The small numerals: the number of eggs found at the point.

On 1 June, we succeeded to collect many eggs, especially *Sweltsa* (Fig. 5, Table 3). Eggs were not found in the upper area (distance: +5~15 cm) from the water edge and comparatively less in the deeper area (-10~-25 cm). While, at the water edge (0~-5 cm), many eggs (163/182) were found and 7 of 9 sample points retained eggs (Table 3). The points with eggs were the calm places that were hit with the swift currents from the main flows (Fig. 5). *Sweltsa* eggs were highly concentrated at a few points in the areas A and E.

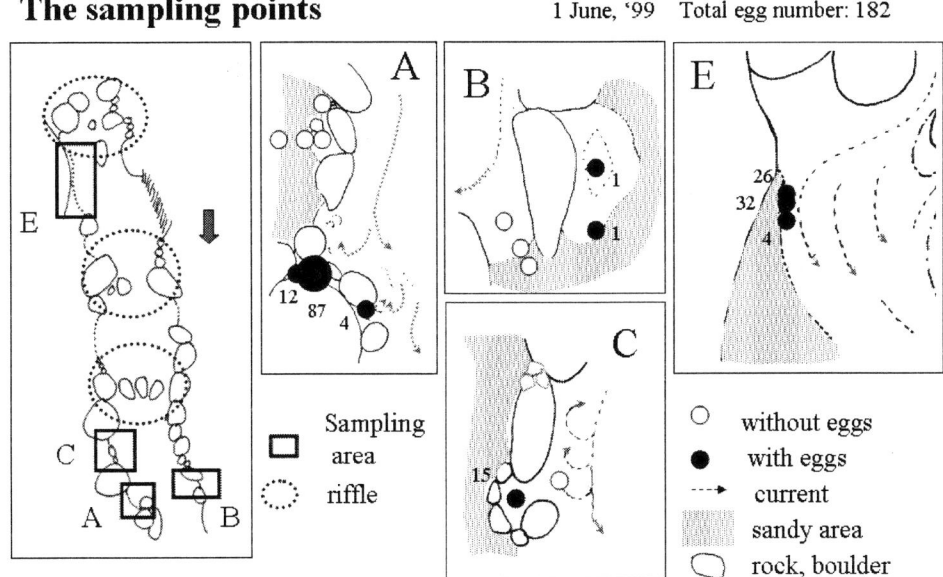

The sampling points   1 June, '99   Total egg number: 182

Figure 5. The sampling and egg-found points on 1 June 1999. The eggs were found at the points that were hit by the current. The small numerals: the number of eggs found at the point.

Table 3. Egg numbers found in three categorized areas in the distance from the water edge on 1 June.

Distance	Total sampling points	Egg number (egg-found point)							
		Sweltsa		Perlodidae		Perlidae		Total	
+5~+15 cm	3	0	(0)	0	(0)	0	(0)	0	(0)
0~-5 cm	198	146	(5)	13	(5)	4	(3)	163	(7)
-10~-25 cm	261	16	(2)	1	(1)	2	(1)	19	(2)

Distance: + upwards from the water edge (land), - downwards (under water)

On 23 June 1999 we found some eggs in the deep areas (Fig. 6, Table 4), although we had some difficulties in sampling. At this time, we could not find many eggs at the water edge (eggs: 13/36, egg-found points: 3/12), while there were more in the deeper areas, (eggs: 20/36, egg-found points: 3/5). Perlid eggs in sandy bottoms showed a tendency to be collected in deeper areas in contrast to *Sweltsa* eggs. Most of the perlid eggs had hatched or had broken anchors. The places where the eggs were found along the water edge were the similar places to those on 1 June 1999.

The sampling points

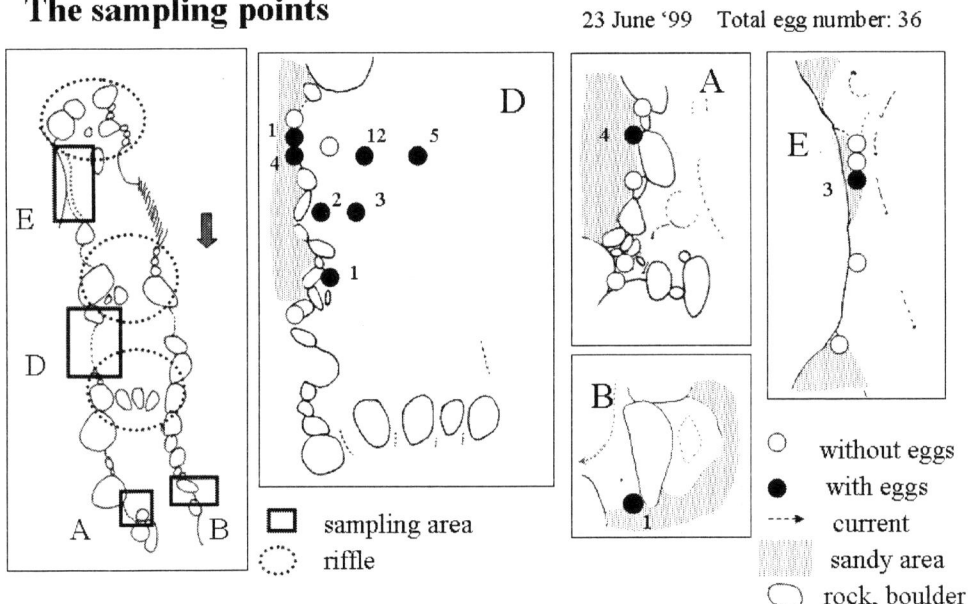

Figure 6. The sampling and egg-found points on 23 June. The eggs were also found in deeper areas. The small numerals: the number of eggs found at the point.

Table 4. Egg numbers found in three categorized areas in the distance from the water edge on 23 June.

Distance	Total sampling points	Egg number (egg-found point)			
		Sweltsa	Perlodidae	Perlidae	Total
0 cm	12	5 (2)	2 (2)	6 (2)	13 (3)
-3~-5 cm	4	0 (0)	0 (0)	3 (2)	3 (2)
-15~-270 cm	5	4 (2)	0 (0)	16 (2)	20 (3)

Distance: + upwards from the water edge (land), - downwards (under water)

Discussion

In Perloidea, most of the eggs have a collar and an anchor for attaching to the substrate (Isobe 1988, 1997). They are considered to be highly adapted to settle on the substrate in the running water. In the family group, however, several species having eggs without attaching structures are known and occur independently of phylogenetic relations. Micropyle position on such eggs including *Sweltsa* suggests the eggs might have lost the attaching ability, such as the ovoid eggs in Archynoperygini (Perlodidae). We focused our study on egg-laying sites and behavior and the egg-settling site.

The females of all taxa observed in this study, including *Sweltsa* with ovoid eggs and the others laying eggs with ability to attach, deposited eggs onto similar places and by similar ways. Both types of eggs were released on the water surface of riffles with broken waves. Such egg-laying sites are highly characteristic among Perloidea. After drifting on or in the swirling water, the two types of eggs are expected to settle into the different places.

We found stonefly eggs for each species approximately reflecting the known egg-laying season. We found many eggs of Sweltsa along the sandy shores. The shores where Sweltsa eggs were found were depositional sites that received flow from upstream riffles where the eggs had been originally deposited. Therefore, we suggest that Sweltsa eggs must generally reach depositional, sandy sites after floating downstream. This was substantiated by our laboratory observation that Sweltsa eggs tend to float on the water surface (Yoshimura et al. 2004). On the other hand, eggs of other species with attachment structures were not found in the sandy shoreline substrates as were the Sweltsa eggs. For example, perlid eggs, which possess a high ability to attach, were more frequently found at the bottom of the primary current of the stream rather than at the shore. When perlid eggs were found in the sands, most of the eggs had already hatched or the anchors were mostly broken, suggesting that these broken eggs were flushed away from the original site of attachment to the sandy shoreline.

Since the sandy shores where the floating Sweltsa eggs are generally very low velocity sites, we expected that recently hatched nymphs of Sweltsa would grow among the sands. Evolutionarily, the ovoid eggs of Sweltsa may have lost the ability to attach, but exploit floating in running water as a tactic to settle into calm places.

Acknowledgments

We would like to acknowledge encouragements from our colleagues in the laboratory at Nara Women's University.

Literature Cited

Hynes, H. B. N. 1974. Observations on the adults and eggs of Australian Plecoptera. Australian Journal of Zoology **29**:37–52.

Isobe, Y. 1988. Eggs of Plecoptera from Japan. Biology of Inland Waters 4:27–39.

Isobe, Y. 1997. Anchors of stonefly eggs. Pages 349–361 *in* P. Landolt, P., and M. Sartori, editors. Ephemeroptera and Plecoptera: Biology-Ecology-Systematics. MTL, Fribourg Switzerland.

Stark, B. P., and S. W. Szczytko. 1982. Egg morphology and phylogeny in Pteronarcyidae (Plecoptera). Annals of the Entomological Society of America **275**:519–529.

Stark, B. P., and S. W. Szczytko. 1984. Egg morphology and classification of Perlodinae (Plecoptera: Perlodidae). Annales de Limnologie. **20**:99–104.

Stark, B. P., and S. W. Szczytko. 1988. Egg morphology and phylogeny in Arcynopterygini (Plecoptera: Perlodidae). Journal of the Kansas Entomological Society **61**:143–160.

Yoshimura, M, Y. Isobe, and T. Oishi. 2004. A preliminary report on the ecological meaning of egg morphology (Plecoptera). Biology of Inland Waters **19**:1–7.

LIFE CYCLES, GROWTH AND PRODUCTION OF STONEFLY POPULATIONS

O. Loskutova

Laboratory of Ecology of Aquatic Organisms, Institute of Biology, Komi Science Center, Russian Academy of Sciences, Syktyvkar, Russia

Abstract

The life cycles of two stonefly species, *Arcynopteryx compacta* (McLachlan) and *Diura nanseni* (Kempny), was studied in a stream in Europe, located in the northeast of European Russia. These large nymphs dominate plecopteran biomass in the rivers of the Pechora Basin, although they are second-rate species in number. Both species have short incubation period of egg and a long nymphal growth period, and a highly synchronized emergence period occurring in early June. The life cycle of these species is univoltine.

Relationship between stonefly body mass and linear size (body length and maximal head width) is established. I obtained data testifying to the prevalence of negative allometry in the growth of studied species. The group growth curves of both species is described by an exponential function. The relationships between body mass and duration of development have been obtained.

Wintertime specific production, monthly specific production for vegetation period and total annual specific production were calculated. Basic production is produced by populations in wintertime under ice, owing to high biomass and relatively high specific production. Population *A. compacta* turned out to be most productive (2.4 g/m^2), production population of *D. nanseni* was smaller (1.3 g/m^2). Calculated P/B coefficients were the following: 4.7 for *D. nanseni* and 5.8 for *A. compacta*.

Key words: Plecoptera; life histories; growth; production; north of Russia.

Introduction

Several scientific papers are devoted to life cycles of stoneflies: Ulfstrand 1968, Harper and Magnin 1969, Brittain 1973, Harper 1973, Lillehammer et al. 1989, Stewart et al. 1990. It was established that part of species has a univoltine (annual) and others have semivoltine (perennial) life cycle. Large-sized species have the longest life cycle, from 3 to 4 years long (Ulfstrand 1968, McDiffett 1970, Nikolaeva 1977). The literature denotes the importance of studying life cycles for the purpose of calculating production (Benke 1984).

In the study of life cycles and stonefly larvae growth, an average body length is suitable only for approximate calculation of larvae age in univoltine species if we do not identify the number of stages within the life cycle (Svensson 1966, 1977, Elliott 1967, Schwarz 1970, Benedetto 1973). Owing to frequent molting, the adjoining stages have little relative difference in size. There is a considerable transgression of features and stages that can not be distinguished (Nikolaeva 1977). Sometimes age is identified by the form of wings expressed via the wing length / head width ratio (Beer-Stiller and Zwick 1995). One method of studying the dynamics of growth of aquatic organisms is to analyze the size structure of species populations. This study can be performed with the help of histograms of the frequency-dimension distribution of organisms.

The present work is not aimed at identification of larvae age stages. Therefore we used larvae body length and head capsule width for description of lifecycles of mass-forming species.

Materials and Methods

Life cycles of stoneflies have been studied at the research station of the Laboratory of Ecology of Aquatic Organisms of the Institute of Biology, located in the middle stream of the Shugor River (64°21'N, 58°28'E) rich in Salmonds (the Pechora river tributary). Larvae of stoneflies were collected from the river every 10 days in 1983 and 1989 and treated with 4% formalin solution. Simultaneously, adult insects were caught. The larvae were measured in laboratory conditions. While judging size composition of stoneflies populations, such parameters as body length, weight and head capsule breadth were taken into account. The head capsule breadth between extreme prominence eye points and the body length from front side of head capsule to the tenth segment end were measured using a microscope. Larvae weight was evaluated with the analytical balance to 0.1 mg accuracy. Before weighing, each larvae was dried on filter paper till moist spots disappeared. Sex affiliation of mature larvae was determined by external genital organs.

Interrelation between body linear size (L – body length, mm) and weight (W, mg) was approximated as the power function: $W=qL^b$, where q and b are equation constants, calculated with empirical data by the least-squares method (Winberg 1971). Interlink between body weight and head capsule breadth was established by analogy: $W=gd_k^a$, where W is an average larvae weight under the given head capsule breadth (mg); g is a weight of larva with a head capsule breadth of 1 mm; d_k is a larvae head capsule breadth; and a is a constant.

Mean specific larvae weight increase rate (C_W) for the time interval $t_2 - t_1$ was defined:

$$C_W = \frac{\ln W_2 - \ln W_1}{t_2 - t_1} \qquad (1)$$

where W_1 and W_2 are larvae weight values to the time t_2 and t_1 (Vinberg 1966). Larvae weight was defined as a mean body weight to the sample-selecting time.

Monthly production was calculated as a product of mean specific production (C_W), mean biomass (B) of population in the given month and this month day number: $P = C_W * B * 30$ (31).

Results and Discussion

Life Cycles. Arcynopteryx compacta (McL.) is the largest stonefly in the Pechora River basin, reaching 21 mm length and 88–178 mg weight. The larvae of *A. compacta* inhabit boulder-pebble bottoms along the entire river length. In the bottom biocenosis of the middle reaches of the Shugor river, this species dominates the biomass, making up ca. 108.7 g/m^2 or 47.2% of total biomass of stoneflies, in the ice-free season. According to this value, it is a second-rate species. Average frequency of larvae in the ice-free season was 5.3 inds./m^2, or 3.8% of total number of stoneflies.

By late May, there are mature nymphs in the river that begin adult emergence. Nymphs accumulate in masses near the water surface, most frequently on stretches with slower current. In the evening (about 21:00), individual larvae come to the river bank through the cracks in the ice. According to our observations of 1983, at this time the larvae slowly moved inland perpendicular to the river. The larvae were languid and some of them seemed torpid. Around midnight, the situation changed. Masses of larvae actively moved from the river towards the forest on snow, 3–5 larvae per 1 m^2, moving speed was 1 m per minute. The distance to the forest was 10–15 m, the larvae hide in the dense grass. That night, 70% of nymphs emergence into adults. Main emergence takes place on June, 3. Adults actively copulated on the river bank.

In the rivers of the Sub-Polar Urals, the nymphs emerge later (in middle June in the lower stream of the Kozhym river and a few days later in the upper stream of the Kozhym). The number of *A. compacta* was 20 - 30 inds./m^2 of riverbank. The adults could be found on the larch trunks 100 m from the river. Females had up 24 - 211 eggs (egg size 0.35 x 0.4 mm) in their abdomens.

In the Shugor River, young drifting larvae were registered already on 15 June. The larvae of first age stage are 1.2–1.7 mm (Lillehammer 1988) in length. Histograms (Fig. 1, 2) show the larvae increase in size during the vegetative season and grow continuously during the winter. It is confirmed by augmenting dimension-age characteristics and occurrence of exuviae in winter samples. At the moment of ice break, the larvae have well-developed wing covers. The larvae are already at the last nymphal instar before emergence into adults. The adults crawls away, the females then fly back to the riverside for depositing their eggs in the early June.

In Fennoscandia, *A. compacta* has a two-year life cycle: the first winter is spent in an egg diapause, hatching taking place the following spring (Lillehammer 1988).

O. Loskutova

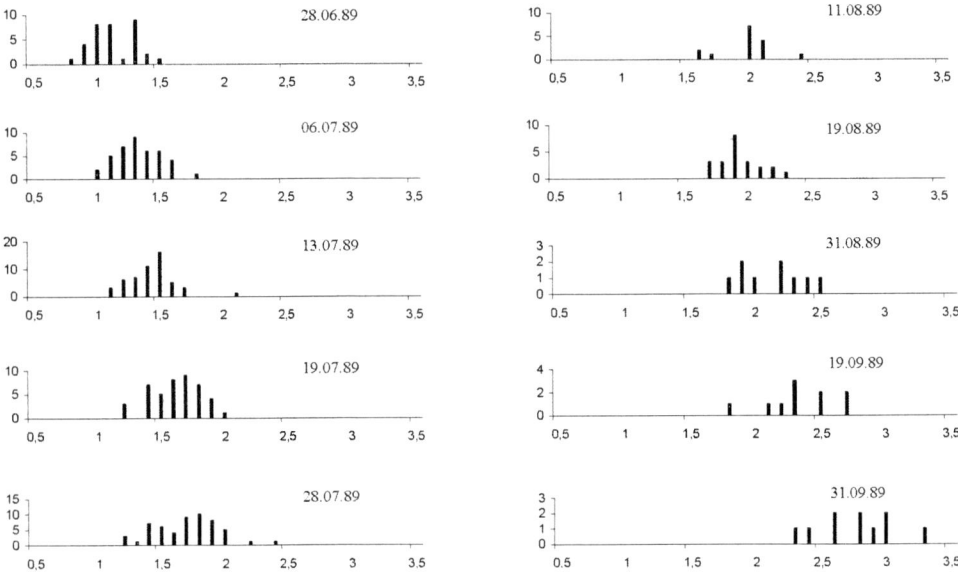

Figure 1. Seasonal changes of the head capsule width of *A. compacta*. Legend: axis of the abscissa – the head capsule width; axis of ordinates – number of individuals.

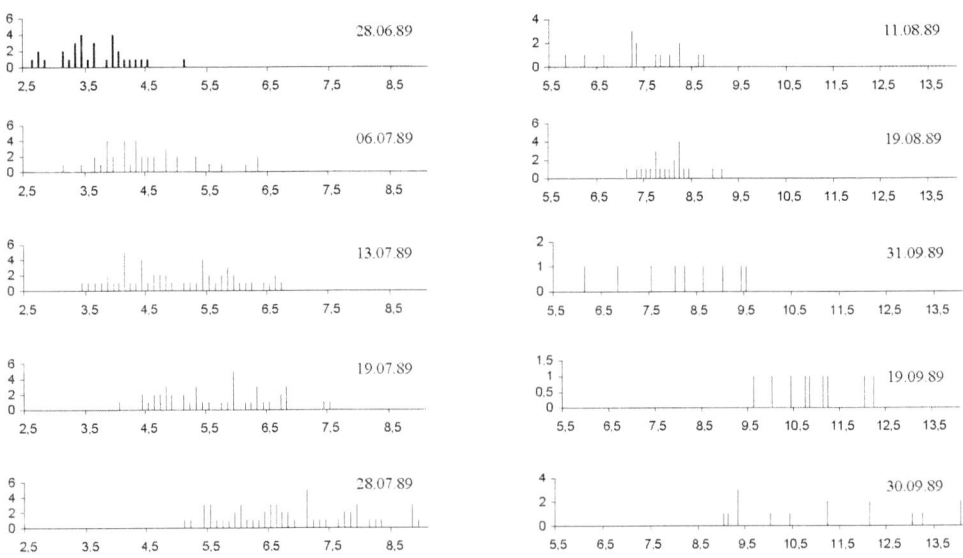

Figure 2. Seasonal changes of the body length of *A. compacta*. Legend: axis of the abscissa – the body length; axis of ordinates – number of individuals.

Group of Species of Genus Diura. We united two species, *Diura bicaudata* and *D. nanseni*, under the same group, because it is hard to identify the young larvae. The larvae of this group hatch at the end of July. Their density at that time was 44.9 inds./m^2, biomass – 16.2 mg/m^2. This group dominates the frequency, making 37.5% of total number of all stoneflies. In August, the larvae of both can be easily identified; *D. bicaudata* is less frequent than *D. nanseni*.

Diura nanseni (Kempny). The mature nymphs of this species are large. Their length approaches 14 mm and body weight ca. 30 mg. The larvae inhabit boulder-pebble river bottoms. In the ice-free period, their density is 21.9 inds./m^2, biomass – 74.9 mg/m^2, or 15.8% of total number and 32.5% of biomass of stonefly larvae.

   Diura males emerge in the early June and females a few days later. During one night and day, 98% of arrived larvae molted into the adult insects. The weight of exuvia is usually 1.5–2 mg. The majority of larvae leave the river to 10 June. Many individuals of both sexes are found on the river banks under the stones. Copulation takes place at that time. Females fly and deposit eggs usually in daytime, in sunny windless weather. The early larval stages with body size ca. 1.0 mm appear in the river in late June to early July. The larval actively grew in the summer (Figs. 3 and 4) and growth continued under the ice. The adult insects fly out soon after the ice breakup.

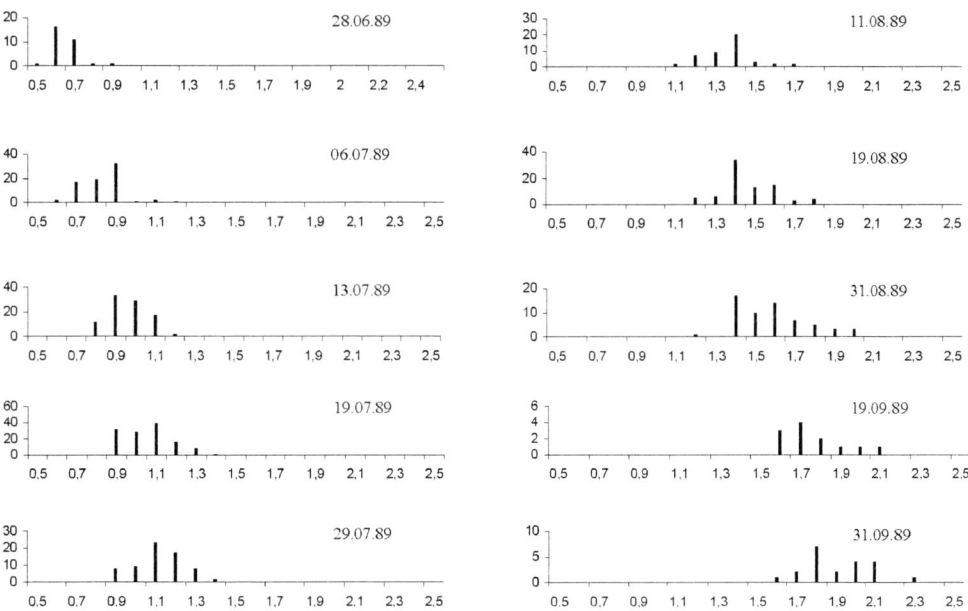

Figure 3. Seasonal changes of the head capsule width of *D. nanseni*. Legend: axis of the abscissa – the head capsule width; axis of ordinates – number of individuals.

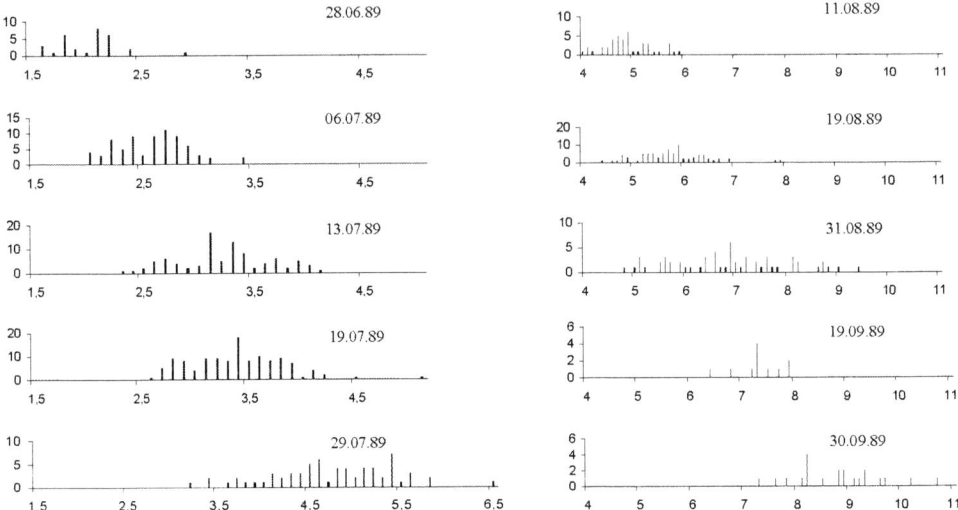

Figure 4. Seasonal changes of the body length of *A. compacta*. Legend: axis of the abscissa – the body length; axis of ordinates – number of individuals.

As *A. compacta D. nanseni* also have diapausal eggs in Norway and the species has a two-year life cycle (Lillehammer 1988). Thus, the annual transformation into the univoltine type of lifecycle is established in both stonefly species.

Growth of Stonefly Larvae. At the present time, for determination of total production, hydrobiologists widely use the method based on the study of growth rate of individuals in the population (Alimov 1981; Balushkina 1987, Golubkov 1987).

Several researchers reveal that the insect larvae have a parabolic growth type (Konstantinov 1958, Golubkov 1979a,b, 1981, 1982, Alimov and Vlasova 1980), described growth by a concave curve. The larvae rapidly inhibit their growth at the last instar and completely cease growing during the adult stage. The larvae usually undergo several instars, but the number of instar is not strictly fixed and can vary depending on growth conditions, sex or individual features of organisms.

The study of growth regularities in several mass-forming species in the Kedrovaya river in the Far East revealed the majority have the parabolic growth type (Bogatov 1994). The stonefly larvae with a temporizing lifecycle have exponential growth type (Teslenko 1992). Other scholars also confirms the possibility of applying the exponential function for description of growth of certain aquatic insects (Winberg 1966, Golubkov 1987, 2000).

One should distinguish between the linear growth and weight growth. The rate of both depends upon the age of organism, its physiological state and environmental conditions (Alimov 1981). The relation between the body weight and linear dimensions is described by the law of allometric growth.

The dependency of larvae body weight upon body length and head capsule width. The relation between body weight and body size of animals is an important ecological-physiological index of their state. This relation is of great value for the applied studies. The preliminary dependency of body weight upon body size of aquatic organisms was used in hydrobiology for a long time. Body weight is calculated by the body linear size if direct weighing is impossible or undesirable (small-sized organisms, experimental studies with feeding the water organisms, reconstruction of fish body weight by body fragments in fish nutrition studies).

For approximation of relation between the body mass and body linear size in different taxa of water organisms, we use exponential equation $W = qL^b$, where W is body weight (mg), L is body length (mm), q and b are constants defined from the empiric data by the method of minimal squares. For this purpose, exponential equation is transferred into the linear form via finding logarithms: log W = logq + b logL, where q is a constant equal to W at L=1; and b - a constant reflecting the level of allometry of organism's form changes during the growth of the organism. If the organism growth without changing its form, then b=3, if the relation of linear sign to body weight decreases, then b>3 (positive allometry), if increases, then b<3 (negative allometry) (Winberg 1971). Interrelation between body mass and linear dimensions in mayflies was studied by S. M. Golubkov (1979) and T. M. Tiunova (1993). For the order of mayflies and stoneflies, a general equation of this dependency ($W=0.026L^{2.84}$) was suggested by Alimov and Vlasova (1980), for three stonefly species from the Far East by Nikolaeva (1984) and for five species by Teslenko (1992).

Our original data on body length and weight in formalin-fixed animals provided us with equation parameters for the larvae of two stonefly species from the Perlodidae family (Table 1). The empirical data for each studied species in the logarithmic system of coordinates, form a line (Figs. 5 and 6). In the obtained equations, the values of the parameter q varied between 0.042 and 0.076, i.e., differed only 1.8 times. The values of exponential indices were below 3. They varied between 2.408 and 2.570, i.e., 1.1 times. The relative errors in parameters negligible. It is seen from Table 1 that the same result was obtained for stonefly species from the Far East (Nikolaeva 1984, Teslenko 1992). Thus, our original and literature data showed that stonefly larvae grow with the negative allometry.

The obtained exponential equations calculated for two predatory species of stoneflies, reliably describe the ratio between the larvae body size and body weight. For two species of Perlodidae, a general equation was derived:

$$\lg W = (-1.2268 \pm 0.0661) + (2.4115 \pm 0.0803)\lg L \text{ or } W = 0.059L^{2.412}.$$

Stonefly larvae make a great share in nutrition of fish in the Pechora tributaries, especially in the ice covered period. Fish stomachs preserve mainly chitinized head capsules of stonefly larvae. That is why it is necessary to know the mass of certain mayfly species when calculating the fish ration. To identify the dependence between body weight and head capsule width, Smock L.A. (1980), E.V. Balushkina (1987) used the following exponential equation:

$$W = gd^a_k,$$

where W is an average larvae weight by certain head capsule weight (mg); g is a weight of larvae with head width of 1 mm; d_k – is larvae head capsule width and a is a constant.

Table 1. Parameters (q, b) of equations of dependencies of body weight (W) upon bode size (L) in stonefly larvae.

Species	n	Limits L, mm	W, mg	$q \pm M_q$	$b \pm M_b$	Source
A. compacta, 1983	185	1.9-21.8	0.7-138.8	0.0415±0.015	2.570±0.171	(1)
A. compacta, 1989	302	2.6-21.8	1.7-116.2	0.076±0.0074	2.397±0.081	(1)
D. nanseni	1068	1.9-16.3	0.1-37.6	0.055±0.016	2.408±0.086	(1)
Allonarcys sachalina	246	3.5-50.0	0.8±1490	0.030±0.0004	2.79±0.005	(2)
Kamimura luteicauda	180	2.5-37.0	0.5-815	0.040±0.005	2.77±0.007	(2)
Stavsolus japonicus	120	1-21.0	0.1-126	0.040±0.007	2.73±0.007	(2)
Plecoptera	-	-	-	0.013-0.04	2.695-3.003	(3)

NB: n – number of measurements; M_q – error of parameter q; M_b – error of parameter b. Sources – (1) – author's original data; (2) – Nikolaeva 1984; (3) – Teslenko 1992.

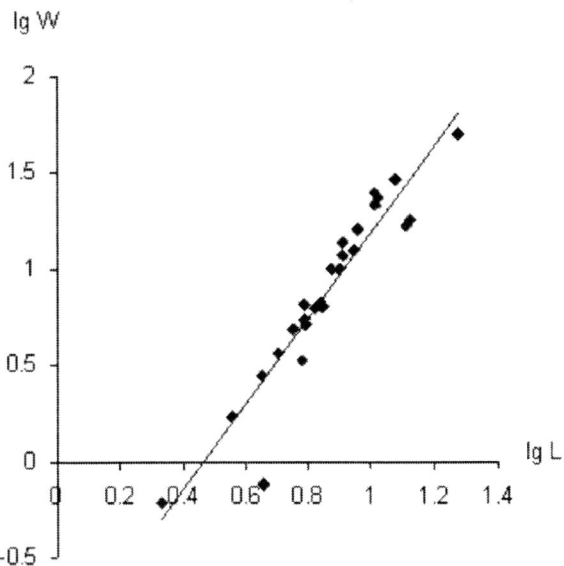

Figure 5. Relation between the lg of body weight and the lg of body length of *Arcynopteryx compacta.*

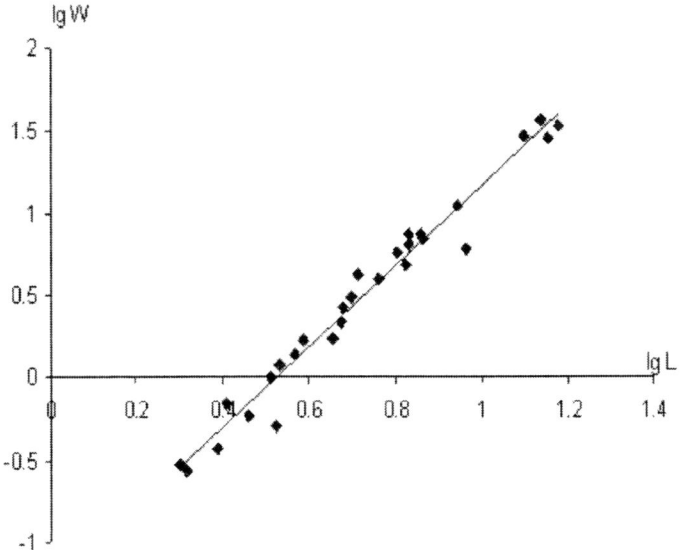

Figure 6. Relation between the lg of body weight and the lg of body length of *Diura nanseni.*

Our work presents equations between average body weight and head capsule width for two stonefly species from the Perlodidae family (Table 2). Dependence of body weight (W) upon head capsule width (d_k) for the Perlodidae family is expressed with the general equation:

$$Lg\ W = (0.4061\pm0.0795) + (1.7670\pm0.3138)Lg\ d_k\ or\ W = 2.54d_k^{1.767}$$

Table 2. Parameters (q, a) of equations of dependencies of body weight (W) upon head capsule width (d_k) in stonefly larvae.

Species	n	Limit		$q\pm M_q$	$a\pm M_a$	Source
A. compacta	487	0.75 – 3.72	1.7 – 116.2	2.5101±0.212	2.1617±0.160	(1)
D. nanseni	1068	0.20 – 2.60	0.3 – 34.0	1.610±0.032	3.197±0.136	(1)
Plecoptera				2.437±0.110	2.287±0.102	(2)

NB: n – number of measurements; W – average larvae mass by the given head capsule width (d_k); M_q – error of parameter q; M_b – error of parameter b. Sources – (1) – author's original data; (2) –Teslenko 1992.

Vital functions of animals, i.e., feeding, metabolism, fertility and growth rate depend upon the body weight. That is why it is necessary to study growth expressed by body weight in animals (Alimov 1981).

The studied species have a univoltine lifecycle, population produced annually only one generation, therefore growth can be investigated via average body weight (Tables 3 and 4). We calculated specific growth rate of stonefly larvae separately during July, August and September, and generally during ice period.

Statistical proceeding of the material gives the following dependencies of body weight upon the duration of larvae growth during vegetation season. Figures 7 and 8 show the curves of group growth of larvae of *Arcynopteryx compacta* and *Diura nanseni*. Curves have exponential nature, that is why we apply exponential function of time to the description of larvae growth:

A. compacta $W = 2.32\ e^{0.0286t}$ (larval group 1.7–29.4 mg);

D. nanseni $W = 0.56e^{0.0345t}$ (larval group 0.3–11.0 mg).

Brittain (1983) also described exponential growth of stonefly species *Capnia atra*, *Nemoura cinerea*, *N. avicularis* and *Diura bicaudata* in a subarctic lake in

Norway.   Average specific growth rate   (determined according to the growth-curve

Table 3. Body length (L), width of head capsule (dk) and weight (W) of *A. compacta.*

Date	n	L, mm	d_k, mm	W, mg
28.06	34	2.6-5.1 (3.63±0.09)	0.75-1.50 (1.1±0.03)	1.7
6.07	40	3.1-6.3 (4.46±0.12)	1.0-1.75 (1.3±0.03)	2.8
13.07	54	3.4-7.8 (5.05±0.13)	1.05-2.10 (1.4±0.03)	3.6
19.07	44	4.0-7.5 (5.62±0.13)	1.15-2.0 (1.6±0.03)	4.9
28.07	57	5.2-9.3 (6.85±0.19)	1.20-2.35 (1.7±0.05)	6.8
11.08	15	5.8-8.7 (7.44±0.22)	1.55-2.35 (1.9±0.05)	9.9
19.08	23	6.9-9.1 (7.89±0.11)	1.65-2.30 (1.9±0.04)	10.0
31.08	9	6.1-9.5 (8.10±0.38)	1.75-2.45 (2.1±0.08)	13.6
19.09	10	7.7-12.2 (10.57±0.41)	1.80-2.70 (2.3±0.09)	23.1
30.09	10	9.3-14.2 (11.95±0.54)	2.25-3.25 (2.7±0.09)	29.4
2.02 female larvae	5	10.0-17.0 (13.9±1.37)	3.00-3.50 (3.24±0.08)	68.5
2.02 male larvae	3	9.0-15.0 (12.5±1.80)	3.00-3.10 (3.00±0.03)	35.0
20.04	13	16.0-21.0 (18.4±0.93)	3.05-3.70 (3.43±0.09)	39.1
31.05 female larvae	9	19.1-21.8 (20.5±0.32)	3.41-3.72 (3.5±0.03)	116.2
31.05 male larvae	8	15.3-17.9 (16.8±0.28)	2.81-3.00 (2.9±0.03)	48.1
1.06 adult females	13	13.2-19.4 (16.8±0.45)	2.80-3.20 (3.0±0.04)	42.3
1.06 adult males	8	12.3-15.6 (13.4±0.47)	2.40-2.70 (2.59±0.05)	28.3

O. Loskutova

Table 4. Body length (L), width of head capsule (d_k) and weight (W) of *D. nanseni.*

Date	n	L, mm	d_k, mm	W, g
28.06	30	1.1-2.9 (2.01±0.05)	0.20-0.85 (0.63±0.01)	0.3
06.07	74	2.0-4.0 (2.59±0.04)	0.60-1.20 (0.84±0.01)	0.7
13.07	93	2.3-4.7 (3.23±0.05)	0.75-1.15 (0.95±0.01)	1.0
19.07	124	2.6-5.1 (3.41±0.04)	0.90-1.40 (1.05±0.01)	1.2
29.07	67	3.2-6.5 (4.74±0.08)	0.90-1.40 1.12±0.02	2.7
11.08	45	4.0-7.4 (5.02±0.09)	1.10-1.70 (1.36±0.02)	3.1
19.08	80	4.4-7.9 (5.76±0.07)	1.20-1.80 (1.47±0.02)	3.9
31.08	60	4.8-9.4 (6.84±0.14)	1.20-2.00 (1.58±0.02)	6.5
19.09	11	6.4-7.9 (7.32±0.27)	1.60-2.10 (1.72±0.03)	7.0
30.09	21	6.8-10.7 (8.73±0.41)	1.60-2.25 (1.89±0.04)	11.0
2.02	14	9.0-11.5 (10.6±0.35)	2.00-3.10 (2.58±0.12)	19.4
31.05	20	10.8-16.3 (13.22±0.34)	2.00-2.60 (2.18±0.04)	20.4
1.06 female imago	11	9.7-14.8 (12.73±0.79)	2.20-2.40 (2.31±0.02)	34.0
1.06 male imago	24	8.4-13.6 (11.73±0.26)	1.70-2.40 (2.03±0.02)	19.9

approximating equation) was higher in *Diura nanseni* larvae and was 0.035 days^{-1}. In the second studied species, specific growth rate was lower, 0.029 days^{-1}.

Tables 3 and 4 show that average larvae body weight in the beginning of ice-free period made in *D. nanseni* 11 mg. in *A .compacta* – 29.4 mg. At the end of ice-free period, body weight was higher: in *D. nanseni* – 20.4 mg and in *A. compacta* – 48.1 mg. The formula (1) gives us the specific growth rate in ice period, which is higher in *A. compacta* – 0.015 days^{-1}, in *D. nanseni* lower, 0.012 days^{-1}. Thus, the larvae of both species continue their growth under ice, but slower. All the facts described above allow us to conclude on the exponential growth of stonefly larvae. They grow all the year round at a high specific rate.

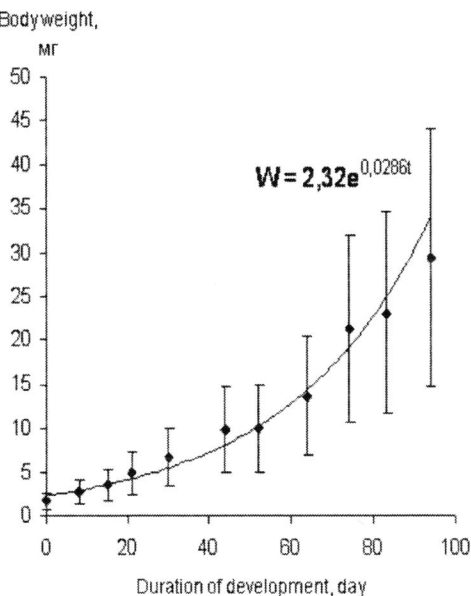

Figure 7. The curves of grow of body weight of larvae *Arcynopteryx compacta* in 1989.

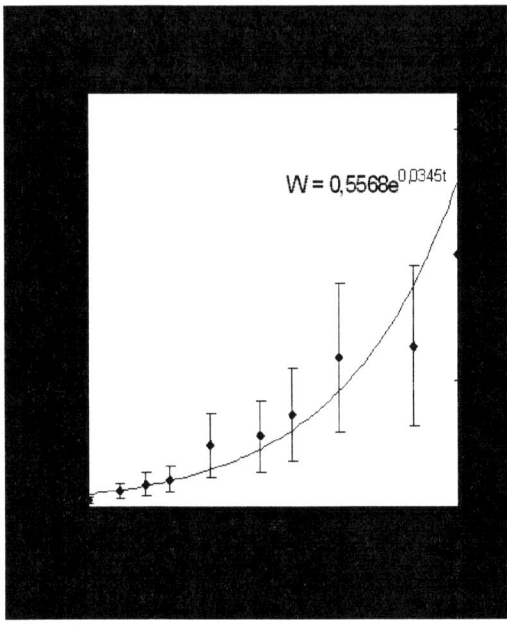

$$W = 0.5568e^{0.0345t}$$

Figure 8. The curves of grow of body weight of larvae *Diura nanseni* in 1989.

Production of Populations of Mass-Forming Stonefly Species. By the exponential type of growth, various-age individual organisms grow at an equal rate. By that, average specific growth of larvae in the population numerically is equal to specific production of the larvae.

We calculated production (P) of two stonefly species separately for each month during the vegetation growth period (June–September) according to average C_w defined by approximation of growth with the help of exponent and average biomass of larvae (B) in each month (Table 5).

Larger larvae of *A. compacta* had higher production indices in summer months than the larvae of *D. nanseni*, but in the ice period their production was considerably higher.

It is seen from Table 5, that population forms larger production in wintertime, owing to high biomass of larvae and to relatively high specific production. It is known that stoneflies are cold-loving insects (Hynes 1970, Brittain 1976, Lillehammer et al. 1989).

Table 5. Production of populations of two stonefly species in the Shugor river in 1983 during vegetation season, mg/m^2.

Species	C_w	June		July		August		September		P
		B	P	B	P	B	P	B	P	
D. nanseni	0.035	4.3	4.52	22.3	24.20	52.5	56.96	61.7	64.79	150.47
A. compacta	0.029	4.9	4.26	9.5	8.54	43.2	38.84	295.2	256.82	308.46

The P/B-coefficient (relation of organic matter annual accretion to the average annual biomass) was higher in populations of a univoltine species *A. compacta*. Our original data are similar to P/B-coefficients of predatory stonefly larvae in the Kedrovaya river in the Far East. The latter varied between 4.4 and 8.1 (Teslenko 1992). These values correspond to the theoretical value for animals developing during one year (Benke 1984, Golubkov 2000).

Acknowledgments

The author thanks Dr. S. Golubkov for the assistance during preparation of the manuscript and meaningful discussions.

Literature Cited

Alimov, A. F. 1981. Functional ecology of fresh-water bivalves. Leningrad, Nauka, 248 pp. Proceedings of the Zoological Institute AS USSR. V. 96 (in Russian)

Alimov, A. F., and V. G. Vlasova. 1980. Growth and production of Trichoptera larvae. Zoological Journal **59**:1483–1493 (in Russian).

Balushkina, E. V. 1987. Functional importance of chironomide larvae for continental water bodies. Leningrad, Nauka, 197 pp. in Proceedings of the Zoological Institute AS USSR. Vol. 143 (in Russian).

Beer-Stiller, A., and P. Zwick. 1995. Biometric studies of some stoneflies and a mayfly (*Plecoptera* and *Ephemeroptera*). Hydrobiologia **299**:169–178.

Benedetto, L. A. 1973. Growth of stonefly nymphs in Swedish Lapland. Entomologisk Tidskrift **94**:15–19.

Benke, A. C. 1984. Secondary production of aquatic insects. Pages 289–322 *in* V. H. Resh, and D. M. Rosenberg, editors. The Ecology of Aquatic Insects. Praeger Publishers, New York.

Brittain, J. E. 1973. The biology and life cycle of *Nemoura avicularis* Morton (*Plecoptera*). Freshwater Biology **3**:199–210.

Brittain, J. E. 1976. The temperature of two Welsh lakes and its effect on the distribution of two freshwater insects. Hydrobiologia. **48**:37–49.

Elliott, J. M. 1967. The Life histories and drifting on the *Plecoptera* and *Ephemeroptera* in a Darmoor stream. Journal of Animal Ecology **36**:343–362.

Golubkov, S. M. 1979a. Growth and exchange of *Cloeon dipterum* L. (Ephemeroptera) larvae under different temperatures. Pages 147–156 *in* Experimental and Field Assays on Biological Bases of Lake Productivity. Leningrad, ZIN AN USSR (in Russian).

Golubkov, S. M. 1979b. Growth rate and body calorie content of single mayfly larvae of the Baetidae family. Page 157–168 *in* Experimental and Field Assays on Biological Bases of Lake Productivity. Leningrad, ZIN AN USSR (in Russian).

Golubkov, S. M. 1981. Parabolic growth type and its efficiency for fresh-water invertebrates. Pages 115–125 *in* Principal Approaches to Freshwater Ecosystems Survey. Leningrad, ZIN AN USSR (in Russian).

Golubkov, S. M. 1982. Relation of growth rates and energy exchange at fresh-water invertebrates: Author's abstract for a Biology Candidate's degree. Minsk, 22 pp. (in Russian).

Golubkov, S. M. 1987. Dynamics of individual production and growth types of invertebrate animals. Page 199–211 *in* Production-Hydrobiological Studies of Water Ecosystems. Leningrad, Nauka (in Russian).

Golubkov, S. M. 2000. Functional Ecology of Aquatic Insects (ed. Alimov). St. Petersburg (in Russian).

Harper, P. P. 1973. Life histories of *Nemouridae* and *Leuctridae* in southern Ontario (*Plecoptera*). Hydrobiologia **41**:309–356.

Harper, P., and E. Magnin. 1969. Cycles vitaux de quelques Plecopteres des Laurentides (insectes). Canadian Journal of Zoology **47**:483–494.

Hynes, H. B. N. 1970. The ecology of stream insects. Annual Review of Entomology **15**:25–42.

Konstantinov, A. S. 1958. Biology of chironomides and their production. *(Biologiya khironomid i ikh razvedeni)*. Proceedings of Saratov Branch of All-Union Research Institute of Lake and River Fishery (in Russian).

Lillehammer, A. 1988. Stoneflies (Plecoptera) of Fennoscandia and Denmark. Fauna Entomologica Scandinavica. Vol. 21, 165 p.

Lillehammer, A., J. E. Brittain, S. J. Saltveit, and P. S. Nielsen. 1989. Egg development, nymphal growth and life cycle strategies in *Plecoptera*. Holarctic Ecology **12**:173–186.

McDiffett, W. F. 1970. The transformation of energy by a Stream detritovore Pteronarcys scotti (*Plecoptera*). Ecology **51**:975–988.

Nikolaeva, E. A. 1977. On the life cycle of *Kamimuria luteicauda* Klapalek (Plecoptera) in the Kedrovaya river. Page 53–63 *in* Fresh-Water Fauna of the Kedrovaya pad' Rreserve. Vladivostok, DVNTS AN USSR (in Russian).

Nikolaeva, E. A. 1984. Intercorrelation between larvae body weight and length of three stonefly species (Plecoptera). Pages 121–127 *in* Freshwater Biology of the Extreme North. Vladivostok, DVNTS AN SSR (in Russian).

Schwarz, P. 1970. Autökologische Untersuchungen zum Lebenzyclus von Setipalpia-Arten (*Plecoptera*). Archiv für Hydrobiologie **67**:141–206.

Smock, L. A. 1980. Relationships between body size and biomass of aquatic insects. Freshwater Biology **10**:375–383.

Stewart, K., B. Hassage, S. Holder, and M. Oswood. 1990. Life Cycles of Six Stonefly Species (Plecoptera) in Subarctic and Arctic Alaska Streams. Annals of the Entomological Society of America. **83**:207–214.

Svensson, P. O. 1966. Growth of nymphs of stream living stoneflies (*Plecoptera*) in northern Sweden. Oikos **17**:197–206.

Svensson, B. 1977. Life cycle, energy fluctuations and sexual differentiation in *Ephemera danica* (*Ephemeroptera*), a streamliving mayfly. Oikos **29**:78–86.

Teslenko, V. A. 1992. The role of Plecoptera larvae in communities of freshwater invertebrates in the small salmon-rich Kedrovaya river. Author's abstract for a Biology Candidate's degree. St. Petersburg, 21 pp. (in Russian).

Tiunova, T. M. 1993. Mayflies of the Kedrovaya river and their ecophysiological characteristics. Vladivostok, Dal'nauka, 194 pp. (in Russian).

Ulfstrand, S. 1968. Life cycles of benthic insects in Lapland streams (Ephemeroptera, Plecoptera, Trichoptera, Diptera, Simuliidae). Oikos **19**:167–190.

Winberg, G. G. 1966. Growth rate and exchange intensity of animals. Modern biology advances. **61**:274–293 (in Russian).

Winberg, G. G. 1971. Linear dimensions and body weight of animals. General Biology Journal. **32**:614–723 (in Russian).

ADULT STONEFLY BEHAVIOR BEFORE AND AFTER MATING

Mayumi Yoshimura

Forest Conservation and Management Group, Kansai Research Center, Forestry and Forest Products Research Institute, Nagaikyutaro 68, Momoyama, Fushimi, Kyoto 612-0855, Japan

Abstract

The postemergence behavior of adults of three species of Perlodidae (*Stavsolus japonicus, Isoperla aizuana, Tadamus kohnonis*) was observed in a cage under seminatural conditions and adult behavior pre- and post-mating was compared. Most of the stoneflies positioned themselves among the branches throughout their adult stage; however, the distribution pattern of individuals was different before and after mating. They tended to be on the ceiling and walls before mating and on the ground after mating. Before mating, individuals on the ceiling and walls were active and their antennas were moving rapidly; however, individuals on the ground did not move. After mating, individuals on the ground were observed walking around, though individuals on the ceiling and walls did not often move. Perhaps individuals found on the walls and ceiling before mating were going to fly away; whereas, females on the floor and under the stones after mating were going to oviposit.

Key words: adult stonefly; behavior; before mating; after mating, mating place; mating time; Perloidea; Stavsolus; Isoperla; Tadamus.

Introduction

Some stonefly species, such as *Microperla (brevicauda)* and *Phasgonophora capitata*, mate just after emergence (Harper 1973b, Hanada et al. 1992), therefore their habitat throughout their adult stage would be close to their emergence site. However other species such as *Taeniopteryx nivalis*, *Isoperla clio, Isoperla transmarina, Nemoura trispinosa* and *Amphnemura delosa* do not have mature eggs at the time of emergence, and egg maturation occurred several days after mating (Sephton and Hynes (1984), Harper 1973a, b). During these several days eating is essential for egg maturation (Harper 1973a, Jop and Szczytko 1984). After emergence they might fly from their emergence stream site to nearby vegetation, like the behavior of *Isoperla signata* (Jop and Szczytko 1984). Harper (1973a) reported that *Amphinemura nigritta* flew into the riparian bushes and trees where they fed quite readily on any available vegetable matter. Because of movement from their emergence site to other vegetation in the area, it is difficult in the field to track their migration behavior after emergence. Only in *Capnia atra* has the entire adult behavior pattern been clarified (Hynes 1976, Zwick 1980).

Recently, several authors have reported on field investigations of the postemergence habitats of a few species (Jackson and Resh 1989, Kuusela and Huusko 1996, Griffith et al. 1998). However, adult migration behavior from emergence to oviposition has not been entirely clarified. The difficulty of observing adult behavior after emergence in the field, necessitated transferring three perlodid species (*Stavsolus japonicus, Isoperla aizuana, Tadamus kohnonis*) to the laboratory under seminatural conditions. We observed their behavior and attempted to predict their adult behavior in the field. Based on these observations and predictions, we discussed the differences in adult behavior before and after mating.

Methods

Around 10 a.m. on 29 March 1999, newly emerged individuals, including eight pairs of *Isoperla aizuana*, seven males and six females of *Stavsolus japonicus*, and seven males and three females of *Tadamus kohnonis*, were collected at Kizu, Kyoto Prefecture (34°44' N, 135°49' E). These stoneflies were brought to the laboratory and kept in a cage (120 × 200 × 120 mm) (Fig. 1) under seminatural light conditions.

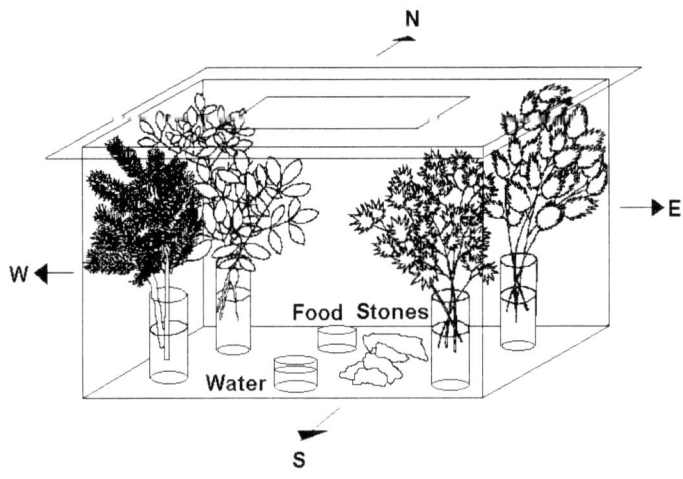

Figure 1. The keeping cage with seminatural light environment.

There were two windows in the laboratory close to the cage, one large, (about 1.8 × 1.8 m) with a southern exposure, located about 0.5 m from the long axis of the cage and one small (about 1.0 × 1.0 m) with an eastern exposure, positioned about 2.5 m from the short axis of the cage. The photoperiod was similar inside and outside the cage except during the time of observations. Branches of four plant species, *Juniperus chinesis* (Chinese juniper), *Nandina domestica* (nandin), *Acer palmatum* (maple) and *Kerria japonica* (Japanese rose), were placed at the corners of the cage and their placement was changed every night after the observations ceased. The cage also contained four small stones, water for ovipositing and honey diluted with water (1:10) soaked into cotton for food.

Day and night from 1 to 30 April, the three species were observed to determine the position of each individual in the cage: on the ceiling, on the walls, between the ceiling and the branches, between the walls and the branches, in the branches, on the ground and under the stones. The direction the wall faced was also recorded for the individuals on the walls at night. Nocturnal observations were done every day between 22:00 and 23:00 under dim-light conditions (120 lux). Diurnal observations were done every second or third day between 12:00 and 13:00 under natural-light conditions. Air temperatures varied from 14°C to 20°C at the nighttime observations and it varied from 16°C to 22°C at the daytime observations. The position of the mating location was also recorded.

The number of individuals at each position was tallied and divided into two groups, those recorded before the first observed mating in the cage and those recorded afterwards. The observations from the various positions were combined into three groups: on the ground and stones, in and around the branches and on the walls or ceiling. The differences in position recorded before versus after mating were analyzed using a Chi-square test among the three groups for each time period, night and day. The differences among the cage positions occupied while mating, and the differences in the direction the wall faced where they were found the night before mating were also analyzed using a Chi-square test or a Binomial test, depending on the sample size. Although these observations were finished on 30 April, the stoneflies were kept in the cage until their death.

Results

Adult Life Cycle. Some emerging individuals of *Isoperla aizuana*, and many exuviae of *Stavsolus japonicus* and *Tadamus kohnonis*, were observed at 10:00 a.m. on the day the individuals were collected. Individuals of *S. japonicus* and *T. kohnonis* had already cast off their skins and were standing motionless under a stone until their body color returned to its original hues. The emerging individuals of *I. aizuana* were found on the vertical sides of large stones or under stones a few meters from the stream edge. During their emergence, they did not react to any small stimuli.

The first matings of *S. japonicus, I. aizuana* and *T. kohnonis* ,observed in the cage, were thirteen, eight and four days after the collection date, respectively. Mating of *S. japonicus* was observed nine times, half of which occurred during the daytime, with the location not restricted to a specific place. Mating of *I. aizuana* was observed eleven times, all of which were at night ($P < 0.0005$, binomial-test), with most of them taking place in and around the branches ($P < 0.05$, binomial-test). Mating of *T. kohnonis* was observed 81 times, two-thirds of which took place at night ($\chi^2_{cal} = 4.5$, $df = 1$, $P < 0.05$, χ^2-test) (Table 1). Nighttime mating was strongly correlated with the areas in and around the branches (χ^2_{cal} = 16, df = 2, P < 0.001, χ^2-test), but there was no preference for a particular mating place in the daytime. Six days after the first observed mating in the cage, eggs were found on the abdomen of *S. japonicus*; for *I. aizuana* it was five days until they were seen, while no individuals with eggs or deposited eggs could be found for *T. kohnonis*.

Table 1. Number of individuals on each position and timing during mating in three species.

		Wall	Ceil-ing	Branches	Wall Branches	Ceiling Branches	Ground	Under Stones	Total
S. japonicus	Night	0	0	1	2	0	2	0	5
	Day	0	0	2	1	0	0	1	4
I. aizuana	Night	0	0	6	2	1	2	0	11
	Day	0	0	0	0	0	0	0	0
T. kohnonis	Night	4	6	16	10	4	7	3	51
	Day	3	6	5	6	4	3	4	30

Adult Distribution Pattern in a Keeping Cage. Distribution pattern of position in the cage in *S. japonicus* was different between before and after mating at night (males: χ^2_{cal} = 25.9, df = 2, N = 189, P < 0.001; females: χ^2_{cal} = 14.6, df = 2, N = 115, P < 0.001; χ^2-test). The number of males on the wall was higher before mating, but it decreased after mating (Fig. 2a). Reversely, the number of males found under the stone and in the branches increased after mating. The number of females on the wall and ceiling was higher before mating, but it decreased after mating (Fig. 2b). Reversely, the number of females under the stone and on the ground increased after mating. This significant difference in the distribution patterns before and after mating was also observed in the daytime (Fig. 2c, d) (males: χ^2_{cal} = 12.9, df = 2, N = 114, P < 0.01; females: χ^2_{cal} = 11.7, df = 2, N = 88, P < 0.01; χ^2-test).

In *I. aizuana*, distribution pattern of position in the cage was different before and after mating at night (males: χ^2_{cal} = 20.2, df = 2, N = 104, P < 0.001; females: χ^2_{cal} = 44.6, df = 2, N = 220, P < 0.001; χ^2-test). The rate of males on the wall was higher

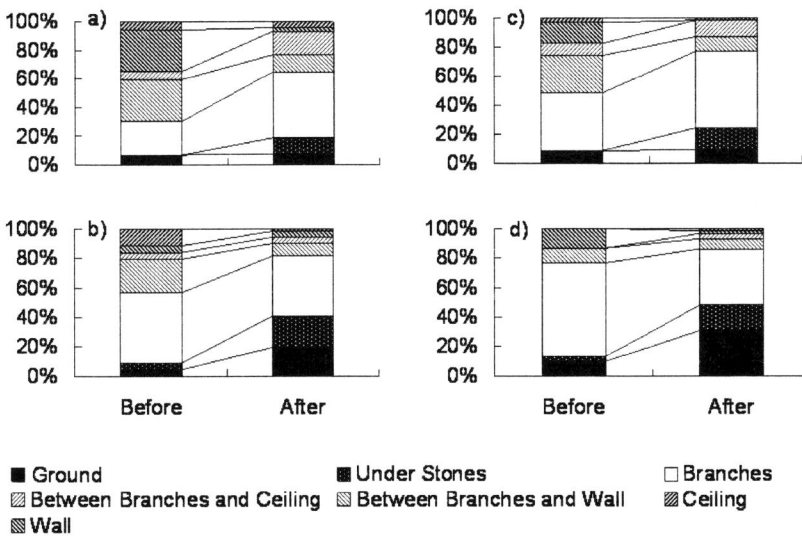

Figure 2. The rate of individuals positioned at each place before and after mating in *Stavsolus japonicus*. a) Males, Night; b) Females, Night; c) Males, Day; d) Females, Day.

before mating, but it decreased after mating (Fig. 3a). Reversely, the rate of males in the branches increased after mating. The rate of females on the wall was higher before mating, but it decreased after mating (Fig. 3b). Reversely, the rate of females in the branches and under the stones increased after mating. This change in the distribution pattern could also be observed during the day (Fig. 3c, d); however, the difference was not significant.

In *T. kohnonis*, there were only a few days before the first observed mating, providing so little data that it was insufficient to delineate a statistically significant difference between cage positions before and after mating. However, there was a significant difference in the female nighttime positioning before and after mating ($\chi^2_{cal} = 14.5$, $df = 2$, $N = 82$, $P < 0.001$; χ^2-test) (Fig. 4).

Directions on the Wall. Males of *S. japonicus* were found on the east wall of the cage significantly more often than elsewhere (males: $\chi^2_{cal} = 9.1$, $df = 3$, $P < 0.05$, χ^2-test) (Fig. 5a); however, there was no difference in the directional preference for females. Males of *I. aizuana* had a statistically significant preference for the south wall ($\chi^2_{cal} = 4.6$, $df = 1$, $P < 0.05$, χ^2-test) (Fig. 5b), and females for the south-east facing wall ($\chi^2_{cal} = 17$, $df = 1$, $P < 0.001$, χ^2-test). *T. kohnonis* (Fig. 5c) tended to be on the south wall, but statistical significance could not be established due to the small number of samples.

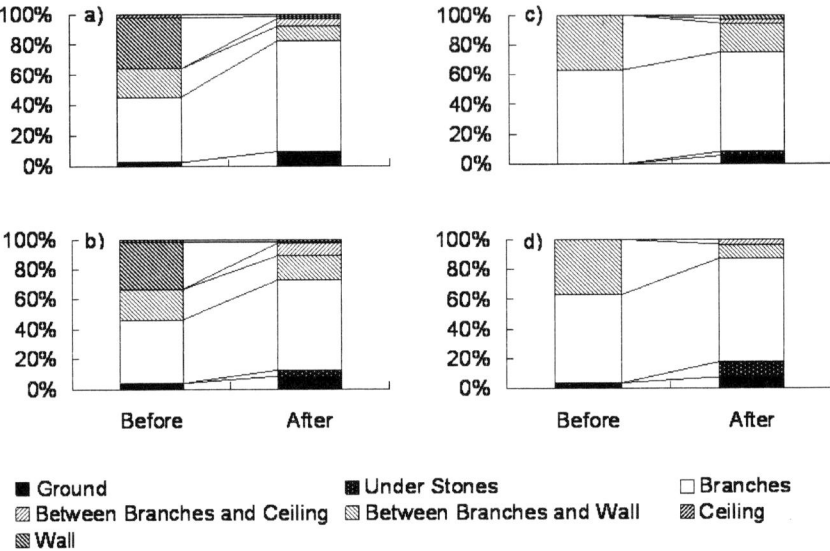

Figure 3. The rate of individuals positioned at each place before and after mating in *Isoperla aizuana*. a) Males, Night; b) Females, Night; c) Males, Day; d) Females, Day.

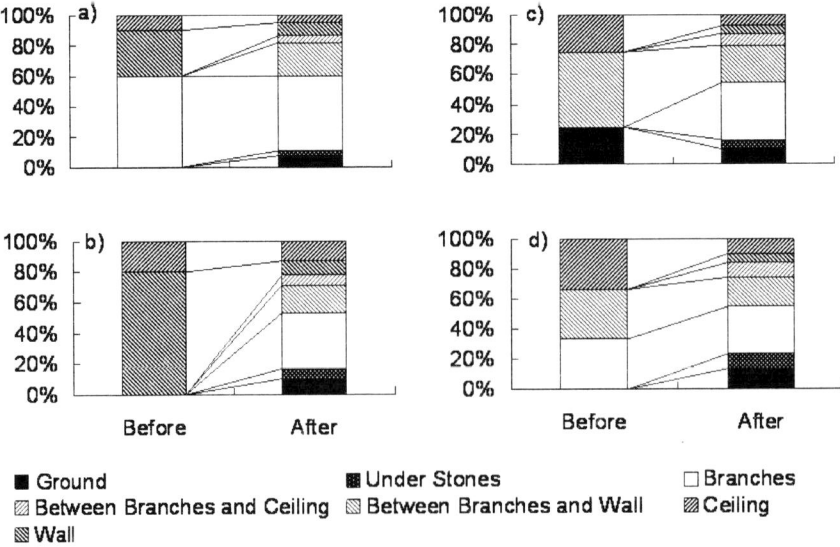

Figure 4. The rate of individuals positioned at each place before and after mating in *Tadamus kohnonis*. a) Males, Night; b) Females, Night; c) Males, Day; d) Females, Day.

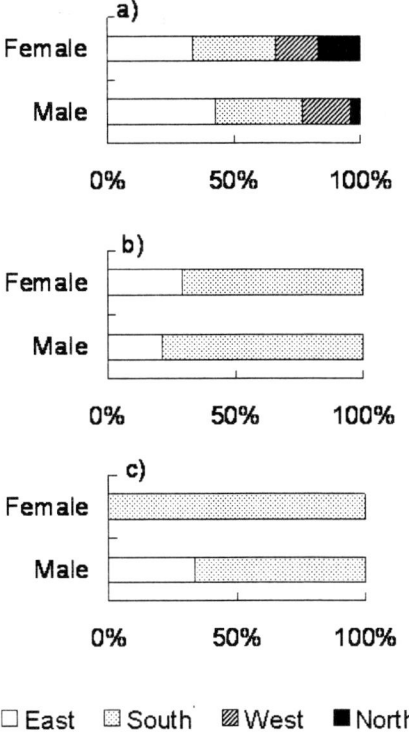

Figure 5. The rate of individuals positioned on the walls facing each direction before mating at night in a) *Stavsolus japonicus*, b) *Isoperla aizuana* and c) *Tadamus kohnonis*.

Adult Behavior in a Keeping Cage. Before mating, individuals on the ceiling and walls were actively walking and moving their antennas rapidly. When the ceiling was removed, individuals on the walls and ceiling began to walk around and tried to walk out of the observation cage. Individuals on the ground, under the stones and in the branches did not move when they were observed. After mating, individuals on the ceiling and walls did not walk around as often and they did not try to fly out even if the ceiling was removed. However, individuals on the ground and under the stones were observed walking around. Egg carrying individuals of *I. aizuana* and *S. japonicus* were observed walking around on the ceiling, the walls and the ground. Most of the individuals in the branches stayed in positions where their wings could be in contact with the leaves; however, if contact was lost between their wings and the leaves, they would move forward in the tree until they reestablished contact.

Discussion

In this study, the position the stoneflies occupied in the cage was different before and after mating and these differences were similar whether observed during the day or night, or whether they were males or females. The rate of individuals on the wall and ceiling was high; before mating, the mating rate was low on the ground and in the branches. Individuals on the wall and ceiling were active, while those on the ground and in the branches were inactive. Individuals on the wall and ceiling came out when the ceiling was removed. Stewart et al. (1969) observed that the individuals released in the laboratory flew up close to the ceiling. They believe this behavior suggests adults fly into trees or high vegetation after emergence. Individuals active on the walls and ceiling in this study would also be considered to be the individuals that are trying to fly away.

After mating, individuals on the ground and under the stones were observed walking around, while individuals on the walls and ceiling moved little. Individuals with eggs were also observed moving around in this study. It is believed active females on the ground and under the stones might be in their ovipositing stage. The rate of individuals on the wall and ceiling was low, whereas that on the ground and under the stones was high. Therefore, after mating, most of the individuals stopped flying and rested in the branches or were moving around the stream to oviposit.

The position of individuals on the walls at night before mating differed amongst the species being studied in this investigation. As reported by Koshima et al. (1981), individuals of *Capnia atra* use sunlight as a compass in order to decide which direction to walk. Brightness seems to be the most important environmental factor used to determine flight direction. In addition, Hayashi et al. (1997) found that individuals of *Sweltsa* sp. tend to fly to the trees while the sun is shining; however, when the light is dim they tend to fly towards the light. During this diurnal investigation there was no strong light; however, moonlight could penetrate through the windows facing east and south and the morning sunlight could shine through the east window. At the time of the field collections for this study, *S. japonicus* emerged earlier than *I. aizuana*, therefore, individuals of *S. japonicus* were showered with eastern light longer than *I. aizuana*. The fact that many of the *S. japonicus* individuals were found on the eastern wall of the observation cage might be related to this emergence timing.

Individuals of *Nemoura trispinosa* fly up to higher altitude after emergence and their mating occurred there in the riparian shrubs or grasses (Harper 1973a). In this study, individuals of *I. aizuana* and *T. kohnonis* also preferred locations in or around the branches as a mating site at night. In addition, these two species tended to mate at night. Individuals of *Claassenia sabulosa* were only active during the two hours just after dusk (Alexander and Stewart 1996), an adaptation which would minimize their risk from predators. However, Zwick (1980) reported the active period of the adult stage in this species is during the daytime and that mating occurs in that time frame. In *Microperla brevicauda* and *Kamimuria tibialis*, the mating and pairing timing was not concentrated in a specific period, but occurred in the daytime (Hanada et al. 1992). Individuals of *S. japonicus* mated both at night and during the day in this study. The timing of mating would be expected to be specific to the species and the mating site. Differences in light preference, mating time and mating site among species would be related to differences in adult behavior. Further investigations should be carried out in order to clarify these differences and the entire adult migration behavior.

Literature Cited

Alexander, K. D., and K. W. Stewart. 1996. Description and theoretical considerations of mate finding and other adult behaviors in a Colorado population of *Claassenia sabulosa* (Plecoptera: Perlodidae). Annals of the Entomological Society of America **89**:290–296.

Griffith, M. B., E. M. Barrows, and S. A. Perry. 1998. Lateral dispersal of adult aquatic insects (Plecoptera, Trichoptera) following emergence from headwater stream in forested Appalachian catchments. Ecology and Population Biology **91**:195–201.

Hanada, S., Y. Isobe, and K. Wada. 1992. Behaviors of adult stoneflies of *Microperla brevicauda* Kawai and *Kamimuria tibialis* (Pictet)(Plecoptera: Insecta). Biology of Inland Waters **7**:1–9.

Harper, P. P. 1973a. Life histories of Nemouridae and Leuctridae in Southern Ontario (Plecoptera). Hydrobiologia **41**:309–356.

Harper, P. P. 1973b. Emergence, reproduction and growth of setipalpian Plecoptera in Southern Ontario. Oikos **24**:94–107.

Hayashi, Y., Y. Isobe, and T. Oishi. 1997. Diel periodicity of emergence of *Sweltsa* sp. (Plecoptera; Chloroperlidae). Pages 52–59 *in* P. Landolt, and M. Sartori, editor. Ephemeroptera and Plecoptera: Biology-Ecology-Systematics. MTL, Fribourg, Switzerland.

Hynes, H. B. N. 1976. Biology of Plecoptera. Annual Review of Entomology **21**:135–153.

Jackson, J. K., and V. H. Resh. 1989. Distribution and Abundance of adult aquatic insects in the forest adjacent to a northern California stream. Entomological Society of America **18**:278–283.

Jop, K., and S. W. Szczytko. 1984. Life cycle and production of *Isoperla Signata* (Banks) in a central Wisconsin trout stream. Aquatic Insects **6**:81–100.

Koshima, S., and T. Hidaka. 1981. Life cycle and behavior of the wingless winter stonefly, *Eocapnia nivalis*. Biology of Inland Waters **2**:19–43.

Kuusela, K., and A. Huusko. 1996. Postemergence migration of stoneflies (Plecoptera) into the nearby forest. Ecological Entomology **21**:171–177.

Sephton, D. H., and H. B. N. Hynes. 1984. The ecology of *Taeniopteryx nivalis* (Fitch) (Taeniopterygidae; Plecoptera) in a small stream in southern Ontario. Canadian Journal of Zoology **62**:637–642.

Stewart, K. W., G. L. Atmar, and B. M. Solon. 1969. Reproductive morphology and mating behavior of *Perlesta placida* (Plecoptera: Perlidae). Annals of the Entomological Society of America **62**:1433–1438.

Zwick, P. 1980. Handbuch der Zoologie, 7 Plecoptera (Steinfliegen). Walter de Gruyter, Berlin.

THE POST GLACIAL DISTRIBUTION OF NEW ZEALAND MAYFLIES

T. R. Hitchings

*Canterbury Museum, Department of Invertebrate Zoology, Rolleston Avenue,
Christchurch 8001, New Zealand*

Abstract

Situated astride the boundary of the Pacific and Austro-Indian tectonic plates, New
Zealand has had a turbulent geological history. Land uplift, earthquakes and
volcanism continue to the present day. Separation from Antarctica and the eastern
coast of Australia in the late Cretaceous resulted in the isolation of the mayfly
population. The survivors evolved into the present fauna, which includes
42 described species in 19 genera and 8 families. All species are endemic. Species
distribution maps, based on more than 8,000 records included herein, illustrate a
limited number of distributional patterns resulting from vegetational modification
and past climatic and geological events. The results support the view that a centre of
dispersal of double gilled leptophlebiid species was in northern New Zealand, but
that glaciation in the south resulted in a different pattern of radiation for some single
gilled Leptophlebiidae, the Nesameletidae and the Rallidentidae.
Keywords: distribution; Ephemeroptera; glaciation; interglacials; mayflies; New
Zealand; Pleistocene; dispersal.

Introduction

Like those from other parts of the world, New Zealand mayflies share a tolerance of
a very narrow range of environmental conditions, especially in the preadult stages.
Thermal tolerance (Quinn et al. 1994), substrate size, flow rates, catchment
development and its impact on water quality (Quinn and Hickey 1990, Harding and
Winterbourn 1995) limit mayfly distributions. Furthermore, no New Zealand species
has radiated into standing waters, except for the mayflies on the occasional rocky
outcrop on a windy lake shore and in one or two lakes at the termination of glaciers.

Herein, I present distribution maps based on the results of extensive field
sampling. However, it is arguable that any patterns shown are not necessarily purely
the result of biogeographical processes (Hitchings 2001). Rather, they may also
reflect the extent of unmodified habitat and/or intensity of collecting effort. The
interpretation of such maps needs to be made with these caveats in mind.

Results and Discussion

Species Widespread Throughout New Zealand. After the end of the glacial period 14 000
bp, New Zealand became largely covered with forest (Leathwick et al. 2003, Fig. 1).

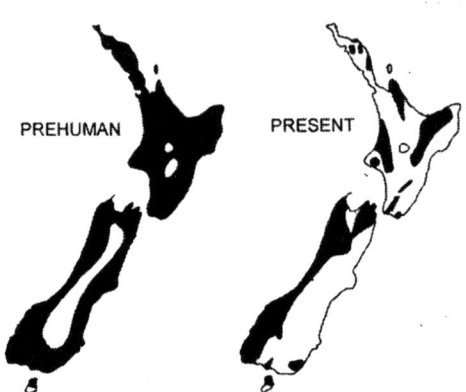

Figure 1. Forest land cover changes, prehuman and present.   (Redrawn from Leathwick et al. 2003).

Before colonization by Polynesians, circa. a thousand years ago, more than four-fifths of New Zealand was forested. Now only approximately a quarter of its surface area remains covered with indigenous forest. In areas that have lost at least 90% of their forest cover, eight species of mayfly are still widely distributed. *Coloburiscus humeralis,* a filter feeder, is a typical example of this group (Fig. 2).

Figure 2. Distribution of *Coloburiscus humeralis*.

C. humeralis shows a preference for, but is not confined to, waterways flowing through forest fragments (Wisely 1962, Harding and Winterbourn 1995). It and seven other species (19%, 4 genera), not only continue to inhabit forested waterways, but have adapted to and are tolerant of conditions in areas that have been converted from forest to agricultural land or become scrub or tussock country. This group appears to be adapting well with environmental change. They could be described as winners in the stakes for evolutionary adaptation. This nucleus of common, well adapted species, has long been recognized (Winterbourn, Rounick and Cowie 1981, Rounick and Winterbourn 1982, Cowie 1985).

Table 1. Ephemeroptera species distributed throughout New Zealand, both in indigenous forest and in areas which have lost at least 90% of forest cover.

Coloburiscus humeralis	*Deleatidium myzobranchia*
Deleatidium autumnale	*Deleatidium vernale*
Deleatidium fumosum	*Neozephlebia scita*
Deleatidium lillii	*Nesameletus ornatus*

Another group of nine species (21%, 7 genera) is widely distributed on both islands, but restricted to the remaining forested regions. It seems probable their range has been reduced by habitat loss arising from the activities of the human population and its commensals.

Table 2. Ephemeroptera species whose range is now largely restricted to the remaining indigenous forest.

Ameletopsis perscitus	*Nesameletus flavitinctus*
Atalophlebioides cromwelli	*Oniscigaster distans*
Austroclima jollyae	*Oniscigaster wakefieldi*
Austroclima sepia	*Zephlebia spectabilis*
Deleatidium cerinum	

An example of a species confined to forest areas is *Ameletopsis perscitus,* the only predatory mayfly among the New Zealand fauna (Fig. 3). Its nearest relatives are *Mirawara* (Australia) and *Chiloporter* (South America).

Whatever the impact of geological uplift and planation, climate change, the rise and fall of sea levels, the advance and retreat of ice sheets and forests, the present wide distribution of these 17 species suggests an adaptability and vagility which has concealed the impact of the Pleistocene.

100 km

Figure 3. Distribution of *Ameletopsis perscitus.*

Background to New Zealand's Present Geography.   The New Zealand Ephemeroptera, along with some other insect orders, are considered to have derived from old Gondwanaland forms whose vicariance came about by continental drift since the Cretaceous, circa. 80 Ma. This has been discussed elsewhere (e.g., Winterbourn 1980). The basement rocks (Fig. 4) of the western portion of the South Island derive from the late Precambrian (circa. 560 ma) to Devonian (345 Ma) Gondwana.

The remainder of New Zealand is more recent. Initially, ancestral New Zealand may have been a substantial continent, half the size of Australia. It has a very large continental shelf, however, this eroded down in the Cretaceous (136 Ma to 65 Ma) to a small low lying archipelago. Many extinctions may have taken place at this time, but conditions would also have been favorable for much speciation. New Zealand's largest genus, *Deleatidium,* a single gilled leptophlebiid, may have begun to radiate in the Oligocene, if not earlier (Fig. 5). At least six species of this genus are now widespread in both islands.

The Miocene saw continuous land, hilly but not mountainous, across a range of latitudes similar to the present (Stevens 1980). It was not until the Pliocene (2 mya) that the sea flooded across a low mid point in the alpine spine. New Zealand became separated into two main islands by a sea channel now 13 km wide at its narrowest point. Cook Strait is now a physical barrier for 17 mayfly species, 7 confined to the North Island (Table 3) and 6 to the South. For example, Cook Strait is a boundary for two species of *Ichthybotus,* a burrower sometimes considered a single genus family with nearest relatives in South America (*Euthyplocia*) (Fig. 6).

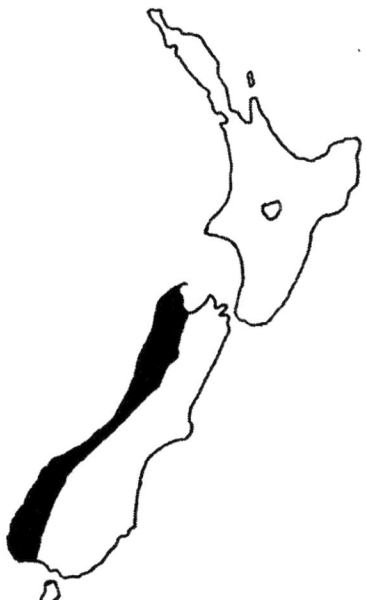

Figure 4. Basement rocks of New Zealand (in black) derived from Gondwana. (Redrawn from Coates and Cox 2002)

Figure 5. In the Oligocene, 35 –5 Ma, the region consisted of numerous small islands shown in black, superimposed on the present land masses. (Redrawn from Coates and Cox 2002).

Table 3. Eleven species in 9 genera restricted to the North Island, mainly in indigenous forests.

Acanthophlebia cruentata	*Isothraulus abditus*
Arachnocolus phillipsi	*Maulus aquilus*
Austronella planulata	*Tepakia caligata*
Deleatidium angustum	*Zephlebia borealis*
Deleatidium magnum	*Zephlebia tuberculata*
Ichthybotus hudsoni	

Ten leptophlebiid species were reported by Towns and Peters (1996) as restricted to the North Island and two to the South Island. In particular, the genus *Zephlebia* was richly represented in the most northern regions but much less so in the south (Fig. 7). These data confirm and refine this conclusion.

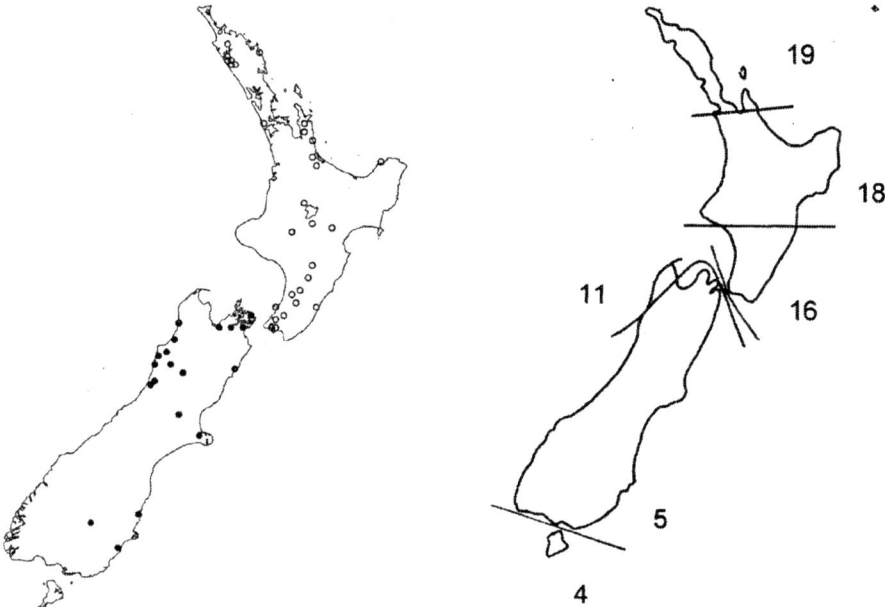

Figure 6. The distributions of *Ichthybotus hudsoni* "*" and *Ichthybotus bicolor* "●".

Figure 7. Variation with latitude of the number of double gilled leptophlebiid species (leptophlebiids excluding the genus *Deleatidium*).

At the time Cook Strait opened, the sea also flooded across other northern points, separating the North Island by at least two further east-west barriers to dispersal (Stevens 1980). These channels can be related, in a general way, to the present distributions of *Zephlebia* and other double gilled species. The gradation of species numbers southwards also supports the view that a centre of leptophlebiid dispersal was in the north. It has been pointed out that the close taxonomic affinities with New Caledonia, shown by some leptophlebiid species, indicate relatively recent past direct links to the north (Towns and Peters 1996, Towns and Hitchings In press).

In the early Pliocene (5–3 ma), earth movements became intense and the pattern of the present mountain ranges became apparent, extending over most of the South Island. These mountains together with smaller highlands in the central North Island were to become icebound during the glacial periods of the Pleistocene (2 ma–8 ka).

The accompanying falls in sea levels were sufficient to connect the two islands. As in the northern hemisphere, glaciers advanced and retreated several times. The seawater barrier to dispersal between the major islands was repeatedly removed and recreated. The most recent advance of glaciers, which is best understood, reached its maximum about 18 000 ya (Fig. 8). On most of the western side of the South Island, the ice sheet extended well out into the sea.

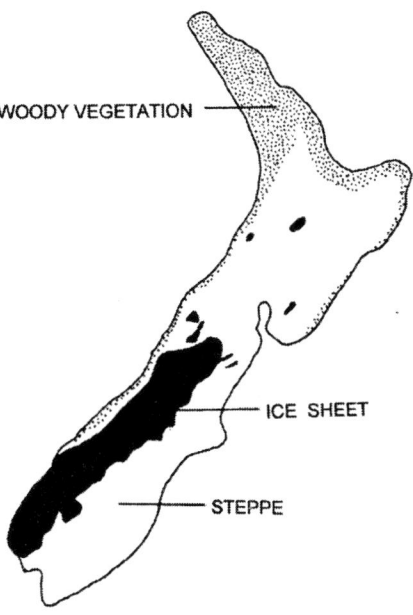

Figure 8. New Zealand 18 000 ya at its final glacial maximum and with lowered shoreline. (Redrawn from Fleming 1980).

On the eastern side much of the mountain mountain and hill country was also ice covered. During the ice advances sea levels fell to about 130 m below the present resulting in an extensive eastern coastal plain (Stevens 1980).

Some Consequences of Pleistocene Glaciation. Although almost of the North Island remained ice free, only the northwest and eastern coast of the South Island remained unglaciated. The distribution pattern shown by eight species of mayfly (Table 4) distributed in the North Island and also in the north west South Island is striking. It can be considered a consequence of the combination of lowered sea levels giving the opportunity for dispersal to the south and the ice sheet barrier confining populations to the northwest portion of the South Island.

Table 4. Eight species with North Island and N.W. South Island distributions.

Mauiulus luma	*Zephlebia inconspicua*
Rallidens mcfarlanei	*Zephlebia nebulosa*
Siphlaenigma janae	*Zephlebia pirongia* (Fig. 9)
Zephlebia dentata	*Zephlebia versicolor*

Some of these species show morphological variation between the North and South Island specimens, which may indicate incipient speciation.

Figure 9. Distribution of *Zephlebia pirongia.* Seven other mayfly species have the same distribution.

In addition to *Ichthybotus bicolor,* an additional five described species are restricted to the South Island (Table 5). Three have mountain distributions centered on the Southern Alps, one on Banks Peninsula and one on and near Stewart Island in the far south.

Table 5. Species restricted to the South Island.

Deleatidium (P) cornutum (Fig. 10)   *Nesameletus austrinus* (Fig. 10)
Deleatidium (P) insolitum   *Nesameletus murihiku*
Ichthybotus bicolor   *Nesameletus vulcanus*

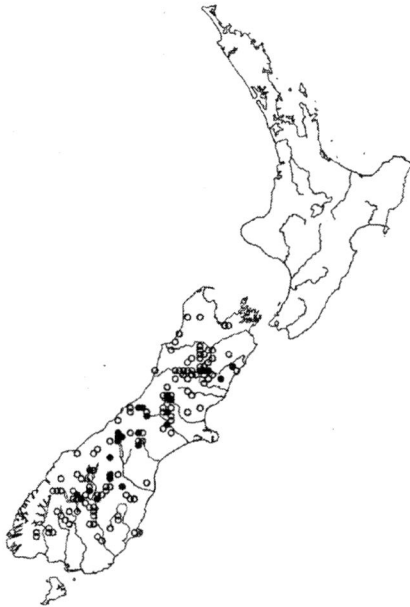

Figure 10. Distributions of the South Island species *Deleatidium (Penniketellum) cornutum* "•" and *Nesameletus austrinus* "∗".

Mountains can be a formidable barrier to mayfly dispersal (Edmunds 1972). There are no records in our New Zealand database of mayflies having been collected at altitudes above 1870 m. During the last glacial maximum, ice extended in the east to the edge of the plains and in much of the west, into the sea. In the east, species could only have survived in streams and rivers of coastal hills and lowlands. On the eastern South Island, such watercourses would have tended to be short, flowing more or less directly into the sea. It is feasible that isolated populations, adapted to cooler, faster water, survived in the lower reaches (Towns and Peters 1996, Staniczek and Hitchings in prep.). About 14 000 ya, retreat of the ice would have enabled some species to reoccupy the upper stretches of rivers and streams.

This has led to some interesting distributional patterns. Lowered sea levels during glacial advances linked Banks Island and Stewart Island to the mainland (Fleming 1980). Banks Island, formed from two volcanoes (12 M–6 M ya) and 80 km from the Southern Alps, was joined to the mainland by glacial outwash. Physical barriers arose during interglacial periods when sea levels rose. On Banks Island, *N. vulcanus* is considered to have speciated from the same stem species, as did the adjacent alpine species *N. austrinu* (Hitchings and Staniczek 2003).

Foveaux Strait, now separating Stewart Island from the South Island by 25 km of open sea, is shallow and became a land bridge during glacial advances.

Figure 11. Distributions of *Nesameletus murihiku* "●" and *N. vulcanus* "*".

Reestablished as an island in interglacial periods, this resulted in the speciation of *N. murihiku* from the same stem species as the now neighbouring *N. flavitinctus*. (Hitchings and Staniczek 2003).

Rallidens mcfarlanei is not a common species but it is distributed throughout the North Island and in the northwest of the South Island. A second species has been found in the east and south of the South Island (Staniczek and Hitchings In prep.). This latter may have arisen from the isolation of a population in the eastern South Island by one or more glaciations. Four relic populations exist, suggesting near extinction during glacial periods. This species seems largely confined to watercourses in foothills and plains. Apparently it was not well adapted to exploit the opportunities provided by retreat of the ice sheet.

Recently two additional *Nesameletus* species have been found in restricted parts of the Southern Alps. It will be interesting to see if their phylogenetics can be related to the changing conditions imposed on the genus by glacial advance and retreat.

I am currently addressing the problem of taxonomy and phylogenetics of the single gilled leptophlebiid genus *Deleatidium*, which contains several undescribed, cold adapted, fast water species confined to the Southern Alps.

Figure 12. Distributions of *Rallidens mcfarlanei* "●" and *R.. sp A* "⁎".

Conclusions

Pleistocene glaciation resulted in a lowering of sea levels sufficiently for the dispersal of eight species from the North into the adjacent portion of the South Island. However the southern ice fields inhibited farther dispersal. Some species, with origins in the north, including the double gilled leptophlebiids, show dispersal and distribution patterns influenced by New Zealand's former archipelagic geography.

Some species, including several single gilled leptophlebiids, restricted to the South Island, have occupied or reoccupied formerly ice covered high country. Speciation may have arisen as a consequence of isolation to restricted areas of the east coast during ice advance or rise in sea levels during interglacials.

Acknowledgments

I would like to thank Dr. N. Hiller, E. N. Jansen, Dr. A. H. Staniczek, Dr. D. R. Towns and Dr. J. B. Ward for helpful advice and practical assistance and Dr. I. M. Henderson for the use of his mapping program. Thanks also to the Directorate of the Canterbury Museum for the use of facilities and to the support of its Mason Foundation to attend the XI International Mayfly-Stonefly Conference 2004.

Literature Cited

Coates, G., and G. Cox. 2002. The Rise and Fall of the Southern Alps. Canterbury University Press, pp. 31, 37.

Cowie, B. 1985. An analysis of changes in the invertebrate community along a southern New Zealand montane stream. Hydrobiologia 120:135–146.

Edmunds, G. F. 1972. Biogeography and Evolution of Ephemeroptera. Annual Review of Entomology 17:21–42.

Fleming, C. A. 1980. The Geological History of New Zealand and Its Life Auckland University Press pp. 94–95.

Harding, J. S., and M. J. Winterbourn. 1995 Effects of contrasting land use on physico chemical conditions and benthic assemblages of streams in a Canterbury (South Island, New Zealand) river system. New Zealand Journal of Marine and Freshwater Research 29:479–492.

Hitchings, T. R. 2001. The Canterbury Museum mayfly collection and database. (Insecta: Ephemeroptera). Records of Canterbury Museum 15:11–32.

Hitchings, T. R., and A. H. Staniczek. 2003. Nesameletidae (Insecta : Ephemeroptera) Fauna of New Zealand 46:11.

Leathwick, J., G. Wilson, D. Rutledge, P. Wardle, F. Morgan, K. Johnston, M. McLeod, and R. Kirkpatrick. 2003. Land Environments of New Zealand, pp. 160–161. David Bateman, Auckland.

Quinn, J. M., and C. W. Hickey. 1990. Magnitude of effects of substrate particle size, recent flooding and catchment development on benthic invertebrates in 88 New Zealand Rivers. New Zealand Journal of Marine and Freshwater Research 24:411–427.

Quinn, J. M., G. L. Steele, C. W. Hickey, and M. L. Vickers. 1994. Upper thermal tolerances of twelve New Zealand stream invertebrate species. New Zealand Journal of Marine and Freshwater Research 28:391–397.

Rounick, J. S., and M. J. Winterbourn. 1982. Benthic faunas of forested streams and suggestions for their management. New Zealand Journal of Ecology 5:140–150.

Staniczek, A. H., and T. R. Hitchings (In preparation). Revision of the genus *Rallidens*.

Stevens, G. R. 1980. New Zealand Adrift. A. H. and A. W. Reed, Wellington. pp. 313, 323, 330.

Towns, D. R., and T. R. Hitchings. (In press). Order Ephemeroptera *in* Gordon, D. P. (ed). The New Zealand Inventory of Biodiversity. Vol. 2. Kingdom Animalia: Chaetognatha and Ecdysozoa. Canterbury University Press, Christchurch.

Towns, D. R., and W. L. Peters. 1996. Leptophlebiidae. (Insecta: Ephemeroptera) Fauna of New Zealand 36, 143 pp.

Winterbourn, M. J. 1980. The Freshwater Insects of Australasia and their affinities. Palaeogeography, Palaeoclimatology, Palaeoecology 31:235–249.

Winterbourn, M. J., J. S. Rounick, and B. Cowie. 1981. Are New Zealand ecosystems really different? New Zealand Journal of Marine and Freshwater Research **15**:321–328.

Wisely, B. 1962. Studies in Ephemeroptera II *Coloburiscus humeralis* (Walker) Ecology and Distribution of the Nymphs. Transactions of the Royal Society of New Zealand, Zoology, **2**(25):209–220.

MAYFLIES OF THE LAKE HOVSGOL REGION, MONGOLIA

S. Enkhtaivan[1] and T. Soldán[2]

[1] Mongolian Academy of Sciences, Geoecology Institute,
Baruun Selbe 13, Ulaanbataar, Tov 211238 Mongolia
[2] Department of Insect Ecology, Institute of Entomology, Czech Academy of Sciences,
Ceske Budejovice CZ-37005, Czech Republic

Abstract

Collections in 1995–1997 and in 2002–2003 from Lake Hovsgol and the surrounding region, especially the eastern tributaries, yielded 56 species, 28 of these representing new Mongolian national records. Unidentified larvae of *Isonychia* sp. and *Ephoron* sp. bring the known mayfly fauna of Lake Hovsgol to at least 58 species. The resulting data may have value for monitoring environmental changes resulting from global warming, permafrost melt and grazing. The study also provides a better understanding of the taxa that may be important for using macroinvertebrates to monitor surface water quality in the region.

Key words: mayfly; tributary streams; Lake Hovsgol; Mongolia

Introduction

Reviews of the published literature and collections during the past several years by the authors and colleagues have demonstrated that at least 79 species of mayflies, or Ephemeroptera, are presently known from Mongolia. Mayflies from Mongolia have been studied from the 1940s with scattered records or new species descriptions in several papers (e.g., D. Braash 1977, 1979, 1980, 1982; V. Landa and T. Soldán 1983; and others). Mostly these studies were based on materials from the Mongolian-Germany biological expedition collected by M. Stubbe and K. Gunther in 1962, 1964 and 1977; the Hungarian-Mongolian expedition collected by Z. Kaszab in 1964, 1965, 1966 and 1968; and the Soviet-Mongolian joint expedition of the Irkutsk and Mongolian State Universities, which started its activities in 1970.

In 1995, The Academy of Natural Sciences began collaborations with the National University of Mongolia and the Mongolian Academy of Sciences in joint biodiversity research expeditions in the Hovsgol Nuur (lake) watershed, Mongolia. As a result, the first Long-Term Ecological Research Site in Mongolia was established in 2002 at Lake Hovsgol.

This study ("the Hovsgol Project") was possible due to funding provided by the Mongolian Academy of Sciences, which received a five-year grant from the Global Environment Facility and implemented by the World Bank for a study entitled, "Dynamics of Biodiversity Loss and Permafrost Melt in Lake Hovsgol."

Lake Hovsgol is a graben of the Baikal Rift System and is the largest lake in Mongolia in terms of the amount of fresh water ($383.3 \, m^3$) and the second by the area ($2760 \, km^2$). Its maximum depth is 262 m, making it the fourth deepest lake in Central Asia. It is the world's fourteenth largest source of fresh water with a virtually undisturbed watershed and a diverse and interesting biota. The outflow of Lake Hovsgol, the River Eg, is a tributary of the Selenge River, the largest source of water entering Baikal (50%); thus, these great lakes are connected (e.g., Kozhova et al. 1989, Baatar, Samiya and Goulden 1997).The climate of the region is distinctly continental: The maximum air temperature is $+35°C$, the minimum is $-45°C$ and annual precipitation is 300–430 mm a year (Atlas of the Lake Hovsgol 1989). The Lake is located on the southern edge of the Siberian taiga, historically with permafrost, such that it is a region of great interest for monitoring climate change. Over the last three decades, average air temperature has increased by $1.8°C$ in this region and permafrost has been melting conspicuously.

Six valleys along the eastern shore of the lake were selected for study beginning in 2002. They included the heavily grazed northeastern valleys of Turag (N 51°18', E 100°48') and Shagnuul south of the town of Hanh; mideastern Noyon and Sevsuul with moderate grazing; and southeastern Dalbay and Borsog (N 50°58', E 100°45') valleys with little or no grazing pressure. In this region, streams usually melt in late May and freeze in mid November.

In this paper, we include species that were collected in 1995–1997 and 2002–2003 with their sample site locality and other information. Also, we are able to report on some other species which were found in the Hovsgol region prior to our study. At least 24 taxa are reported here from Mongolia for the first time.

Material and Methods

The materials from 1995–1997 were collected by J. Morse, J. Gelhaus, B. Hayford, A. Prather, E. Hunter, B. Namkhaidorj and C. Goulden and identified by T. Soldán in 2001. Mayfly larvae and adults were collected by me during May through September, 2002–2003, as part of the summer field work of the Hovsgol-GEF Project. Larvae were collected by use of the Rapid Bioassessment Protocol from upper, middle and lower sites in each of the six study streams each June, July and August in 2002–2003. Adults were collected primarily with sweep nets and with Malaise traps continuously set in the forest edge and on the riverbank of each study valley.

A few specimens were reared in the field. Ice is not available in the Mongolian countryside. Nevertheless, I was successful in rearing mature larvae and subimagoes to adulthood by packing rearing-vials in mosses in a cooler.

Light trapping proved unsatisfactory probably due to the extended daylight period in summer, cold crepuscular temperatures and/or inactivity or lack of attraction of the insects.

Specimens were preserved in 75% ethanol and studied with a Wild® M5 dissecting microscope.

Sample Collected Localities

1. G710: Hovsgol Aimag, Hovsgol Nuur area, W side ±48 km N of Khatgal, 50.41.14.N, 100.14.46.E; elev. ±1650 m; ice-free pool; 10 June 1996;
2. G713: Hovsgol Aimag, Hovsgol Nuur area, S side Alag Tsar River (near lake); 22 km ENE of Khatgal, 50.30.31.N, 100.24.12.E; elev.1615 m; 11–12 June 1996;
3. G714: Hovsgol Aimag, Hovsgol Nuur area, S side Alag Tsar River (near lake); 29 km ENE of Khatgal, 50.26.26.N, 100.23.30.E; elev.1693 m; 12 June 1996; in isolated small pool;
4. G715: Hovsgol Aimag, Hovsgol Nuur area, S side steppe pools along Alag Tsar River (near lake); 22 km ENE of Khatgal, 50.30.42.N, 100.23.44.E; elev.1615 m; 12 June 1996;
5. G716: Hovsgol Aimag, Hovsgol Nuur area, E side; Borsogiin gol (river), at bridge; 50.55.32.N, 100.46.19.E; elev.1640 m; 19 June 1996; Malaise trap;
6. G718: Hovsgol Aimag, Hovsgol Nuur area, E side; Borsogiyn gumnag am (stream); 50.57.19.N, 100.45.15.E; elev.1646 m; 14 June 1996; Malaise trap;
7. G719: Hovsgol Aimag, Hovsgol Nuur area, E side; S of Dalbayn Khyr; 50.59.22.N, 100.43.01.E; elev.1695 m; 14 June 1996; Malaise trap;
8. G721: Hovsgol Aimag, Hovsgol Nuur area, N side; Jargalant gol (stream), 25km NW of Khankh; 51.38.59.N, 100.31.48.E; isolated snowmelt pool; elev.1724 m; 15–16 June 1996; Malaise trap;
9. G722: Hovsgol Aimag, Hovsgol Nuur area, N side; Ikh Khoroo gol (river), 37 km NW of Khankh; benthic from river; 51.34.26.N, 100.28.48.E; elev.1721 m; 16 June 1996;
10. G724: Hovsgol Aimag, Hovsgol Nuur area, E side; Sevsuuliyn gol (river), 58 km S of Khankh; 51.09.42.N, 100.45.21.E; elev.1626 m; isolated pools along the river, 17 June 1996;
11. G725a: Hovsgol Aimag, Hovsgol Nuur area, E side; Noyon gol (stream), 51 km S of Khankh; 51.12.44.N, 100.46.06.E; elev.1669 m; from cobble in river; 18 June 1996;
12. G725b: Hovsgol Aimag, Hovsgol Nuur area, E side; Noyon gol (stream), 51 km S of Khankh; 51.12.44.N, 100.46.06.E; elev.1669 m; small pool along river; 18 June 1996;
13. G731: Hovsgol Aimag, steppe, 1 km N Moron (lat/long) of Moron 49.38.N, 100.10.E; elev.1290m; 21 June 1996;
14. G733a: Hovsgol Aimag, Selenge River, 14 km W of Ikh Uul (Bayan-Uhaa, lat/long.) of Bayan Uhaa; 48.33.N, 98.40.E; elev.1040 m; 21–22 June 1996;

15. G733b: Hovsgol Aimag, Selenge River, 14 km W of Ikh Uul (Ih-Uul, Yihe Uula Suma or Selenge; lat/long. Selenge); 49.26.N, 101.28.E; elev.1040 m; 21 June 1996;
16. G746: Hovsgol Aimag, Hovsgol Nuur area, Alag Tsar River (near lake); 22 km ENE of Khatgal, 50.4813194N, 100.4001850E; elev.1675 m; 13 July 1997;
17. G747: Hovsgol Aimag, Hovsgol Nuur area, E side, 50.5961823N, 100.4829633E; elev.1620 m; Heg Tsar cove; 14 July 1997;
18. G750: Hovsgol Aimag, Hovsgol Nuur area, E side, S of Dalbayn Khyr; Borsogo Cove; 50.9948283N, 100.7123029E; elev.1655 m; 15 July 1997;
19. G751: Hovsgol Aimag, Hovsgol Nuur area, E side, Borsogiyn gol (river main); 50.9253591N, 100.7552880E; elev.1710 m; 15 July 1997;
20. G752: Hovsgol Aimag, Hovsgol Nuur area, E side, Borsogiyn gol (tributary stream); 50.9351385N, 100.7595098E; elev.1700 m; 15 July 1997;
21. G753: Hovsgol Aimag, Hovsgol Nuur area, E side, Borsogiyn gol (tributary stream); 50.9542412N, 100.7538289E; elev.1640 m; 16 July 1997;
22. G756: Hovsgol Aimag, Hovsgol Nuur area, E side, Noyon gol (river); 51.2100917N, 100.7784033E; elev.1620 m; 17 July 1997;
23. G757: Hovsgol Aimag, Hovsgol Nuur area, E side, Turagyn gol (river); 51.2873608N, 100.8270800E; elev.1640 m; 17 July 1997;
24. G759: Hovsgol Aimag, Hovsgol Nuur area, E side, Toyn gol (river); 51.4449084N, 100.8178800E; elev.1700 m; 18 July 1997;
25. G761: Hovsgol Aimag, Hovsgol Nuur area, N side, Ikh Horoo gol (river); 51.5889215N, 100.4610175E; elev.1775 m; 19 July 1997;
26. G763: Hovsgol Aimag, Hovsgol Nuur area, N side, Jargalant gol (river); 51.6497219N, 100.5295426E; elev.1765 m; 20 July 1997;
27. G765: Hovsgol Aimag, Hovsgol Nuur area, NE side, Khankh River; 51.5015942N, 100.7691336E; elev.1705 m; 21 July 1997;
28. G767: Hovsgol Aimag, Hovsgol Nuur area, E side, Shagnuul gol; 51.2567353N, 100.8504474E; elev.1700 m; 21 July 1997;
29. H414: Hovsgol Aimag, Hovsgol Nuur area, east shore/ Baturant; 50.52.24N, 100.3444113E; elev. ???? m; 2 July 1995;
30. H415: Hovsgol Aimag, Hovsgol Nuur area, east shore/ Garvan; 50.52.24N, 100.3444110E; 2 July 1995;
31. H419: Hovsgol Aimag, Hovsgol Nuur area, west shore/ Garvan; 50.55.08N, 100.14.41930E; 3 July 1995;
32. H420: Hovsgol Aimag, Hovsgol Nuur area, Hoar Us; 50.56.30N, 100.14.37.102E; 3 July 1995;
33. H427: Hovsgol Aimag, Hovsgol Nuur area, east shore/ Juragol; 51.17.43N, 100.47.17935E; 7 July 1995;
34. H428: Hovsgol Aimag, Hovsgol Nuur area, east shore/ Juragol; 51.17.43N, 100.47.17100E; 5 July 1995;
35. H446: Hovsgol Aimag, Hovsgol Nuur; 7 July 1995;

36. H447: Hovsgol Aimag, Hovsgol Nuur area, east shore; 50.58.46N, 100.42.25930E; 7 July 1995;
37. H453: Hovsgol Aimag, Hovsgol Nuur area, east shore; Modon Hui; 50.59.26N, 100.32.33636E; 7 July 1995;
38. H464: Hovsgol Aimag, Hovsgol Nuur area, Hankh compound; 51.30.33.4N, 100.39.21.0E; 8–16 July 1995;
39. H465: Hovsgol Aimag, Hovsgol Nuur area, Bayan gol; 51.36.48.0N, 100.36.01.3E; 10 July 1995;
40. H467: Hovsgol Aimag, Hovsgol Nuur area, Hankh; 12 July 1995;
41. H468: Hovsgol Aimag, Hovsgol Nuur area, near mouth Hankh gol; 12 July 1995;
42. H477: Hovsgol Aimag, Hovsgol Nuur area, east shore of lake between Hankh and Bayan gol; 13 July 1995;
43. H479: Hovsgol Aimag, Hovsgol Nuur area, Jargalant gol; 51.38.59.0N, 100.31.46.4E; 14–15 July 1995;
44. H484: Hovsgol Aimag, Hovsgol Nuur area, Locaha Island; Southeast shore; 50.57.50N, 100.32.31E; 18 July 1995;
45. H494: Hovsgol Aimag, Hovsgol Nuur area, bridge Egiin gol Hatgal; 50.25.0N, 100.08.55E; 21 July 1995;
46. H495: Hovsgol Aimag, Hovsgol Nuur area, Hatgal; 1 km S. Bridge; 21 July 1995;
47. MP: Hovsgol Aimag, Hovsgol Nuur area, W of Lake, S of SSW of Ar Khilent Lake; 50.45.19N, 100.13.47E; 10 June 1996;
48. En: Hovsgol Aimag, Bugsey gol (river); 50.45.19N, 100.13.47E; 2 Aug 1997;
49. Go: Hovsgol Aimag, Hovsgol Nuur area, Ikh Khoroo gol (between bridge crossing and Lake Shore); 51.57.26.4N, 100.48.108E; 20 July 1999;
50. Hovsgol Aimag, Hovsgol Nuur area, mouth Khoroo gol;15 July 1995;
51. Hovsgol Aimag, Hovsgol Nuur area, Borsog gol;
52. Hovsgol Aimag, Hovsgol Nuur area, Dalbay gol;
53. Hovsgol Aimag, Hovsgol Nuur area, Sevsuul gol;
54. Hovsgol Aimag, Hovsgol Nuur area, Noyon gol;
55. Hovsgol Aimag, Hovsgol Nuur area, Shagnuul gol;
56. Hovsgol Aimag, Hovsgol Nuur area, Turag gol;
57. Hovsgol Aimag, Hovsgol Nuur area, 6km WNW of Tosontsengel sum, elev 1400 m, 18 June 1968, light trap.
58. Hovsgol Aimag, Hovsgol Nuur area, Tesiin gol, 22 km W of Tsetserleg sum, elev 1820 m, 22 June 1968.
59. Hovsgol Aimag, Hovsgol Nuur area, Delgermoron river., 8 km of Burenkhaan sum, elev 1450 m, 14–20 June 1968.
60. Hovsgol Aimag, Hovsgol Nuur area, Egiin gol., 8 km of Alag Erdene sum, elev 1600 m, 17 July 1968, light trap.
61. Hovsgol Aimag, Hovsgol Nuur area, Tesiin gol, Alag Mort, 42 km NE of Khalzan Sogootiin Davaa, elev 1900 m, 14 July 1968.

In the following list of species, the dates of capture for specimens collected as part of the Hovsgol GEF project at sites #51–56 are provided after the site number. Dates for other collections are noted in the list above. Specimens captured during 1995–1997 were identified by T. Soldán in November 2001. All other specimens were identified by S. Enkhtaivan. Species that are newly recorded for Mongolia are indicated with an asterisk (*).

Ameletidae

1. *Ameletus inopinatus* Eaton, 1887 — #1-1L, #6-1L, #8-4L, #9-17L, #17-1I, #18-2L, #21-13L, #31-1L, #35-1I, #37-1L+4I, #39-1L, #42-34L, #43-1I, #44-2L, #47-20L, #51-? I (23 Aug. 2002).
   Ameletus sp. — #6-9L, #39-6L, #45-1I(F).

Baetidae

2. *Acentrella fenestratus* (Kazlauskas 1963) — #4-3L, #5-17L, #16-1L, #21-1L.
   Acentrella "putoranica" Kluge (nomen nudum) — #25-8L+9I.
   Acentrella "putoranica" Kluge (in litt.) — #49-4L.
3. *Acentrella sibiricus* (Kazlauskas 1963) — #45-2I.
4. *Baetiella tuberculatus* (Kazlauskas 1963) — #23-2L.
5. *Baetis bicaudatus* Dodds, 1923 — #16-2L, #19-6L, #21-5L, #51-I (23 Aug 2002), #55-I (02 Sep. 2002, 24 Aug. 2003).
6. *Baetis feles* Kluge, 1980 — #20-9L.
7. *Baetis fuscatus* [(Linnaeus 1761)] — #14-6L, #23-6L, #57-1F, #58-1F/1M, #59-3F/1M. Previously recorded from Mongolia by Baykova and Varychanova (1978) and Landa and Soldán (1983).
   [*Baetis*] *bioculatus* [Linnaeus 1758];
   [*Baetis*] *venustulus* [Eaton 1885];
8. *Baetis pseudothermicus* Kluge, 1983 — #9-2L, #11-5L, #21-5L,#26-4L, #43-4L, #47-9L,#49-1L+1I; previously recorded for Mongolia by Landa and Soldán (1983).
   Baetis sp. — #15-3I, #16-1I, #21-5I, #23-2I, #24-1L, #26-3I, #38-1I; #42-2L, #43-2I; previously recorded for Mongolia by Landa and Soldán (1983).
9. *Baetis ussuricus* Kluge, 1983 — #14-1L; previously recorded for Mongolia by Landa and Soldán (1983).
10. *Baetis vernus* Curtis, 1834 — #14-6L,#15-3L, #16-6L,#22-28L, #25-16I, #28-5I, #38-30L, #49-1L, #60-1M; previously recorded for Mongolia by [Author, date?];
    [*Baetis*] *finitimus* Eaton, 1871;
    [*Baetis*] *tenax* Eaton, 1870;

11. *Baetopus asiaticus* Soldán, 1978 — #61-2M/1F; previously recorded for Mongolia by Soldán (1978) and Landa and Soldán (1983).
12. *Baetopus montanus* Soldán, 1978 — #61-11M/10F; previously recorded for Mongolia by Soldán (1978) and Landa and Soldán (1983).
13. *Centroptilum kazlauskasi* Kluge, 1963 — #54-1I (15–25 Aug. 2002).
14. *Centroptilum luteolum* (Müller 1776) — #53-1I (23 Aug. 2003).
15. *Cloeon (Cloeon) dipterum* Linnaeus, 1761 — #52-1I (01 Aug. 2003);
    Cloeon sinense (Walker 1853);
16. *Cloeon (Procloeon) bifidum* Bengtsson, 1912 — #51-2I (23-31 Aug. 2002, Malaise trap).
17. *Cloeon (Procloeon) pennulatum* (Eaton 1870) — #23-1, #51-1L (23 Aug. 2002); previously recorded for Mongolia by [Baykova and Varychanova (1978)].

Caenidae

18. *Brachycercus harrisella* Curtis, 1834 — #14-6L; previously recorded for Mongolia by Kluge (1991);
    Eurycaenis harrisella (Curtis, 1834): Bengtsson, 1917.
19. *Caenis horaria* (Linnaeus 1758) — #7-1L; previously recorded for Mongolia by Landa and Soldán (1983).
    Caenis dimidiata Stephens, 1835;
    Caenis lactella Eaton, 1884;
20. *Caenis jungi* Braasch, 1980 — #60-1M; previously recorded for Mongolia by Landa and Soldán (1983).
21. *Caenis robusta* Eaton, 1884 — #56-1L (24 Aug. 2002); previously recorded for Mongolia by Braasch (1982); Landa and Soldán (1983); newly recorded for the Hovsgol region;
    Caenis incus Bengtsson, 1912;
    Caenis ulmeri Brodsky, 1930;
    Caenis sp. — #40-1L.

Ephemerellidae

22. *Drunella cryptomeria* Imanishi, 1937 — #14-2L, #15-1L;
    Ephemerella bicornis Gose, 1980;
23. *Drunella submontana* Brodsky, 1930 — #59-1M/2F subimagoes; previously recorded for Mongolia by Landa and Soldán (1983).
24. *Drunella triacantha* Tshernova, 1949 — #14-3L, #15-4L;
    Ephemerella tenax Tshernova, 1952;
    Ephemerella eroensis Gose, 1980;
25. *Ephemerella aurivillii* Bengtsson, 1908 — #8-16L, #9-7L, #14-1I;
    Ephemerella aronii Eaton, 1908;

Ephemerella norda McDunnough, 1924;
Ephemerella concinnata Traver, 1934;
Ephemerella taeniata Tshernova, 1952;
Ephemerella maxima Allen, 1971;
Ephemerella ezoensis Gose, 1985;
26. *Ephemerella mucronata* (Bengtsson 1909) — #14-7L; previously recorded for Mongolia by Braasch (1982); Landa and Soldán (1983).
    Chitonophora kreignoffi Ulmer, 1920;
    Chitonophora unicolorata Ikonomov, 1961;
  Ephemerella sp. — #16-33L, #22-15L.
  Ephemerella sp. 1 — #14-2L, #15-1L.
  Ephemerella sp. 2 — #22-5L.
27. *Torleya ignita* (Poda 1761) — #14-7L,#15-2L, #17-4L, #23-30L+1I, #24-7L,#25-1L, #27-7I, #51-5L/7I (23 Aug. 2002), #59-1M/3F subimagoes; previously recorded for Mongolia by Baykova and Varychanova (1978), Braasch (1982) and Landa and Soldán (1983).
28. *Torleya nuda* Tshernova, 1949 — #16-18L, #23-19L,#24-11L+1I, #25-56I+30L, #28-4I, #38—1L, #43-1L, #46-1L, #48-5I, #49-1L, #51-3M (24-27 Jul. 2002, 27 Aug. 2003), #55-1M (14 Aug. 2002);
    Ephemerella thymali Tshernova, 1952;
    Ephemerella verrucosa Kluge, 1980;
29. *Uracanthella lenoki* Tshernova, 1952 — #14-1L;
    Uracanthella markevithsi Belov, 1979;

Ephemeridae

30. *Ephemera orientalis* McLachlan, 1875 — #59-1F; previously recorded from Mongolia by Baykova and Varychanova (1978), Braasch (1982) and Landa and Soldán (1983).
31. *Ephemera sachalinensis* Matsumura, 1934 — #13-1I, #14-28L.
32. *Ephemera strigata* Eaton, 1892 — #59-1F; previously recorded for Mongolia by Landa and Soldán (1983).
33. *Ephemera transbajkalica* Tshernova, 1973 — #14-3L.

Heptageniidae

34. *Cinygmula cava* Ulmer, 1927 — #9-7L,#25-1L,#26-5L, #32-8L, #39-33L, #43-5L, #47-21L; previously recorded for Mongolia by Braasch (1979), Braasch (1986);
    Cinygmula guentheri Braasch, 1979;
    Cinygmula altaica Tshernova, 1949;
35. *Cinygmula kurenzovi* Bajkova, 1962 — #45-5I, #60-3M/1F; previously recorded for Mongolia by Landa and Soldán (1983);
    Cinygmula kaszabi Landa and Soldán, 1983;

36. *Cinygmula putoranica* (Kluge 1980) — #16-9L+1I, #23-1L, #51-10I (25 Aug. 2002), #55-3I (14 Aug. - 02 Sep. 2002).
    Cinygmula sp. — #2-1L,#8-2L, #9-16L, #16-19L, #22-2L.
37. *Ecdyonurus* (*Afghanurus*) *aspersus* Kluge, 1980 — #23-1I, #51-I (16-27 Jul. 2002 Malaise trap, 26 Aug. 2002).
38. *Ecdyonurus* (*Afghanurus*) *inversus* Kluge, 1980 — #24-12L+1I, #28-14L+9I.
39. *Ecdyonurus* (*Afghanurus*) *joernensis* Bengtsson, 1909 — #17-9L, #23-34L,#27-3I, #46-2L; previously recorded for Mongolia by Braasch (1986);
    [*Ecdyonurus*] *flavomaculatus* [Aro 1928];
    [*Heptagenia*] *mongolicus* Bajkova and Varychanova, 1978;
    [*Heptagenia*] *dentatus* Braasch, 1979;
    [*Ecdyonurus*] *stubbei* Braasch, 1979;
40. *Ecdyonurus* (*Afghanurus*) *simplicioides* (McDunnough 1924) — #14-4L.
41. *Ecdyonurus* (*Afghanurus*) *vicinus* (Demoulin 1964) — #14-??, #15-??.
    Ecdyonurus sp. - #22-2L.
42. *Epeorus* (*Belovius*) *pellucidus* (Brodsky 1930) — #14-4L, #23-2L, #51- 8I (16-25 Aug. 2002, Malaise trap); previously recorded for Mongolia by Braasch (1979)
    Cinygmula smirnovi Tshernova, 1978;
    Epeorus tshernovae Braasch, 1979;
43. *Epeorus anatolii* Sinitshenkova, 1981 — #23-8L, #51-4I (23–31 Aug. 2002, Malaise trap), #56-3I (24 Aug. 2003);
    Epeorus rautiani Sinitshenkova, 1982;
44. *Epeorus sinitshenkovae* Tshernova, 1981 — #55-1I (14 Aug. 2002); previously recorded for Mongolia by Baykova and Varychanova (1978).
    Epeorus ninae Kluge, 1995;
45. *Heptagenia flava* Rostock,1877 — #14-2I, #52-1I (24–27 Jul. 2002), #54-1(15 Aug. 2002); previously recorded for Mongolia by Braasch 1979 and Landa and Soldán (1983);
    Ephemera citrina Hummel, 1825;
    Heptagenia bipunctata Esben-Petersen, 1916;
    Heptagenia arsenjevi Tshernova, 1952;
46. *Heptagenia sulphurea* Müller, 1776 — #14-3L, #15-1L; previously recorded for Mongolia by Braasch (1979); Braasch (1986);
    Ephemera helvola Subz., 1776;
    Ephemera leucophthalma Strom 1783;
    Ephemera ferruginea Gmelin 1790;
    Baetis costalis Curtis, 1834;
    Heptagenia sulphurea f. *dalecarlica* Bengtsson, 1912;
    Baetis elegans Curtis, 1834
    Baetis straminea Curtis, 1834
    Baetis cyanops Pictet, 1843

Heptagenia soldatovi Tshernova, 1952
Heptagenia f. *dalecarlica* Bengtsson, 1912
Subsp. *Heptagenia sulphurea. albicauda* Kluge, 1987
47. *Rhithrogena bajkovae*   Sowa, 1973 – #15-12I, #56-2M (24 Aug. 2003);
previously recorded for Mongolia by Landa and Soldán (1983).
       [*Rhithrogena*] *quadrinotata* Sinitshenkova, 1982;
       Rhithrogena gr. *lepnevae* — #14-1L.
48. *Rhithrogena lepnevae* Brodsky, 1930 — #23-24L+1I, #55-1I (24 Aug. 2003),
       #59-20M/30F; previously recorded for Mongolia by Bajkova
       andVarychanova (1978), Braasch (1979) and Landa and Soldán (1983);
       Rhithrogena unicolor Tshernova, 1952;
       Rhithrogena binotata Sinitsh, 1982;
49. *Rhithrogena sibirica* Brodsky, 1930 — #14-3I, #45-4I, #59-2M, #60-1M;
       previously recorded for Mongolia by Braasch (1979a) and Landa and
       Soldán (1983).
Rhithrogena sp. — #14-1I+1F, #59-1M/3F.

Isonychiidae

Isonychia sp. — #14-10L.

Leptophlebiidae

50. **Paraleptophlebia strandii* Eaton, 1901 — #4-2L,#10-5L, #23-3I+1L,#27-4I,
       #52-1I (05 Aug. 2003), #54-3L (15 Aug. 2002, 24 Aug. 2003), #55-I
       (14 Aug. – 02 Sep. 2002 Malaise trap), #56-10L (24 Aug. 2002);
       Leptophlebia lunata Tshernova, 1928;

Metretopodidae

51. *Metretopus alter* Bengtsson, 1930 — #51-3I (21–26 Aug. 2002); previously
       recorded for Mongolia by [Author, date?].
       Metretopus sp. — #14-14L, #16-6L+1I, #22-5L, #27-1I.
52. **Metretopus tertius* Tiunova, 1999 — #14-11L.

Polymitarcyidae

Ephoron sp. — #14-7L.

Siphlonuridae

53. **Siphlonurus* cf. *chankae* Tshernova, 1952 — #16-1L.
54. **Siphlonurus chankae* Tshernova, 1952 — #16-5L, #19-24L, #33-28L.
55. **Siphlonurus immanus* Kluge, 1985 — #51-5M/5F (14 Jul. - 26 Aug. 2002).

56. *Siphlonurus lacustris* Eaton, 1870 — #61-1F; previously recorded for Mongolia
     by Braasch (1982) and Landa and Soldán (1983);
  Siphlurus zetterstedii Bengstson 1909;
  Siphlonurus pyrenaicus Navas 1930;
  Siphlonurus nuessleri Jacob, 1972;
 Siphlonurus sp. — #3-3L,#4-38L, #6-3L, #7-2L,#10-31L, #12-11L, #16-
     4L, #21-2L, #27-2L+1I, #29-6L, #30-1L, #34-4L, #36-4L, #41-5L,
     #45-1I(F).

Discussion and Conclusion

If climate change continues at the same rate and trend as now, what will happen to region's biodiversity, including its mayfly fauna? The answer to this question can only be provided by long-term monitoring of temperature, permafrost and fauna. Such monitoring is very compatible with the goals of a Long-Term Ecological Research Site like this one at Lake Hovsgol, Mongolia. The data provided in this paper establishes a baseline for mayfly diversity against which future monitoring can be compared.

Does excessive grazing decrease favorable habitats for insects and contribute to biodiversity loss? The different grazing intensities in the six rivers of this study and differences in their mayfly taxa richness may provide some important insight to this question. Figure 1 shows that the highest mayfly richness occurs in Borsog River, a stream that has been affected least by humans and their cattle in recent years.

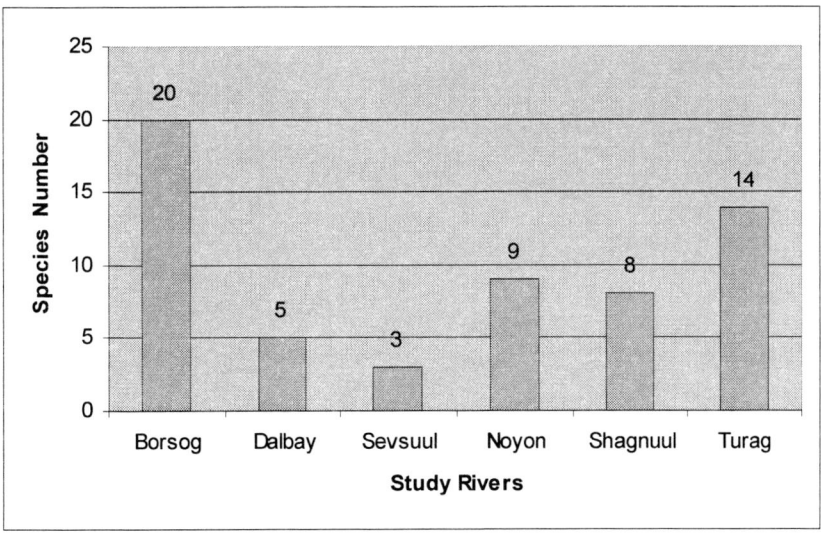

Figure 1. Comparison of six valleys by mayfly taxa richness.

Other streams have some kind of over grazing and intensive livestock. However, grazing intensity is not otherwise well-correlated with mayfly diversity in these other streams: Turag is one of the most heavily grazed valleys but has the next highest number of mayfly species, whereas Dalbay is one of the least heavily grazed valleys but has nearly the lowest number of mayfly species. Clearly, mayfly taxa richness cannot be explained solely by grazing density.

Despite increasing temperature and human influence, we expect that additional intensive sampling in the Hovsgol region will uncover additional national and regional records and undescribed species because the mayfly species fauna of the Hovsgol region and Mongolia are still very incompletely known. Thus, more careful study is needed to understand mayfly habitat requirements before humans disturb the ecosystem further.Including mayflies and other macroinvertebrates in a national water monitoring program in Mongolia would help obtain that knowledge. This biological monitoring will more-likely happen if better illustrated keys can be made available to Mongolian scientists and water quality regulators.

The following 28 species are newly recorded for Mongolia: *Ameletus inopinatus, Acentrella fenestratus, Acentrella sibiricus, Baetiella tuberculatus, Baetis bicaudatus, Baetis feles, Centroptilum kazlauskasi, Centroptilum luteolum, Cinygmula putoranica, Cloeon* (C.) *dipterum, Cloeon* (Procloeon) *bifidum, Ecdyonurus* (Afghanurus) *aspersus, Ecdyonurus* (Afghanurus) *inversus, Ecdyonurus* (Afghanurus) *simplicioides, Ecdyonurus* (Afghanurus) *vicinus, Epeorus anatolii, Ephemera sachalinensis, Ephemera transbaikalica, Drunella cryptomeria, Drunella triacantha, Uracanthella lenoki, Ephemerella aurivillii, Ephemerella nuda, Paraleptophlebia strandii, Metretopus tertius, Siphlonurus chankae, Siphlonurus immanus* and *Siphon* cf. *chankae.*

Acknowledgments

This study was made possible by World Bank GEF funding provided to the Mongolian Academy of Sciences for a study entitled "Dynamics of Biodiversity Loss and Permafrost Melt in Lake Hovsgol National Park, Mongolia." Thanks to Drs. Clyde Goulden, John Morse, Jon Gelhaus and Barbara Hayford for their valuable advice and support. Thanks also to Drs. Pat McCafferty (and his students), Riley Nelson and Richard Baumann for help with taxonomic identifications. S. Enkhtaivan is grateful to the Scholarship Subcommittee of the Permanent Committee of the International Conferences on Ephemeroptera for providing a Ryzord Sowa scholarship and to the USA NSF Biotic Surveys and Inventories grant (DEB-0206674) that helped her to participate in this meeting. Thanks to colleagues and family for continuing encouragement.

Literature Cited

Baatar, D., R. Samiya, and C. E. Goulden. 1997. Long-Term Ecological Research in Mongolia. Pages 67–75 *in* King, H-B., S. P. Hamburg, and Y. Hsia, editors. Long-Term Ecological Research In East Asia-Pacific Region. Taiwan Forestry Research Institute.

Braasch, D. 1977. *Rithrogena piechockii* n.sp. aus der Mongolei (Eph. Heptageniidae). Ergebnisse der Mongolisch-Deutschen Biologischen Expeditionen seit 1962. Entomologische Abhandlungen (Dresden) **21**(9):140–142, illust. (with English and Russian summary).

Braasch, D. 1979. Die Eintagsfliegen (Ephemeroptera, Heptageniidae) der Mongolisch-Deutschen Biologischen Expeditionen seit 1964 und 1977. Ergebnisse der Mongolisch-Deutschen Biologischen Expeditionen seit 1962. Entomologische Abhandlungen (Dresden) **23**(5):65–76.

Braasch, D. 1980. *Ecdyonurus klugei* n.sp. (Ephemeroptera, Heptageniidae) aus der Mongolei. Ergebnisse der Mongolisch-Deutschen Biologischen Expeditionen seit 1962. Entomologische Abhandlungen (Dresden) **24**(3):41–43.

Braasch, D. 1982. Die Eintagsfliegen der Mongolisch-Deutschen Biologischen Expeditionen 1964 (Ephemeroptera). Dt.Entom.Z., N. F. 29, Heft 1-3, Seite 43-47.

Braasch, D. 1986. Eintagsfliegen aus der Mongolischen Volksrepuplik (Ephemeroptera, Heptageniidae). Entomologische Abhandlungen (Dresden) **30**(2):77–80.

Kozhova, O. M., O. Shagdarsuren, A. Dashdorj, and N. Sodnom (eds). 1989. Atlas of Lake Hövsgöl. Cartographic Ministry of USSR, Moscow (in Russian).

Landa, V., and T. Soldán. 1983. Ephemeroptera from Mongolia. Entomologica Hungarica Rovartani kozlemenyek. **44**(2):189–204.

Soldán, T. 1978. Two new species of *Baetopus* (Ephemeroptera, Baetidae) from Mongolia with a special reference to related genera. Vestnik Ceskoslovenske Spolecnosti Zoologicke, Svazek XLII-Cislo3-1978-Str. 209–214.

GENETIC DIVERSITY IN HEADWATER-SPECIFIC MAYFLIES BASED ON THE MITOCHONDRIAL 16S RRNA GENE SEQUENCES

Koji Tojo

Department of Biology, Faculty of Science, Shinshu University, Asahi, Matsumoto, Nagano 390-8621, Japan

Abstract

Headwaters are unique components of catchments as they usually support a taxonomically and ecologically unique fauna. This paper, focuses on an endemic Japanese dipteromimid mayflies, which have possibly very limited dispersal ability, and are restricted to headwaters. Its ecological relationships (its dispersal ability) and genetic variation were discussed and compared with some other mayfly species living in other water systems. Dipteromimid mayflies showed greater interpopulation genetic distances than those of any other of the mayfly species examined.

Key words: mayfly; Dipteromimidae; genetic property; genetic diversity; headwater area.

Introduction

Headwaters, the streams that make up the beginnings of rivers, are unique components of catchments as they usually support a taxonomically and ecologically unique fauna. However, since headwaters are generally narrow and have a fluctuating course, they have received little attention from ecologists and their benthic fauna, including insects, is poorly studied (Ota and Takahashi 1999). This paper, focuses on an endemic Japanese dipteromimid mayflies *Dipteromimus tipuliformis* and *D. flavipterus*, which have possibly very limited dispersal ability, and are restricted to headwaters (Tojo and Matsukawa 2003). Its ecological relationships (its dispersal ability) and genetic variation (genetic distance within and between populations in the mitochondrial 16S rRNA gene sequences) were discussed and compared with some other mayfly species (related ameletid and siphlonurid mayflies, and mayflies, e.g., ephemerid, polymitarcyid mayflies) living in other water systems.

Materials and Methods

Two dipteromimid mayflies, *Dipteromimus tipuliformis* MaLachlan (33 individuals from 13 localities) and *D. flavipterus* Tojo and Matsukawa (3 individuals from one locality), were used in this study (Fig. 1). Sequence data of 27 *D. tipuliformis* and all *D. flavipterus* individuals of which were referred from Tojo and

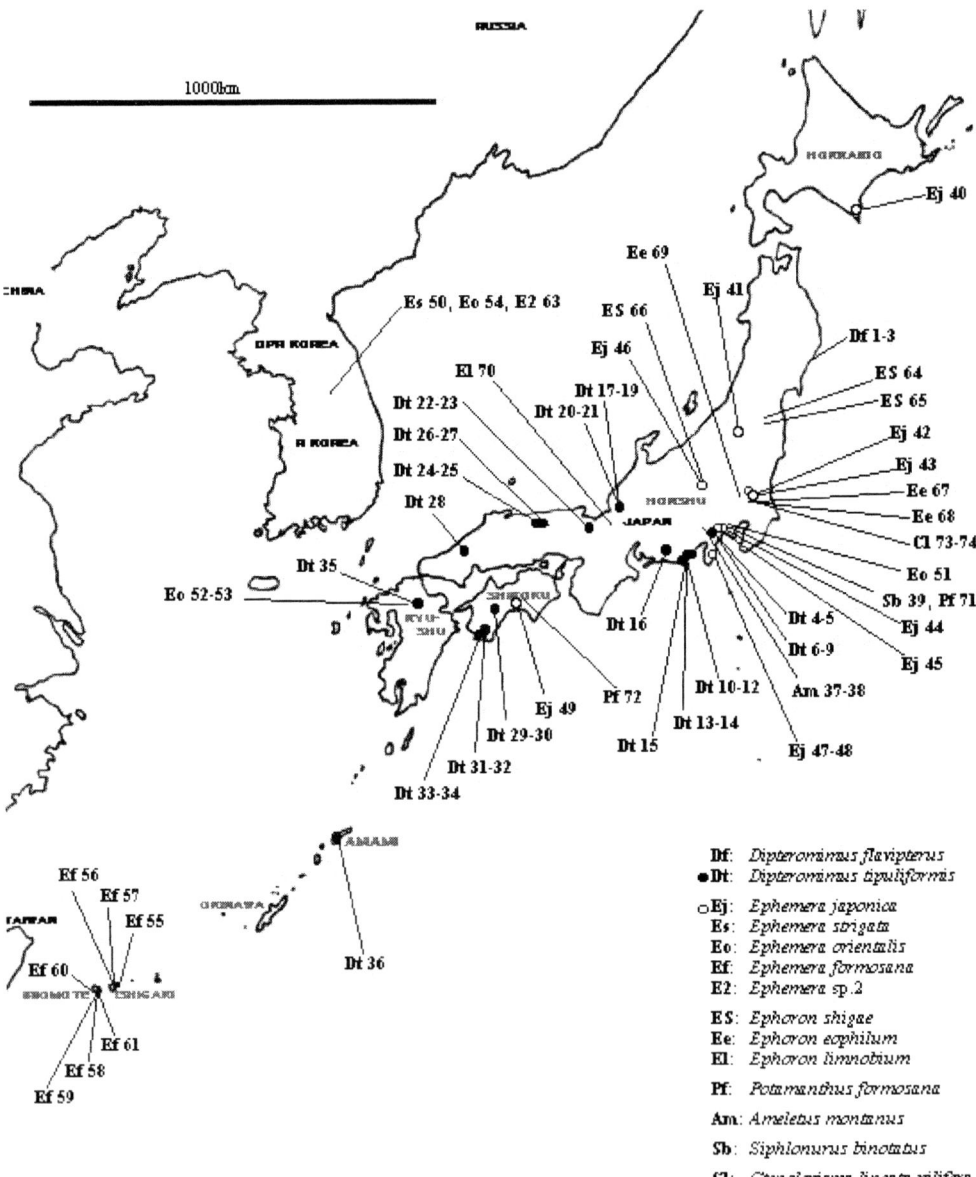

Figure 1. Sampling localities of mayflies examined.

Matsukawa (2003: DDBJ/AB110236-110266). Some related mayflies of Ameletidae and Siphronuridae and some burrowing mayflies of Ephemeridae, Polymitarcyidae and Potamanthidae, which have been studied for molecular analyses, were used as control for the analysis of genetic diversity. A sequence data of *Ephemera* sp. (from Immenburg, Bonn, Germany) was referred from GenBank (AF266047; Misof et al. 2001) and two individuals of silverfish were used as outgroup. Adults and nymphs were fixed with pure ethanol for molecular examination.

DNA Analyses by Sequencing of the Mitochondrial 16S rRNA Region. DNA was extracted from the specimens and purified using the DNeasy Tissue Kit (QIAGEN, Hilden). The 16S rRNA genes were amplified by a PCR method using as the forward primer 5'-TTACGCTGTTATCCCTAA-3' and the reverse primer 5'-CGCCTGTTTATCAAAAACAT-3'. PCR products were purified with Microcon Kit (MILLIPORE, Massachusetts). The purified DNA was sequenced directly by an automated method using the DYEnamic ET Terminator Cycle Sequencing Kit (Amersham Biosciences, New Jersey) on an automated sequencer (ABI PRISM 377 Genetic Analyzer; Perkin Elmer/Applied Biosystems, California).

Sequence data were aligned using Clustal W (Thompson et al. 1994, 1999) and phylogenetic analyses were performed by the neighbor-joining (NJ) method (Saitou and Nei, 1987), implemented using PHYLIP version 3.57 (Felsenstein 1995). The NJ analyses employed using Kimura's two-parameter method (Kimura 1980) and confidences of branches were assessed by 1,000 bootstrap resamplings. The significant difference in relationships between pairwise genetic distance (uncorrected p-distance) of individuals and geographic distance of their collection site was tested by Mantel test.

Results and Discussion

Dipteromimid mayflies, *Dipteromimus tipuliformis* and *D. flavipterus* were used for examining genetic relationships with related ameletid, siphlonurid mayflies and controlled burrowing mayflies, and silverfishes as outgroups. The neighbor-joining (NJ) dendrogram derived from Kimura's (1980) distance matrix from aligned sequences is shown in Figs. 2 and 3. The monophyly of the dipteromimid mayflies (and other examined mayfly groups) could be strongly supported (the bootstrap proportion [BP] of Dipteromimidae = 82%; [BP] of *D. tipuliformis* = 96%; [BP] of *D. flavipterus* = 100%). As for the in-group, *D. tipuliformis* has high intraspecific variations and some local populations are distinguished on the basis of genetic differentiation (the subclusters were well separated geographically; Figs. 2 and 3). On the other hand, other mayfly groups: Ephemeridae and Polymitarcyidae could be also strongly supported as monophyly at species level (e.g., [BP] of *Ephemera japonica* =84%; [BP] of *E. formosana* = 100%; [BP] of *E. orientalis* =97%; [BP] of *Ephoron shigae* and *E. eophilum* = 100% all), but they have not so high intraspecific variations than dipteromimid mayflies.

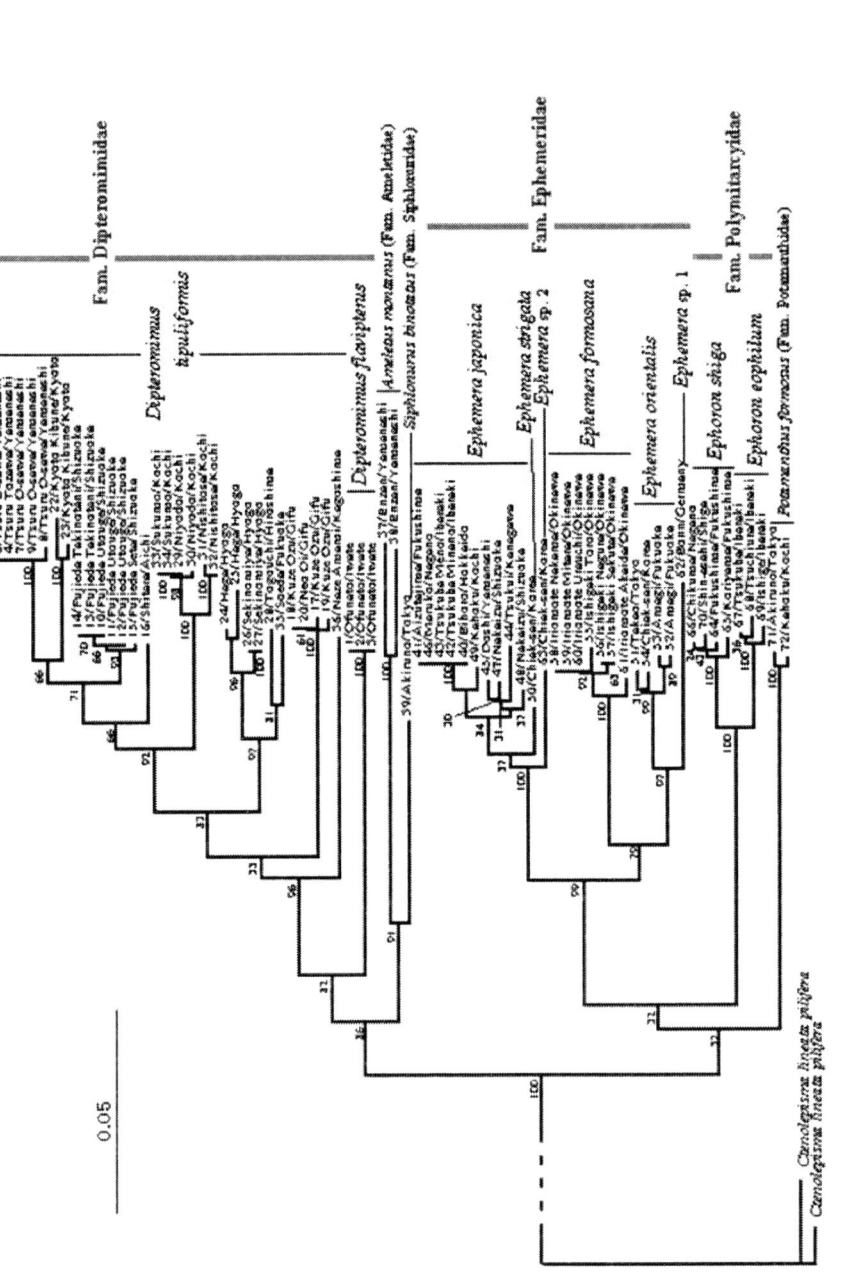

Figure 2. Neighbor-joining (NJ) dendrogram using mitochondrial 16S rRNA gene sequences of mayflies, based on Kimura's 2-parameter genetic distance matrix (Kimura 1980), with two silverfish specimens of *Ctenolepisma lineata pilifera* as outgroups. Bootstrap values for 1000 replicates are indicated at major nodes. Sample number and population information are given in Fig. 1.

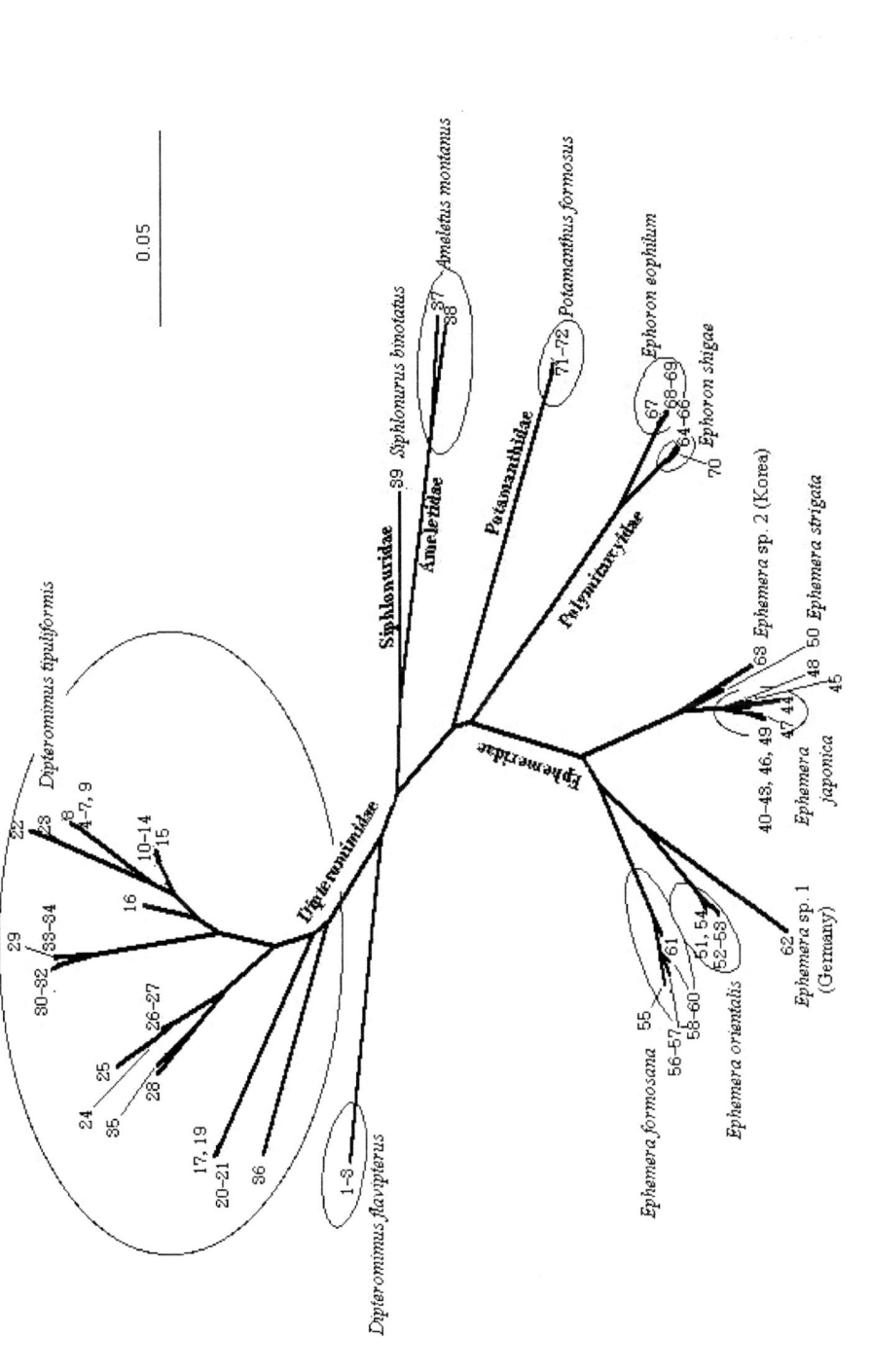

Figure 3. Unrooted NJ dendrogram of the mitochondrial 16S rRNA gene sequences from 72 mayfly specimens (see Fig. 1), based on the matrix as same as that of Fig. 2.

Figure 4 shows the relationships between pairwise genetic distance (uncorrected p-distance) of individuals and geographic distance of their sites in *D. tipuliformis* and *E. japonica*.   Dipteromimid mayflies, especially *D. tipuliformis* showed greater intraspecific (interpopulation) genetic distances than those of *E. japonica* and any other of the mayfly species examined (not shown).   Dipteromimid mayflies, ephemerid mayflies and the other mayflies examined have univoltine life cycles.   We do not consider differences of the interpopulation's genetic variations as originated from these life cycle differences.   The dipteromimid mayflies, however, are endemic to headwater areas.   These species possibly have very limited dispersal ability.   This may be characteristic of headwater aquatic insects (not only dipteromimid mayflies). Other mayfly species and the other aquatic insects living in headwaters should be compared with those living in other water systems (cf. Waples 1995).

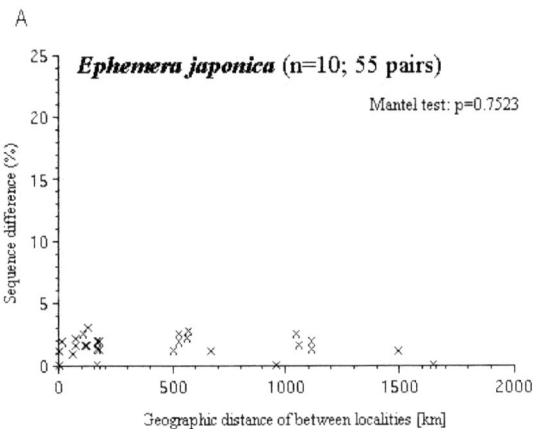

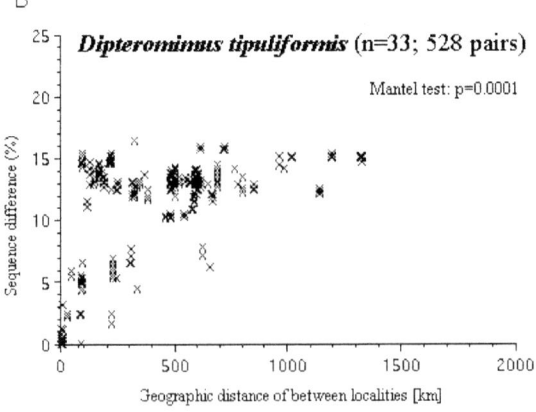

Figure 4. Relationships between pairwise genetic distance (uncorrected p-distance) of individuals and geographic distance of their collection sites. A: *Ephemera japonica* (N = 10; 55 pairs); B: *Dipteromimus flavipterus* (N = 33; 528 pairs).

Acknowledgments

I acknowledge the valuable suggestions and support of Drs. R. Kuranishi, F. Hayashi, T. Tsutsumi, M. Hatakeyama, J. M. Lee and M. D. Picker. I am also indebted to Drs. H. Ichiyanagi and Y. Yamaguchi, and Mr/s. K. Matsukawa, T. Torii, G. Yoshinari, N. Kawase and K. Nio, for their cooperation in collecting materials. This study was supported by a Grant-in Aid for Scientific Research from the Japanese Society for the Promotion of Science (JSPS: no. 16510180) to R. Kuranishi and by a JSPS postdoctoral fellowship to author.

Literature Cited

Felsenstein, J. 1995. Confidence limits on phylogenies: An approach using the bootstrap. Evolution **39**:783–791.

Kimura, M. 1980. A simple method for estimating evolutionary rates of base substitution through comparative studies of nucleotide sequences. Journal of Molecular Evolution **16**:111–120.

Misof, B., A. M. Rickert., T. R. Buckley, G. Fleck., and K. P. Sauer. 2001. Phylogenetic signal and its decay in mitochondrial SSU and LSU rRNA gene fragments of Anisoptera. Molecular Biology and Evolution **18**:27–37.

Ota, T., and G. Takahashi. 1999. Erosion Control and Ecological Management of Mountain Streams. University of Tokyo Press, Tokyo (in Japanese).

Saitou, N., and M. Nei. 1987. The neighbor-joining method: A new method for reconstructing phylogenetic trees. Molecular Biology and Evolution **4**:406–425.

Thompson, J. D., D. G. Higgins, and T. J. Gibson. 1994. CLUSTAL W: improving the sensitivity of progressive multiple sequence alignment through sequence weighting, position specific gap penalties and weight matrix choice. Nucleic Acid Research **22**:4673–4680.

Thompson, J. D., F. Plew, and O. Poch. 1999. A comprehensive comparison of multiple sequence alignment programs. Nucleic Acid Research **27**:2682–2690.

Tojo, K., and K. Matsukawa. 2003. A description of the second species of the family Dipteromimidae (Insecta, Ephemeroptera), and genetic relationship of two dipteromimid mayflies inferred from mitochondrial 16S rRNA gene sequences. Zoological Science **20**:1249–1259.

Waples, R. S. 1995. Evolutionary significant units and the conservation of biological diversity under the endangered species act. Pages 8–27 *in* J. Nielsen, editor. Evolution and the aquatic ecosystem: defining unique units in population conservation. American Fisheries Society, California.

OLD SPECIES OF NEOTROPICAL PLECOPTERA

Claudio G. Froehlich

*Departamento de Biologia, Universidade de São Paulo,
14040-901 Ribeirão Preto, SP, Brazil*

Abstract

Several types of stoneflies described in the nineteenth and early twentieth century were examined. *Gripopteryx tesselata* Brauer is confirmed as a good species. *Onychoplax limbatella* Klapálek, characterized by the very small subgenital plate, has the *Anacroneuria* type postfrontal line; the generic position hangs on the finding of males. *Perla luteicollis* Walker is not a *Macrogynoplax,* but should be included in *Anacroneuria;* while males are not found; the female type has an unusually large subgenital plate, but the postfrontal line is of the *Anacroneuria* type. *Perla morio* Pictet may not be an *Anacroneuria*; the type of the postfrontal line could not be ascertained, but it presents a field of small conical hairs on T10, as in other Neotropical genera but not in *Anacroneuria*, and the subgenital plate has no trace of a hammer. *Anacroneuria atrifrons* Klapálek from the Amazon is a good species of which *A. montera* Stark is a junior synonym; the syntypes from the Hamburg Museum, from the State of Sao Paulo, belong to *A. stanjewetti* Froehlich.

Keywords: Plecoptera; Neotropical genera.

Introduction

A number of species of stoneflies described in the nineteenth and in the first half of the twentieth century remain dubious because the original descriptions are too incomplete. To aggravate the situation for some species, the types have been lost and the names should be considered forgotten. Other species, however, have extant types and I have examined a few of these in order to try to clarify their position, not always satisfactorily. I have also examined briefly three Navás' species of *Neoperla,* the types and single specimens of which are deposited in DEI: *N. collaris* Navás 1932, a female from the State of Rio de Janeiro, *N. genualis* Navás 1932, stated to be a male by Navás, but lacking the abdomen, and *N. melzeri* Navás 1932, a female. The last two are from S. Martinho, Mato Grosso, a place that could not be located. All three are *Anacroneuria*.

Results and Discussion

Tupiperla tesselata (Brauer)

(Fig. 1)

Gripopteryx tesselata Brauer 1859: 51.
Tupiperla tesselata; Froehlich 1998: 21.

In my revision of *Tupiperla* (Froehlich 1998), I considered *T. tesselata* to be a good species, removing it from the synonymy of *T. gracilis* (Burmeister 1839) (Illies 1963). The evidence, however, was not strong; so in a visit to the Vienna Museum in 2001, I succeeded in loosening the specimen from the card to which it was glued and cleared the terminalia. The shape of the female subgenital plate (Fig. 1) confirmed my decision.

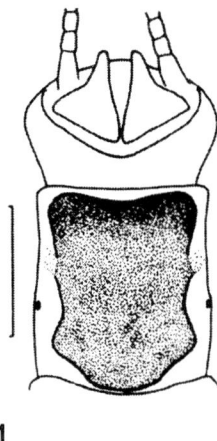

1

Figure 1. *Tupiperla tesselata* (Brauer), ♀ holotype. Subgenital plate, ventral. Scale bar = 0.5 mm.

Onychoplax limbatella Klapálek

(Figs. 2–4)

Onychoplax limbatella Klapálek 1916: 56. — Stark 2001: 419.

The genus *Onychoplax* is characterized chiefly by the small female subgenital plate. The single species, *Onychoplax limbatella*, is known only by the female type, deposited in the Vienna Museum. It has a number of labels, one of which states "vidit Pictet 1842", therefore after the publication of Pictet's book but on the other hand, showing that the specimen has been collected a very long time ago. On another label is written "Par O", which could indicate a locality, as already commented by Klapálek (1916), but where? Klapálek gives the provenance as Brazil?

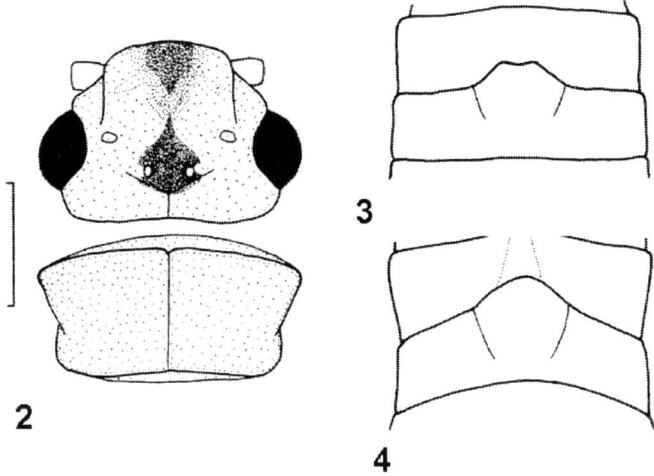

Figs. 2–4. *Onychoplax limbatella* Klapálek, ♀ holotype. 2, head and pornotum; 3-4, sterna 8 and 9, before and after clearing. Scale bars = 1 mm.

The postfrontal line forms an open V, as in most stoneflies. The color pattern of the head (Fig. 2), with its brown-black middle markings contrasting with the yellowish lateral parts, is distinctive. The forewings are brownish behind R-R$_1$, somewhat darker in basal half and along R$_1$. C, Sc, R and R$_1$ to anastomosis, as well as the costal and subcostal areas, are yellowish. The rest of the veins are brownish. The subgenital plate is very small; before clearing it showed an apical notch (Fig. 3), which disappeared after clearing (Fig. 4). St. 9 is simple, lacking hair patches or sclerotized areas.

Remarks. The colour pattern of the head and of the forewings and the small subgenital plate are distinctive. The V-shaped postfrontal line separates it from the eastern South American genera *Enderleina*, *Kempnyia* and *Macrogynoplax*. A simple St. 9 is found also in *Klapalekia* (Stark 1991) and, apparently, judging by Illies' (1964) figures, in the Andean genera (*Inconeuria*, *Kempnyella*, *Nigroperla* and *Pictetoperla*). In *Anacroneuria*, on the other hand, St. 9 always has differentiated hair patches. The possibility exists, therefore, that *Onychoplax* is not Brazilian at all, but comes from some Andean area.

Anacroneuria luteicollis (Walker 1852), stat.nov.

(Figs. 5–7)

Perla luteicollis Walker 1852: 154.
Macrogynoplax luteicollis; Klapálek 1916: 74.

This is a dark species with an orange pronotum. The postfrontal line forms an open V (Fig. 5). The forewing length is 13 mm. The female subgenital plate is very large (Figs. 6 and 7). Sternum 9 has a pair of hair patches, with the short hairs directed forwards (Figs. 6 and7).

Remarks. Perla luteicollis, from Venezuela, known only from the female type deposited in the Natural History Museum, London, was transferred by Klapálek to *Macrogynoplax*, but its assignment to the genus was not firm (Stark & Zwick 1989). The present revision shows that it surely does not belong to this genus by the lack of the unpigmented mesal portions of the eyes and by the shape of the postfrontal line. It stands near *Anacroneuria*, deviating by the size of the subgenital plate and the structure of sternum 9. Nevertheless, I am transferring it to the latter, pending on finding males. The colour pattern recalls that of some *Enderleina*, what should be due to convergence.

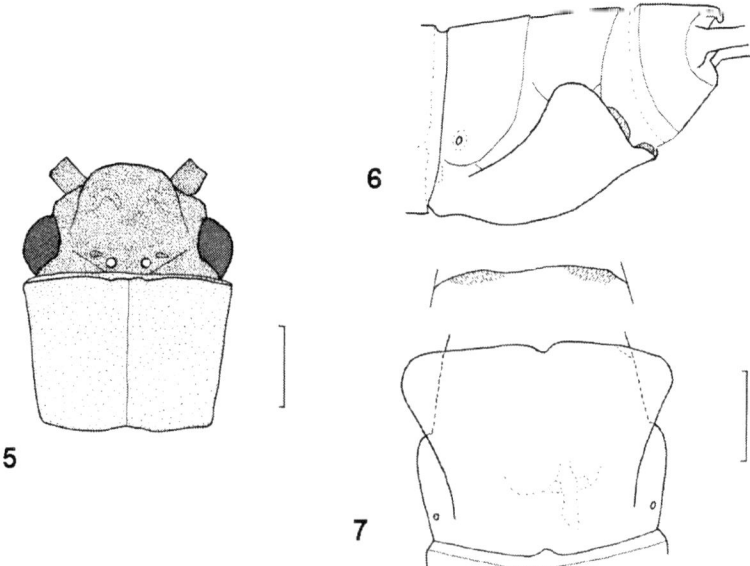

Figures 5–7. *Anacroneuria luteicollis* (Walker), ♀ holotype. 5, head and pronotum; 6, terminalia lateral before clearing; 7, sterna 8 and 9 after clearing. Scale bars = 1 mm.

Anacroneuria morio (Pictet)

Perla (*Perla*) *morio* Pictet 1841: 272.
Anacroneuria morio; Zwick 1972: 1162.

The male type of *Perla morio* Pictet is kept in the Berlin Museum, in alcohol. It is not in good conditions, as already stated by Zwick (1972), who transferred it to *Anacroneuria*, although with some reservations. It is a very dark small species. The shape of the postfrontal line could not be ascertained. The subgenital plate is evenly rounded, with no trace of a hammer. Tergum 10 has a small number of peg-like hairs, as in *Kempnyia* and some *Macrogynoplax* and *Enderleina*, but not in *Anacroneuria*.

Remarks. *Perla morio*, from Colombia, may not be an *Anacroneuria* but its position is uncertain. I am not removing it from this genus for lack of firm characters.

Anacroneuria atrifrons Klapálek

(Figs. 8–11)

Anacroneuria atrifrons Klapálek 1922: 89.
Anacroneuria montera Stark 1998: 41, nov. syn.

This is a brown species with a distinctive head colour pattern: the anterior frons to the M-line and two spots at the sides of the ocelli, that include the tentorial scars, are pale brown. The rest of the head, lappets included, is dark brown (Fig. 8). The forewing length is 9 mm.

The hammer is an elliptical truncated cone ca. 0.08 mm long (Fig. 9). The penial armature (Figs. 10 and 11) has a down-sloping tip and a low keel that in dorsal view appears as a pair of opposing parenthesis. The hooks are a little thickened in their distal halves, but end in pointed tips.

Remarks. The species is known by the male type, deposited in the MNHL. It was collected on the Amazon, at São Paulo de Olivença, by Bates. The Hamburg Museum has 3, "Kotypen" (1 ♂ 1 ♀ and 1 lacking the abdomen, but probably a ♀ by the size), not cited in Klapálek's (1922) posthumous article but in which he gives the female wing expanse, indicating that he was taking the syntypes into account. These syntypes belong to another species, considered below.

Anacroneuria montera Stark 1998, from Iquitos, in the Peruvian Amazon, is a synonym of this species. The male agrees well in color, size and penial characters with *A. atrifrons*.

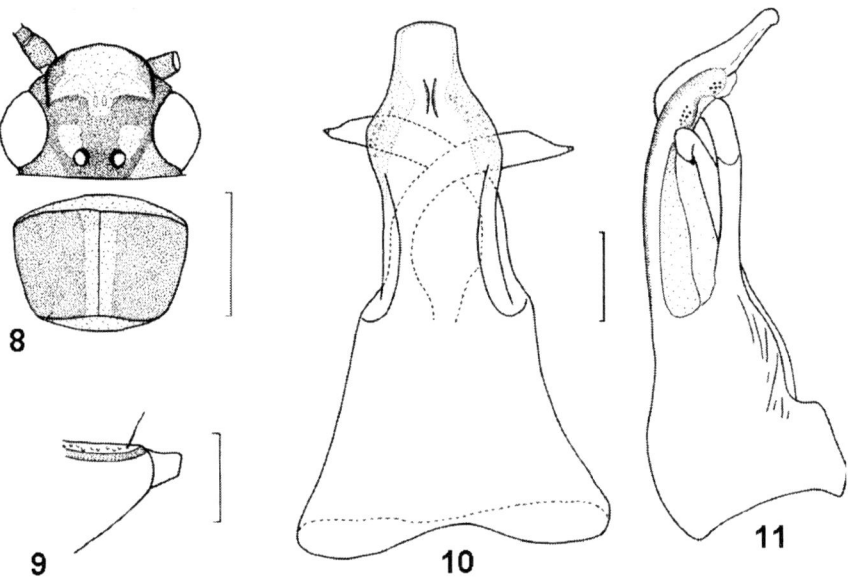

Figures 8–11. *Anacroneuria atrifrons* Klapálek, ♂ holotype. 8, head and pronotum, scale bar = 1mm; 9, hammer, one hair drawn on St9 to show the length, scale bar = 0.2 mm; 10-11, penial armature, dorsal and lateral, scale bar = 0.1 mm.

Anacroneuria stanjewetti Froehlich

(Figs. 12–15)

Anacroneuria stanjewetti Froehlich 2002: 93.
Anacroneuria atrifrons Klapálek, i.l. (Hamburg Museum).

General colour brown, male darker than female. Frons dark between postfrontal and M-lines, the latter lighter; anterior frons and genae a lighter brown; parietalia much lighter. In one female, a light wedge extends backwards from M-line (Fig. 12) Pronotum brown. Legs brown, femora yellowish basally. Wing membrane infuscated, veins brown, but C. Sc and R-R$_1$ paler. Cerci in male ringed.

Male. Forewing length, 8.3 mm. Hammer a short truncate cone. The penial armature (Figs. 13 and 14) narrows abruptly near apex, forming shoulders. Dorsal keel in dorsal view appears as a pair of lines converging but not touching anteriorly. Hooks regularly curved, tips blunt. Ventral vesicles small.

Female. Forewing length, 10–11.4 mm. Subgenital plate with 4 subequal lobes (Fig. 15). Sternum 9 with moderately dense hair patches, distal margin almost straight.

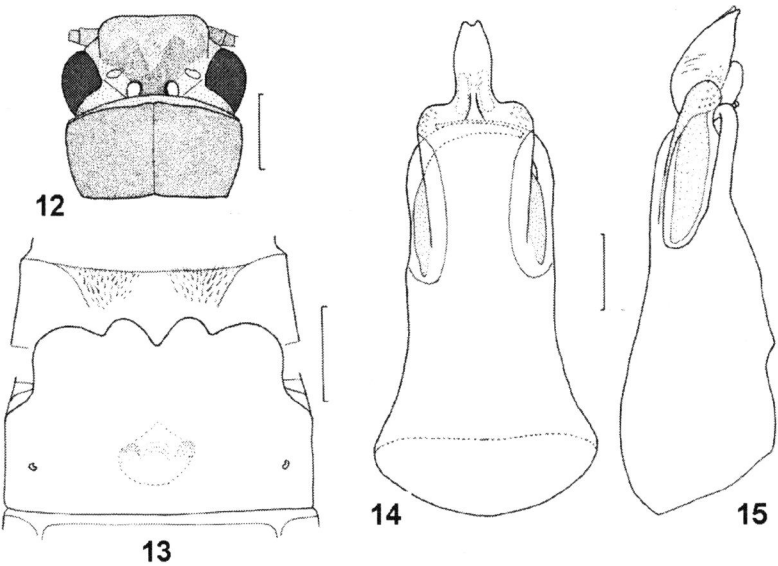

Figures 12–15. *Anacroneuria stanjewetti* Froehlich. 12, head and pronotum, scale bar = 0.5 mm; 13-14, male, penial armature dorsal and lateral, scale bar = 0.1 mm; 15, female, sterna 8 and 9, scale bar = 0.5 mm.

Remarks. The present material differs somewhat from the type series from the State of Santa Catarina and the provinces of Misiones and Entre Ríos in Argentina, being a little smaller and darker. In the male, the penial armature has smaller ventral vesicles and a larger apical piece. In the female, the 4 lobes of the subgenital plate end at about the same level, while in specimens from Santa Catarina and Argentina the lateral lobes are larger.

These differences, however, are in my opinion compatible with variations to be expected within the range of one species.

Acknowledgments

My special thanks to Peter Zwick for allowing me to work in his laboratory and for obtaining beforehand the needed specimens from several museums; to him and his wife Heide for their splendid hospitality; to the curators of the museums involved, for the lending of types; and to the Brazilian Council of Scientific and Technological Development (CNPq) for Fellowship N° 301247/1996-0.

Literature Cited

Brauer, F. 1868. Familie Perlidae. Pages 51–52 *in* Neuropteren. Reise der Österreichischen Fregatte Novara um die Erde in den Jahren 1857, 1858, 1859. Zoologischer Theil, 2. Band (1:A) **4**:1–105, pls I–II.

Burmeister, H. 1839. Handbuch der Entomologie, Plecoptera. 2 (2):863–881.

Froehlich, C. G. 1998. Seven new species of Tupiperla (Plecoptera: Gripopterygidae) from Brazil, with a revision of the genus. Studies on Neotropical Fauna and Environment **33**:19–36.

Froehlich, C. G. 2002. Anacroneuria mainly from southern Brazil and northeastern Argentina (Plecoptera: Perlidae). Proceedings of the Biological Society of Washington **115**:75–107.

Illies, J. 1963. Revision der südamerikanischen Gripopterygidae (Plecoptera). Mitteilungen der Schweizerischen Entomologischen Gesellschaft **36**:145–248.

Illies, J. 1964. Südamerikanische Perlidae (Plecoptera), besonders aus Chile und Argentinien. Beiträge zur Neotropischen Fauna **3**:207–233.

Klapálek, F. 1916. Subfamilia Acroneuriinae Klp. Časopis České Spolecnosti Entomologické **13**:45–84.

Klapálek, F. 1922. Plécoptères nouveaux. Quatrième partie. Annales de la Societé Entomologique de Belgique **62**:89–95.

Navás, L. 1932. Plecópteros. Pages 58–63 *in* Insectos suramericanos. Revista de la Real Academia de Ciencias, Madrid, 29.

Pictet, F.-J. 1841. Histoire Naturelle Générale et Particulière des Insectes Névroptères. Famille des Perlides. J. Kessmann, Genève.

Stark, B. P. 1991. Redescription of Klapalekia augustibraueri (Klapálek) (Plecoptera: Perlidae). Aquatic Insects **13**(3):189–192.

Stark, B. P. 2001. A synopsis of Neotropical Perlidae (Plecoptera). Pages 405–422 *in* E.Domínguez, editor. Trends in Research in Ephemeroptera and Plecoptera. Kluwer Academic/Plenum, New York.

Stark, B. P., and I. Sivec. 1998. Anacroneuria of Peru and Bolivia (Plecoptera: Perlidae). Scopolia **40**:1–64.

Stark, B. P., and Zwick, P. 1989. New species of Macrogynoplax from Venezuela and Surinam (Plecoptera: Perlidae). Aquatic Insects **11**:247–255.

Walker, F. 1852. Perlides. Pages 136–192 *in* Catalogue of the specimens of neuropterous insects in the collection of the British Museum. Part I. (Phryganides – Perlides). 192 pp.

Zwick, P. 1972. Die Plecopteren Pictets und Burmeisters, mit Angaben über weitere Arten (Insecta). Revue Suisse de Zoologie **78**:1123–1194.

ZOOGEOGRAPHIC AFFINITIES OF SOUTHWESTERN USA PLECOPTERA

Steven J. Cary[1] and Gerald Z. Jacobi[2]

[1] New Mexico State Parks, Santa Fe, New Mexico 87505 USA
[2] Jacobi and Associates, Santa Fe, New Mexico 87505 USA

Abstract

Zoogeographic affinities were delineated for the 75 Plecoptera species recorded from Arizona and New Mexico. Nearly half (37) were characterized as widespread western species; 18 were southwestern species; 17 were Rocky Mountain species; and three were widespread eastern species. In four major physiographic subdivisions of the study area, stonefly species arrange themselves in ways that reveal dispersal corridors, dispersal barriers and refugia that operated during late Pleistocene pluvial and post-pluvial Holocene environments. As a bridge between Nearctic and Neotropical regions, the southwestern Plecoptera fauna was shown to be distinct in its species composition, taxonomic representation and level of endemism.

Key words: Plecoptera, stoneflies, Southwestern United States, zoogeography, refugia, dispersal, pluvial, Pleistocene, Holocene.

Introduction

The American Southwest is a rewarding area for the study of island biogeography. Forested, well-watered uplands, some reaching into Canadian life zones, are separated by broad arid valleys in Lower and Upper Sonoran zones. Plant and animal species inhabiting these mesic uplands are separated by many miles from sister populations with which they are not currently in reproductive contact. Brown (1971) showed that the distribution of small mammals in montane islands in the Great Basin reflected no colonization between mountains since the end of the Pleistocene.

Stream-bottom insects, Plecoptera in particular, have been a popular subject of biogeographical analysis (Illies 1965, Nelson and Baumann 1987, Hynes 1988; Stewart and Ricker 1997, Houseman and Baumann 1997, Huntsman et al. 1999, Nelson 2004). Many stonefly species are closely tied to specific aquatic habitats (Baumann 1979) and adults generally do not wander far from their host streams (Griffith et al. 1998, Hughes et al. 1999, Myers et al. 2001, Briers et al. 2002). Life histories that make them good indicators of aquatic conditions also make them desirable subjects of geographic analysis. Changes in climate during the Pleistocene had the result of moving Plecoptera habitats all over North America. Regional studies often find interesting Pleistocene histories for Plecoptera, with Southwestern species of particular interest (Stewart et al. 1974, Jacobi and Cary 1986, 1996, Sargent et al. 1991) because most occur in montane habitats that are discontinuous

133

and were subjected to expansion, contraction and movement during the Pleistocene. Such movements, some approaching continental scale (Ross and Yamamoto 1967, Ross et al. 1967), gave rise to the notion of geographic affinity, or origin, of individual species. Before regional faunas were well known, Ricker (1964) suggested that Arizona and New Mexico functioned as a refuge for stoneflies during Pleistocene glaciations. Stewart et al. (1974) proposed that pluvial conditions stimulated southward dispersal of cordilleran species, southwestward dispersal of eastern species and southeastward dispersal of western species. Jacobi and Cary (1986) filled in data gaps for southwestern New Mexico and added detail to the Arizona/New Mexico Plecoptera refugium proposed by Ricker. Sargent et al. (1991) characterized pluvial and post-pluvial Plecoptera movements in northern Mexico and concluded that pluvial conditions in the region were sufficiently moist to permit interbasin dispersal of Nearctic species southward into northern parts of Chihuahua, Sonora and Baja California. Relict populations of southwestern species still survive there in disjunct montane populations, as they do in the American Southwest.

Current knowledge of Arizona and New Mexico stoneflies indicates a collective fauna of at least 75 species, including local endemic species (Baumann and Olson 1984, Stanger and Baumann 1993, Baumann and Jacobi 2002). Enough information is now available to add detail to prior zoogeographic analyses as they pertain to this region. This region played key roles in the dispersal and isolation of Plecoptera during pluvial and postpluvial times. Because of this, several questions can be asked. Where were the dispersal corridors? Where were the dead ends and bottlenecks? Which species made successful transitions from pluvial to postpluvial conditions? What were the areas of isolation and evolution in the region? Is there any evidence of ongoing dispersal?

Study Area

Arizona and New Mexico offer diverse features of geology, physiography and aquatic habitat, all of which were in place by the Late Pleistocene (Fig. 1). In this semiarid landscape, orographically enhanced precipitation favored uplands with moisture; moist uplands then served as centers of Plecoptera diversity. Major upland areas have variable degrees of connection or separation from the Rocky Mountain cordillera to the north and the Mexican Sierra Madre cordillera to the south (Fig. 1). The Continental Divide bisects the study area from north to south, steering runoff to the Mississippi River, Rio Grande, or Colorado River. There are multiple watercourses of varying sizes, substrates and elevations. For purposes of this analysis we divide the study area into four parts, each built around an important aggregation of uplands, mountains and peaks.

Southern Rocky Mountains. Three elements of the southern Rocky Mountains comprise the highest upland complex in the study area. The Sangre de Cristo Mountains extend south from Colorado and terminate in north-central New Mexico near Taos and Santa Fe. Topping out in tundra at over 3960 m elevation, this linear

Figure 1. Major topographic and hydrologic features in study area.

spine sustained alpine valley glaciers in the Late Wisconsin period. These peaks distribute runoff eastward via the Canadian River to the Mississippi River. The San Juan Mountains extend from southwestern Colorado into north-central New Mexico, where they approach 3660 m elevation. South of the southern San Juans, the Jemez Mountains reach 3500 m elevation. The southern San Juans, the Jemez and the Sangre de Cristos shed runoff toward the Rio Grande. The southbound Rio Grande follows a north-south trench dividing the Sangre de Cristo Mountains on the east from the southern San Juan and Jemez Mountains on the west.

Mogollon Plateau. The second most prominent upland is the Mogollon Plateau, which begins south of the Grand Canyon in north-central Arizona. It extends from the volcanic San Francisco Peaks (3660 m) southeastward past Flagstaff along the Mogollon Rim through the White Mountains and Escudilla Peak (3500 m) in east-central Arizona. This extinct volcanic field reaches into New Mexico north of Silver City, where peaks exceeding 3050 m elevation dominate the landscape. The Mogollon Plateau complex has a dominant east-west axis that makes it unique in the region. The Continental Divide passes across its eastern limb and its abrupt eastern edge drains to the Rio Grande. Some watercourses drain southeast into the closed Mimbres Basin. Most runoff from the Mogollon massif flows north or south at first, before joining the Little Colorado River or the Gila River which eventually turn west toward the Colorado River. The Gila River flows through Phoenix and joins the Colorado River about 80 km above the Gulf of California. The Gila is perennial for much of its length and its east-to-west flow is unique in the region.

Sacramento Mountains. The Sacramento Mountains complex of south-central New Mexico represents a third important upland area. Of its major components, the north-south trending limestone Sacramento Mountains exceed 2900 m elevation. The east-west oriented Capitan Mountains are of igneous origin and exceed 3050 m elevation. The core of this complex is Sierra Blanca Peak (3658 m). Its erosion-resistant igneous rhyolites help it maintain its status as the southernmost glaciated peak and the highest peak in the United States for its latitude. Its flanks are made of limestones heavily influenced by karst features like caves and springs. This is the main upland separating the Rio Grande from its major US tributary—the Pecos River. The steep west face drains to the topographically closed Tularosa Basin. The gentler east slopes drain through Roswell and the Pecos River, which joins the Rio Grande in Texas.

Sky Islands. The final noteworthy upland complex is the Sky Islands—a series of lesser uplands lying at southern latitudes between Las Cruces, New Mexico and Tucson, Arizona. In this basin-and-range landscape, broad lowlands are bordered by mountain ranges of varying heights. Major mountain ranges in this region on the Arizona side include the Chiricahua Mountains, Huachuca Mountains, Pinaleno Mountains, Santa Catalina Mountains and Santa Rita Mountains, all exceeding 2600 m. Important uplands on the New Mexico side include the Animas Mountains and the Organ Mountains, both over 2600 m. Between these and lower ranges lie basins, many of which are topographically closed and some which supported pluvial lakes during the Late Pleistocene. This region lies generally west of the Rio Grande and south of the west-flowing Gila River. It contains a few major watercourses flowing northward from Mexico, including the San Pedro River and Santa Cruz River.

Basins also are important in regional Plecoptera biogeography. Two large, adjacent, topographically closed basins in south-central New Mexico are notable. The Tularosa Basin is the site of White Sands National Monument. During the Pleistocene it collected enough water to support a freshwater pluvial lake. The adjacent Jornada del Muerto earned its name from early Spanish colonists who sought to cross its vast, waterless emptiness. Each has a north-south orientation. No major watercourses flow in or out of these basins. Together they create a large, inhospitable region for Plecoptera.

Paleoclimates and Paleoenvironments

Paleoecological reconstructions depend on clues to paleoclimates; precipitation and hydrologic regimes are particularly important for stoneflies in the semiarid Southwest. Late Pleistocene and Holocene climates have been inferred based on a wealth of evidence including fossil pollen and geologic deposits. Most importantly, Van Devender and Spaulding (1979) summarized the key role played in reconstruction efforts by plant and animal macrofossils preserved in woodrat (*Neotoma* spp.) middens.

Prevailing regional climate during the last (Wisconsin) glaciopluvial episode was cooler and wetter than today. The atmospheric circulation pattern thought to be responsible for that climate is the winter storm track. During the late Pleistocene, Pacific winter storms typically came onshore south of 36° N latitude, south of the high peaks of the Sierra Nevada (Spaulding and Graumlich 1986). These moisture laden storms delivered rain and snow to the interior Southwest, including the study area. Consequences of that circulation pattern can be seen, to a lesser degree, in the modern atmospheric circulation pattern driven by the eastern Pacific Ocean sea surface temperature phenomenon known as the El Niño Southern Oscillation (ENSO) (Molles and Dahm 1990). ENSO also delivers a strong flow of moist air onshore across southern California and into Arizona and New Mexico. When this flow encounters major uplands like the Mogollon Rim, it leaves behind significant moisture in the form of rain and snow. Melting of those mountain snows supports aquatic habitats for the large number of Plecoptera that emerge in late winter and early spring (Jacobi and Cary 1986 and 1996).

Fossil pollen recovered from cored lake sediments above 2200 m on the Mogollon Rim in central Arizona document a mid Wisconsin (37,000 to 25,000 years ago) flora consistent with a climate that was cooler and wetter than today (Jacobs 1985, Anderson 1993). Similar results were obtained for middle and late Wisconsin times from analysis of packrat middens in the Grand Canyon (Mead and Phillips 1981). Fossil pollen retrieved from sediment cores in the semiarid San Agustin Plains in west central New Mexico document a lake 18,000 years ago, surrounded by spruce, pine and sage (Markgraf et al. 1984), compared to today's piñon pine and juniper savanna.

Contents of late Wisconsin (ca. 12,000 years ago) woodrat assemblages from Picacho Peak, between Tucson and Phoenix and near the Gila River, indicated a climate with cool, dry summers and mild, moist winters. Compared to modern conditions, Van Devender et al. (1991) estimated mean air temperatures to have been 3.6 to 5.0° C lower in January and 8.4° C lower in July. Precipitation was estimated to be 50 percent greater annually and 110 percent greater in winter. Similar results were found for the lower Grand Canyon (Van Devender et al. 1977), the Sacramento Mountains (Van Devender et al. 1984), the San Andres and Fra Cristobal Mountains of south-central New Mexico (Elias 1987) and the Guadalupe Mountains of southeastern New Mexico and west Texas (Harris 1993).

A variety of aquatic environments characterized the region during late Wisconsin time. Cave calcite deposits in the eastern foothills of the Sierra Nevada document Late Pleistocene water tables 5 m to 9 m higher than today (Szabo et al. 1994). Higher water tables generated greater spring and seep discharges, documented by travertine and other geologic deposits in southern Nevada and the Grand Canyon (Quade and Pratt 1989, Szabo 1990). Late Wisconsin lowland aquatic environments were dynamic and diverse, including streams, marshes and lakes (Quade 1986). Sedge-marsh habitats bordered rivers and lakes in the New Mexico Bootheel (Smartt 1977). Evidence further suggests that pluvial basins were neither uniform nor static;

mosaics of aquatic habitats may have characterized lake margins at different localities and at different times. Modern occurrence of low-vagility stonefly populations in widely separated aquatic habitats in semiarid southwestern New Mexico implies that ecological conditions once were suitable for widespread Plecoptera populations (Jacobi and Cary 1996).

Ecological impacts of Pleistocene climates in the Southwest have been expressed in terms of altitudinal displacement of vegetation zones compared to modern conditions. Estimates for lowering of vegetation zones in the region range from 1700 m at high elevations to 350 m at low elevations (Van Devender and Spaulding 1979). However, low elevations in the Mohave and Sonoran deserts near the mouth of the Colorado River supported desert vegetation below 300m elevation during the Late Wisconsin (Cole 1986; Spaulding 1983, 1990). The resulting picture is one of differential life zone compression: down-slope displacement of plant communities was greatest at high altitudes, but the amount of displacement decreased at elevations approaching sea level.

Pluvial conditions that expanded aquatic habitats from southern California to New Mexico apparently were less influential in Trans-Pecos Texas. Evidence from late Pleistocene woodrat middens suggest that the pluvial climate may not have been sufficiently moist to permit montane plants to cross intervening lowlands near El Paso (Van Devender et al. 1987) or Big Bend (Wells 1966). The large number of uplands between the Pacific Ocean and west Texas may have extracted enough moisture from passing storms that little was left for areas east of the Sacramento and Guadalupe Mountains. This placed an eastern boundary, of sorts, on the region of pluvial influence.

The shift from cool, wet Late Wisconsin pluvial conditions to warm, dry Holocene interpluvial conditions was not a synchronous, region-wide event. Radiocarbon dating of lake sediment sequences from pluvial Lake Cochise, near Willcox in southeast Arizona, documented two periods of pluvial, lake-filling climates in that region: one during late-glacial time (ca. 13,500 years ago) and a second ca. 9000 years ago (Waters 1989). Weather conditions were probably not the same during both events. The first occurred as a result of enhanced winter precipitation while the second occurred in response to enhanced winter and summer precipitation (Spaulding and Graumlich 1986).

In summary, climate and physiography made the pluvial American Southwest a moist, but nonetheless, complex place. Valley glaciers decorated high peaks in northern New Mexico, while the Sacramento Mountains, Mogollon Mountains, White Mountains and San Francisco Peaks probably supported snowfields through much of the year. Pluvial temperature and moisture regimes raised groundwater tables and enhanced flow from springs and seeps. Pluvial aquatic environments were not static; lakes, marshes, springs, seeps, rivers and wetlands underwent episodes of expansion and shrinkage. Aquatic species in the area may have moved around considerably, adjusting to new conditions as storm tracks shifted and moisture influxes waxed or waned over decades and centuries. Not only were Late Wisconsin aquatic environments widespread, varied and dynamic over space and time, but the shift to warmer dryer Holocene conditions was neither instantaneous nor identical everywhere. Dispersal corridors may have opened and closed repeatedly during the last pluvial episode. At any given moment, dispersal corridors may have been open in one area and but closed in others.

Ways in which Plecoptera arranged themselves in pluvial space and how they responded to Holocene changes must have varied from species to species, but some generalizations seem plausible. Sargent et al. (1991) explained that mountain glaciers and cooler temperatures caused stonefly habitats and populations to shift downward in elevation. In the Southwest, this downward shift would have been enabled by expanded stream environments in what are now waterless lowland deserts. Downward shifts in elevation by Plecoptera were likely directed downstream along host watercourses because of preferences for dispersal along stream axes rather than perpendicular to them (Griffith et al. 1998, Hughes et al. 1999, Briers et al. 2002). Expanded springs and groundwater-fed streams in pluvial time probably improved rates of interbasin dispersal by shortening lateral distances between suitable aquatic habitats.

Methods

Biogeographic Affinities. To characterize the geographic affinity of each Plecoptera species, we reviewed available sources of distributional information. Comprehensive US distribution maps available at the website (Kondratieff and Baumann 2000) were particularly useful. We also used historical published data (Ricker 1964, Stark et al. 1975, Stewart et al. 1974, Jacobi and Baumann 1983, Baumann and Jacobi 1984), our recent collection data and unpublished information not yet incorporated into the NPWRC website, including Jacobi et al. (2005) and Spindler (DEQ, Phoenix, Arizona, unpublished data).

Zoogeographic affinity for each species was identified as one of four types. This enabled comparison of the geographic origins and faunal compositions of various portions of the study area.

(1) Rocky Mountain species are those limited to the prominent north-south cordilleran chain that extends from New Mexico north through Canada and into Alaska. Examples of species with Rocky Mountain distributions include *Malenka coloradensis* and *Bolshecapnia milami.* Some species in this class (e.g., *Allocapnia pilosa*) have very local ranges, but always within or near the spine of the Rocky Mountains.

(2) Widespread western North American species are those that are broadly distributed across several major western uplands including the Rockies, the Sierra Nevada, or Pacific Coast Ranges. Species from several stonefly families exemplify this type of distribution: *Eucapnopsis brevicauda, Prostoia besametsa, Suwallia pallidula, Taenionema pacificum, Skwala americana* and *Claassenia sabulosa,*

(3) Eastern North American species have core ranges in parts of North America east of the Mississippi River, but their overall distributions extend west across the Great Plains to the Rocky Mountain Front Range. Three species fall into this category: *Acroneuria abnormis, Taeniopteryx parvula* and *Perlesta decipiens.*

(4) Southwestern species generally have ranges restricted to Arizona, New Mexico or Mexico. Some extend radially into regions north of the Great Basin (Nevada, Utah, Colorado), west into California or south into Mexico. Species exhibiting this type of distribution include *Taenionema jacobii, Amphinemoura mogollinca and* several Capniidae, e.g., *Capnura fibula, Capnia decepta* and *Mesocapnia werneri.*

Similarity Index. A similarity index (SI) was constructed to quantify the degree of similarity between Plecoptera faunas of different portions of the study area and with faunas of neighboring areas. Faunal similarity (SI) between two areas was calculated as:

$$SI = 2T_c / (T_a + T_b)$$

where T_a = number of species reported from area A; T_b = number of species reported from area B; and T_c = number of species common to areas a and b.

Results

The Southwestern Plecoptera Fauna. Collectively, New Mexico and Arizona support 75 stonefly species. This fauna is comprised of 19 Capniidae, 11 Chloroperlidae, 5 Leuctridae, 13 Nemouridae, 5 Perlidae, 15 Perlodidae, 2 Pteronarcyidae and 5 Taeniopterygidae (Table 1).

Table 2 shows the New Mexico fauna is most similar to faunas of the neighboring states of Colorado and Utah, with SI values greater than 0.70. At the north end of the Rocky Mountain chain, more than 3500 km away, lies the Yukon Territory. Plecoptera of New Mexico and Yukon Territory had an SI value of 0.42. In contrast, Plecoptera of New Mexico's eastern neighbors of Texas, Oklahoma and Kansas, had SI values with New Mexico of less than 0.10. The Arizona fauna is most similar to New Mexico, with an SI value of 0.54. Arizona Plecoptera had SI values between 0.29 and 0.35 with its neighbors of Utah, Nevada and Chihuahua, Mexico. It was even lower with California, 0.13 (Table 2).

Zoogeographic Affinities. Of the total regional fauna of 75 species, only six are common to all four subregions. Zoogeographic affinities of the Arizona and New Mexico species are given in Table 1 and Fig. 2. The largest contribution to the bistate fauna was made by 37 species (almost 50% of the total) that were widespread across the western half of the North American continent. Many occurred from New Mexico west to California and north to the Arctic Circle.

Table 1. Plecoptera occurrence by origin in subregions—Sacramento Mountains, Mogollon Plateau, Sky Islands and Southern Rocky Mountains—of Arizona and New Mexico; with zoogeographic affinities: Western North America (W NA), Southwestern (SW), Rocky Mountain (R) and Eastern North America (E NA).

	Sacramento Mtns.	Mogollon Plateau	Sky Islands	S. Rocky Mtns.
Capniidae				
Bolshecapnia milami (Nebeker and Gaufin)				R
Capnia californica Claassen		SW	SW	
Capnia caryi Baumann and Jacobi		SW		
Capnia coloradensis Claassen				R
Capnia confusa Claassen	W NA	W NA		W NA
Capnia decepta Banks	SW	SW	SW	SW
Capnia gracilaria Claassen	W NA	W NA		W NA
Capnia uintahi Gaufin		SW		
Capnia vernalis Newport				W NA
Capnura fibula Claassen		SW		SW
Capnura wanica Frison	SW	SW		SW
Eucapnopsis brevicauda Claassen	W NA	W NA	W NA	W NA
Isocapnia crinita (Needham and Claassen)				R
Isocapnia vedderensis (Ricker)	R	R		R
Mesocapnia arizonensis (Baumann and Gaufin)		SW	SW	
Mesocapnia frisoni (Baumann and Gaufin)	SW	SW	SW	
Mesocapnia werneri (Baumann and Gaufin)		SW	SW	
Utacapnia logana (Nebeker and Gaufin)				SW
Utacapnia poda (Nebeker and Gaufin)				R
Chloroperlidae				
Alloperla pilosa Needham and Claassen	R			R
Paraperla frontalis Banks	W NA			W NA
Plumiperla diversa (Frison)				W NA
Suwallia pallidula (Banks)	W NA	W NA	W NA	W NA
Suwallia starki Alexander and Stewart				W NA
Sweltsa borealis (Banks)				W NA
Sweltsa coloradensis (Banks)		W NA	W NA	W NA
Sweltsa hondo Baumann and Jacobi				SW
Sweltsa lamba (Needham and Claassen)				W NA
Triznaka pintada (Ricker)	W NA			W NA
Triznaka signata (Banks)				R

Table 1 (*continued*)

	Sacramento Mtns.	Mogollon Plateau	Sky Islands	S. Rocky Mtns.
Leuctridae				
Paraleuctra jewetti Nebeker and Gaufin				R
Paraleuctra occidentalis (Banks)				W NA
Paraleuctra rickeri Nebeker and Gaufin				R
Paraleuctra vershina Gaufin and Ricker				W NA
Perlomyia utahensis Needham and Claassen				W NA
Nemouridae				
Amphinemoura apache Baumann and Gaufin			SW	
Amphinemoura banksi Baumann and Gaufin	R			R
Amphinemoura. mogollonica Baumann and Gaufin	SW	SW	SW	SW
Amphinemoura venusta (Banks)			SW	
Malenka californica (Claassen)				W NA
Malenka coloradensis (Banks)	R	R	R	R
Malenka flexura (Claassen)				R
Podmosta delicatula (Claassen)				W NA
Prostoia besametsa (Ricker)		W NA		W NA
Zapada cinctipes (Banks)				W NA
Zapada frigida (Claassen)	W NA			W NA
Zapada haysi (Ricker)	W NA			W NA
Zapada oregonensis (Claassen)				W NA
Perlidae				
Acroneuria abnormis (Newman)				E US
Anacroneuria wipukupa Baumann and Olson		SW		
Claassenia sabulosa (Banks)	W NA			W NA
Hesperoperla pacifica (Banks)		W NA	W NA	W NA
Perlesta decipiens (Walsh)			E US	
Perlodidae				
Cultus aestivalis (Needham and Claassen)				W NA
Diura knowltoni (Frison)				W NA
Isogenoides colubrinus (Hagen)				R
Isogenoides elongatus (Hagen)	R	R		R
Isogenoides zionensis (Hanson)		SW		SW
Isoperla fulva (Claassen)				W NA
Isoperla jewetti Szczytko and Stewart			SW	
Isoperla longiseta (Banks)				R
Isoperla mormona Banks	W NA	W NA		W NA

Table 1 (*continued*)

	Sacra-mento Mtns.	Mogol-lon Plateau	Sky Islands	S. Rocky Mtns.
Perlodidae (Continued)				
Isoperla phalerata (Smith)				W NA
Isoperla quinquepunctata (Banks)		W NA	W NA	W NA
Isoperla sobria (Hagen)				W NA
Kogotus modestus (Banks)				R
Megarcys signata (Hagen)				R
Skwala americana (Klapalek)		W NA		W NA
Pteronarcyidae				
Pteronarcella badia (Hagen)			W NA	W NA
Pteronarcys californica Newport				W NA
Taeniopterygidae				
Doddsia occidentalis (Needham and Claassen)				W NA
Taenionema jacobii Stanger and Baumann		SW	SW	
Taenionema pacificum (Banks)	W NA	W NA		W NA
Taenionema pallidum (Banks)	W NA			W NA
Taeniopteryx parvula Banks				E US
Total for Subregion	21	27	18	63
Total Number of Species	75			
Zoogeographic Affinity				
Western North America (W NA)	37	12	11	6	37
Southwestern (SW)	18	4	13	10	7
Rocky Mountain (R)	17	5	3	1	17
Eastern North America (E US)	3	0	0	1	2

Table 2. Similarity indices comparing Plecoptera fauna of New Mexico and Arizona with those of other US and Mexico states and Yukon Territory.

State	# taxa	Similarity Index with NM	Similarity Index with AZ
Texas	26	$(2*3)/(72+26) = 6/98 = 0.06$	$(2*2)/(32+26) = 4/58 = 0.07$
Oklahoma	57	$(2*1)/(72+57) = 2/129 = 0.02$	$(2*1)/(32+57) = 2/89 = 0.02$
Kansas	21	$(2*2)/(72+21) = 4/93 = 0.04$	$(2*0)/(32+21) = 0/53 = 0.00$
Colorado	86	$(2*61)/(72+86) = 122/158 = \mathbf{0.77}$	$(2*20)/(32+86) = 40/118 = \mathit{0.34}$
Utah	81	$(2*56)/(72+81) = 112/153 = \mathbf{0.73}$	$(2*20)/(32+81) = 40/113 = \mathit{0.35}$
New Mexico	72		$(2*28)/(32+72) = 56/104 = \mathbf{0.54}$
Arizona	32	$(2*28)/(72+32) = 56/104 = \mathbf{0.54}$	
Nevada	37	$(2*21)/(72+37) = 42/109 = \mathit{0.39}$	$(2*11)/(32+37) = 22/69 = \mathit{0.32}$
California	161	$(2*32)/(72+161) = 64/233 = 0.27$	$(2*13)/(32+161) = 26/193 = 0.13$
Chihuahua	9	$(2*5)/(72+9) = 10/81 = 0.12$	$(2*6)/(32+9) = 12/41 = \mathit{0.29}$
Sonora	2	$(2*2)/(72+2) = 4/74 = 0.05$	$(2*2)/(32+2) = 4/34 = 0.12$
Baja Calif.	11	$(2*5)/(72+11) = 10/82 = 0.12$	$(2*5)/(32+11) = 10/43 = 0.23$
Yukon Terr.	71	$(2*30)/(72+71) = 60/143 = \mathbf{\mathit{0.42}}$	$(2*8)/(32+71) = 16/103 = 0.16$

Arizona and New Mexico host 18 species (24%) whose geographic distributions are largely restricted to the Southwest quadrant of North America (CA, NV, UT, CO, AZ, NM). Twelve of these (16%) are centered in Arizona and New Mexico, with overall ranges that may reach into neighboring states of the US and Mexico. Two species (3%) have distributions that are chiefly Mexican: *Amphinemoura venusta* and *Amphinemoura apache*.

The third most numerous group, with 17 species in the study area (23%), was composed of species associated with the Rocky Mountains. The Cordilleran chain reaches from Yukon Territory south to New Mexico, but does not include the Great Basin or the Pacific Coast Ranges of the United States.

The zoogeographic region that makes the smallest contribution to the Arizona/New Mexico fauna is eastern North America. Three species are broadly distributed and most prevalent east of the Great Plains: *Acroneuria abnormis*, *Taeniopteryx parvula* and *Perlesta decipiens*. Their westernmost distribution outposts are in the central and southern Rocky Mountains or in the Southwest. They represent a tiny fraction of North America's diverse eastern Plecoptera fauna and only 4% of the southwestern regional fauna.

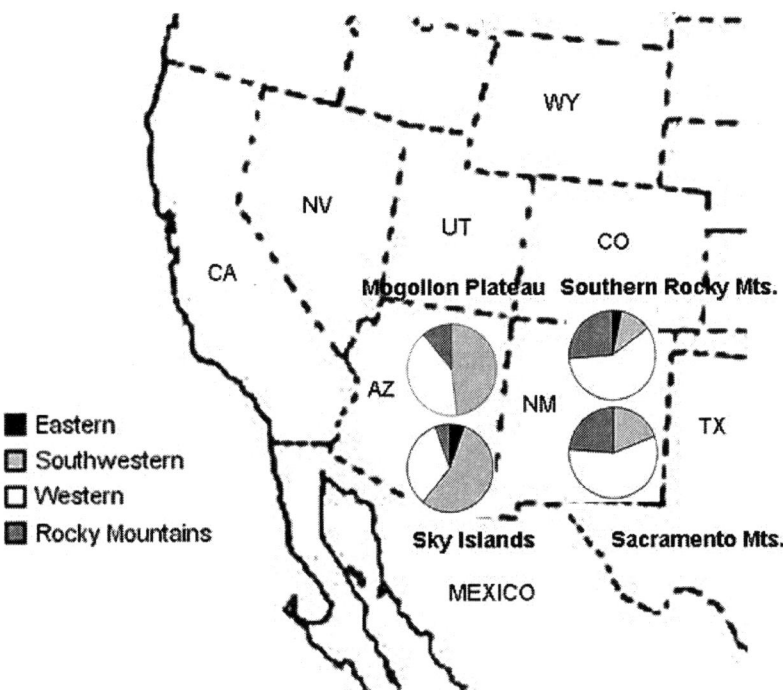

Figure 2. Zoogeographic affinities in Arizona and New Mexico by subregion.

The relative importance of species with different zoogeographic affinities (widespread western, eastern, Rocky Mountains and southwestern) comprising the Plecoptera fauna of the different subregions of the study area is shown in Table 1. The Southern Rocky Mountains of north-central New Mexico, with 63 species, has the richest Plecoptera fauna of the four subregions in the study area. The most depauperate subregion was the Sky Islands, with only 18 species.

Faunal contributions exhibited two patterns (Fig. 2). The Southern Rocky Mountains and the Sacramento Mountains, though different in absolute numbers, were similar because they consisted chiefly of widespread western and Rocky Mountain species, with fewer southwestern species. The second pattern was exhibited by the Mogollon Plateau and the Sky Islands. There was a primary southwestern influence, secondary western influence and a weak Rocky Mountain influence.

Capniidae, and to a lesser extent, Perlodidae, Nemouridae and Chloroperlidae, dominated the composition of the Arizona/New Mexico Plecoptera (Table 3 in bold). The southwestern contribution was dominated by Capniidae; the western by Chloroperlidae, Nemouridae and Perlodidae; and the Rockies by Capniidae and Perlodidae.

Table 3. Zoogeographic affinities of Plecoptera families found in Arizona and New Mexico.

Family	Western	Southwestern	Rockies	Eastern	Total
Capniidae	4	**10**	**5**	0	19
Chloroperlidae	**8**	1	2	0	11
Leuctridae	3	0	2	0	5
Nemouridae	**7**	3	3	0	13
Perlidae	2	1	0	2	5
Perlodidae	**8**	2	**5**	0	15
Pteronarcyidae	2	0	0	0	2
Taeniopterygidae	3	1	0	1	5
Total	37	18	17	3	75

Discussion

Composition of the Southwestern Fauna. The Arizona-New Mexico Plecoptera fauna is a blend of species with different geographic affinities. Against a rich background of widespread western stoneflies, Rocky Mountain species from the north mix with southwestern species in varying proportions throughout the study area. Several endemics emphasize opportunities for isolation and differentiation in the Southwest's diverse landscapes. Stoneflies of eastern North America make a numerically minor contribution, but their presence is noteworthy so far from their presumed place of origin.

The influence of Rocky Mountain and widespread western species in the southwestern fauna is demonstrated by comparing stoneflies in New Mexico with those in Yukon Territory, 3500 km apart at opposite ends of the Rocky Mountain cordillera. The north-westernmost province of Canada, Yukon Territory has 71 stonefly species (Stewart and Ricker 1997), similar in number to New Mexico's 72 species (Jacobi et al. 2005). Remarkably, New Mexico and Yukon Territory have 30 stonefly species in common (SI = 0.42). New Mexico shares almost as much of its stonefly fauna with the distant Yukon Territory as it does with neighboring Arizona (SI = 0.54). Clearly there is an assemblage of western species as content at 36° N latitude as they are at 63° N latitude.

Almost one-third of the Arizona–New Mexico fauna consists of species that are narrowly distributed in the Southwest. Geographic influence of this southwestern component drops to about 10% of the fauna north of 37° N latitude. What it lacks in geographic extent is balanced by its uniqueness, manifest in several endemic species. Moreover, the large proportion of Capniidae, a largely winter-emerging group, in the modern southwestern fauna is evocative of that fauna. More than half of all

Capniidae recorded in the study area have southwestern affinities and more than half of the southwestern faunal group are capniids (Table 3). Predominantly winter moisture regimes of pluvial climates were likely responsible for the abundance and diversity of Capniidae in the Southwest. Their modern representatives remain localized in the Southwest because of their high degree of site loyalty and generally low vagility, which make them slow dispersers.

Perspective from Individual Upland Complexes. Southern Rocky Mountains—North-central New Mexico represents the southern terminus of the Rocky Mountain cordillera and the Rocky Mountain Plecoptera fauna is relatively intact there, 64 species. Strong connections with the Colorado Rockies to the north and considerable terrain above 3050 m elevation ensure that the fauna is dominated by stoneflies typical of the western United States (59%) and the Rocky Mountains (26%). In this complex are *Bolshecapnia milami* and *Paraleuctra jewetti*, species at the extreme south end of their ranges and quite disjunct from sister populations. The southwestern faunal contribution to this region consists of 7 species (11%), more than might be expected. *Sweltsa hondo* is a local endemic. In *Acroneuria abnormis* and *Taeniopteryx parvula*, moderate sized rivers in the southeast foothills of the region have a small, but noteworthy influence from the eastern US. This is unique in New Mexico and the Southwest.

Sacramento Mountains Complex—The Sacramento Mountains have a Plecoptera fauna of 21 species to date. This assemblage is largely a depauperate version of the Southern Rocky Mountains fauna. All stoneflies in the Sacramentos also occur in the Southern Rocky Mountains, 250 km due north. Even *Alloperla pilosa*, a high-altitude species sparsely scattered in the higher Rockies, is found here (Jacobi et al. 2005). The Sacramentos are effectively the southern terminus of the Rocky Mountains fauna. Despite its southerly location, the Sacramento fauna has a small Southwestern influence, represented by only three species: *M. frisoni*, *C. decepta* and *A. mogollonica*. No eastern species have yet been collected in larger waters of the eastern footslopes. Also, no endemic stoneflies have yet been discovered here.

Mogollon Plateau—The Mogollon Plateau has a larger fauna than the Sacramento Mountains (27 species). Although it is at the same latitude as the Sacramentos, its faunal composition differs. Only three Rocky Mountain species (11% of the fauna) have penetrated the area: *Isocapnia vedderensis*, *Isogenoides elongatus* and *Malenka coloradensis*. Widespread western species make up 41% of the local Plecoptera, but the larger component (48%) is of southwestern affinity. The area hosts three southwestern endemics: *Taenionema jacobii*, *Capnia caryi* and *Anacroneuria wipukupa*. This can be attributed in part to a greater diversity of habitats spanning more than 3050 m elevation. No species of eastern affinity has yet been found in this subregion. With the Sky Islands, the Mogollon Plateau represents the core area for southwestern fauna.

Sky Islands—The Sky Islands region has a very distinctive fauna. The total fauna is smaller than other portions of the study area, but it is uniquely dominated by species of southwestern affinity. Ten of the 18 species recorded to date have southwestern affinities. *Amphinemoura venusta* is essentially a Mexican species that barely reaches into the US. *A. apache* appears to be a local endemic in the Chiricahua Mountains. Widespread western species contribute a third of the fauna, less than in the other three subregions of the study area. There is only one Rocky Mountain species (*Malenka coloradensis*). The Sky Islands share many species with the Mogollon Mountains (SI = 0.58). One difference is the presence in the Sky Islands of three species representative of larger waters: *Perlesta decipiens*, *Isoperla jewetti* and *Pteronarcella badia*. This might be expected because the Sky Islands have considerable lowlands with moderate to large waters.

The American Southwest as a Plecoptera Refugium. One stonefly's refuge is another's demise; that is, each species has its own requirements. What constitutes a refuge also depends on the conditions from which refuge is sought. Modern, interpluvial conditions in the southwest are relatively warm and dry compared to pluvial conditions. The modern regional climate exhibits a bimodal precipitation regime that, on average, delivers moderate moisture in winter and summer, but is dry in spring and autumn. Under modern, semiarid conditions, most lowland areas are too warm and dry to support stoneflies. Modern stonefly refuges are in the mountains, which are characterized by cool temperatures and more abundant aquatic habitats. Within the study area, the Southern Rocky Mountains are the best regional refuge today because they support the most species. The Mogollon Plateau and Sacramento Mountains are secondary refuges; their advantages include southerly location. Even the many minor isolated mountain ranges that reach 2600 m elevation are part of the modern regional refugium because most harbor at least one stonefly species (Jacobi and Cary 1996). Plecoptera species richness declines downstream from alpine habitats. No stoneflies are recorded from either the Canadian River or the Pecos River as they exit New Mexico into Texas. The Rio Grande has one species, *Isoperla jewetti*, as it enters Texas. An old record of *Pteronarcella badia* is from the Colorado River where it exits Arizona into Mexico. This single stonefly, as with *I. jewetti* in the Rio Grande, is probably at risk due to hydrologic modifications and other perturbations (Jacobi et al. 2005).

Pluvial Refugia. A different scenario existed under pluvial conditions. The Southern Rocky Mountains can be viewed as a jumping-off point for Plecoptera that were forced out by the advent of valley glaciers and colder glaciopluvial conditions or lured out as pluvial climates created favorable conditions in previously unsuitable areas. Their ability to move would depend on availability of suitable habitats downstream or across modest interbasin divides. Downstream dispersal from the Southern Rockies favored routes east along the Canadian River, south along the Pecos River and Rio Grande, and west along the Colorado River. Dispersal perpendicular to major drainages, though slower, would have been possible using semicontinuous upland alignments such as the Continental Divide and the divide between the Rio Grande and Pecos River.

Sacramento Mountains—South of north-central New Mexico, the Rocky Mountain Plecoptera fauna fragments into impoverished remnants that survive in isolated, smaller uplands. The Sacramento Mountain complex seems to have been the major refuge for Rocky Mountain species heading south during pluvial episodes and a suitable disembarkation point for Plecoptera headed north under interpluvial conditions. It served these functions because it had perennial waters spanning 1525 to 3050 m elevation. In addition, there were two ways in and out for Plecoptera. A wet route was available down the Pecos River and up into the Sacramentos via the Rio Hondo and Rio Peñasco near Roswell (Fig. 1). The Sacramentos also are connected to the Southern Rockies via mesas and small uplands at elevations that exceed 2135 m. *Alloperla pilosa* and *Triznaka pintada*, which are known from the Sacramentos and the high Sangre de Cristos, but not from elsewhere in the Pecos River drainage, may have taken the highland route down and back via the Sandia Mountains, Manzano Mountains, Mesa de Los Humanos and Gallinas Mountains. Since the Sacramento Mountains Plecoptera fauna is essentially a subset of the Southern Rocky Mountains fauna, it effectively marks the southern terminus of that fauna.

Mogollon Plateau—In comparison, the Mogollon Plateau was a less convenient destination for southward dispersing Rocky Mountain species. Aquatic insects with low vagility, such as Plecoptera, could disperse from the Southern Rockies to the Mogollon Mountains in one of two general ways. The wet route is down the San Juan River to the Colorado River, down the Colorado to the Little Colorado River and then upstream into the high country. A potential dry route existed by jumping sequentially along the Continental Divide from the Jemez Mountains to Mt. Taylor, the Zuni Mountains, the Gallo and San Francisco Mountains and thence to the Mogollon Plateau. This combination of wet and overland routes apparently was successful for only three Rocky Mountain species: *Isocapnia vedderensis*, *Malenka coloradensis* and *Isogenoides elongatus*. The Sky Islands support only one Rocky Mountain species (*Malenka coloradensis*).

Lower Colorado River, Gila River and pluvial lakes—One region that seems to have been a major refuge for Plecoptera during pluvial conditions was the area extending from the Colorado River lowlands on the west, eastward along the Gila River into New Mexico, perhaps even across the insignificant Continental Divide as far east as the Rio Grande. Aquatic habitats in this region included big glacier-fed waters (the Colorado River), moderate rivers (the Gila River) and south-draining tributaries from the White Mountains (the San Francisco, Verde and Salt rivers). Some stoneflies have disjunct distributions straddling the lower Colorado River: one population on the Mogollon Plateau and a second in California, i.e., *Mesocapnia werneri* and *Capnia californica*. If modern populations of these capniids are back-shifted down-slope to estimate their pluvial ranges, the result suggests they shared habitat in the lower Colorado River or nearby tributaries and pluvial lakes. Cohorts of *C. californica* and *M. werneri* became separated at the end of the latest pluvial episode due to desertification of Colorado River lowlands and induced dispersal upward, to the east and west, into their disjunct modern locations (Nelson 2004). Also consider a species pair: Californian *Capnia guilianii* and recently described *Capnia caryi* from the Mogollon Plateau. In describing *C. caryi*, Baumann and Jacobi (2002) noted that its closest relative may be *C. guilianii*. For these two species, one could hypothesize a common origin in the lower Colorado River, perhaps during an older pluvial episode. The split may have taken place during the subsequent interpluvial episode, after which the two disjunct populations stayed reproductively isolated long enough to differentiate into separate species.

The lowland pluvial refuge in southern Arizona and southwestern New Mexico also included Gila River tributaries draining north from Mexico like the San Simon, Santa Cruz and San Pedro rivers, as well as low-gradient streams and marshes associated with basin-filling lakes near Wilcox, Arizona and Lordsburg and Playas, New Mexico. Pluvial lakes themselves may not have been optimal habitat for Plecoptera, but the lakes were fed and drained by springs, streams and rivers that would have been good habitat around the margins. By back-shifting modern populations of *Mesocapnia arisonensis* down-slope to presumed pluvial altitudes, the result suggests this species thrived in Gila River lowlands and tributaries. If its modern preference for low gradient, marshy streams has not changed since pluvial times, then spring-fed tributaries around pluvial lake margins might have been home base for this insect in pluvial times.

Middle Colorado River—Middle reaches of the Colorado River, including the segment that runs through the Grand Canyon, were also a probable refuge for regional Plecoptera. Evidence suggests that the Grand Canyon supported abundant springs (Van Devender et al. 1987, Mead and Philips 1981). Now it is joined by major tributaries such as the Virgin River heading in Utah, the San Juan River heading in the San Juan Mountains of Colorado and the Little Colorado River originating on the north slope of the Mogollon Plateau. Modern Plecoptera habitats in the Grand Canyon are limited by its modern arid climate and flow manipulation below Glen Canyon Dam and there are few recent Plecoptera records from that area. Nevertheless, two stoneflies exhibit modern distributions that suggest they occupied the middle Colorado River region during pluvial times. *Capnia uintahi* and *Isogenoides zionensis* currently occupy upland areas that they would have arrived at by simple upstream dispersal from the Colorado River when Late Pleistocene pluvial conditions gave way to Holocene warming and drying.

Dispersal Corridors. Major physiographic features in the region would seem to strongly favor north-south dispersal. Because Plecoptera disperse more effectively along a stream axis than perpendicular to it, the best dispersal corridors are rivers themselves. The three largest rivers (Colorado, Rio Grande, Pecos) all flow generally from north to south. For a dispersal corridor to be effective, it must connect places of suitable habitat. The Pecos River was an effective dispersal corridor because it functioned as a link between habitats in the Sangre de Cristo Mountains and the Sacramento Mountains. The Colorado River was an effective dispersal route for species west of the Continental Divide because it linked habitats in the Colorado Rockies and Great Basin with habitats in the Mogollon Plateau. The Rio Grande was a principal north-south corridor as well, but it may have had limited effectiveness because it does not connect with much Plecoptera habitat in its southern reaches.

The Canadian River was a likely dispersal route for east-west movement between the Sangre de Cristo Mountains and the Great Plains. Moreover, *M. frisoni* has been documented in New Mexico's eastern plains (Jacobi and Cary, unpublished data). Stewart et al. (1974) noted the occurrence of southwestern *Mesocapnia frisoni* as far east as the Texas Panhandle. Three other eastern species (*Taeniopteryx parvula*, *Acroneuria abnormis* and *Perlesta decipiens*) have been documented in New Mexico (Jacobi et al. 2005) and are known from other western populations (Kondratieff and Baumann 2000). These recent records support the former existence of one or more dispersal routes connecting eastern and western population centers during pluvial times. The likely corridors for such dispersal would have been the major rivers that flow from the Rocky Mountains east to the Mississippi, thereby connecting eastern and western habitats. The Canadian River, Arkansas River, Red River and their numerous tributaries all offered potential dispersal routes. *A. abnormis* and *T. parvula* could have reached their New Mexico locations via these routes. The bigger challenge now is to explain the presence of *P. decipiens* in the lower Colorado River and the middle Rio Grande.

Dispersal Barriers. East-west dispersal of Plecoptera within the study area seems to have been very limited. The Sacramento Mountains and the Mogollon Plateau are significant and diverse uplands at the same latitude, only 320 km apart, yet their fauna are different. The small contribution of southwestern species to the Sacramentos and the small contribution of Rocky Mountain species to the Mogollon Plateau suggest that Plecoptera inhabiting these two uplands have never been in good communication. The greatest barrier to Plecoptera dispersal in the study area lies between these two uplands. That barrier consists of four parallel geologic structures, from east to west: the Continental Divide, the Rio Grande trench, the Jornada del Muerto and the Tularosa basin.

South of the Sacramento Mountains, a combination of limestone and gypsum geology, rapidly declining elevation and weakened pluvial climatic influence may have created a dead end for southbound Plecoptera. Passage to the Guadalupe Mountains in Texas, 100 km away, necessitates crossing a dry mesa at 1530 m. Once there, the Guadalupe Mountains top out at only 2590 m. The Guadalupe

Mountains have limestone and gypsum geology with associated chemical and physical attributes. Unlike the relatively moist Sacramento Mountains, the Guadalupe Mountains have only one perennial stream and few perennial springs. Availability of permanent flow in limestone terrain has been shown to have greater effect on the benthic macroinvertebrate community than other physical or chemical parameters (Smith and Wood 2002). One endemic Trichoptera (caddisfly) is known from the Guadalupe Mountains (Harris and Moulton 1993). No stoneflies have yet been found there, but additional research may document them in the middle or upper reaches of McKittrick Canyon. Based on elevation, life zones and remnant aquatic habitats, the Guadalupes may have supported some Plecoptera during pluvial times. If so, postpluvial changes may have eliminated them. South of the Guadalupe Mountains, the Delaware Basin seems fully capable of preventing southward dispersal of any stoneflies. Pluvial climates lost their effectiveness east of the Front Range of Sacramento Mountains and the Guadalupe Mountains. West Texas has several uplands with potential stonefly habitat, such as the Chisos Mountains of Big Bend region, but no Plecoptera have yet been reported from there.

Endemism. Existence of the distinctive Southwestern Plecopterofauna is the result of three factors: the east-west oriented Mogollon Mountains complex, the east-to-west flowing Gila River and effectiveness of Late Pleistocene climates in creating pluvial conditions in that area. Several species seem to be centered in the Gila watershed. The Gila River is well suited to harbor Plecoptera under a variety of climatic conditions because the watershed contains abundant and diverse habitats from near sea level at its junction with the Colorado River to above 3000 m elevation on the Mogollon Plateau. The Gila River system is able to isolate stonefly species from sister populations because of its unusually long, often inhospitable journey to the Colorado River. Stoneflies of small coldwater streams can migrate downstream under pluvial conditions, while species of larger, low gradient reaches can find refuge upstream during interpluvial conditions, all without leaving the Gila River watershed.

This view is consistent with evolution and endemism in other obligate aquatic groups. There is remarkable endemism among fish in the Gila River watershed. Of the 10 fish than were native to the Gila River in New Mexico, eight were endemic and restricted to that river system (Sublette et al. 1990). The native endemic ichthyofauna is even richer in the Gila River and lower Colorado River (Minckley 1973). This degree of endemism and differentiation among fish could only result from long-term isolation of aquatic habitats.

High rates of endemism are now becoming apparent among freshwater aquatic invertebrates in the Gila River system. The Gila River and adjacent Mogollon Plateau appear to be incubators for new Plecoptera species: *Taenionema jacobii, Capnia caryi, Amphinemoura apache* and *Anacroneuria wipukupa.* The endemic mayfly (Ephemeroptera) fauna may be richer still; 10 species are known to be endemic to the Gila watershed in New Mexico and Arizona (Lugo-Ortiz and

McCafferty 1995, McCafferty et al. 1997). The freshwater aquatic invertebrates in this region also include at least two endemic beetles (Murvosh 1993, Barr and Shepard 1993) and an endemic oligochete (Fend and Brinkhurst 2000).

Conclusion

We have offered some revisions and upgrades to the concept of a Plecoptera refuge in the American Southwest. The southwestern refuge was probably large, diverse and effective. We suggest that most modern southwestern stoneflies did not have to travel far to find suitable habitats under pluvial, then postpluvial conditions. High elevation mountain chains, like the Sangre de Cristos, which had glaciers in upper valleys, probably supported Plecoptera populations from lower ice margins (Koshima 1984) far downstream (Milner et al. 2001). With the large amount of topographic relief in the region, many species could adjust hundreds of meters in altitude, if necessary, while staying in their host watersheds and moving less than 160 horizontal km.

East of the Continental Divide, north-south dispersal prevailed along the Pecos River and Rio Grande. Many Rocky Mountain and widespread western stoneflies found refuge in the southern Rocky Mountains and the Sacramento Mountains. East-west dispersal corridors between the Great Plains and southern Rocky Mountains allowed three eastern stoneflies to extend their ranges into the Southwest.

West of the Divide, north-south dispersal was less effective in connecting the Rocky Mountains with the Mogollon Plateau. The Gila River, along with middle and lower reaches of the Colorado River, formed an effective refuge for southwestern species. Today this distinctive southwestern fauna is characterized by several endemic species. The large number of resident Capniidae is evidence of the key role played by pluvial climate patterns in creating Plecoptera habitat in the region.

Acknowledgments

M. Donna Jacobi graciously assisted in creating the tables and figures. The authors thank Patty Spindler, Arizona Department of Environmental Quality, for sharing the agency's data.

Literature Cited

Anderson, R. S. 1993. A 35,000 year vegetation and climatic history from Potato Lake, Mogollon Rim, Arizona. Quaternary Research **40**:351–359.

Barr, C. B., and W. D. Shepard. 1993. Survey for *Heterelmis stephani* Brown (Insecta: Coleoptera: Elmidae) in Madera Canyon and other localities in the Santa Rita Mountains, Arizona. Final Report prepared for U. S. Forest Service.

Baumann, R. W. 1979. Nearctic stonefly genera as indicators of ecological parameters (Plecoptera:Insecta). Great Basin Naturalist **39**(3):241–244.

Baumann, R. W., and G. Z. Jacobi. 1984. Two new species of stoneflies (Plecoptera) from New Mexico. Proceedings of the Entomological Society of Washington **86**(1):147–154.

Baumann, R. W., and G. Z. Jacobi. 2002. *Capnia caryi*, an interesting new species of stonefly from the American Southwest (Plecoptera: Capniidae). Western North American Naturalist **62**(4):484–486.

Baumann, R. W., and C. A. Olson. 1984. Confirmation of the stonefly genus *Anacroneuria* (Plecoptera: Perlidae) from the Nearctic region with the description of a new species from Arizona. Southwestern Naturalist **29**(4):489–492.

Briers, R. A., H. M. Cariss, and J. H. R. Gee. 2002. Dispersal of adult stoneflies (Plecoptera) from upland streams draining catchments with contrasting land-use. Archiv fur Hydrobiologie **156**(4):627–644.

Brown, J. H. 1971. Mammals on mountaintops: nonequilibrium insular biogeography. American Naturalist **105**(945):467–478.

Cole, K. L. 1986. The lower Colorado River valley: a Pleistocene desert. Quaternary Research **25**:392–400.

Elias, S. A. 1987. Paleoenvironmental significance of Late Quaternary insect fossils from packrat middens in south-central New Mexico. Southwestern Naturalist **32**(3):383–390.

Fend, S. V., and R. O. Brinkhurst. 2000. New species of *Rhynchelmis* (Clitellata, Lumbriculidae), with observations on the Nearctic species. Hydrobiologia **428**(1):1–59.

Griffith, M. B., E. M. Barrows, and S. A. Perry. 1998. Lateral dispersal of adult aquatic insects (Plecoptera, Trichoptera) following emergence from headwater streams in forested Appalachian catchments. Annals of the Entomological Society of America. **91**(2):195–201.

Harris, A. H. 1993. Wisconsinan pre-pleniglacial biotic change in southeastern New Mexico. Quaternary Research **40**:127–133.

Harris, S. C., and S. R. Moulton. 1993. New species of *Ochrotrichia (Ochrotrichia)* from the southwestern United States and northern Mexico. (Trichoptera: Hydroptilidae). Journal of the New York Entomological Society **101**(4):542–549.

Houseman, R. M., and R. W. Baumann. 1997. Zoogeographic affinities of the stoneflies (Plecoptera) of the Raft River Mountains, Utah. Great Basin Naturalist **57**(3):209–219.

Hughes, J. M., P. B. Mather, A. L. Sheldon, and F. W. Allendorf. 1999. Genetic structure of the stonefly, *Yoraperla brevis*, populations: the extent of gene flow among adjacent montane streams. Freshwater Biology **41**(1):1365–1427.

Huntsman, B. O., R. W. Baumann, and B. C. Kondratieff. 1999. Stoneflies (Plecoptera) of the Black Hills of South Dakota and Wyoming, USA: distribution and zoogeographic affinities. Great Basin Naturalist **59**(1):1–17.

Hynes, H. B. N. 1988. Biogeography and origins of the North American stoneflies (Plecoptera). Memoirs of the Entomological Society of Canada **144**:31–37.

Illies, J. 1965. Phylogeny and zoogeography of the Plecoptera. Annual Review of Entomology **10**:117–140.

Jacobi, G. Z., and R. W. Baumann. 1983. Winter stoneflies (Plecoptera) of New Mexico. Great Basin Naturalist **43**(4):585–591.

Jacobi, G. Z., and S. J. Cary. 1986. New records of winter stoneflies (Plecoptera) from southwestern New Mexico, with notes on habitat preferences and zoogeographical origins. Southwestern Naturalist **31**(4):503–510.

Jacobi, G. Z., and S. J. Cary. 1996. Winter stoneflies (Plecoptera) in seasonal habitats in New Mexico, USA. Journal of the North American Benthological Society **15**(4):690–699.

Jacobi, G. Z., S. J. Cary, and R. W. Baumann. 2005. An updated list of the Stoneflies (Plecoptera) of New Mexico, USA. Entomological News **116**(1)29–34.

Jacobs, B. F. 1985. A middle Wisconsin pollen record from Hay Lake, Arizona. Quaternary Research **24**:121–130.

Kondratieff, B. C., and R. W. Baumann. 2000. State Plecoptera list:hhtp://www.npwrc.gov/resource/distr/insects/sfly/nm/toc/htm.

Koshima, S. 1984. A novel cold tolerant insect found in a Himalayan glacier. Nature **310**:225–227.

Lugo-Ortiz, C. R., and W. P. McCafferty. 1995. Annotated inventory of the mayflies (Ephemeroptera) of Arizona. Entomological News **106**:131–140.

Markgraf, V., J. P. Bradbury, R. M. Forester, G. Singh, and R. S. Sternberg. 1984. San Agustin Plains, New Mexico: age and paleoenvironmental potential reassessed. Quaternary Research **22**:336–343.

McCafferty, W. P., C. R. Lugo-Ortiz, and G. Z. Jacobi. 1997. Mayfly fauna of New Mexico. Great Basin Naturalist **57**(4):283–314.

Mead, J. I., and A. M. Philips, III. 1981. The Late Pleistocene and Holocene fauna and flora of Vulture Cave, Grand Canyon, Arizona. Southwestern Naturalist **26**(3):257–288.

Milner, A. M., J. E. Brittain, E. Castella, and G. E. Petts. 2001. Trends of macroinvertebrate community structure in glacier-fed rivers in relation to environmental conditions: a synthesis. Freshwater Biology **46**(12):1833–1847.

Minkley, W. L. 1973. Fishes of Arizona. Arizona Game and Fish Department. Sun Printing Co., Phoenix, Arizona.

Molles, C. M., Jr., and C. N. Dahm. 1990. A perspective on El Niño and La Niña: global implications for stream ecology. Journal of the North American Benthological Society **9**(1):68–76.

Murvosh, C. M. 1993. Microdistribution of the water penny *Psephenus montanus* (Coleoptera: Psephenidae), with notes on life history and zoogeography. Southwestern Naturalist **38**(2):119–126.

Myers, M. J., F. A. H. Sperling, and V. H. Resh. 2001. Dispersal of two species of Trichoptera from desert springs: conservation implications for isolated vs. connected populations. Journal of Insect Conservation **5**(3):207–215.

Nelson, C. R. 2004. Systematics of the *Capnia californica* species group, including a morphological phylogeny, zoogeography, and description of *Capnia kersti*, new species (Plecoptera: Capniidae). Annals of the Entomological Society of America **97**(1):97–104.

Nelson, C. R., and R. W. Baumann. 1987. The stonefly genus Capnura (Plecoptera: Capniidae) in North America: systematics, phylogeny, and zoogeography. Transactions of the American Entomological Society **113**:1–28.

Quade, J. 1986. Late Quarternary environmental changes in the upper Las Vegas valley, Nevada. Quarternary Research **26**:340–357.

Quade, J., and W. L. Pratt. 1989. Late Wisconsin groundwater discharge environments of the southwestern Indian Springs Valley, southern Nevada. Quaternary Research **31**:351–370.

Ricker, W. E. 1964. Distribution of Canadian stoneflies. Gewässer Abwässer Heft **34/35**:50–71.

Ross, H. H., G. L., Rotramel, J. E. H. Martin, and J. F. McAlpine. 1967. Postglacial colonization of Canada by its subboreal winter stoneflies of the genus *Allocapnia*. Canadian Entomologist **99**:703–712.

Ross, H. H., and T. Yamamoto. 1967. Variations in the winter stonefly *Allocapnia granulata* as indicators of Pleistocene faunal movements. Annals of the Entomological Society of America **60**(2):447–458.

Sargent, B. J., R. W. Baumann, and B. C. Kondratieff. 1991. Zoogeographic affinities of the nearctic stonefly (Plecoptera) fauna of Mexico. Southwestern Naturalist **36**(3):323–331.

Smartt, R. A. 1977. The ecology of Late Pleistocene and Recent *Microtus* from south-central and southwestern New Mexico. Southwestern Naturalist **22**(1):1–19.

Smith, H., and P. J. Wood. 2002. Flow permanence and macroinvertebrate community variability in limestone spring systems. Hydrobiologia **487**(1):45–58.

Spaulding, W. G. 1983. Late Wisconsin macrofossil records of desert vegetation in the American Southwest. Quaternary Research **19**:256–264.

Spaulding, W. G. 1990. Vegetation dynamics during the last deglaciation, southeastern Great Basin, U.S.A. Quaternary Research **33**:188–203.

Spaulding, W. G., and L. J. Graumlich. 1986. The last pluvial climatic episode of southwestern North America. Nature (London) **320**:441–444.

Stanger, J. A., and R. W. Baumann. 1993. A revision of the stonefly genus *Taenionema* (Plecoptera: Taeniopterygidae). Transactions of the American Entomological Society **119**:171–229.

Stark, B. P., T. A. Wolff, and A. R. Gaufin. 1975. New records of stoneflies (Plecoptera) from New Mexico. Great Basin Naturalist **35**(1):97–99.

Stewart, K. W., R. W. Baumann, and B. P. Stark. 1974. The distribution and past dispersal of southwestern United States Plecoptera. Transactions of the American Entomological Society **99**:507–546.

Stewart, K. W., and W. E. Ricker. 1997. Stoneflies (Plecoptera) of the Yukon. Pages 201–222 *in* H. V. Danks and J. A. Downes, editors. Insects of the Yukon. Biological Survey of Canada (Terrestrial Arthropods). Ottawa.

Sublette, J E., M. D. Hatch, and M. Sublette. 1990. The Fishes of New Mexico. University of New Mexico Press. Albuquerque.

Szabo, B. J. 1990. Ages of travertine deposits in eastern Grand Canyon National Park, Arizona. Quaternary Research **34**:24–32.

Szabo, B. J., P. T. Kolesar, A. C. Riggs, I. J. Winograd, and K. R. Ludwig. 1994. Paleoclimatic inferences from a 120,000-yr calcite record of water-table fluctuation in Browns Room of Devils Hole, Nevada. Quaternary Research **41**:59–69.

Van Devender, T. R., J. L. Betancourt, and M. Wimberly. 1984. Biogeographic implications of a packrat midden sequence from the Sacramento Mountains, south-central New Mexico. Quaternary Research **22**:344–360.

Van Devender, T. R., G. L. Bradley, and A. H. Harris. 1987. Late Quaternary mammals from the Hueco Mountains, El Paso and Hudspeth counties, Texas. Southwestern Naturalist **32**(2):179–195.

Van Devender, T. R., and W. G. Spaulding. 1979. Development of vegetation and climate in the southwestern United States. Science **204**:701–710.

Van Devender, T. R., A. M. Phillips, III, and J. I. Mead. 1977. Late Pleistocene reptiles and small mammals from the lower Grand Canyon of Arizona. Southwestern Naturalist **22**(1):49–66.

Van Devender, T. R., J. I. Mead, and A. M. Rea. 1991. Late Quaternary plants and vertebrates from Picacho Peak, Arizona. Southwestern Naturalist **36**(3):302–314.

Waters, M. R. 1989. Late Quaternary lacustrine history and paleoclimatic significance of pluvial Lake Cochise, southeastern Arizona. Quaternary Research **32**:1–11.

Wells, P. V. 1966. Late Pleistocene vegetation and degree of pluvial climatic change in the Chihuahuan Desert. Science **153**:970–975.

A CHECKLIST OF THE BURROWING MAYFLY FAMILY EPHEMERIDAE

J. M. Hwang[1], Y. J. Bae[1] and W. P. McCafferty[2]

[1]*Department of Biology, Seoul Women's University*
Seoul 139-774, Korea
[2]*Department of Entomology, Purdue University*
West Lafayette, Indiana 47907, USA

Abstract

A checklist of the Ephemeridae is provided including 96 species, 7 genera and two subfamilies (Ephemerinae: *Ephemera* Linnaeus and *Afromera* Demoulin; Hexageniinae: *Denina* McCafferty, *Hexagenia* Walsh, *Litobrancha* McCafferty, *Eatonigenia* Ulmer and *Eatonica* Navas). Original and selected secondary reference sources and distributions are provided.

Key words: Ephemeroptera; Ephemeridae; Ephemerinae; Hexageniinae; checklist.

Introduction

The burrowing mayfly family Ephemeridae is nearly cosmopolitan in distribution being absent only in Australia and the oceanic islands. They inhabit fine substrates such as gravel, sand, silt, or mud in lotic and lentic habitats. The family presently includes 96 species in 7 genera and two subfamilies (Ephemerinae: *Ephemera* Linnaeus and *Afromera* Demoulin; Hexageniinae: *Denina* McCafferty, *Hexagenia* Walsh, *Litobrancha* McCafferty, *Eatonigenia* Ulmer and *Eatonica* Navas) (McCafferty, 2004).

The largest subfamily Ephemerinae includes *Ephemera* and *Afromera*, each of which includes 68 and seven species, respectively. The largest portion of *Ephemera* species (34 spp.) is known in Southeast Asia with a majority (30 spp.) recorded in Southern China. The Ephemeridae and related burrowing mayfly families were studied mainly by McCafferty (McCafferty 1973, 1991, McCafferty and Gilles 1979). The generic and familial catalogs of Ephemeridae were provided by McCafferty (1973, 1991, 2004), Hubbard (1978, 1990), McCafferty and Gilles (1979), etc. and the classification follows a recent phylogenetic classification by McCafferty (2004).

Although members of the family have been studied relatively frequently due to large body size, common occurrence and ecological importance in freshwater habitats, the species taxonomy and generic revisions are still needed in particular on the species groups in the Oriental Region. In the present paper, we provide an up-to-date checklist of the species and higher taxa of Ephemeridae with their original and selected secondary reference sources and known distributional data for further revisionary studies of the family.

Checklist

Subfamily Ephemerinae Latreille

Genus *Ephemera* Linnaeus, 1758
1. *Ephemera annandalei* Chopra, 1937
   Ephemera annandalei Chopra, 1937 (in Hafiz, 1937: 360) (Ms, F).
   Distribution: India, Burma.
2. *Ephemera axillaris* Navas, 1930
   Ephemera axillaris Navas, 1930: 138 (M).
   Distribution: China.
3. *Ephemera blanda* Traver, 1932
   Ephemera blanda Traver, 1932: 104 (L).
   Distribution: USA.
4. *Ephemera compar* Hagen, 1875
   Ephemera compar Hagen, 1875: 578; Eaton, 1883-88: 65 (M).
   Distribution: USA.
5. *Ephemera consors* Eaton, 1891
   Ephemera consors Eaton, 1891: 412 (M).
   Distribution: India.
6. *Ephemera danica* Müller, 1764
   Ephemera danica Müller, 1764 (cited from Walker, 1853); Eaton, 1871: 72 (M, F); Eaton, 1883-88: 61 (M, F, Ms).
   Distribution: England, France, Switzerland, Germany.
7. *Ephemera diffusa* Chopra, 1937
   Ephemera diffusa Chopra, 1937 (in Hafiz, 1937: 355) (M, Ms, Fs).
   Distribution: India.
8. *Ephemera distincta* Hubbard, 1982
   Ephemera distincta Hubbard, 1982: 192 (M).
   Distribution: India.
9. *Ephemera duporti* Lestage, 1921
   Ephemera duporti Lestage, 1921: 216 (M, F, Ms, Fs).
   Distribution: Vietnam.
10. *Ephemera exspectans* (Walker), 1860
    Potamanthus expectans Walker, 1860: 198 (Fs).
    Ephemera exspectans (Walker): Eaton, 1871: 74 (Fs); Hafiz, 1937: 366 (M).
    Distribution: India, Burma.
11. *Ephemera flaveola* Walsh, 1862
    Ephemera flaveola Walsh, 1862: 377 (M, F, Ms, Fs).
    Distribution: USA.

12. *Ephemera formosana* Ulmer, 1919
    Ephemera formosana Ulmer, 1919: 6 (M, F, Ms, Fs); Kang and Yang, 1994: 394 (L).
    Distribution: China, Japan, Taiwan.
13. *Ephemera fulvata* Navas, 1935
    Ephemera fulvata Navas, 1935: 100 (M, F).
    Distribution: India.
14. *Ephemera glaucops* Pictet, 1843
    Ephemera glaucops Pictet, 1843: 132 (M, F, Ms).
    Distribution: Algeria, France, Germany, Italy, Portugal, Russia, Sweden, Switzerland.
15. *Ephemera guttulata* Pictet, 1843
    Ephemera guttulata Pictet, 1843: 135; Eaton, 1883-88: 66 (M, F).
    Distribution: Canada, USA.
16. *Ephemera hainanensis* Zhang, Gui and You, 1995
    Ephemera hainanensis Zhang, Gui and You, 1995; 71 (M, F, Ms, Fs).
    Distribution: China.
17. *Ephemera hasalakensis* Hubbard, 1983
    Ephemera (Ephemera) hasalakensis Hubbard, 1983: 387 (M, F, Ms, Fs).
    Distribution: Sri Lanka.
18. *Ephemera hongjiangensis* Zhang, Gui and You, 1995
    Ephemera hongjiangensis Zhang, Gui and You, 1995: 71 (M, F, Ms, Fs).
    Distribution: China.
19. *Ephemera hsui* Zhang, Gui and You, 1995
    Ephemera hsui Zhang, Gui and You, 1995: 72 (M, F, Ms, Fs).
    Distribution: China.
20. *Ephemera immaculata* Eaton, 1871
    Ephemera immaculata Eaton, 1871: 74 (M); Hafiz, 1937: 363 (M).
    Distribution: India.
21. *Ephemera innotata* Navas, 1922
    Ephemera innotata Navas, 1922: 54 (M).
    Distribution: Vietnam.
22. *Ephemera japonica* McLachlan, 1875
    Ephemera japonica McLachlan, 1875: 169 (M); Eaton, 1883-88: 74 (M, F); Tshernova, 1973: 222 (M, L).
    Distribution: China, Japan, Far East Russia.
23. *Ephemera javana* Navas, 1936
    Ephemera javana Navas, 1936 (cited from Ueno, 1969); Ueno, 1969: 233 (M).
    Distribution: Thailand.
24. *Ephemera jianfengensis* Zhang, Gui and You, 1995
    Ephemera jianfengensis Zhang, Gui and You, 1995: 69 (M, F, Ms, Fs).
    Distribution: China.

25. *Ephemera kirinensis* Hsu, 1931
    Ephemera kirinensis Hsu, 1931-32: 40 (M, Fs); Hsu, 1936-37a: 287 (M, F, Fs).
    Distribution: China.
26. *Ephemera koshunensis* Matsumura, 1931
    Ephemera koshunensis Matsumura, 1931:1468 (F).
    Distribution: Japan.
27. *Ephemra lankensis* Hubbard, 1983
    Ephemra (*Ephemera*) *lankensis* Hubbard, 1983: 386 (M, F, Ms, Fs).
    Distribution: Sri Lanka.
28. *Ephemera lineata* Eaton, 1870
    Ephemera lineata Eaton, 1870: 1; Eaton, 1883-88: 63 (M, F, Ms); Tshernova, 1973:222 (M, L).
    Distribution: Throughout Europe (not recorded in Italy), USSR, China.
29. *Ephemera longiventris* Navas, 1917
    Ephemera longiventris Navas, 1917: 9 (F).
    Distribution: Vietnam.
30. *Ephemera lota* Navas, 1933
    Ephemera lota Navas, 1933: 6; Hsu, 1936-37a: 289 (Fs).
    Distribution: China.
31. *Ephemera maoyangensis* Zhang, Gui and You, 1995
    Ephemera maoyangensis Zhang, Gui and You, 1995: 68 (M, F, Ms, Fs).
    Distribution: China.
32. *Ephemera media* Ulmer, 1936
    Ephemera media Ulmer, 1935-36: 204 (M, F, Fs).
    Distribution: China.
33. *Ephemera mooiana* McCafferty, 1971
    Ephemera mooiana McCafferty, 1971a: 60 (M, L).
    Distribution: South Africa.
34. *Ephemera nadinae* McCafferty and Edmunds, 1973
    Ephemera (*Aethephemera*) *nadinae* McCafferty and Edmunds, 1973: 306 (L).
    Distribution: India.
35. *Ephemera nathani* Hubbard, 1982
    Ephemera nathani Hubbard, 1982: 194 (Ms, Fs).
    Distribution: India.
36. *Ephemera nigroptera* Zhou, 1998
    Ephemera nigroptera Zhou, 1998: 139 (M, F).
    Distribution: China.
37. *Ephemera orientalis* McLanclan, 1875
    Ephemera orientalis McLanclan, 1875: 168 (M); Tshernova, 1973: 225 (M, F, Ms, L).
    Distribution: China, Japan, Korea, Mongolia, Russia.

38. *Ephemera parnassiana* Demoulin, 1958
    Ephemera parnassiana Demoulin, 1958: 226 (M, F).
    Distribution: Greece.
39. *Ephemera paulea* Grandi, 1955
    Ephemera paulea Grandi, 1955: 202 (M, F).
    Distribution: Italia.
40. *Ephemera pictipennis* Ulmer, 1924
    Ephemera pictipennis Ulmer, 1924: 28 (M).
    Distribution: China.
41. *Ephemera pictiventris* McLanclan, 1904
    Ephemera pictiventris McLanclan, 1904 (cited from Wu, 1935); Hsu,
        1936-37a: 293 (M, F, Fs).
    Distribution: China.
42. *Ephemera pieli* Navas, 1934
    Ephemera pieli Navas, 1934: 5; Hsu, 1936-37a: 295 (Ms).
    Distribution: China.
43. *Ephemera postica* (Banks), 1914
    Rhoenanthus posticus Banks, 1914: 613 (M).
    Ephemera postica (Banks): Hubbard, 1983: 384 (M, F, Ms).
    Distribution: Sri Lanka.
44. *Ephemera pramodi* Dubey, 1971
    Ephemera pramodi Dubey, 1971: 525 (M, F).
    Distribution: Nepal.
45. *Ephemera pulcherrima* Eaton, 1892
    Ephemera pulcherrima Eaton, 1892: 185 (Fs); Hsu, 1936-37b: 433 (M,
        F, Fs).
    Distribution: Burma, China, Hong Kong, India.
46. *Ephemera punctata* (Navas), 1922
    Nirvius punetatus Navas, 1922: 56 (M).
    Ephemera punctata (Navas): Lestage, 1922: 253.
    Distribution: Malaysia.
47. *Ephemera purpurata* Ulmer, 1919
    Ephemera purpurata Ulmer, 1919: 8 (M, Ms); Ueno, 1969: 234 (F).
    Distribution: China.
48. *Ephemera quadriguttata* Lestage, 1927
    Ephemera quadriguttata Lestage, 1927: 93 (M).
    Distribution: Vietnam.
49. *Ephemera remensa* Eaton, 1891
    Ephemera remensa Eaton, 1891: 410 (M, F).
    Distribution: India.
50. *Ephemera rufomaculata* Zhou and Zheng, 2003
    Ephemera rufomaculata Zhou and Zheng, 2003: 666 (M, F, L).
    Distribution: China.

51. *Ephemera sachalinensis* Matsumura, 1911
    Ephemera sachalinensis Matsumura, 1911: 6; Tshernova, 1973: 226 (M,
      F, Ms, L).
    Distribution: China, Japan, Korea, Russia.
52. *Ephemera sauteri* Ulmer, 1912
    Ephemera sauteri Ulmer, 1912: 369 (M, F, Fs); Kang and Yang, 1994:
      396 (L).
    Distribution: China, Taiwan.
53. *Ephemera separigata* Bae, 1995
    Ephemera separigata Bae, 1995: 159 (M, F, L).
    Distribution: Korea.
54. *Ephemera serica* Eaton, 1871
    Ephemera serica Eaton, 1871: 75 (M, F, Ms).
    Distribution: China, Hong Kong, Vietnam.
55. *Ephemera shengmi* Hsu, 1937
    Ephemera shengmi Hsu, 1936-37b: 440 (M, Ms); Tshernova, 1973: 228
      (M, L).
    Distribution: China, Russia.
56. *Ephemera simulans* Walker, 1853
    Ephemera simulans Walker, 1853: 536 (M); Eaton, 1883-88: 67 (M, F,
      Ms).
    Distribution: Canada, USA.
57. *Ephemera soanica* Ali, 1967
    Ephemera soanica Ali, 1967: 75 (L).
    Distribution: Pakistan.
58. *Ephemera spilosa* Navas, 1936
    Ephemera spilosa Navas, 1936: 117; Hsu, 1937-38: 53 (M, F).
    Distribution: China.
59. *Ephemera strigata* Eaton, 1892
    Ephemera strigata Eaton, 1892 (cited from Tshernova, 1973);
      Tshernova, 1973: 228 (M, L).
    Distribution: China, Japan, Korea, Mongolia, Russia.
60. *Ephemera supposita* Eaton, 1883
    Ephemera supposita Eaton, 1883-88: 72 (M).
    Distribution: China, India, Sri Lanka.
61. *Ephemera transbajkalica* Tshernova, 1973
    Ephemera transbajkalica Tshernova, 1973: 229 (M, F).
    Distribution: Russia.
62. *Ephemera traverae* Spieth, 1938
    Ephemera traveri Spieth, 1938: 5 (M, F).
    Distribution: USA.

63. *Ephemera varia* Eaton, 1883
    Ephemera varia Eaton, 1883-88: 69 (M, F, Ms).
    Distribution: USA.
64. *Ephemera vulgate* Linnaeus, 1758
    Ephemera vulgate Linnaeus, 1758: 546; Eaton, 1883-88: 59 (M, F).
    Distribution: England, Switzerland, Portugal, France, Russia.
65. *Ephemera wanquanesis* Zhang, Gui and You, 1995
    Ephemera wanquanesis Zhang, Gui and You, 1995: 70 (M, F, Ms).
    Distribution: China.
66. *Ephemera wuchowensis* Hsu, 1937
    Ephemera wuchowensis Hsu, 1937-38: 54 (Ms, Fs).
    Distribution: China.
67. *Ephemera yaosani* Hsu, 1937
    Ephemera yaosani Hsu, 1937-38: 55 (Ms, Fs)
    Distribution: China.
68. *Ephemera zettana* Kimmins, 1937
    Ephemera zettana Kimmins, 1937: 431 (M, F).
    Distribution: India.

Genus *Afromera* Demoulin

69. *Afromera aequatorialis* (Kimmins), 1956
    Ephemera aequatorialis Kimmins, 1956: 86 (M, F).
    Afromera aequatorialis (Kimmins): Kimmins, 1960: 355.
    Distribution: Congo, Madagascar, Togoland, Uganda.
70. *Afromera congolana* Demoulin, 1955
    Afromera congolana Demoulin, 1955: 293 (M).
    Distribution: Congo, Madagascar.
71. *Afromera evae* Gillies, 1979
    Afromera evae McCafferty and Gillies, 1979: 172 (M, F, L).
    Distribution: Gambia.
72. *Afromera gilliesi* Elouard, 1986
    Afromera gilliesi Elouard, 1986a: 170 (M).
    Distribution: Guinee.
73. *Afromera natalensis* (Barnard), 1932
    Ephemera natalensis Barnard, 1932: 210 (M); Kimmins, 1949: 828 (Ms,
      F).
    Afromera natalensis (Barnard): Demoulin, 1955: 295 (M).
    Distribution: Uganda, Madagascar.
74. *Afromera siamensis* (Ueno), 1969
    Ephemera siamensis Ueno, 1969: 235 (M, Ms, Fs).
    Ephemera (*Dicrephemera*) *siamensis* Ueno: McCafferty and Edmunds,
      1973: 302 (M, F, L).
    Afromera siamensis (Ueno): McCafferty and Gillies 1979: 169.
    Distribution: Thailand.

75. *Afromera troubati* Elouard, 1986
    Afromera troubati Elouard, 1986a: 171 (M).
    Distribution: Mali, Niger.

Subfamily Hexageniinae McCafferty

Genus *Denina* McCafferty
    76. *Denina dubiloca* McCafferty, 1987
        Denina dubiloca McCafferty, 1987: 472.
        Distribution: unknown.
        Note: This species is fossil mayfly and deposited in the Paleontological
            Institute of the USSR Academy of Sciences (McCafferty 1987).
Genus *Hexagenia* Walsh
    77. *Hexagenia albivitta* (Walker), 1853
        Baetis albivitta Walker, 1853: 566 (M).
        Hexagenia albivitta (Walker): Eaton, 1871: 64 (M); Spieth, 1941: 270
            (M, L).
        Distribution: Argentina, Brazil, Colombia, Costa Rica, Guyana, Mexico,
            Paraguay, Uruguay.
    78. *Hexagenia atrocaudata* McDunnough, 1924
        Hexagenia atrocaudata McDunnough, 1924: 92 (M, F); Traver, 1931:
            616 (L).
        Distribution: Canada, USA.
    79. *Hexagenia benedicta* Navas, 1922
        Hexagenia benedicta Navas, 1922: 55 (F).
        Distribution: Brazil.
    80. *Hexagenia bilineata* (Say), 1824
        Baetis bilineata Say, 1824: 303 (F).
        Hexagenia bilineata (Say): Walsh, 1863: 199; Spieth, 1941: 242
            (M, F, L).
        Distribution: Canada, Mexico, USA.
    81. *Hexagenia callineura* Banks, 1914
        Hexagenia callineura Banks, 1914: 613 (F).
        Distribution: Colombia, Ecuador.
    82. *Hexagenia limbata* (Serville), 1829
        Ephemera limbata Serville, 1829 (cited from Walker, 1853).
        Hexagenia limbata (Serville): Walsh, 1863: 197; Spieth, 1941: 245
            (M, F).
        Distribution: Canada, Mexico, USA.
    83. *Hexagenia mexicana* Eaton, 1885
        Hexagenia mexicana Eaton, 1883-88: 50 (M).
        Distribution: Colombia, Ecuador, Mexico, Peru.

84. *Hexagenia orlando* Traver, 1931
    Hexagenia orlando Traver, 1931: 608 (M, F).
    Distribution: USA.

85. *Hexagenia rigida* McDunnough, 1924
    Hexagenia rigida McDunnough, 1924: 90 (M, F); Spieth, 1941: 267 (M, L).
    Distribution: Canada, USA.

Genus *Litobrancha* McCafferty

86. *Litobrancha recurvata* (Morgan), 1913
    Hexagenia recurvata Morgan, 1913: 395 (M, F).
    Litobrancha recurvata (Morgan): McCafferty, 1971b: 45 (M, L).
    Distribution: Canada, USA.

Genus *Eatonigenia* Ulmer, 1939

87. *Eatonigenia chaperi* (Navas), 1935
    Hexagenia chaperi Navas, 1935: 99 (M).
    Eatonigenia chaperi (Navas): Ulmer, 1939: 479; McCafferty, 1973: 55
        (M, F, L).
    Distribution: China, Indonesia (Java and Borneo), Thailand.

88. *Eatonigenia indica* (Chopra), 1924
    Hexagenia indica Chopra, 1924: 416 (F).
    Eatonigenia indica (Chopra): McCafferty, 1973: 57.
    Distribution: India.

89. *Eatonigenia seca* McCafferty, 1973
    Eatonigenia seca McCafferty, 1973: 57 (M, F).
    Distribution: Thailand.

90. *Eatonigenia trirama* McCafferty, 1973
    Eatonigenia trirama McCafferty, 1973: 59 (L).
    Distribution: India.

Genus *Eatonica* Navas

91. *Eatonica crassi* McCafferty, 1971
    Eatonica crassi McCafferty, 1971a: 57 (M, F); Elouard, 1986b: 88 (M).
    Distribution: Sudan, Tanzania, Malawi.

92. *Eatonica denysae* Elouard and Sartori, 1998
    Eatonica denysae Elouard and Sartori, 1998: 2 (Ms, F, Fs).
    Distribution: Madagascar.

93. *Eatonica josettae* Demoulin, 1969
    Eatonica josettae Demoulin, 1969:5 (Fs).
    Distribution: Madagascar.

94. *Eatonica luciennae* Elouard and Oliarinony, 1998
    Eatonica luciennae Elouard and Oliarinony, 1998: 4 (F, Ms, Fs).
    Distribution: Madagascar.

95. *Eatonica patriciae* Elouard, 1986
    Eatonica patriciae Elouard, 1986b: 89 (M, F).
    Distribution: Guinee.

96. *Eatonica schoutedeni* (Navas), 1911
    Ephemera schoutedeni Navas, 1911: 222 (F).
    Eatonica schoutedeni (Navas): Navas, 1913: 181.
    Distribution: Ethiopia, Uganda, Tanzania.

Acknowledgments

This work was supported by the research project "Ecotophia 21" from the Ministry of Environment of Korea in 2005.

Literature Cited

Ali, S. R. 1967. The mayfly nymphs (Order: Ephemeroptera) of Rawalpindi District. Pakistan Journal of Science **19**:73–86.

Bae, Y. J. 1995. Ephemera separigata, a new species of Ephemeridae (Insecta: Ephemeroptera) from Korea. Korean Journal of Systematic Zoology **11**:159–166.

Banks, N. 1914. New neuropteroid insects, native and exotic. Proceedings of the Academy of Natural **Science** of Philadelphia **66**:608–618.

Barnard, K. H. 1932. South Africa mayflies (Ephemeroptera). Transactions of the Royal Society of South Africa pp. 201–259.

Chopra, B. 1924. The fauna of an island in the Chilka Lake. The Ephemeroptera of Barkuda Island. Records of the Indian Museum **26**:415–422.

Demoulin, G. 1955. *Afromera* gen. Nov., Ephemeridae de la faune ethiopienne (Ephemeroptera). Bulletin et Annales de al Societe Royale d'Entomologie de Belgique **91**:291–295.

Demoulin, G. 1958. Mission E. Janessens en Grece, 1957 3 note. Ephemeroptera. Bulletin et Annales de al Societe Royale d'Entomologie de Belgique **94**:226–228.

Demoulin, G. 1969. Le genre Eatonica Navas (Ephemeroptera, Ephemeridae) existe-t-il a Madagascar. Bulletin et Annales de al Societe Royale d'Entomologie de Belgique **45**:1–9.

Dubey, O. P. 1971. Torrenticole insects of the Himalaya. VI. Descriptions of nine new species of Ephemeridae from the Northwest Himalaya. Oriental Insects **5**:521–548.

Eaton, A. E. 1870. On some new British species of Ephemeridae. Transactions of the Royal Entomological Society of London pp. 1–8.

Eaton, A. E. 1871. A monograph on the Ephemeridae. Transactions of the Royal Entomological Society of London pp. 1–164.

Eaton, A. E. 1883–88. A revisional monograph of recent Ephemeridae or mayflies. Transactions of the Royal Entomological Society of London, 2nd Ser. Zool. **3**:1–352.

Eaton, A. E. 1891. Notes on some native Ephemeridae in the Indian Museum. Journal of the Asiatic Society of Bengal **4**:407–413.

Eaton, A. E. 1892. New species of Ephemeridae from the Tenasserim Valley. Transactions of the Linnean Society of London pp. 185–190.

Elouard, J. M. 1986a. Ephemeroptera from West Africa: The Genus *Afromera* (Ephemeridae). Revue d'Hydrobiologie Tropicale **19**:169–176.

Elouard, J. M. 1986b. Ephemeres d'Afrique de l'Ouest: le genre Eatonica (Ephemeridae). Revue d'Hydrobiologie Tropicale **19**:87–92.

Elouard, J. M., R. Oliarinony, and M. Sartori. 1998. Biodiversité aquatique de Madagascar. 9. Le genre *Eatonica* Navas (Ephemeroptera, Ephemeridae). Bulletin de la Societe Entomologique Suisse **71**:1–9.

Grandi, M. 1955. Contributi allo studio degli Efemeroidei italiani. Boll. Ist. Ent. Univ. Bologna. pp. 202–212.

Hafiz, H. A. 1937. The Indian Ephemeroptera (Mayflies) of the sub-order Ephemeroidea. Records of the Indian Museum pp. 351–361.

Hagen, H. 1875. Report on the Pseudoneuroptera and Neuroptera collected by Lieut. W. L. Carpenter in 1873 in Colorado. Ann. Rpt., U. S. Geol. Surv. of the Territories for 1873, Part 3, pp. 578–583.

Hsu, Y. C. 1931–1932. Two new species of Mayflies from China (order Ephemeroptera). Peking Natural History Bulletin **6**:39–41.

Hsu, Y. C. 1936–1937a. The mayflies of China (Order Ephemeroptera). Peking Natural History Bulletin **11**:287–296.

Hsu, Y. C. 1936–1937b. The mayflies of China (Order Ephemeroptera). Peking Natural History Bulletin **11**:433–440.

Hsu, Y. C. 1937–1938. The mayflies of China. Peking Natural History Bulletin **12**:53–56.

Hubbard, M. D. 1982. Two new species of *Ephemera* from South India (Ephemeroptera: Ephemeridae). Pacific Insects **2**:192–195.

Hubbard, M. D. 1983. Ephemeroptera of Sri Lanka: Ephemeridae. Systematic Entomology **8**:383–392.

Hubbard, M. D. 1990. Mayfies of the world. A catalog of the family and genus group taxa (Insecta: Ephemeroptera). Sandhill Crane Press, Gainesville, Florida. pp. 1–119.

Hubbard, M. D. and W. L. Peters. 1978. A catalogue of the Ephemeroptera of the Indian Subregion. Oriental Insects Supplement **9**:1–42.

Kang, S. C., and C. T. Yang. 1994. Ephemeroptera of Taiwan (Ephemeroptera). Chinese Journal of Entomology **14**:391–399.

Kimmins, D. E. 1937. Some new Ephemeroptera. Annals and Magazine of Natural History **19**:430–440.

Kimmins, D. E. 1949. Ephemeroptera from Nyasaland, with descriptions of new species. Annals and Magazine of Natural History Ser. 12, **1**:825–836.

Kimmins, D. E. 1956. New species of Ephemeroptera from Uganda. Bulletin of the British Museum (Natural History) Entomology **4**:71–87.

Kimmins, D. E. 1960. Notes on East African Ephemroptera, with descriptions of new species. Bulletin of the British Museum (Natural History) Entomology **9**:337–355.

Lestage, J. A. 1921. Les Ephémères indo-chinoises. Annales de la Société Entomologique de Belgique **61**:211–222.

Lestage, J. A. 1922. Notes sur le genre Nirvius Navas (=*Ephemera* L.) [Ephemeroptera]. Bulletin de la Société Entomologique de France. **16**:253–254.

Lestage, J. A. 1927. Une Ephemera nouvelle du Tonkin et tableau des espèces de la faune orientale. Annales de la Société Entomologique de France 96:93–100.

Linnaeus, C. 1758. Systema naturae per regna tria naturae secundum classes, ordines, genera, species, cum characteribus, differentiis, locis. Ed. Decimal reformata. Vol. 1. Laur. Salvii, Holmiae, pp. 1–823.

Matsumura, S. 1911. Erster Beitrag zur Insekten-Fauna von Sachalin. Journal of the College of Agriculture Tohoku Univ. **4**:1–145.

Matsumura, S. 1931. Ephemerida. Pages 1465–1480 *in* 6,000 Illustrated Insects of the Japanese Empire. Tokyo (in Japanese).

McCafferty, W. P. 1971a. New burrowing mayflies from Africa (Ephemeroptera: Ephemeridae). Journal of the Entomological Society of Southern Africa **34**:57–62.

McCafferty, W. P. 1971b. New genus of mayflies from easten North America (Ephemeroptera: Ephemeridae). Journal of the New York Entomological Society **79**:45–51.

McCafferty, W. P. 1973. Systematic and zoogeographic aspects of Asiatic Ephemeridae (Ephemeroptera). Oriental Insects **7**:49–67.

McCafferty, W. P. 1987. New fossil mayfly in Amber and its relationships among extant Ephemeridae (Ephemeroptera). Annals of the Entomological Society of America **80**:472–474.

McCafferty, W. P. 1991. Toward a phylogentic classification of the Ephemeroptera (Insecta): A commentary on systematics. Annals of the Entomological Society of America **4**:343–360.

McCafferty, W. P. 2004. Higher classification of the burrowing mayflies (Ephemeroptera: Scapphodonta). Entomological News **115**:84–92.

McCafferty, W. P., and G. F. Edmunds Jr. 1973. Subgeneric classification of *Ephemera* (Ephemeroptera). Pan-Pacific Entomologist **49**:300–307.

McCafferty, W. P., and M. T. Gilles. 1979. The African Ephemeridae (Ephemeroptera). Aquatic Insects **1**:169–178.

McDunnough, J. R. 1924. New Canadian Ephemeridae with notes, II. Canadian Entomologist **56**:90–98.

McLachlan, R. 1875. A sketch of our present knowledge of the Neuropterous fauna of Japan (excluding Odonata and Trichoptera). Transactions of the Entomological Society **2**:167–190.

Morgan, A. H. 1913. A contribution to the biology of mayflies. Annals of the Entomological Society of America **6**:371–413.

Navás, L. 1911. Deux Ephémérides (Ins. Neur.) nouveaux du Congo Belge. Annales de la Société Scientifique de Bruxelles **35**(pt. 1):221–224.

Navás, L. 1913. Algunos órganos de la alas de los insectos. II International Congress of Entomology, Oxford, 1912, **2**:178–186.

Navás, L. 1917. Névroptères de l'Indo-Chine. 2e série. Revue illustrée d'Entomologie: Rennes **7**:8–17.

Navás, L. 1922. Efemerópteros nuevos o poco conocidos. Boletin de la Sociedad Entomológica de España **5**:54–63.

Navás, L. 1930. Epemeropteros. Insectos Del Museo De Paris. pp. 137–140.

Navás, L. 1933. Névroptères et insectes voisins. Chine et pays environnants. 4ᵉ Série. Notes d'Entomologie Chinoise **1**(9):1–23.

Navás, L. 1934. Névroptères et insectes voisins. Chine et pays environnants. Sixième Série. Notes d'Entomologie Chinoise, Musée Heude **1**(14):1–10.

Navás, L. 1935. Décadas de insectos nuevos. Década 27. Brotéria (Ciências Naturais) **31**:97–107.

Navás, L. 1936. Névroptères et insectes voisins. Chine et pays environnants. 9ᵉ Série, suite. Notes d'Entomologie Chinoise, Musée Heude **3**(7):117–132.

Pictet, P. J. 1843. Histoire Naturelle, generale et particuliere des Insectes Neuropteres. Familledes Ephemerines. Geneva and Paris. pp. 1–300.

Say, T. 1824. Narrative on an expedition to the source of the St. Peters River, Lake Winnepeek, Lake of the Woods, etc. Performed in the year 1823, by order of the Hm. J. C. Calhoun, Secretary of War, under the command of Stephan H. Ling, Major U.S.T.E. by W. M. Keating. Philadelphia. **2**:268–378.

Spieth, H. T. 1938. Taxonomic studies on Ephemerida, I: Description of new North American species. American Museum of Natural History **1002**:1–11.

Spieth, H. T. 1941. Taxonomic studies on the Ephemeroptera. II. The genus *Hexagenia*. American Midland Naturalist **26**:233–280.

Traver, J. R. 1931. Seven new southern species of the mayfly genus *Hexagenia*, with notes on the genus. Annals of the Entomological Society of America **24**:591–621.

Traver, J. R. 1932. Mayflies of North Carolina. Journal of the Elisha Mitchell Scientific Society **47**:85–161.

Tshernova, O. A. 1973. Palearctic species of the Genus Ephemera L. (Ephemeroptera, Ephemeridae). Entomologicheskoe Obozrenie **52**:223–233.

Ueno, M. 1969. Mayflies (Ephemeroptera) from various regions of Southeast Asia. Oriental Insects **3**:221–238.

Ulmer, G. 1912. H. Sauter's Formosa-Ausbeute. Ephemeriden. Entomologische Mitteilungen **1**:369–375.

Ulmer, G. 1919. Neue Ephemeropteren. Archiv fur Naturgeschichte (A) **85**:1–80.

Ulmer, G. 1924. Einige alte und neue Ephemeropteren. Sonder-Abdruck Aus Konowia. pp. 23–37.

Ulmer, G. 1935–36. Neue Chinesische Ephemropteren, Nesbst Urersicht uber die bisher aus China bekannten arten. Peking Natural History Bulletin. **10**:201–215.

Ulmer, G. 1939. Eintagsfliegen (Ephemeropteren) von den Sunda-Inseln. Archiv für Hydrobiologie Supplement **16**:443–692.

Walker, F. 1853. Ephemerinae. List of the specimens of neuropterous insects in the collection of the British Museum, Part III (Termitidae- Ephemeridae). pp. 533–585

Walker, F. 1860. Characters of undescribed neuroptera in the collection of W.W. Saunders, Esq. Transactions of the Royal Entomological Society of London **5**:176–199.

Walsh, B. D. 1862. List of the Pseudoneuroptera of Illinois contained in the cabinet of the writer, with descriptions of over forty new species, and notes on their structural affinities. Proceedings of the Academy of Natural Sciences Philadelphia **13**:361–402.

Walsh, B. D. 1863. Observations on certain N. A. Neuroptera. by H. Hagen, M. D., of Koenigsburg, Prussia; translated from the original French M.S., and published by permission of the author, with notes and descriptions of about twenty new N. A. species of Pseudoneuroptera. Proceedings of the Entomological Society of Philadelphia **2**:167–272.

Wu, C. F. 1935. Order VII. Ephemeroptera. Catalogus Insectorum Sinensium **1**:247–253.

Zhang, J., H. Gui, and D. You. 1995. Studies on the Ephemeridae (Insecta: Ephemeroptera) of China. Journal of Nanjing Normal University (Nature Science). **18**:68–76.

Zhou, C., Gui, H. and Su, C. 1998. New species of genus *Ephemera* (Ephemeroptera: Ephemeridae). Entomologia Sinica **5**:139–142.

Zhou, C., and L. Y. Zheng. 2003. Two synonyms and one new species of the genus *Ephemera* from China (Ephemeroptera, Ephemeridae). Acta Zootaxonomica Sinica **28**:665–668.

STONEFLIES OF GLACIER NATIONAL PARK AND FLATHEAD RIVER BASIN, MONTANA

Robert L. Newell[1], R. W. Baumann[2] and J. A. Stanford[3]

[1,3] *The University of Montana, Flathead Lake Biological Station, Polson, Montana 59860 USA*
[2] *Brigham Young University, Provo, Utah 84602 USA*

Abstract

One hundred stonefly species records were located in the Glacier National Park and Flathead River basin of western Montana. Some 58 species were documented in the Flathead River, 58 from Glacier Park, 95 and 74 species respectively from west and east of the continental divide. Common and uncommon species were documented along with high elevation, lentic and hyporheic species.

Key words: Plecoptera; stoneflies; Montana; Glacier National Park; Flathead River.

Introduction

Descriptions of Montana stoneflies first appeared in Needham and Claassen (1925) and other early papers, Ricker (1952), Petty (1965), Lehmkuhl (1966), Nebeker (1966), Newell (1970) and Baumann (1970). Gaufin et al. (1972) published the first systematic account of the stoneflies of Montana. This work listed 119 species from the entire state. Gaufin's presence at the University of Montana in the late 1960s stimulated an interest in Montana stoneflies that continues today, Robson (1968), Stanford (1975), Howe (1974), Hauer et al. (1980), Perry (1984), Hauer et al. (1989), Jones et al. (1988), Imbert (1990), Gangemi (1991), Varrelman (1992), Case (1995), Wicklum (1998) and Hauer et al. (2000). Roemhild (1986) compiled a list of 110 Montana species and provided some ecological and specific distributional information for each species. In addition many systematic studies listed Montana stonefly records, e.g., Alexander et al. (1999), Baumann et al. (1977), Lyon et al. (1997), Nelson et al. (1989), Stanger et al. (1993), Stark et al. (2001), Stewart et al. (2002), Szczytko et al. (1979) and Van Wieren et al. (2001).

Methods

This document lists all Plecoptera taxa collection records from Glacier National Park (GNP) and the Flathead River Basin, upstream from the confluence with the Clark Fork River in Sanders County. The study area includes all of Flathead and Lake Counties, portions of Glacier, Lincoln, Missoula and Sanders counties in NW Montana. The classification system used in this paper follows that outlined by Stark et al. (1986).

Published and unpublished reports and theses were examined for records of any stonefly records from the study area. Reference collections at the University of Montana and Brigham Young University were also surveyed. Stonefly taxa recorded from the study area were extracted along with information on distribution in the Flathead River system, Glacier National Park, relationship to the continental divide, altitude collected, dates adults collected and number of collection records. All collection records were compared with known distributional records for each taxon. This information is listed in Table 2.

Results

A total of 100 species, representing 45 genera and 9 families were recorded (Table 1). As a comparison, Gaufin et al. (1972) recorded 119 species from all of Montana. The following list enumerates the number of species from each family:

Nemouridae- 16

Taeniopterygidae- 3

Capniidae- 20

Leuctridae- 10

Peltoperlidae- 1

Pteronarcidae-2

Perlodidae- 23

Perlidae- 4

Chloroperlidae- 21

Glacier National Park (GNP)

GNP is comprised of 4140 km^2 or 1.4 million acres. The U.S./Canadian border is its northern boundary. Its western boundary is the north fork of the Flathead River and its southern boundary is the middle fork of the Flathead River. The continental divide nearly bisects the park in a northwest-southeast direction. The area of GNP, west of the continental divide, drains into the Columbia River system, the area north east of the continental divide drains into Hudson Bay and the south eastern portion drains into the Missouri River system.

Some 89 species were collected in GNP with some small areas having a diverse stonefly fauna (Table 2). McDonald Creek (west of the divide) and its tributaries had 66 species. The Many Glacier region (east of the divide, 78 km^2 in area), had 65 species over an elevational gradient of 1480–2580 m, m.s.l. Several lakes in this area have been studied for their Plecoptera fauna. Lehmkuhl (1966) found nine species of stoneflies living in three lakes of this area. The nearby amictic Iceberg Lake is the type locality for *Bolshecapnia spenceri*. Twelve species of stonefly adults have been collected along the shores of this small cirque lake. Wilbur Creek flows out of Iceberg Lake for a distance of only about 9 km. A total of 33 species of stoneflies were found in this small stream.

Some 95 species of stoneflies were recorded from west of the continental divide and 74 species were found east of the divide. Four species collected, *Amphinemura banksi, Bolshecapnia spenceri, Isoperla longiseta* and *Zapada glacier* were only found east of the divide and 25 species collected were only found west of the divide (Table 2).

Flathead River

The main stem Flathead River is comprised of three forks, the north, middle and south forks. The south fork flows out of the Bob Marshall Wilderness area and is impounded by Hungry Horse Dam. The three forks and the main stem comprise about 365 km (219 mi) of river and drain an area of 22,241 km^2 in southeast British Columbia and western Montana. The main stem river eventually empties into 496 km^2 Flathead Lake. A total of 58 species were found in the three forks and main stem above Flathead Lake (Stanford, 1975) and others. Two species are known to occur in Flathead Lake (*Capnia confusa* and *Suwallia salish*).

Six kilometers below the lake outlet, the river is impounded by Kerr Dam, which regulates the upper 3 m of Flathead Lake and creates flow fluctuations in the lower river and seasonal changes in lake levels. The river then flows for 100 km before merging with the Clark Fork River. Only five species have been collected in the lower Flathead River (*Pteronarcys californica, Taenionema pacificum, Claassenia sabulosa, Isoperla fulva* and a Nemourid).

General Distribution Data

The most infrequently reported species in the study area were (Table 2):

Amphinemura banksi   *Utacapnia columbiana*
Capnia cheama   *Isoperla longiseta*
Isocapnia hyalite   *Megaleuctra kincaidi*

The most frequently reported species in decreasing order were (Table 2):

Sweltsa fidelis
Sweltsa coloradensis
Suwallia pallidula
Zapada columbiana
Eucapnopsis brevicauda
Suwallia lineosa

Those species collected at extreme elevation were (Table 2):

Bolshecapnia spenceri- 2216 m   *Zapada frigida-* 2195 m
Lednia tumana- 2195 m   *Megarcys signata-* 2121 m
Paraperla wilsoni- 2195 m

Collection records noted that over half of all species (51) were collected along lake shores. Lentic habitats that yielded emerging stoneflies were: Lake McDonald, Flathead Lake, Iceberg Lake, Grinnell Lake, Swiftcurrent Lake, and Lake Josephine.

Stoneflies recorded from hyporheic habitats (pumped wells) included: *Alloperla severa, Capnia confuse, Claassenia sabulosa, Diura knowltoni, Hesperoperla pacifica, Isocapnia crinita, I. grandis, I. integra, I. vedderensis, Isoperla fulva, Kathroperla, Paraperla frontalis,* and *P. wilsoni.*

Table 1. Systematic listing of the stoneflies (Order Plecoptera) of the study area.

Family Nemouridae

    Genus *Amphinemura*
        Amphinemura banksi Baumann & Gaufin
    Genus *Lednia*
        Lednia tumana (Ricker)
    Genus *Malenka*
        Malenka californica (Claassen)
        Malenka flexura (Claassen)
    Genus *Podmosta*
        Podmosta decepta (Frison)
        Podmosta delicatula (Claassen)
    Genus *Prostoia*
        Prostoia besametsa (Ricker)
    Genus *Soyedina*
        Soyedina potteri Baumann & Gaufin
    Genus *Visoka*
        Visoka cataractae (Neave)
    Genus *Zapada*
        Zapada cinctipes (Banks)
        Zapada columbiana (Claassen)
        Zapada cordillera (Baumann & Gaufin)
        Zapada frigida (Claassen)
        Zapada glacier (Baumann & Gaufin)
        Zapada haysi (Ricker)
        Zapada oregonensis (Claassen)
Table 1 (*continued*)

Family Taeniopterygidae

    Genus *Doddsia*
        Doddsia occidentalis (Banks)
    Genus *Taenionema*
        Taenionema pacificum (Banks)
        Taenionema pallidum (Banks)

Family Capniidae

    Genus *Bolshecapnia*
        Bolshecapnia milami (Nebeker & Gaufin)
        Bolshecapnia sasquatchi (Ricker)
        Bolshecapnia spenceri (Ricker)
    Genus *Capnia*
        Capnia cheama Ricker
        Capnia confusa Claassen
        Capnia gracilaria Claassen
        Capnia nana Claassen
        Capnia petila Jewett
        Capnia sextuberculata Jewett
    Genus *Eucapnopsis*
        Eucapnopsis brevicauda Claassen
    Genus *Isocapnia*
        Isocapnia crinita (Needham & Claassen)
        Isocapnia grandis (Banks)
        Isocapnia hyalita Ricker
        Isocapnia integra Hanson
        Isocapnia vedderensis (Ricker)
    Genus *Mesocapnia*
        Mesocapnia oenone (Neave)
    Genus *Utacapnia*
        Utacapnia columbiana (Banks)
        Utacapnia distincta (Frison)
        Utacapnia poda (Nebeker & Gaufin)
        Utacapnia trava (Nebeker & Gaufin)

Table 1 (*continued*)

Family Leuctridae

    Genus *Despaxia*
        Despaxia augusta (Banks)
    Genus *Megaleuctra*
        Megaleuctra kincaidi Frison
        Megaleuctra stigmata (Banks)
    Genus *Paraleuctra*
        Paraleuctra forcipata (Frison)
        Paraleuctra jewetti Nebeker & Gaufin
        Paraleuctra occidentalis (Banks)
        Paraleuctra projecta (Frison)
        Paraleuctra vershina Gaufin & Ricker
    Genus *Pomoleuctra*
        Pomoleuctra purcellana (Neave)
    Genus *Perlomyia*
        Perlomyia utahensis Needham & Claassen

Family Peltoperlidae

    Genus *Yoraperla*
        Yoraperla brevis (Banks)

Family Pteronarcidae

    Genus *Pteronarcys*
        Pteronarcys californica Newport
    Genus *Pteronarcella*
        Pteronarcella badia (Hagen)

Family Perlodidae

    Genus *Cultus*
        Cultus aestivalis (Needham & Claassen)
        Cultus pilatus (Frison)
        Cultus tostonus (Ricker)
    Genus *Diura*
        Diura knowltoni (Frison)
    Genus *Isogenoides*
        Isogenoides colubrinus (Hagen)
        Isogenoides elongatus (Hagen)

Table 1 (*continued*)

Genus *Isoperla*
    Isoperla fulva Claassen
    Isoperla fusca Needham & Claassen
    Isoperla longiseta Banks
    Isoperla mormona Banks
    Isoperla petersoni Needham & Christenson
    Isoperla quinquepunctata (Banks)
    Isoperla sobria (Hagen)
    Isoperla sordida (Banks)
Genus *Kogotus*
    Kogotus modestus (Banks)
    Kogotus nonus (Needham & Claassen)
Genus *Megarcys*
    Megarcys signata (Hagen)
    Megarcys subtruncata (Needham & Claassen)
    Megarcys watertoni (Ricker)
Genus *Pictetiella*
    Pictetiella expansa (Claassen)
Genus *Setvena*
    Setvena bradleyi (Smith)
Genus *Skwala*
    Skwala americana (Frison)
    Skwala curvata (Hanson)

Family Perlidae

Genus *Calineuria*
    Calineuria californica (Banks)
Genus *Claassenia*
    Claassenia sabulosa (Banks)
Genus *Doroneuria*
    Doroneuria theodora (Needham & Claassen)
Genus *Hesperoperla*
    Hesperoperla pacifica (Banks)

Table 1 (*continued*)

Family Chloroperlidae

Genus *Alloperla*
    Alloperla medveda Ricker
    Alloperla serrata Needham & Claassen
    Alloperla severa (Hagen)
Genus *Kathroperla*
    Kathroperla perdita Banks
Genus Paraperla
    Paraperla frontalis (Banks)
    Paraperla wilsoni Ricker
Genus *Plumiperla*
    Plumiperla diversa (Frison)
Genus *Suwallia*
    Suwallia autumna (Hoppe)
    Suwallia dubia (Frison)
    Suwallia forcipata (Neave)
    Suwallia lineosa (Banks)
    Suwallia pallidula (Banks)
    Suwallia salish Alexander & Stewart
    Suwallia starki Alexander & Stewart
Genus *Sweltsa*
    Sweltsa albertensis (Needham & Claassen)
    Sweltsa borealis (Banks)
    Sweltsa coloradensis (Banks)
    Sweltsa fidelis (Banks)
    Sweltsa revelstoka (Jewett)
Genus *Triznaka*
    Triznaka signata (Banks)
Genus *Utaperla*
    Utaperla sopladora Ricker

Table 2. Distributional data of the stoneflies in the study area, the Flathead River, Glacier National Park (GNP), East and West of the continental Divide, altitude collected, dates adults collected and number of collection records.

Species	Flat. R.	GNP *	West **	East* *	ALT(m)	dates-adult	No. coll.#
Amphinemura banksi		x		x			1
Lednia tumana		x	x	x	2195–1677	3-VIII to16-X	9
Malenka californica	x	x	x	x	890–1623	11-VII to 26-X	19
Malenka flexura		x	x	x	883–2057	21-IV to11-XII	64
Podmosta decepta		x	x	x	1006–1589	21-VI to 17-VII	15
Podmosta delicatula		x	x	x	1220–1519	1-VII to 4-VIII	13
Prostoia besametsa	x	x	x	x	963–1170	12-IV to 13-VIII	47
Soyedina potteri	x	x	x	x	1049–1858	12-IV to 24-VII	5
Visoka cataractae	x	x	x	x	975–2200	28-III to 27-VII	38
Zapada cinctipes	x	x	x	x	794 –1460	20-II to17-VII	54
Zapada columbiana	x	x	x	x	882–2195	3-I to 6-VIII	72
Zapada cordillera	x		x		939–1104	25-III to 4-IV	4
Zapada frigida		x	x	x	882–2195	4-IV to 9-VIII	26
Zapada glacier		x		x	1525–1830	9-VII to 3-VIII	12
Zapada haysi	x	x	x	x	1095–2100	11-IV to 24-VII	32
Zapada oregonensis		x	x	x	882–1525	26-IV to 22-VII	24
Doddsia occidentalis	x	x	x	x	960–1350	18-III to 9-VII	39
Taenionema pacificum	x	x	x	x	774–1487	26-III to 20-XI	22
Taenionema pallidum	x	x	x	x	957–1800	27-IV to 30-VII	59
Bolshecapnia milami	x	x	x		963–1050	13-III to 18-III	9
Bolshecapnia missiona[a]	x	x	x	x	1050–1890	6-III to 26-III	5
Bolshecapnia spenceri		x		x	1858–2216	27-VII to 3-VIII	4
Capnia cheama	x		x		1190	25-III	1
Capnia confusa	x	x	x	x	882–1439	27-II to 17-VII	33
Capnia gracilaria	x	x	x		963–1006	24-II to 14-V	14
Capnia nana	x	x	x		963–1090	30-I to 28-III	33
Capnia petila	x	x	x	x	990–1975	16-II to 28-VI	19
Capnia sextuberculata	x	x	x		964–1090	19-II to 28-III	19
Eucapnopsis brevicauda	x	x	x	x	963–2057	8-V to 14-VII	71
Isocapnia crinita	x	x	x		948–1026	20-III to 17-V	7
Isocapnia grandis	x	x	x	x	957–1525	10-III to 8-V	9
Isocapnia hyalita		x	x		1086	17-V	1
Isocapnia integra	x	x	x	x	963–2195	24-II to 27-VII	9

Table 2 (*continued*)

Species	Flat. R.	GNP *	West **	East* *	ALT(m)	dates-adult	No. coll.#
Isocapnia vedderensis	x	x	x		963–1026	21-IV to 11-VI	14
Mesocapnia oenone		x	x		975–1158	6-IX to 25-X	6
Utacapnia columbiana	x		x		1095	5-II to 20-III	2
Utacapnia distincta	x	x	x		945–1020	21-II to 15-III	18
Utacapnia poda	x	x	x		964–1198	27-III to 22-IV	5
Utacapnia trava	x	x	x	x	948–1525	10-II to 2-VI	13
Despaxia augusta		x	x		882–1350	31-VIII to 11-XI	17
Megaleuctra kincaidi		x	x		1049	13-V	1
Megaleuctra stigmata			x		883	15-IV-17-VII	4
Paraleuctra forcipata	x	x	x	x	882–2100	5-V to 27-VII	30
Paraleuctra jewetti		x	x	x	1975	28-VI	2
Paraleuctra occidentalis	x	x	x	x	882–1858	4-IV to 10-VII	54
Paraleuctra projecta		x	x	x	895–1770	8-VI to 25-VI	2
Paraleuctra vershina	x	x	x	x	882–1615	17-V to 24-VII	68
Pomoleuctra purcellana		x	x	x	882–1975	9-V to 28-VIII	27
Perlomyia utahensis	x	x	x	x	882–1525	26-IV to 10-VII	31
Yoraperla brevis		x	x	x	778–1982	2-VI to 28-VII	39
Pteronarcys californica		x	x	x	774–1140	17-IV to 4-VIII	9
Pteronarcella badia	x	x	x	x	945–1490	4-VI to 24-VII	21
Cultus aestivalis	x		x		804–1830	4-VI to 20-VII	7
Cultus pilatus		x	x	x	774–1950	16-IV to 25-VII	9
Cultus tostonus	x	x	x	x	948–1384	5-VI to 1-VIII	7
Diura knowltoni	x		x		948–1198	20-IV to 18-V	10
Isogenoides colubrinus	x	x	x	x	948–1008	20-VI to 7-VII	8
Isogenoides elongatus			x		774	17-IV to 25-IV	2
Isoperla fulva	x	x	x	x	795–1490	25-IV to 20-VIII	35
Isoperla fusca	x	x	x	x	1370–1950	6-VII to 13-VIII	13
Isoperla longiseta		x		x	1384	10-VII	1
Isoperla mormona			x		794–800	30-VI to 19-VII	4
Isoperla petersoni		x	x		963–1103	15-VIII to 9-X	4
Isoperla quinquepunctata	x	x	x	x	794–1487	26-V to 24-VII	17
Isoperla sobria	x	x	x	x	963–1800	2-VI to 27-IX	15
Isoperla sordida		x	x	x	1010–2100	13-VIII	7

Species	Flat. R.	GNP *	West **	East *	ALT(m)	dates-adult	No. coll.#
Kogotus modestus	x	x	x	x	939–1525	4-IX to 14-IX	20
Kogotus nonus		x	x	x	963–1525	21-VIII to 24-X	7
Megarcys signata		x	x	x	939–2121	25-VI to 28-VIII	12
Megarcys subtruncata			x		959	21-VI	2
Megarcys watertoni		x	x	x	778–2100	13-VI to 27-VIII	55
Setvena bradleyi		x	x	x	883–1952	5-VI to 8-VIII	22
Skwala curvata		x	x	x	963–1525	26-IV to 17-V	6
Skwala americana	x	x	x	x	778–1341	21-III to 3-VIII	18
Calineuria californica		x	x	x	940–1490	22-VI to6-VIII	22
Claassenia sabulosa	x	x	x		778–1280	11-VII to 22-IX	17
Doroneuria theodora	x	x	x	x	959–1341	21-VIII to 2-IX	17
Hesperoperla pacifica	x	x	x	x	778–2100	13-V to 23-VIII	36
Alloperla medveda		x	x	x	975–1982	6-VII to 28-VII	26
Alloperla serrata		x	x	x	963–2100	17-VI to 10-VIII	36
Alloperla severa	x	x	x	x	959–1487	18-VI to 4-VIII	64
Kathroperla perdita	x	x	x	x	883–1783	15-IV to 28-VI	29
Paraperla frontalis	x	x	x	x	778–1975	24-V to 9-VIII	55
Paraperla wilsoni		x	x	x	882–2195	16-V to 9-VIII	12
Plumiperla diversa	x	x	x	x	794–1490	15-VI to 6-VIII	18
Suwallia autumna		x	x	x	882–1490	13-VII to 12-X	12
Suwallia dubia		x	x	x	1370–1525	20-VII to 22-VII	4
Suwallia forcipata		x	x		963–1073	29-VIII to 25-IX	7
Suwallia lineosa	x	x	x	x	959–1951	23-VI to 1-X	71
Suwallia pallidula	x	x	x	x	794–1950	23-VI to 8-X	79
Suwallia salish			x		882	1-VII	2
Suwallia starki	x	x	x	x	963–1490	15-VII to 30-VIII	6
Sweltsa albertensis		x	x	x	963–1525	2-VI to 15-VII	32
Sweltsa borealis	x	x	x	x	882–1858	15-V to 23-VII	60
Sweltsa coloradensis	x	x	x	x	794–2100	17-V to 19-IX	98
Sweltsa fidelis	x	x	x	x	883–1975	1-VI to 15-VIII	157
Sweltsa revelstoka		x	x	x	882–2134	7-VII to 10-IX	60
Triznaka signata	x	x	x	x	925–1384	5-VI to 25-VII	19
Utaperla sopladora	x	x	x	x	963–1490	16-VI to 22-IX	15
TOTALS - 100	58	89	95	74			

[a] Formerly *Bolshecapnia sasquatichi*, see Baumann and Potter 2007.
*= GNP, Glacier National Park
**= East and West of the Continental Divide
#= Number of collection records

Discussion

The stonefly fauna, 100 species, of Glacier National Park and the Flathead River Basin, Montana had never been concisely reported prior to this work. Gaufin et al. (1972) was the last to document all of the stonefly records from Montana proper and he listed 119 species. A large volume of stonefly research and aquatic invertebrate research in general has resulted in a thoroughly studied geographic area. This thorough range of studies has created a record of stonefly distribution, flight patterns, and altitude and drainage information greatly increasing our knowledge of stonefly ecology. The possible number of Lentic stoneflies is a possible area for further research. The knowledge of the hyporheic stonefly fauna continues to grow. This research should make additional studies of stonefly ecology in this geographic area easier and more involved.

Literature Cited

Baumann, R. W., and D. S. Potter. 2007. What is *Bolshecapnia sasquatchi* Ricker? Plus a new species of *Bolshecapnia* from Montana (Plecoptera: Capniidae). Illiesia, **3**(15):157–162.

Gangemi, J. T. 1991. Comparative effects of wildfire and timber harvest on periphyton and zoobenthos in two Northern Rocky Mountain streams. M.S. Thesis, University of Montana, Missoula.

Gaufin, A. R., W. E. Ricker, M. Miner, and R. A.Hays. 1972. The stoneflies (Plecoptera) of Montana. Transactions of the American Entomological Society **98**:1–161.

Hauer, F. R., J. A. Stanford, J. J. Giersch, and W. H. Lowe. 2000. Distribution and abundance patterns of macroinvertebrates in a mountain stream: an analysis along multiple environmental gradients. Vereinigung für Theoretische und Angewante Limnologie **27**:1485–1488.

Hauer, F. R., J. A. Stanford, and R. Steinkraus. 1989. The zoobenthos of the lower Flathead River: The effect of Kerr Dam operation. FLBS Report 108-89, Flathead Lake Biological Station, The University of Montana, Polson, Montana.

Hauer, F. R., E. G. Zimmerman, and J. A. Stanford. 1980. Preliminary investigations of distributional relationship of aquatic insects and genetic variation of a fish population in the Kintla Drainage, Glacier National Park. Pages 71–84 *in* Proceedings of the Second Conference on Scientific Research in the National Parks. 26–30 November 1979, National Park Service and American Institute of Biological Sciences, San Francisco, California.

Howe, P. J. 1974. A study of the community gradient of the benthic insects of Wilbur Creek in Glacier National Park, Montana. M.S. Thesis, The University of Montana, Missoula.

Imbert, J. B. 1990. An ecological study of a regulated palouse prairie stream. M.S. Thesis, The University of Montana, Missoula, Montana.

Jones, T. S., and V. H. Resh. 1988. Movements of adult aquatic insects along a Montana (USA) springbrook. Aquatic Insects 10(2):99–104.

Lehmkuhl, D. M. 1966. a study of the littoral invertebrates of three mountain lakes in Glacier National Park, Montana. M.S. Thesis, The University of Montana, Missoula, Montana.

Lyon, M. L., and B. P. Stark. 1997. Alloperla (Plecoptera:Chloroperlidae) of Western North America. Entomological News 108(5):321–334.

Nebeker, A. 1966. The taxonomy and ecology of the family Capniidae of the Western United States. Ph.D. Dissertation, University of Utah.

Needham, J. G., and P. W. Claassen. 1925. A monograph of the Plecoptera or stoneflies of America north of Mexico. Thomas Say Foundation of the Entomological Society, Publication No. 2, pp.1–397.

Nelson, C. R., and R. W. Baumann. 1989. Systematics and distribution of the winter stonefly genus *Capnia* (Plecoptera:Capniidae) in North America. The Great Basin Naturalist 49(3):289–363.

Newell, R. L. 1970. Checklist of some aquatic insects from Montana. Proceedings of the Montana Academy of Science 30:45–56.

Perry, S. A. 1984. Comparative ecology of benthic communities in natural and regulated areas of the Flathead and Kootenai rivers, Montana. Ph.D. Dissertation, North Texas State University, Denton, Texas.

Petty, W. C. 1965. Distribution of the Stoneflies (Plecoptera) of Northwestern Montana. National Science Foundation, Undergraduate Fellowship, University of Utah.

Ricker, W. E. 1952. Systematic studies in Plecoptera. Indiana University Publications in Science Series, No. 18, 200 pp.

Robson, E. B. 1968. The Limnology of a Glacial Stream in Montana. M.S. Thesis, University of Utah, Salt Lake City.

Roemhild, G. 1986. Aquatic insects of Montana. Montana State University, Bozeman, Montana, pp.108.

Stanford, J. A. 1975. Ecological studies of Plecoptera in the Upper Flathead and Tobacco Rivers, Montana. Ph.D. Dissertation, University of Utah.

Stanford, J. A., J. V. Ward, and B. K. Ellis. 1994. Ecology of the alluvial aquifers of the Flathead River, Montana. Pages 367–390 *in* J. Gibert, D. L. Danielopol, and J. A. Stanford, editors. Groundwater Ecology. Academic Press, San Diego.

Stanger, J. A., and R. W. Baumann. 1993. A revision of the stonefly genus *Taenionema* (Plecoptera:Taeniopterygidae). Transactions of the American Entomological Society 119(3):171–229.

Stark, B. P., S. W. Szczytko, and R. W. Baumann. 1986. North American stoneflies (Plecoptera): systematics, distribution, and taxonomic references. Great Basin Naturalist 46:383–397.

Stark, B. P., and J. W. Kyzar. 2001. Systematics of Nearctic *Paraleuctra*, with description of a new genus (Plecoptera:Leuctridae). Tijdscript voor Entomologie 144:119–135.

Stewart, K. W., and B. P. Stark. 2002. Nymphs of North American Stonefly Genera (Plecoptera). Caddis Press, Columbus, Ohio.

Szczytko, S. W., and K. W. Stewart. 1979. The genus *Isoperla* (Plecoptera) of western North America; holomorphology and systematics, and a new stonefly genus *Cascadoperla*. Memoirs of the American Entomological Society No. 32, 120 pp.

Varrelman, S. K. 1992. Influence of water-level regulation on littoral macrozoobenthos of Flathead Lake, Montana. M.S. Thesis, The University of Montana, Missoula.

Varrelman, S. K., and C.N. Spencer. 1991. Preliminary investigation and effects of water-level regulation on nearshore benthic invertebrates of Flathead Lake compared to Lake McDonald, Northwest Montana. Proceedings of the Montana Academy of Science **51**:85–102.

Van Wieren, B. J., B. C. Kondratieff, and B. P. Stark. 2001. A revision of the North American species of *Megarcys* Klapalek (Plecoptera:Perlodidae). Proceedings of the Entomological Society of Washington **103**:409–427.

Wicklum, D. 1998. Effects of fish on lacustrine invertebrate community and seston dynamics. Ph.D. Dissertation, The University of Montana, Missoula, Montana.

A SYNOPSIS OF THE AFROTROPICAL TRICORYTHIDAE

Helen M. Barber-James

Department of Freshwater Invertebrates, Albany Museum, Grahamstown 6139, South Africa and the Entomology Department, Rhodes University, Grahamstown

Abstract

The Tricorythidae of the Afrotropical Region is currently composed of five described genera, three of which are thought to be restricted to Madagascar (*Madecassorythus* Elouard and Oliarioniny, *Ranorythus* Oliarinony and Elouard, and *Spinirythus* Oliarinony and Elouard), one which is restricted to Africa (*Dicercomyzon* Demoulin), and one which is thought to be distributed on both landmasses (*Tricorythus* Eaton). Based on sexual dimorphism, manifest in the relative eye size of mature male and female nymphs and adults and on the structure of the genitalia of adult males, it is proposed that there are two additional genera in Africa, as yet undescribed. One of these genera is represented by a species currently placed in *Tricorythus* (*T. discolor* [Burmeister]). Several other undescribed species within South Africa have been identified as belonging to the group. A second lineage is represented by *Tricorythus tinctus* Kimmins, from Uganda, the only currently described species. There are also several more undescribed species of this group widespread in Africa.

Key words: Africa; Madagascar; Afrotropical; Tricorythidae; Ephemeroptera; mayfly.

Introduction

Currently, ten African species are placed in the genus *Tricorythus*, all described as adults, but with only three also known in the nymphal stage. A further nine species of *Tricorythus* are described from Madagascar (Oliarinony et al. 1998b), all described as adults. Only one species, *Tricorythus tinctus* Kimmins 1956 has a sound nymphal–adult association (Corbet 1960). For *Tricorythus discolor* (Burmeister) 1839, the nymph was described nearly 100 years later (Barnard 1932). Barnard (1932) also described adult *Tricorythus reticulatus* with Crass (1947) describing the presumed nymph.

Little is known of the other six so-called *Tricorythus* species from Africa, each of which have been scantily illustrated and described, often from female subimagos (Ulmer 1916, 1930; Navás 1936). Even the type species, *T. varicauda* Eaton (1868), is not adequately described. Demoulin (1954b) summarized all that was known about the Tricorythidae at that time. Material of each of these species needs to be examined and nymphal-adult associations made from collections of fresh material.

The taxa being considered in this paper are distributed within the Afrotropical region. This is defined by Crosskey and White (1977) to include Africa south of the Sahara Desert, conveniently delimited by the 254 mm (10 inches) rainfall isohytes (Fig.1), and the Malagasy region. Although Egypt falls out of this region, it is included in this paper as the type species of *Tricorythus* was collected there. This is interesting as it represents the northern-most distribution of the family Tricorythidae, as defined by McCafferty and Wang (2000). The presence of the genus in Egypt suggests possible colonization by the Egyptian species via the Nile and suggests the type species is not the ancestral species. Other members of the family are known from Asia (Ulmer 1913; Sroka and Soldán, these proceedings), suggesting a Gondwanan origin of the family. The currently known genera of Afrotropical Tricorythidae are summarized in Table 1.

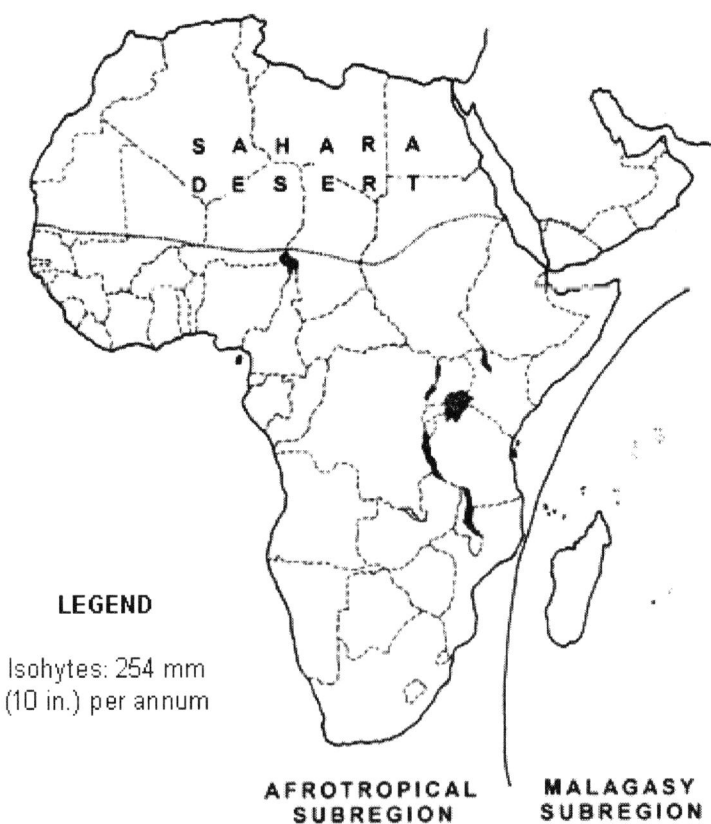

Figure 1. The Afrotropical region.

Table 1. Summary of current genera in the Tricorythidae of the Afrotropical region.

Genus	No. of species	Distribution
Madecassorythus Elouard and Oliarioniny, 1997	4 spp	Madagascar
Spinirythus Oliarinony and Elouard, 1998	3 spp	Madagascar
Ranorythus Oliarinony and Elouard, 1997	2 spp	Madagascar
Tricorythus Eaton, 1868	19 spp	Madagascar and Africa
Dicercomyzon Demoulin, 1954a	4 spp	Africa

Historical Background to Classification of Afrotropical Tricorythidae

Eaton (1868) placed *Tricorythus varicauda* from Egypt, (originally described by Pictet [1843] as *Caenis*), into his newly created genus *Tricorythus* as the type species for the genus. Only the adult male was described and this was done in inadequate detail. Before the family Tricorythidae was erected, Burmeister (1839) described three species, which he placed in the genus *Oxycypha* (two of these were subsequently placed in *Caenis* by Jacob [1974]). Burmeister's (1839) third *Oxycypha* species, *O. discolor,* was placed in *Tricorythus* by Eaton (1884), after having undergone a series of generic name changes prior to this (Appendix A). Burmeister's poor description was of a female subimago, with the only locality information given as the Cape of Good Hope. Eaton (1884) evidently re-examined Burmeister's material, giving a slightly expanded description of the dry female subimago. Esben-Petersen (1913) described a male of *T. discolor* from Tulbagh (western Cape) and two female subimagos from M'fongosi, Zululand, which he assumed belonged to the same species. Ulmer (1921) further expanded the description of the female from Burmeister's material. Lestage (1924) stated that Ulmer's description agreed with the attributes of *T. discolor* as described by Esben-Petersen (1913). Esben-Petersen (1920) briefly described another unnamed species, again a female subimago, from the Free State, which he maintained was different to *T. discolor.* Lestage (1924) suggested that Esben-Petersen's (1920) female subimago needed to be reassessed in the light of Ulmer's (1921) elaboration on the known information for *T. discolor*, though nothing further was concluded.

The first *Tricorythus* nymph described was *T. discolor*, from the western Cape by Barnard (1932), who redescribed both sexes of the adult at the same time as describing the nymph. Lestage (1942) noted the similarity between Barnard's nymph and the nymph of the Javanese *T. jacobsoni* described by Ulmer (1939), except for the absence of maxillary palps in *T. jacobsoni*. Recent reexamination of Barnard's material indicates that *T. discolor* is not a *Tricorythus* at all, but represents an undescribed genus, many specimens of which have been collected from all over southern Africa since then. This will be discussed further in a section below. Sroka and Soldán (these proceedings) have erected a new genus to accommodate *T. jacobsoni* and have described a further six new species within that genus from Asia.

Barnard (1932) described a female subimago as *Tricorythus reticulatus*, basing the description on the irregular network of veins in the wings. The type of this is a shrivelled female subimago on a pin. Crass (1947) described a male that he thought was *Tricorythus reticulatus*, though in little detail, and tentatively designated an unassociated nymphal exuvium as representing the nymph. These were from the Lions River, in KwaZulu-Natal, an area approximately five degrees of latitude north of the site where Barnard's adult was collected (east of Swellendam, in the Western Cape). There is nothing to indicate that Crass' material is related to that described by Barnard. There is no record of where Crass' material was lodged, and it is not amongst the material housed at the Natal Museum or Albany Museum, so there are no types to compare with. However, material from the Mooi River, geographically close to the Lions River, fits his description and has allowed closer observation of this species.

Navás (1936) erected the genus *Neurocaenis* (basing this on a female subimaginal specimen, and placed within the family Caenidae at that stage) to accommodate certain African species with a particular kind of wing venation. He decided this wing venation was distinct from what had been described as typical *Tricorythus* wing venation. *Tricorythus fuscata* was the type species for *Neurocaenis*. A further five African species were included in *Neurocaenis* (Appendix 1) as was the Oriental species *T. jacobsoni*. After examining nine new species of *Tricorythus* from Madagascar, Oliarinony et al. (1998b) found the venation was intermediate between that described for *Neurocaenis* and *Tricorythus* and thus decided on the synonymy of the two genera, with the name *Tricorythus* taking precedence due to seniority.

The genus *Dicercomyzon* Demoulin (1954a) has a less complex history than *Tricorythus*. It was first described and named on its very characteristic nymphal stage (Demoulin 1954a) and the first adults were described by Kimmins (1957). These have quite distinct male genitalia with less distinct wing venation.

Madecassorythus (Elouard and Olinarionony 1997) was first described from adult material with nymphs subsequently described by Oliarinony et al. (2000). As nymphs, specimens of *Madecassorythus* are very similar to those of *Tricorythus*. The differences lie more clearly with the adults, particularly in the structure of male genitalia. *Madecassorythus* (Fig. 2B) has completely divided penes and long auxiliary processes (terminology used by McCafferty and Wang (2000)—also referred to as gonostyles by Elouard and Olinarionony (1997). *Tricorythus* on the other hand (Figs. 2G and H) has fused penes and no auxiliary processes. McCafferty and Wang (2000) suggested that *Madecassorythus* and *Dicercomyzon* belong together in a subfamily Dicercomyzoninae based on similarities in the male genitalia (Figs. 2A and B), but the description of *Madecassorythus* nymphs by Oliarinony et al. (2000) suggests a closer association of *Madecassorythus* with *Tricorythus*.

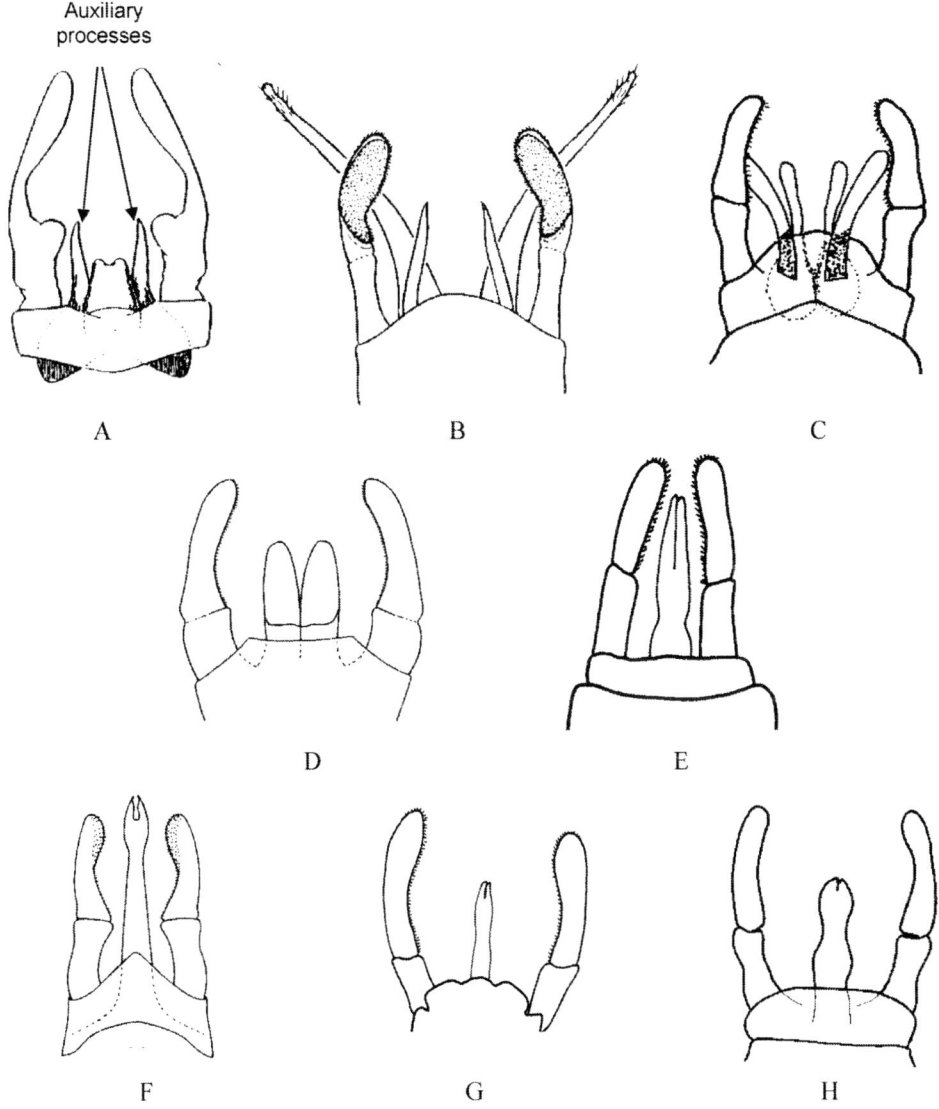

Figure 2. Male genitalia representing each genus of the Tricorythidae. (A) *Dicercomyzon* (redrawn from Kimmins 1957). (B) *Madecassorythus* (redrawn from Elouard and Oliarinony 1997). (C) *Spinirythus* (redrawn from Oliarinony 1998). (D) *Ranorythus* (redrawn from Oliarinony and Elouard 1997). (E) *T. discolor* group (material from Mukhutswi stream, Mpumalanga, South Africa). (F) *T. tinctus* group (material from the Sabie River, Kruger National Park, South Africa). (G) *T. reticulatus* (material from Mooi River, KwaZulu-Natal). (H). *Tricorythus ambinintsoae* (Madagascan sp.) (redrawn from Oliarinony et al. 1998b).

Ranorythus (Oliarinony and Elouard 1997) and *Spinirythus* (Oliarinony and Elouard 1998, in Oliarinony et al. 1998a) have only been formally described in the adult stage. The species in each of these two genera have distinctive male genitalia. Like *Madecassorythus*, *Spinirythus* has divided penes and auxiliary processes (Fig. 2C). *Ranorythus* however, has broad, partially fused penes and no auxiliary processes (Fig. 2D). The nymphs of the species in these two genera have been documented by Oliarinony (1998). The general nymphal form and mouthparts showing a remarkable resemblance to those of the other African and Malagasy Tricorythidae, except for *Dicercomyzon*.

A member of the genus *Tricorythafer* (Lestage 1942), originally described from the Congo by Needham (1920) as *Caensopsis futigans*, was included in the Tricorythidae by Demoulin and Edmunds (1954). McCafferty and Wang (2000) synonymised *Tricorythafer* with *Tricorythodes* Ulmer, placing these in the family Leptohyphidae. This family is considered to have a Nearctic and Neotropical distribution. Thus the generic and family placement of Needham's species "*futigans*" is unclear and will be treated in a separate paper after examination of the type material.

The current genera and species of Afrotropical Tricorythidae are summarized in Appendix A, which includes synonyms and known distributional ranges.

Observations from Recent Studies

Currently grouped within the genus *Tricorythus*, there is a distinct lineage of which only one species, *T. tinctus* Kimmins (1956), has to date been described from Uganda. Both the adult and nymph (the latter described by Corbet [1960]) of this species are known. There are several other undescribed species that clearly belong to this group. These are widely distributed in Africa from the Ivory Coast (Elouard, *personal communications*) and the Congo (Demoulin 1957), and several collected more recently from the Kabompo River (Zambia), Cunene River (Namibia) and tributaries of the Limpopo River system (South Africa). This group of species is clearly different to those described as *Tricorythus* s.s., and deserves to be elevated to a generic ranking. The distinctive characters include the greatly produced incisors on the mandibles (Fig. 3) and the male genitalia (Fig. 2F), which have penes slightly longer than the claspers, fused from the base for about four-fifths of their length. These are consistent characters in all of the new species examined, though nymphal–adult associations are not known for all species. There are clear specific differences between the species from each area. The actual description of this new genus and each new species will be done in a future paper.

There is a second African group, also currently placed within the group traditionally called *Tricorythus*. The species in this group have clear sexual dimorphism, with males smaller than females, and with eyes much bigger than in the female (Fig. 4A). This characteristic is clearly visible in both the nymphs and adults.

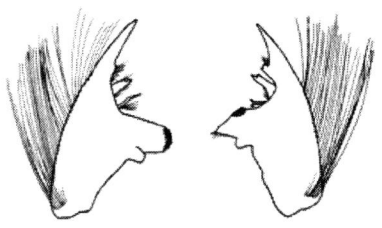

Figure 3. Dorsal view of left and right mandibles of a species contained in the genus the "*tinctus*" group. Material from the Sabie River, Kruger National Park, South Africa.

Reexamination of some of the material used by Barnard (1932) in his expanded description of *T. discolor* has revealed that those specimens also have such dimorphism, suggesting that *T. discolor* is in fact not a member of *Tricorythus* after all. In *Tricorythus*, the males and females have similarly small eyes as have members of the *T. tinctus* group (Fig. 5). However, the sexual dimorphism with large eyes in the male is also seen in the genera *Ranorythus* and *Spinirythus* (Figs. 4B, C). Interestingly, it has also been noted in a new genus described from Asia (Sroka and Soldán, these proceedings). The African specimens with these characteristics need to be placed in a genus of their own, leaving *T. reticulatus* as the only southern African *Tricorythus* species, together with the remaining species from further north in Africa (Appendix A).

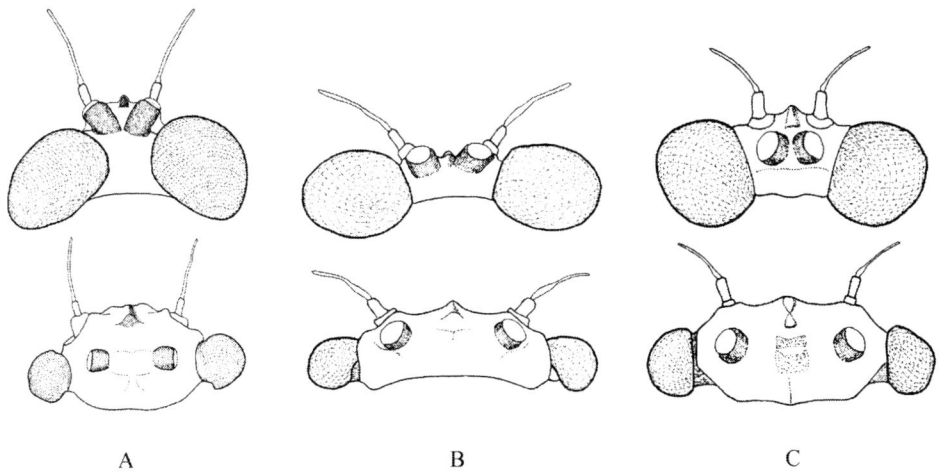

A   B   C

Figure 4. Heads of adult males (above) and females (below) of genera which have male eyes larger than female eyes (adults illustrated, but also evident in the nymphs). (A) *T. discolor* group (material from Mukhutswi stream, Mpumalanga, South Africa). (B) *Ranorythus* (redrawn from Oliarinony and Elouard 1997). (C) *Spinirythus* (redrawn from Oliarinony 1998). *Dicercomyzon* not included.

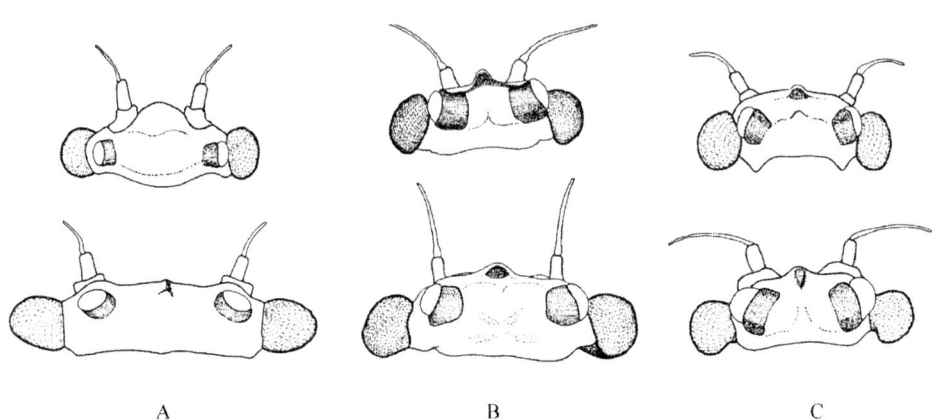

Figure 5. Heads of adult males (above) and females (below) of genera which have male eyes similar in size to female eyes (adults illustrated, but also evident in the nymphs). (A) *Tricorythus* sp. (Madagascar) (redrawn from Oliarinony et al. 1998b). (B) *Tricorythus reticulatus* (material from Mooi River, KwaZulu-Natal). (C) *T. tinctus* group (material from the Sabie River, Kruger National Park, South Africa).

Apart from male eye size differences in some species and the extended mandibles evident in the *T. tinctus* group, the nymphs of all genera except *Dicercomyzon* are very similar. In order to clarify the relationships between these seemingly similar nymphs, one has to look at their adults. Certain of the groups have very unusual and distinctive male genitalia (Figs. 3A–H), which clearly distinguish them as different genera. Wing venation does not seem very helpful as it is very variable between specimens within a species. Table 2 provides a useful summary of the salient morphological features of the adult males of each genus.

Discussion

The information gathered during this study demonstrates the presence of two new Tricorythidae genera in Africa (Table 3). While the adult characteristics of *Dicercomyzon* lie within the variation shown by the other genera of African, Malagasy and the Asian Tricorythidae, the nymphs of *Dicercomyzon* species show many distinctive morphological features (Demoulin 1954a, Kimmins 1957). These include a very flattened body form with broad, flattened tibiae; highly derived mouthparts; uniquely fibrillate gills; and a unique arrangement of ventral abdominal setae, which apparently form a suction disc. Until recently, the Tricorythidae contained two other genera, which have been given their own family status (McCafferty and Wang 2000), these being Machadorythidae and Ephemerythidae. At this point, *Dicercomyzon* clearly stands out from the other members of the

Tricorythidae in its nymphal stage, and further consideration needs to be given to decide whether it also deserves its own family status.

Table 2. Diagnostic features of adult males in the Afrotropical Tricorythidae genera.

	Eyes	Auxiliary processes (Gonostyle)	Penes	Length of auxiliary processes	Spines at base of penes
Dicercomyzon	♂ larger than ♀	present	Fused for most of length	Longer than penis	absent
Madecassorythus	♂ larger than ♀	present	Completely separated	Much shorter than penis	present
Spinirythus	♂ larger than ♀	present	Completely separated	Only slightly shorter than penis	present
Ranorythus	♂ larger than ♀	absent	Fused for half of length	No auxiliary processes	absent
Tricorythus	♂ similar to ♀	absent	Fused except near apex	No auxiliary processes	absent
"discolor" group	♂ larger than ♀	absent	Fused for half of length	No auxiliary processes	absent
"tinctus" group	♂ similar to ♀	absent	Fused except near apex	No auxiliary processes	absent

Table 3. Revised estimate of number of genera and species of Tricorythidae in Africa and Madagascar.

Genus	No. of species	Distribution
Dicercomyzon	4 spp	Africa
Madecassorythus	4 spp	Madagascar
Spinirythus	3 spp	Madagascar
Ranorythus	2 spp	Madagascar
" discolor" group (new genus)	3 spp	Africa
" tinctus" group (new genus)	5 spp	Africa
Tricorythus (*varicauda* group)	17 spp	Madagascar and Africa

With the realization that *T. discolor* belongs in a genus distinct from *Tricorythus*, it is apparent that no nymphs of "true *Tricorythus*" have been associated with adults and described, apart from the description of *T. reticulatus*. This, however, was poorly described by Crass (1947) and was not associated with the type material of that species described by Barnard (1932). Nymphs and reared adult males of *T. reticulatus* need to be found and associated in the western Cape to confirm whether Crass' nymphs and adults from KwaZulu-Natal are the same species as those found in the western Cape. To get a better understanding of the relationships between the genera, it is essential to collect nymphs of the type species of *Tricorythus* (*T. varicauda*) from Egypt. It is also important for the other African species currently described as adults to have the nymphal stage correlated with them.

One may question how the seemingly obvious character of eye size between males and females can have been overlooked in the earlier descriptions of *T. discolor*. Esben-Petersen's (1913) drawing and notes of the species are very vague. Burmeister (1839) described only the female subimago and in his description of his genus *Oxycypha* he says "compound eyes small, simple ...", so due to lack of sufficient material he missed the important sex-related eye dimorphism. Examining some of the material used by Barnard 1932 (Hex River, Western Cape), the adult male has distinctly larger eyes than the females, which has also not been mentioned in the literature. Looking at nymphs from the same collection, there are several female nymphs, but only a few males, with the bigger eyes. Until recently, this dimorphism has remained unnoticed, despite the fact that the genus is widespread in South Africa. The observation of male-female sexual dimorphism was first reported by Barber-James (1995) in an unidentified species from the northeast Cape.

It is interesting to note that all descriptions of female specimens by the various authors have apparently been of subimagos. Closer observation of many female specimens in the Albany Museum collection has revealed that none have shed the subimaginal skin on their wings. Males do, however, go through a moult from subimago to imago. Whether the females do not moult at all or just retain the subimaginal skin on their wings is not clear. It is known, for example, that some of the Oligoneuriidae shed the subimaginal cuticle from the body but not from the wings (Edmunds and McCafferty 1988). Detailed studies of the cuticle from the body of each of the Tricorythidae genera will have to be undertaken to confirm what is happening in this family. The weak, reticulate venation observed in some specimens may be due to the fact they are subimagos. Crass (1947) observes that the reticulate appearance of the wings in his male specimens of *T. reticulatus* is less obvious than in the female.

Selected characters from both nymphal and adult stages of known or described species from each genus (Table 4) were assessed to produce a matrix of character states, which are considered either plesiomorphic or derived (Table 5). These are by no means comprehensive, but access to material of some of the genera to closely investigate further characters is needed before producing a more comprehensive set of characters.

Table 4. Fourteen characters of the adult and nymphal stages of Afrotropical Tricorythidae. Polarity is determined against a hypothetical outgroup which is entirely plesiomorphic (0 = ancestral state; 1,2 = derived states). Multistate characters are ordered.

Character	Character States

Adults

1. ♂ genitalia (penes)
   - completely separated = 0
   - partially fused = 1
   - fused (apex notched) = 2

2. Auxiliary process on ♂ genitalia
   - Absent = 0 present = 1

3. Spines at base of penes
   - Absent = 0 present = 1

4. ♂ eyes size relative to ♀
   - same size = 0 bigger = 1

5. Median cercus
   - present = 0 absent = 1

Nymphs

6. Arrangement of setae on fore-femur
   - scattered = 0 in a pattern = 1

7. Ratio of fore-femur length/width
   - >1.5 (relatively longer) = 0
   - <1.5 (relatively wider) = 1

8. Claw denticle distribution
   - many along length of claw = 0
   - specialized = 1

9. Gill structure
   - lamellate with fibrillar tufts beneath = 0
   - fibrillate only = 1

10. Maxillary palps
    - well developed = 0
    - reduced = 1
    - absent = 2

11. Mandible - incisors
    - incisors small and even = 0
    - outer and inner incisors well developed but approx even in size = 1
    - outer incisor greatly produced = 2

12. Mandible - setae
    - without fringe of setae = 0
    - lateral margin fringed with setae = 1

13. Labium
    - glossae and paraglossae not reduced or fused = 0
    - glossae reduced = 1
    - glossae and paraglossae fused = 2

14. Shape of hypopharyngeal superlingua
    - not produced laterally = 0
    - produced laterally = 1

Table 5. Matrix of selected character states for seven taxa in the Afrotropical Tricorythidae. For definition of characters refer to Table 4. "–" indicates character state uncertain at present.

Character

Taxon	1	2	3	4	5	6	7	8	9	10	11	12	13	14	15
Dicercomyzon	2	1	0	1	1	0	1	0	1	0	0	0	1	1	2
Madecassorythus	0	1	1	0	0	1	0	1	0	1	1	1	2	0	1
Spinirythus	0	1	1	1	0	–	0	–	0	1	1	1	2	0	1
Ranorythus	1	0	0	1	0	1	0	1	0	1	1	1	2	0	2
Tricorythus	2	0	0	0	0	1	0	1	0	1	1	1	2	0	0
"discolor" group	1	0	0	1	0	1	0	1	0	1	1	1	2	0	1
"tinctus" group	2	0	0	0	0	1	0	1	0	1	2	1	2	0	0

The selected characters were analyzed using the program HENNIG86 (Farris 1988) using the exhaustive tree-finding process known as implicit enumeration (i.e.). All multistate characters were ordered by default by HENNIG86. A hypothetical plesiomorphic outgroup was chosen. Of the nine equally parsimonious trees produced, one of these is represented in Fig. 6. In all of the trees, *Dicercomyzon* remained clearly separated from the other groups. Three other clades grouped in pairs as in Fig. 6, though the relationships of these pairs varied. Thus, this analysis indicates a close relationship between *Ranorythus* and the new genus represented by *T. discolor*, between *Madecassorythus* and *Spinirythus*, and between *Tricorythus* (characters from *T. reticulatus* and the Madagascan *Tricorythus* species) and the new genus represented by *T. tinctus*. Molecular work and comparative studies of the eggs of each group, and accumulation of more morphological feature will help to refine this preliminary analysis.

Conclusions

Much works needs to be done on the African Tricorythidae to resolve subtle differences between the nymphs of as yet undescribed species in the two new genera. It is also necessary for many species currently described as adults only to have the nymphal stage correlated with them. Scanning electron microscopy is underway in collaboration with Professor Elda Gaino of Perugia University to compare the eggs of as many species within each genus as possible. Molecular analysis is being carried out in association with Dr. Michael Monaghan at the Natural History Museum, London. Such additional information will help to test the phylogenetic relationships between the different genera that are proposed here.

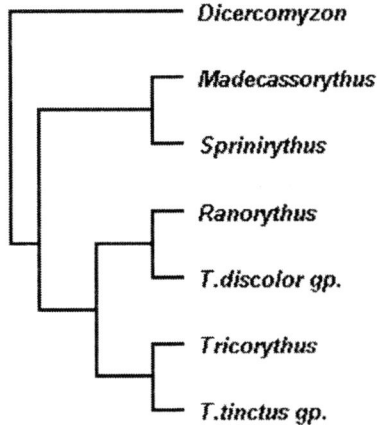

Figure 6. Possible phylogenetic relationships between the genera of the Afrotropical Tricorythidae based on a cladistic analysis of the character matrix in Table 5. Length = 22; ci = 77; ri = 68.

Acknowledgments

The NRF has generously supported the cost of my attending this conference. Dr. Fred Gess is thanked for his help with translating difficult old German scripts. Nick James helped with the French. Dr. Jean-Marc Elouard has very generously shared information on West African and Malagasy Tricorythidae, which he has studied. Sylvia de Moor is thanked for her help with the illustrations of eyes. Dr. Ferdy de Moor and Professor Martin Villet made useful comments on the manuscript. The Directorate of Museums and Heritage Resources, Eastern Cape, are thanked for providing research facilities.

Literature Cited

Barber-James, H. M. 1995. A preliminary survey of the Ephemeroptera of the north east Cape rivers, South Africa. Pages 97–109 *in* L. D.Corkum, and J .H. Ciborowsky, editors. Current Directions in Research on Ephemeroptera. Canadian Scholars Press, Toronto.

Barnard, K. H. 1932. South African may-flies (Ephemeroptera). Transactions of the Royal Society of South Africa **20**:201–259.

Burmeister, H. 1839. Handbuch der Entomologie II. Besondere Entomologie 2. Gymnognatha. 2. Neuroptera. Berlin. 757–1004.

Corbet, P. S. 1960. Larvae of certain East African Ephemeroptera. Revue de zoologie et de botanique africaines **61**(1–2):119–129.

Crass, R. S. 1947. The mayflies (Ephemeroptera) of Natal and the Eastern Cape. Annals of the Natal Museum **11**:37–110.

Crosskey, R.W., and G. B. White. 1977. The Afrotropical Region. A recommended term in zoogeography. Journal of Natural History **11**:541–544.

Demoulin, G. 1954a. Description préliminaire d'un type larvaire nouveau d' Ephéméroptères Tricorythidae du Congo belge. Institut royal des Sciences natruelles de Belgique **30**(6):1–4.

Demoulin, G. 1954b. Recherches critiques sur les Ephéméroptères Tricorythidae d'Afrique et d'Asia. Bulletin et Annales de la societé d'entomologie de Belgique **90**(9–10):264–277.

Demoulin, G. 1957. Le type larvaire probable des *Tricorythus* Eaton (Ephemeroptera Tricorythidae). Insitute Royal des Sciences naturelles de Belgique **33**(19):1–4.

Demoulin, G., and G. F. Edmunds. 1954. Note à propos des nom et quteurs de deux genres d'Éphéméroptères. Bulletin et Annales de la Société d'entomologie de Belgique **90**(1–2):46–48.

Eaton, A. E. 1868. An outline of the re-arrangement of the genera of Ephemeridæ. Entomologist's Monthly Magazine **5**:82–91.

Eaton, A. E. 1883–1888. A revisional monograph of Recent Ephemeridae or mayflies. Transactions of the Linnean Society of London, Zoology Second Series Vol. 3:1–352. pl. 1–65.

Edmunds Jr., G. F., and W. P. McCafferty. 1988. The Mayfly Subimago. Annual Review of Entomology **33**:509–529.

Elouard J.-M., and R. Oliarinony. 1997 - Biodiversité aquatique de Madagascar. 6- *Madecassorythus* un nouveau genre de Tricorythidae définissant la nouvelle sous-famille des Madecassorythinae (Ephemeroptera, Pannota). Bulletin de la Société entomologique de France. **102**(3):225–232.

Esben-Petersen, P. 1913. Ephemeridae from South Africa. Annals of the South African Museum **10**:177–187.

Esben-Petersen, P. 1920. New Species of Neuropterous Insects from South Africa. Annals of the South African Museum **17**:499–505.

Jacob, U. 1974. Zur Kenntnis zweier *Oxycypha* – Arten Hermann Burmeisters (Ephemeroptera, Caenidae). Reichenbachia **15**:93–97.

Kimmins, D. E. 1956. New Species of Ephemeroptera from Uganda. The Bulletin of the British Museum (Natural History) **4**(2):71–87.

Kimmins, D. E. 1957. New species of *Dicercomyzon* Demoulin (Ephemeroptera, fam. Tricorythidae). Bulletin of the British Museum (Natural History) Entomology **6**(5):129–136.

Lestage, J. A. 1924. Les Ephémères de l'Afrique du Sud. Catalogue critique and systèmatique des espèces connues et description de trois genres nouveaux et de sept espèces nouvelles. Revue zoologique africaine **12**(3):316–351.

Lestage, J. A. 1942. Contribution á l'étude des Ephéméroptères. XVIII. Notes critiques sur les anciens Caenidiens d' Afrique et sur l'indépendance de l'évolution tricoryththido-caenidienne. Bulletin du Musée royal d'Histoire naturelle de Belgique **48**:1–20.

McCafferty, W. P., and T. Q.Wang. 2000. Phylogenetic systematics of the major lineages of pannote mayflies (Ephemeroptera: Pannota). Transactions of the American Entomological Society **126**:9–101.

Navás, L. 1936. Insectes du Congo belg. Série IX. Revue de zoologie et de botanique africanes **28**(3):333–368.

Needham, J. G. 1920. African Stone-flies and May-flies collected by the American Museum Congo Expedition. Bulletin of the American Museum of Natural History **43**:35–40.

Oliarinony, R. 1998. Systèmatique et biogeography des Tricorythidae malagaches, (Ephéméroptères: Pannotes). Mèmoire de D.E.A. Option "Ecologie et Environment".Université d'Antananarivo, Madagascar. vol. 1 text 60 pp, vol. 2 figs 61 pp.

Oliarinony R., and J.-M. Elouard. 1997 - Biodiversité aquatique de Madagascar. 7. *Ranorythus* un nouveau genre de Tricorythidae définissant La nouvelle sous-famille des Ranorythinae (Ephemeroptera Pannota). Bulletin de la Société entomologique de France **102**(5):439–447.

Oliarinony R., J.-M. Elouard, and N. H. Raberiaka. 1998a. Biodiversité aquatique de Madagascar. 8. *Spinirythus* un nouveau genre de Tricorythidae (Ephemeroptera Pannota). *Bulletin de la Société entomologique de France* **103**(3):237–244.

Oliarinony R., J.-M. Elouard, and N. H. Raberiaka. 1998b – Biodiversité aquatique de Madagascar. 19. Neuf nouvelles espèces de *Tricorythus* Eaton (Ephemeroptera Pannota, Tricorythidae). Revue francaise d'entomologie **20**(3):73–90.

Oliarinony, R., M. Sartori, and J.-M. Elouard. 2000. Première description des larves et des oeufs du genre malgache *Madecassorythus* (Ephemeroptera, Tricorythidae). Mitteillung der Schweizerischen Entomologischen Gesellschaft / Bulletin de la Société Entomologique Suisse **73**:69–378.

Pictet, F. J. 1843–1845. Histoire naturelle générale et particulière des insects névroptères, Famillie des éphémérines 1–319.

Sroka, P., and T. Soldán. 2006. Revision of the family Tricorythidae (Ephemeroptera) in the Oriental region. Mayfly-Stonefly Conference Proceedings, University of Montana, Biological Station.

Ulmer, G. 1913. Ephemeriden aus Java, gesammelt von Edw. Jacobson. Notes from the Leyden Museum **35**:102–120.

Ulmer, G. 1915/16. Ephemeropteren von Aquatorial-Afrika. Archiv für Naturgeschichte 81(A) **7**:1–19.

Ulmer, G. 1921. Über einige Ephemeropteren - Typen älterer Autoren. Archiv für Naturgeschichte 87(A) **6**:229–267.

Ulmer, G. 1930. Entomological Expedition to Abyssinia, 1926–27; Trichptera and Ephemeroptera. Annals and Magazine of Natural History **10**(6):479–511.

Ulmer G. 1939. Eintagsfliegen (Ephemeropteren) von den Sunda-Inseln. Archiv für Hydrobiologie, Supplement **16**:443–692.

Appendix A

Checklist of Afrotropical Tricorythidae as they currently stand.

● current valid species names
○ subordinate names

Family **TRICORYTHIDAE** Lestage, 1942

1. Genus *Dicercomyzon* Demoulin, 1954 [Africa]
    ● *Dicercomyzon costale* Kimmins, 1957 [Tanzania, Malawi, South Africa]
    ● *Dicercomyzon femorale* Demoulin, 1954 [Congo, Ghana]
    ● *Dicercomyzon sjösterdti* (Ulmer, 1910) [Tanzania, Ghana]
        ○ *Caenis sjösterdti* Ulmer, 1909 (orig.)
        ○ *Tricorythus sjösterdti* Lestage, 1918 (comb.)
        ○ *Dicercomyzon sjösterdti* Demoulin 1954 (comb.)
        ○ *Dicercomyzon marginatum* Kimmins, 1957 (syn.)
    ● *Dicercomyzon verrierae* Demoulin, 1964 [Guinea]

2. Genus *Madecassorythus* Elouard and Oliarioniny, 1997 [Madagascar]
    ● *Madecassorythus hertui* Elouard and Oliarioniny, 1997 [Madagascar]
    ● *Madecassorythus linae* Elouard and Oliarioniny, 1997 [Madagascar]
    ● *Madecassorythus ramanankasinae* Elouard and Oliarioniny, 1997 [Madagascar]
    ● *Madecassorythus raphaeli* Oliarioniny and Sartori, 2000 [Madagascar]

3. Genus *Ranorythus* Oliarinony and Elouard, 1997 [Madagascar]
    ● *Ranorythus violettae* Oliarinony and Elouard, 1997 [Madagascar]
    ● *Ranorythus langrani* Elouard and Oliarinony, 1997 [Madagascar]

4. Genus *Spinirythus* Oliarinony and Elouard, 1998 [Madagascar]
    ● *Spinirythus colasi* Elouard and Oliarinony, 1998 [Madagascar]
    ● *Spinirythus martini* Oliarinony and Elouard, 1998 [Madagascar]
    ● *Spinirythus rosae* Oliarinony and Raberiaka, 1998 [Madagascar]

5. Genus *Tricorythus* Eaton, 1868 [Africa and Madagascar]
    ● *Tricorythus abyssinica* Ulmer, 1930 [Ethiopia]
        ○ *Neurocaenis abyssinica* Demoulin, 1954 (comb.)
        ○ *Tricorythus abyssinica* Oliarinony, Elouard and Raberiaka, 1998 (comb.)
    ● *Tricorythus ambinintsoae* Oliarinony and Elouard, 1998 [Madagascar]

- *Tricorythus discolor* (Burmeister, 1839) [South Africa]
  - *Oxycypha discolor* Burmeister, 1839 (orig.)
  - *Cloeon discolor* Walker, 1853 (comb.)
  - *Caenis discolor* Eaton, 1871 (comb.)
  - *Tricorythus discolor* Eaton, 1884 (comb.)
  - *Neurocaenis discolor* Demoulin, 1954 (comb.)
  - *Tricorythus discolor* McCafferty and de Moor, 1995 (comb.)
- *Tricorythus fyae* Oliarinony and Raberiaka, 1998 [Madagascar]
- *Tricorythus fuscata* (Navás, 1936) [Congo]
  - *Neurocaenis fuscata* Navás, 1936 (orig.)
  - *Tricorythus fuscata* Oliarinony, Elouard and Raberiaka, 1998 (comb.)
- *Tricorythus goodmani* Elouard and Oliarinony, 1998 [Madagascar]
- *Tricorythus jeanne* Oliarinony and Elouard, 1998 [Madagascar]
- *Tricorythus lanceolatus* Kimmins, 1960 [Uganda]
- *Tricorythus latus* Ulmer, 1916 [Congo, Sudan]
  - *Tricorythurus latus* Lestage, 1942 (comb.)
  - *Tricorythus latus* Demoulin, 1954 (comb.)
- *Tricorythus longus* Ulmer, 1916 [Congo, Sudan, Uganda]
  - *Caenis regia* Navás, 1932 (orig.)
  - *Caenis collarti* Navás, 1933 (comb.)
- *Tricorythus pierrei* Elouard and Oliarinony, 1998 [Madagascar]
- *Tricorythus poincinsi* Navás, 1926 [Kenya]
  - *Neurocaenis poincinsi* Demoulin, 1954 (comb.)
  - *Tricorythus poincinsi* Oliarinony, Elouard and Raberiaka, 1998 (comb.)
- *Tricorythus reticulatus* Barnard, 1932 [South Africa]
  - *Neurocaenis reticulata* Demoulin, 1954 (comb.)
  - *Tricorythus reticulatus* McCafferty and de Moor, 1995 (comb.)
- *Tricorythus rolandi* Oliarinony and Raberiaka, 1998 [Madagascar]
- *Tricorythus sylvestris* Oliarinony and Elouard, 1998 [Madagascar]
- *Tricorythus tinctus* Kimmins, 1956 [Uganda]
- *Tricorythus variabilis* Oliarinony and Raberiaka, 1998 [Madagascar]
- *Tricorythus varicauda* (Kollar and Pictet, 1843) [Egypt]
  - *Caenis varicauda* Kollar and Pictet, 1843 (orig.)
  - *Tricorythus varicauda* Eaton, 1868 (comb.)
- *Tricorythus vulgaris* Raberiaka and Oliarinony, 1998 [Madagascar]

THE FAMILY BAETIDAE FROM JAPAN

Toshihito Fujitani

College of Integrated Arts and Science, Osaka Prefecture University, Sakai, Osaka 599-8531, Japan

Abstract

A revised list of the family Baetidae (Ephemeroptera) from Japan is presented including 11 genera and 39 species with a valid name. Baetid genera that occur in Japan are *Acentrella* Bengtsson, *Alainites* Waltz and McCafferty, *Baetiella* Uéno, *Baetis* Leach, *Centroptilum* Eaton, *Cloeon* Leach, *Labiobaetis* Novikova and Kluge, *Nigrobaetis* Novikova and Kluge, *Procloeon* Bengtsson, *Promatsumura* Hubbard and *Tenuibaetis* Kang and Yang.

Key words: Baetidae; genera; species; Japan.

Introduction

The family Baetidae occurs throughout the world except for New Zealand (Edmunds et al. 1976). To date, about 40 genera have been established (Lugo-Ortiz and McCafferty 1999) and more than 500 species are described in this family (Dudgeon 1999).

Concerning Baetidae from Japan, 27 species were described by Takahashi (1924), Matsumura (1931), Uéno (1931a, 1969), Imanishi (1937) and Gose (1965, 1980a, b, c). New records of 9 species from Japan were provided by Matsumura (1904), Takahashi (1929), Horasawa (1931), Uéno (1931b, 1969), Kobayashi (1992), Nakasone et al. (1998), Ishiwata (2000) and Fujitani et al. (2004). Kobayashi (1987) and Fujitani (2002) distinguished 13 species and two species as nymphs, respectively, and gave them alphabetical provisional names.

Ishiwata (2001a) provided a checklist of Ephemeroptera in Japan, in which Baetidae included 38 species with a valid name belonging to 7 genera: *Acentrella* Bengtsson, *Baetiella* Uéno, *Baetis* Leach, *Centroptilum* Eaton, *Cloeon* Leach, *Procloeon* Bengtsson and *Matsumuracloeon* Hubbard. Fujitani et al. (2003a, b) showed the occurrence of the following genera in Japan: *Nigrobaetis* Novikova and Kluge, *Alainites* Waltz and McCafferty, *Labiobaetis* Novikova and Kluge and *Tenuibaetis* Kang and Yang, which Fujitani et al. (2003a) raised from subgenus. They also transferred 14 species originally described in *Baetis* to these genera (Fujitani et al. 2003a, b). Herein, a revised checklist of Baetidae from Japan is provided. I include 39 species with a valid name belonging to 11 genera. I also provide notes on the species given alphabetical provisional names by Kobayashi (1987) and Fujitani (2002). In synonymic list of the species originally described from regions except Japan, I cited the original descriptions in addition to the papers and reports referring to material or descriptions from Japan. Omission of sex or life stages indicates that the cited papers and reports only recorded the species.

Family Baetidae

Genus *Nigrobaetis* Novikova and Kluge, 1987
  Baetis Leach, 1815: Müller-Liebenau, 1969, 1 (the *niger* group, the *gracilis* group in part); Müller-Liebenau, 1974, 163 (the *niger* group, the *gracilis* group).
  Baetis (*Nigrobaetis*) Kazlauskas, 1972, 337 nom. nud., in Waltz et al., 1994, 34; Novikova and Kluge, 1987, 8 (type species: *Ephemera niger* Linné 1761, by original designation); Novikova and Kluge, 1994, 623 (in part).
  Nigrobaetis: Waltz et al., 1994, 34; Waltz and McCafferty, 1997, 138; Fujitani et al., 2003a, 122; Fujitani et al., 2003b, 128.
  Baetis (*Margobaetis*) Kang and Yang, in Kang et al., 1994, 11 [type species: *Baetis* (*Margobaetis*) *mundus* Chang and Yang, 1994, by original designation].
  Nigrobaetis chocoratus (Gose, 1980)
  Baetis chocoratus Gose, 1980a, 79, nymph; Gose, 1985, 21, nymph; Kobayashi, 1987, 54, nymph; Yamasaki, 1987, 85; Kobayashi, 1989, 62; Tanida (ed.), 1991, 14, nymph; Ishiwata, 2001a, 66, nymph; Ishiwata, 2001b, 180; Ishiwata, 2002, 14.
  Baetis (*Nigrobaetis*) *chocoratus*: Novikova and Kluge, 1994, 628, nymph.
  Alainites chocoratus: Waltz and McCafferty, 1997, 135.
  Nigrobaetis chocoratus: Fujitani, 2002, 114; Fujitani et al., 2003a, 123; Fujitani et al., 2003b, 128.
  Nigrobaetis sacishimensis (Uéno, 1969)
  Baetis sacishimensis Uéno, 1969, 222, male; Gose, 1980b, 122, male; Ishiwata, 2001a, 67; Ishiwata and Kobayashi, 2003, 306, male.
  Alainites sacishimensis: Waltz et al., 1994, 34, male.
  Nigrobaetis sacishimensis: Fujitani et al., 2003a, 123; Fujitani et al., 2003b, 128.

In Japan, *Nigrobaetis* sp. D, *Nigrobaetis* sp. I, *Nigrobaetis* sp. N and *Nigrobaetis* sp. P were originally distinguished in nymph and only given alphabetical provisional names in *Baetis* by Kobayashi (1987). They were transferred to *Nigrobaetis* by Fujitani et al. (2003b).

Genus *Alainites* Waltz and McCafferty, 1994
  Baetis Leach, 1815: Müller-Liebenau, 1969, 174 (the *gracilis* group in part); Müller-Liebenau, 1974, 34 (the *muticus* group).
  Baetis (*Nigrobaetis*): Novikova and Kluge, 1994, 623 (in part).

Alainites Waltz and McCafferty, in Waltz et al., 1994, 34 (type species: *Baetis muticus* Linnaeus, 1758, by original designation); Waltz and McCafferty, 1997, 13; Fujitani et al., 2003a, 124; Fujitani et al., 2003b, 128.

Baetis (*Acerbaetis*) Kang and Yang, in Kang et al., 1994, 35 [type species: *Baetis* (*Acerbaetis*) *clivosus* Chang and Yang, 1994, by original designation].

Alainites atagonis (Imanishi, 1937)

Baetis atagonis Imanishi, 1937, 337, male, female; Gose, 1980b, 122, male; Ishiwata, 2001a, 66.

Alainites atagonis: Waltz and McCafferty, 1997, 135; Fujitani et al., 2003a, 124; Fujitani et al., 2003b, 128.

Alainites florens (Imanishi, 1937)

Baetis florens Imanishi, 1937, 336, male; Gose, 1980a, 79, nymph; Gose, 1980b, 122, male; Gose, 1985, 20, nymph; Ishiwata, 2001a, 66.

Baetis (*Nigrobaetis*) *florens*: Novikova and Kluge, 1994, 628, male, nymph.

Alainites florens: Waltz and McCafferty, 1997, 135; Fujitani et al., 2003a, 124; Fujitani et al., 2003b, 128.

Alainites muticus (Linnaeus, 1758)

Baetis pumilus (Burmeister, 1839): Horasawa, 1931, 121, male.

Baetis muticus (Linnaeus, 1758): Ishiwata, 2001a, 67.

Alainites muticus: Waltz et al., 1994, 34, list.

Alainites yoshinensis (Gose, 1980)

Baetis yoshinensis Gose, 1980a, 79, nymph; Gose, 1980b, 122, male; Gose, 1985, 21, nymph; Kobayashi, 1987, 54, nymph; Yamasaki, 1987, 86; Kobayashi, 1989, 61; Kuranishi and Kuhara, 1994, 1212; Ishiwata, 2001a, 68; Ishiwata, 2001b, 180, male, nymph; Ishiwata, 2002, 15; Ishiwata and Kobayashi, 2003, 304, male, nymph.

Baetis (*Nigrobaetis*) *yoshinensis*: Novikova and Kluge, 1994, 629, male, nymph.

Alainites yoshinoensis [!]: Waltz and McCafferty, 1997, 135.

Alainites yoshinensis: Fujitani, 2002, 114; Fujitani et al., 2003a, 124; Fujitani et al., 2003b, 128.

Genus *Labiobaetis* Novikova and Kluge, 1987

Baetis Leach, 1815: Müller-Liebenau, 1969, 150 (the *atrebatinus* group); Edmunds Jr. et al., 1976, 158 (in part); Morihara and McCafferty, 1979, 137 (in part) (the *propinquus* group); Müller-Liebenau, 1984, 253 (the *molawiensis* group); Waltz and McCafferty, 1990, 775 (in part).

Baetis (*Labiobaetis*) Novikova and Kluge, 1987, 8 (type species: *Baetis atrebatinus* Eaton, 1870, by original designation).

Baetis (*Mullerbaetis*) Kang and Yang, in Kang et al., 1994, 32 [type species: *Baetis* (*Mullerbaetis*) *molawiensis* Müller-Liebenau, 1984, by original designation].

Labiobaetis: McCafferty and Waltz, 1995, 19; Gattolliat, 2001, 97; Fujitani et al., 2003a, 126; Fujitani et al., 2003b, 129; previously synonymized with *Pseudocloeon* Klapálek, 1905 by Lugo-Ortiz et al., 1999, 1 (in part).

In Japan, *Labiobaetis* sp. G and *Labiobaetis* sp. Q were originally distinguished in nymph and only given alphabetical provisional names in *Baetis* by Kobayashi (1987). They were transferred to *Labiobaetis* by Fujitani et al. (2003b).

Genus *Tenuibaetis* Kang and Yang, 1994

Baetis (*Tenuibaetis*) Kang and Yang, in Kang et al., 1994, 26 [type species: *Baetis* (*Tenuibaetis*) *pseudofrequentus* Müller-Liebenau, 1985, by original designation]; previously synonymized with *Baetiella* Uéno, 1931 by Waltz and McCafferty, 1997, 136.

Tenuibaetis: Fujitani et al., 2003a, 129; Fujitani et al., 2003b, 130.

Tenuibaetis pseudofrequentus (Müller-Liebenau, 1985)

Baetis pseudofrequentus Müller-Liebenau, 1985, 98, nymph; Kobayashi, 1987, 55, nymph; Kobayashi, 1989, 61; Ishiwata, 2001a, 67; Ishiwata and Kobayashi, 2003, 304, nymph.

Baetiella pseudofrequenta: Waltz and McCafferty, 1997, 136; Tong and Dudgeon, 2000, 144, male, female.

Tenuibaetis pseudofrequentus: Fujitani et al., 2003b, 131.

In Japan, *Tenuibaetis* sp. E and *Tenuibaetis* sp. H were originally distinguished in nymph and only given alphabetical provisional names in *Baetis* by Kobayashi (1987). They were transferred to *Tenuibaetis* by Fujitani et al. (2003b).

Genus *Baetis* Leach, 1815

Baetis Leach, 1815, 137 (type species: *Ephemera fuscata* Linneaus, 1716, emended from *Ephemera bioculata* by Kimmins, 1964, 146); Müller-Liebenau, 1969, 1 (the *alpinus* group, the *lutheri* group, the *pavidus* group, the *rhodani* group, the *vernus* group, the *fuscatus* group, the *buceratus* group); Edmunds Jr. et al., 1976, 158 (in part); Morihara and McCafferty, 1980, 139 (the *rhodani* group, the *vernus* group, the *fuscatus* group, species not assignable to any species groups); Tshernova et al., 1986, 128 (in part); Waltz and McCafferty, 1997, 137 (the *rhodani* group, the *vernus* group, the *fuscatus* group, species having not been assigned to any species groups).

Baetis (*Baetis*): Novikova and Kluge, 1987, 8.

Baetis acuminatus Gose, 1980

Baetis acuminatus Gose, 1980b, 122, male; Ishiwata, 2001a, 66; Fujitani et al., 2003b, 131.

Baetis bicaudatus Dodds, 1923

Baetis bicaudatus Dodds 1923, 71, female, nymph; Ishiwata et al., 2000, 71, nymph; Ishiwata, 2001a, 66; Fujitani et al., 2003b, 131; Konno et al., 2003, 141; Konno, 2004, 21.

Baetis sp. L: Kobayashi, 1987, 54, nymph; Kobayashi, 1989, 59; Tanida (ed.), 1991, 1, nymph.

Baetis celcus Imanishi, 1937

Baetis celcus Imanishi, 1937, 335, male, female; Gose, 1980b, 122, male; Ishiwata, 2001a, 66; Fujitani et al., 2003b, 131.

Baetis flexifemora Gose, 1980

Baetis flexifemora Gose, 1980b, 122, male; Kobayashi, 1992, 19, male; Fujitani et al., 2003b, 131.

Baetis frexifemora [!]: Ishiwata, 2001a, 67; Ishiwata, 2002, 15.

Baetis fuscatus (Linnaeus, 1761)

Ephemera fuscata Linnaeus, 1761, 376, imago.

Baetis bioculatus (Linnaeus, 1758): Matsumura, 1904, 160, male; Matsumura, 1915, 37, male; Uéno, 1928, 50, nymph; Uéno, 1931b, 220; Uéno, 1935, 64, male, nymph.

Baetis fuscatus: Ishiwata, 2001a, 67.

Baetis hyugensis Gose, 1980

Baetis hyugensis Gose, 1980a, 79, nymph; Gose, 1980b, 122, male; Gose, 1985, 21, nymph; Ishiwata, 2001a, 67; Fujitani et al., 2003b, 131.

Baetis iriomotensis Gose, 1980

Baetis sp. (No. 1): Uéno, 1969, 224, female.

Baetis iriomotensis Gose, 1980b, 122, female; Ishiwata, 2001a, 67, list; Fujitani et al., 2003b, 131.

Baetis sahoensis Gose, 1980

Baetis sahoensis Gose, 1980a, 79, nymph; Gose, 1980b, 122, male; Gose, 1985, 21, nymph; Kobayashi, 1987, 54, nymph; Yamasaki, 1987, 85; Kobayashi, 1989, 59; Tanida (ed.), 1991, 1, nymph; Kuranishi and Kuhara, 1994, 1212; Ishiwata, 2001a, 67; Ishiwata, 2001b, 180, nymph; Ishiwata, 2002, 15; Fujitani, 2002, 114; Fujitani et al., 2003b, 131.

Baetis shinanonis Uéno, 1931

Baetis shinanonis Uéno, 1931a, 93, male, female; Uéno, 1935, 64, nymph; Ishiwata, 2001a, 67.

Baetis taiwanensis Müller-Liebenau, 1985

Baetis taiwanensis Müller-Liebenau, 1985, 94, nymph; Fujitani et al., 2004, 40, male, female, male subimago, female subimago, nymph.

Baetis sp. S1: Fujitani, 2002, 114.

Baetis takamiensis Gose, 1980

Baetis takamiensis Gose, 1980a, 79, nymph; Gose, 1980b, 122, male; Gose, 1985, 79, nymph; Ishiwata, 2001a, 67; Fujitani et al., 2003b, 131.

Baetis thermicus Uéno, 1931

Baetis thermicus Uéno, 1931b, 92, male, female; Imanishi, 1937, 333, male, female; Imanishi, 1940, 219, nymph; Uéno, 1950, 130, male; Gose, 1962, 18, nymph; Gose, 1980b, 79, male; Kobayashi, 1987, 55, nymph; Yamasaki, 1987, 86; Kobayashi, 1989, 59; Tanida (ed.), 1991, 1, nymph; Kuranishi and Kuhara, 1994, 1212; Ishiwata, 2001a, 67; Ishiwata, 2001b, 180, male, nymph; Fujitani, 2002, 114; Ishiwata, 2002, 15; Ishiwata and Kobayashi, 2003, 304, male, nymph; Fujitani et al., 2003b, 131.

Baetis totsukawaensis Gose, 1980

Baetis totsukawensis, Gose, 1980a, 79, nymph; Gose, 1980b, 122, male; Gose, 1985, 21, nymph; Ishiwata, 2001a, 67.

Baetis tsushimensis Gose, 1980

Baetis tsushimensis Gose, 1980a, 79, nymph; Gose, 1980b, 122, male; Gose, 1985, 21, nymph; Ishiwata, 2001a, 67; Fujitani et al., 2003b, 131.

Baetis uenoi Gose, 1980

Baetis sp. (No. 2): Ueno, 1969, 224, female.

Baetis uenoi Gose, 1980b, 122, female; Ishiwata, 2001a, 68; Fujitani et al., 2003b, 131.

Baetis yamatoensis Gose, 1965

Baetis yamatoensis Gose, 1965, 218, male; Gose, 1973, 1; Gose, 1980a, 79, nymph; Gose, 1980b, 122, male; Gose, 1985, 21, nymph; Ishiwata, 2001a, 68; Fujitani et al., 2003b, 131.

In Japan, *Baetis* sp. F, *Baetis* sp. J, *Baetis* sp. M and *Baetis* sp. O were originally distinguished in nymph and only given alphabetical provisional names by Kobayashi (1987) and *Baetis* sp. M1 by Fujitani (2002).

Genus *Acentrella* Bengtsson, 1912

Acentrella Bengtsson, 1912, 110 (type species: *Acentrella lapponicus* Bengtsson, 1912, by original designation); Bogoescu and Tabacaru, 1957, 483; Waltz and McCafferty, 1987, 559; McCafferty and Waltz, 1990, 774; Park et al., 1996, 56; Ishiwata, 2001a, 65; Ishiwata, 2001b, 178; Ishiwata, 2002, 14.

Baetis Leach, 1815: Edmunds Jr. and Traver, 1954, 238 (in part); Grandi, 1957, 119; Müller-Liebenau, 1965, 111 (in part); Müller-Liebenau, 1969, 81 (the *lapponicus* group); Edmunds Jr. et al., 1976, 158 (in part); Morihara and McCafferty, 1980, 139 (the *lapponicus* group).

Baetis (*Acentrella*): Novikova and Kluge, 1987, 16.

Acentrella gnom (Kluge, 1983)

Pseudocloeon gnom Kluge, 1983, 74, male, female, nymph.

Acentrella gnom: Waltz and McCafferty, 1987, 560; Kobayashi, 1992, 18, nymph; Ishiwata, 2000, 75, nymph; Ishiwata, 2001a, 65; Ishiwata, 2001b, 179, nymph; Ishiwata, 2002, 14.

Acentrella lata (Müller-Liebenau, 1985)

Pseudocloeon latum Müller-Liebenau, 1985, 100, nymph; Nakasone et al., 1998, 166.

Acentrella lata: Ishiwata, 2001a, 65; Ishiwata and Kobayashi, 2003, 304, nymph.

Acentrella sibirica (Kazlauskas, 1963)

Pseudocloeon sibiricum Kazlauskas, 1963, 319, nymph.

Acentrella sibirica: Ishiwata, 2000, 75, nymph; Ishiwata, 2001a, 65; Ishiwata, 2002, 14.

Acentrella suzukiella Matsumura, 1931

Acentrella suzukiella Matsumura, 1931, 1471, male; Ishiwata, 2001a, 65.

Genus *Baetiella* Uéno, 1931

Baetiella Uéno, 1931b, 220 (type species: *Acentrella japonica* Imanishi, 1930, by original designation); Uéno, 1935, 65; Edmunds Jr. and Traver, 1954, 238; Waltz and McCafferty, 1987, 561; Park et al., 1996, 58; Waltz and McCafferty, 1997, 136 (in part); Ishiwata, 2001a, 65; Ishiwata, 2001b, 179; Ishiwata, 2002, 14.

Pseudocloeon Matsumura, 1931, 1473 (type species not designated) (in part); Gose, 1980c, 211; Gose, 1985, 21.

Baetis (*Baetiella*): Novikova and Kluge, 1987, 16.

Neobaetiella Müller-Liebenau, 1985, 103 (type species: *Neobaetiella uenoi* Müller-Liebenau, 1985, by original designation).

Matsumuracloeon Hubbard, 1989, 388 (type species: *Pseudocloeon aino* Matsumura, 1931a, by monotypy).

Baetiella japonica (Imanishi, 1930)

?*Acentrella* (sp. nov.?): Uéno, 1928, 51, nymph.

Acentrella japonica Imanishi, 1930, 263, male, female, male subimago, nymph.

Baetiella japonica: Uéno, 1931b, 220, male, nymph; Uéno, 1935, 65, male; Imanishi, 1940, 226, nymph; Waltz and McCafferty, 1987, 563; Yamasaki, 1987, 85; Kuranishi and Kuhara, 1994, 1212; Ishiwata, 2001a, 65; Ishiwata, 2001b, 179, male, nymph; Ishiwata, 2002, 14.

Pseudocloeon aino Matsumura, 1931, 1437, female.

Baetiella nosegawaensis Gose, 1965, 218, male.

Pseudocloeon japonica [!]: Gose, 1980c, 211, male, nymph; Gose, 1985, 21, nymph.

Pseudocloeon japonicum: Müller-Liebenau, 1985, 107, nymph.

Baetiella aino: Waltz and McCafferty, 1987, 563.

Matsumuracloeon aino: Hubbard, 1989, 388.

Baetiella bispinosa (Gose, 1980)

Baetiella sp. 1: Tanida, 1974, 162.

Pseudocloeon bispinosus Gose, 1980c, 211, nymph.

Baetiella bispinosa: Waltz and McCafferty, 1987, 563; Tong and Dudgeon, 2000, 143, nymph; Ishiwata, 2001a, 65; Ishiwata and Kobayashi, 2003, 304, nymph.

Genus *Centroptilum* Eaton, 1869

Centroptilum Eaton, 1869, 131 (type species: *Ephemera luteola* Müller, 1776, by original designation): Gose, 1962, 18; Gose, 1980c, 212; Gose, 1985, 21; Tshernova et al., 1986, 120; Ishiwata, 2001a, 68; Ishiwata, 2001b, 181; Ishiwata, 2002, 15.

Cloeon (*Centroptilum*): Novikova and Kluge, 1994, 63.

Centroptilum rotundum Takahashi, 1929

Centroptilum rotundum Takahashi, 1929, 64, female; Uéno, 1931b, 220, male; Gose, 1980c, 212, nymph; Gose, 1985, 21; Ishiwata, 2001, 68.

Genus *Cloeon* Leach, 1815

Cloeon 1815, 137 (type species: *Ephemera diptera* Linnaeus, 1761, by original designation); Uéno, 1935, 64; Gose, 1962, 18; Gose, 1980b, 77; Gose, 1985, 21; Tshernova et al., 1986, 120; Ishiwata, 2001a, 68; Ishiwata, 2001b, 181; Ishiwata, 2002, 15.

Cloeon (*Cloeon*): Novikova and Kluge, 1994, 64.

Cloeon dipterum (Linnaeus, 1761)

Ephemera diptera Linnaeus, 1761, 377, imago.

Cloeon dipterum: Matsumura, 1904, 160, female; Uéno, 1935, 64, nymph; Gose, 1962, 18; Gose, 1980b, 77; Gose, 1985, 20, nymph; Ishiwata, 2001a, 68; Ishiwata, 2001b, 181; Ishiwata, 2002, 15.

Cloeon kyotonis Matsumura, 1931

Cloeon kyotonis Matsumura, 1931, 1472, female; Ishiwata, 2001a, 68.

Cloeon maikonis Takahashi, 1924

Cloeon maikonis Takahashi, 1924, 372; female; Ishiwata, 2001a, 68.

Cloeon marginale (Hargen, 1858)

Cloe marginale Hagen, 1858, 477, female subimago.

Cloeon marginale: Uéno, 1969, 226, male, female; Ishiwata, 2001a, 68.

Cloeon okamotoi Takahashi, 1924

Cloeon okamotoi Takahashi, 1924, 372, female; Ishiwata, 2001a, 68.

Cloeon ryogokuense Gose, 1980

Cloeon ryogokuense Gose, 1980b, 77; Gose, 1985, 21; Ishiwata, 2001a, 68; Ishiwata, 2002, 15.

Cloeon tamagawanum (Matsumura, 1931)

Procloeon tamagawanum Matsumura, 1931, 1472, male.

Promatsumura tamagawanum: Hubbard, 1988.

Cloeon tamagawanum: Ishiwata, 2001a, 68

Genus *Procloeon* Bengtsson, 1915

Pseudocloeon Bengtsson, 1914, 218 (type species: *Cloeon bifidum* Bengtsson, 1912 by original designation).

Procloeon Bengtsson, 1915, 34 (type species: *Cloeon bifidum* Bengtsson, 1912 by original designation): replacement name for *Pseudocloeon* Bengtsson, 1914; Tshernova et al., 1986, 120; Ishiwata, 2001a, 68; Ishiwata, 2001b, 181; Ishiwata, 2002, 16.

Cloeon (*Procloeon*): Novikova and Kluge, 1994, 74.

Procloeon bimaculatum (Eaton, 1885)

Cloeon bimaculatum Eaton, 1885, 182, male, female; Uéno, 1969, 226, female.

Procloeon bimaculatum: Ishiwata, 2001a, 69.

Genus *Promatsumura* Hubbard, 1988

Procloeon Matsumura, 1931, 1473 (type species not designated).

Promatsumura: Hubbard, 1988, 240 (type species: *Procloeon nipponicum* Matsumura, 1931): replacement name for *Procloeon* Matsumura, 1931; Ishiwata, 2001a, 69.

Promatsumura nipponicum Matsumura, 1931

Procloeon nipponicum Matsumura, 1931, 1472, female.

Promatsumura nipponicum: Hubbard, 1988, 240; Ishiwata, 2001a, 69.

Acknowledgments

I cordially thank Professor Dr. K. Tanida, College of Integrated Arts and Science, Osaka Prefecture University for his supervision on my study. I am deeply grateful to Professor Dr. M. Ishii, Dr. T. Hirowatari and Mr. N. Hirai, Entomological Laboratory, Graduate School of Agriculture and Biological Science, Osaka Prefecture University for their critical and useful comments. My gratitude is extended to Mr. N. Kobayashi, Institute of River Biology Co. Ltd., and Mr. S. Ishiwata, Kanagawa Environmental Research Center, for kindly providing me with invaluable taxonomic information on Baetidae. This study was partly supported by the fund to T. Fujitani from the Ecology and Civil Engineering Society, Japan (No. 2: 1999), to K. Tanida from the Ministry of Education, Culture, Sports, Science and Technology, Japan (No. 08874106) and from Foundation of River and Water Environment Management, Japan (No. 13-1-4-19: 2000–2002).

Literature Cited

Bengtsson, S. 1912. Neue Ephemeriden aus Scherden. Entomologisk Tidskrift **33**:107–117 (in German).

Bengtsson, S. 1914. Benerkungen über die nordischen Arten der Gattung *Cloëon* Leach. Entomologisk Tidskrift **35**:210–220 (in German).

Bengtsson, S. 1915. Eine Namensänderung. Entomologisk Tidskrift **36**:34 (in German).

Bogoescu, C., and I. Tabacaru. 1957. Etude comparée des nymphes d'*Acentrella* et de *Pseudocloeon*. Considération phylogénétiques concernant la famille Baëtidae (Ephemeroptera). Beiträge zur Entomologique 7:483–491 (in French).

Dodds, G. S. 1923. Mayflies from Colorado. Transactions of the American Entomological Society **49**:93–116.

Dudgeon, D. 1999. Tropical Asian streams. Hong Kong University Press, Hong Kong.

Eaton, A. E. 1869. On *Centroptilium*, a new genus of the Ephemeridae. The Entomologist's Monthly Magazine **6**:131–132.

Eaton, A. E. 1885. A revisional monograph of recent Ephemeridæ or mayflies. The Transactions of the Linnean Society of London. Series 2. Zoology **3**:1–346.

Edmunds Jr., G. F., and J. R. Traver. 1954. An outline of a reclassification of the Ephemeroptera. Proceedings of the Entomological Society of Washington **56**:236–240.

Edmunds Jr., G. F., S. L. Jensen, and L. Berner. 1976. The mayflies of North and Central America. University of Minnesota Press, Minneapolis, USA.

Fujitani, T. 2002. Species composition and distribution patterns of *Baetis* and its related genera (Baetidae: Ephemeroptera) in a Japanese stream. Hydrobiologia **485**:111–121.

Fujitani, T., T. Hirowatari, and K. Tanida. 2003a. Genera and species of Baetidae in Japan: *Nigrobaetis*, *Alainites*, *Labiobaetis* and *Tenuibaetis* n. stat. (Ephemeroptera). Limnology **4**:121–129.

Fujitani, T., T. Hirowatari, and K. Tanida. 2003b. Nymphs of *Nigrobaetis*, *Alainites*, *Labiobaetis*, *Tenuibaetis* and *Baetis* from Japan (Ephemeroptera: Baetidae): diagnoses and keys for genera and species. Pages 127–133 *in* E. Gaino, editor. Research Update on Ephemeroptera and Plecoptera. Universitá di Perugia, Italy.

Fujitani, T., T. Hirowatari, and K. Tanida. 2004. The first record of *Baetis taiwanensis* Müller-Liebenau from Japan, with description of imago and subimago (Ephemeroptera: Baetidae). Entomological Science **7**:39–46.

Gattolliat, J. L. 2001. Six new species of *Labiobaetis* Novikova & Kluge (Ephemeroptera: Baetidae) from Madagascar with comments on the validity of the genus. Annales de Limnologie **37**:97–123.

Gose, K. 1962. Ephemeroptera. Pages 6–24 *in* M. Tsuda, editor. Aquatic Entomology. Hokuryûkan Publisher, Tokyo (in Japanese.)

Gose, K. 1965. Description of two new species of Baetidae from Japan (Ephemeroptera). Kontyû **33**:218–220.

Gose, K. 1973. Benthic community of Yoshino River, Japan. Study on Bioproductivity of Benthos in Yoshino River **5**:1–7 (in Japanese).

Gose, K. 1980a. The mayfly of Japanese 6. Aquabiology **2**:76–79 (in Japanese).

Gose, K. 1980b. The mayfly of Japanese 7. Aquabiology **2**:122–123 (in Japanese).

Gose, K. 1980c. The mayfly of Japanese 8. Aquabiology **2**:286–288 (in Japanese).

Gose, K. 1985. Ephemeroptera. Pages 7–32 *in* T. Kawai, editor. An Illustrated Book of Aquatic Insects of Japan. Tokai University Press, Tokyo (in Japanese).

Grandi, M. 1957. Contributi allo studio degli Ephemeroidei Italiani. XXI. Intorno ai generi *Acentrella* Bgtssn. e *Baetis* Leach. Bollettino dell'Istituto di Entomologia della Università di Bologna **22**:119–124 (in Italian).

Hagen, H. A. 1858. Synopsis der Neuroptera Ceylons. Verhandlungen des Zoologisch-botanischer Verein in Wien 8:471–488 (in Latin).

Horasawa, I. 1931. Short notes on Japanese Ephemeridae (2). Kontyû **5**:121–124 (in Japanese).

Hubbard, M. D. 1988. *Promatsumura*, replace name for *Procloeon* Matsumura, 1931 (Ephemeroptera: Baetidae) with designation of type-species. Insecta Mundi **2**:240.

Hubbard, M. D. 1989. *Matsumuracloeon*, a replacement name for *Pseudocloeon* Matsumura, 1931 (Ephemeroptera: Baetidae). Florida Entomologist **72**:388.

Imanishi, K. 1930. Mayflies from Japanese torrents I. New mayflies of the genera *Acentrella* and *Ameletus*. Transactions of the Natural History Society of Formosa **20**:263–267.

Imanishi, K. 1937. Mayflies from Japanese torrents VIII. Notes on the genera *Paraleptophlebia* and *Baetis*. Annotations Zoologicae Japonenses **16**:330–338.

Imanishi, K. 1940. Ephemeroptera of Manchoukuo, Inner Mongolia, and Chôsen. Pages 169–263 in T. Kawamura, editor. Report of the Limnological Survey of Kwantung and Manchoukuo. Kyoto University Press (in Japanese).

Ishiwata, S. 2000. Notes on mayflies in Kanagawa Prefecture. Natural History Report of Kanagawa Prefecture **21**:73–82 (in Japanese).

Ishiwata, S. 2001a. A checklist of Japanese Ephemeroptera. Pages 55–84 *in* Y. J. Bae, editor. The 21st Century and Aquatic Entomology in East Asia. Proceedings of the 1st symposium of AESEA.

Ishiwata, S. 2001b. Mayflies of Chiba Prefecture, Japan. Checklist, diagnoses and keys. Journal of the Natural History Museum and Institute, Chiba **6**:163–200 (in Japanese).

Ishiwata, S. 2002. Mayflies of Kanagawa Prefecture, Japan. Kanagawa Chûho, **138**:1–46 (in Japanese).

Ishiwata, S., and N. Kobayashi. 2003. Ephemeroptera. Pages 296–331 *in* M. Nishida, N. Shikatani, and S. Shokita, editors. The Flora and Fauna of Inland Waters in the Ryukyu Islands. Tokai University Press, Tokyo (in Japanese).

Ishiwata, S., T. M. Tiunova, and R. B. Kuranishi. 2000. The mayflies (Insecta: Ephemeroptera) collected from the Kamchatka Peninsula and the North Kuril Islands in 1996–1997. Pages 67–75 *in* T. Komai, editor. Results of Recent Research on Northeast Asian Biota, Natural History Research, Special Issue.

Kang, S. C., H. C. Chang, and C. T. Yang. 1994. A revision of the genus *Baetis* in Taiwan (Ephemeroptera, Baetidae). Journal of Taiwan Museum **47**:9–44.

Kazlauskas, R. 1963. New and little-known mayflies (Ephemeroptera) from the USSR. Entomological Review **42**:582–589 (in Russian).

Kazlauskas, R. 1972. Neues über das System der Einstagfliegen der Familie Baetidae (Ephemeroptera). Pages 337–338 *in* XIII International Congress of Entomology Moscow, 2–9 August, 1968 Proceedings (in German).

Kimmins, D. E. 1964. *Baetis* [Leach, 1815] (Insecta, Ephemeroptera): proposed designation of a type-species under the plenary powers. Z. N. (S). 1620. Bulletins of Zoological Nomenclature **21**:146–147.

Kluge, N. J. 1983. New and little known mayflies of the family Baetidae (Ephemeroptera) from the Primorye. Entomologiceskoe obozrenie **61**:65–79 (in Russian).

Kobayashi, N. 1987. *Baetis* species as biological indices of water quality. Pages 41–52 *in* M. Yasuno, and T. Iwakuma, editors. Proceedings of Symposium "Problem and Perspective of Aquatic Biological Index". National Institute of Environmental Studies, Tokyo (in Japanese).

Kobayashi, N. 1989. Genus Baetis, with consideration to their systematic relationships and distribution patterns. Pages 53–67 *in* A. Shibatani and K. Tanida, editors. Recent Progress in Aquatic Entomology in Japan, With Special Reference to Speciation and "*Sumiwake*". Tôkai University Press, Tokyo, Japan (in Japanese).

Kobayashi, N. 1992. Ephemeroptera. Pages 18–24 *in* H. Moriya, editor. Benthic Fauna of Sagamihara City. Sagamihara City Government (in Japanese).

Konno, Y. 2004. Lotic aquatic insects in the Azusagawa River, in the Kamikochi Valley of Chubu Sangaku National Park. Japanese Journal of Limnology **65**:21–26 (in Japanese).

Konno, Y., H. Nishimoto, H. Maruyama, T. Torii, and S. Ishiwata. 2003. Lotic aquatic insects in the alpine zone of Daisetsuzan National Park, Hokkaido. Japanese Journal of Limnology **64**:141–144 (in Japanese).

Kuranishi, R., and N. Kuhara. 1994. Benthic fauna of Akan National Park. The Nature of Akan National Park, 1993. V. Fauna of Akan National Park. Maeda Ippoen Foundation, Japan (in Japanese).

Leach, E. A. 1815. Entomology. Brewster's Edinburgh Encyclopedia, **9**:57–172.

Linnaeus, C. 1761. Fauna Suecia. Edition II. No. 1476 (in Latin).

Lugo-Ortiz, C. R., and W. P. McCafferty. 1999. Global diversity of the mayfly family Baetidae (Ephemeroptera): a generic perspective. Trends in Entomology **2**:45–54.

Lugo-Ortiz, C. R., W. P. McCafferty, and R. D. Waltz. 1999. Definition and reorganization of the genus *Pseudocloeon* (Ephemeroptera: Baetidae) with new species description and combinations. Transactions of the American Entomological Society **125**:1–37.

Matsumura, S. 1904. Guide to 1,000 species of insects from Japan Empire. Keiseisha Shoten, Tokyo (in Japanese).

Matsumura, S. 1915. Insect taxonomy. Volume 1. Keiseisha Publisher, Tokyo (in Japanese).

Matsumura, S. 1931. 6000 illustrated insects of Japan Empire. Tôko Shoin, Tokyo (in Japanese).

McCafferty, W. P., and R. D. Waltz. 1990. Revisionary synopsis of the Baetidae (Ephemeroptera) of North and Middle America. Transaction of the American Entomological Society **116**:769–799.

McCafferty, W. P., and R. D. Waltz. 1995. *Labiobaetis* (Ephemeroptera: Baetidae): new status, new North American species, and related new genus. Entomological News **106**:19–28.

Morihara, D. K., and W. P. McCafferty. 1979. Systematics of the *propinquus* group of Baetis species (Ephemeroptera: Baetidae). Annals of the Entomological Society of America **72**:130–135.

Morihara, D. K., and W. P. McCafferty. 1980. The *Baetis* larvae of North America (Ephemeroptera: Baetidae). Transactions American Entomological Society **105**:139–221.

Müller-Liebenau, I. 1965. Revision der von Simon Bengtsson aufgestellten *Baetis*-Arten (Ephemeroptera). Opuscula entomologica **30**:79–123 (in German).

Müller-Liebenau, I. 1969. Revision of European species of the genus *Baetis* Leach, 1815 (Insecta, Ephemeroptera). Gewässer und Abwässer **48/49**:1–214 (in German).

Müller-Liebenau, I. 1974. Baetidae from southern part of France, Spain and Portugal (Insecta, Ephemeroptera). Gewässer und Abwässer **53/54**:7–42 (in German).

Müller-Liebenau, I. 1984. New genera and species of the family Baetidae from West-Malaysia (River Gombak) (Insecta: Ephemeroptera). Spixiana **7**:253–284.

Müller-Liebenau, I. 1985. *Baetis* from Taiwan with remarks on *Baetiella* Ueno, 1931 (Insecta, Ephemeroptera). Archiv für Hydrobiologie 104:93–110.

Nakasone, K., H. Mitsumoto, M. Yonamine, T. Kishimoto, E. Higa, and T. Omija. 1998. Studies on suspended material content in sediment and benthic macro-invertebrate fauna of river in U. S. military base. Annual report of Okinawa Prefectural Institute of Health and Environment news **12**:161–167 (in Japanese).

Novikova, E. A., and N. Y. Kluge. 1987. Systematics of the genus *Baetis* (Ephemeroptera: Baetidae) with description of a new species from Middle Asia. Vestnik Zoologii **1987**:8–19 (in Russian).

Novikova, E. A., and N. Y. Kluge. 1994. Mayflies of the subgenus *Nigrobaetis* (Ephemeroptera, Baetidae, *Baetis* Leach, 1815). Entomologiceskoe obozrenie **73**:623–644 (in Russian).

Park, S. Y., Y. J. Bae, and I. B. Yoon. 1996. Revision of the Baetidae (Ephemeroptera) of Korea (1). Historical review, *Acentrella* Bengtsson and *Baetiella* Uéno. Entomological Research Bulletin (Korean Entomological Institute, Korea University) **22**:55–66.

Takahashi, Y. 1924. Some new species of mayflies from Japan. Zoological magazine **36**:377–380 (in Japanese).

Takahashi, Y. 1929. A new Japanese may-fly, *Centroptilium rotundum* n. sp. Lansania **1**:63–64 (in Japanese, with English description).

Tanida, K. 1974. Preliminary reports on the aquatic insects of the Ryukyus with some ecological remarks I. Ecological Study on Natural Conservation of the Ryukyu Islands **1**:161–174 (in Japanese).

Tanida, K., editor. 1991. Aquatic Insects of Shiga Prefecture, Japan. Shingakusha Publisher, Kyoto (in Japanese).

Tong, X., and D. Dudgeon. 2000. *Baetiella* (Ephemeroptera: Baetidae) in Hong Kong, with description of a new species. Entomological News **143**:111–118.

Tshernova, O. A., N. J. Kluge, N. D. Sintshenkova, and B. B. Belov. 1986. Order Ephemeroptera. Pages 99–142 *in* P. A. Lehr, editor. Identification of Insects of Far East USSR. Vol. 1. Leningrad Press, Leningrad (in Russian).

Uéno, M. 1928. Some Japanese mayfly nymphs. Memoirs of the college of science, Kyoto Imperial University, Series B. **4**:19–63.

Uéno, M. 1931a. Einige neue Ephemeroptera und Plecoptera aus Mittel-Japan. Annotations Zoologicae Japonenses **13**:91–99 (in German).

Uéno, M. 1931b. Contribution to the knowledge of Japanese Ephemeroptera. Annotation Zoologicae Japonense **16**:189–231.

Uéno, M. 1935. Reports on benthic fauna of Minami-Azumi region, with special attention to Azusagawa River. Pages 9–116 *in* D. Miyadi, editor. Benthic Fauna and Kamikôchi Valley and Azusagawa River. Iwanami Publisher, Tokyo (in Japanese).

Uéno, M. 1950. Ephemeroptera. Pages 120–130 *in* T. Ishii et al., editors. Illustrated Book of Insects of Japan. Hokuryûkan Publisher, Tokyo (in Japanese).

Uéno, M. 1969. Mayflies (Ephemeroptera) from various regions of Southeast Asia. Oriental Insects **2**:221–228.

Waltz, R. D., and W. P. McCafferty. 1987. Systematics of *Pseudocloeon, Acentrella, Baetiella*, and *Liebebiella*, new genus (Ephemeroptera: Baetidae). Journal of New York Entomological Society **95**:553–568.

Waltz, R. D., and W. P. McCafferty. 1990. Revisionary synopsis of the Baetidae (Ephemeroptera) of North and Middle America. Transaction of American Entomological Society **116**:769–799.

Waltz, R. D., and W. P. McCafferty. 1997. New generic synonymies in Baetidae (Ephemeroptera). Entomological News **108**:134–140.

Waltz, R. D., W. P. McCafferty, and A. Thomas. 1994. Systematics of *Alainites* n. gen., *Diphetor, Indobaetis, Nigrobaetis* n. stat., and *Takobia* n. stat. (Ephemeroptera, Baetidae). Bulletin de la Société d'histoire Naturelle de Toulouse **130**:33–36.

Yamasaki, T. 1987. Mayflies of Tamagawa River system and their distribution. Pages 81–120 *in* R. Ishikawa, T. Yamasaki, J. Kojima and S. Uchida, editors. Analytic Studies on the Distribution of Some Insect-Groups in the Tamagawa River System and Its Upper Reaches. Tokyu Foundation for Better Environment, Tokyo (in Japanese).

A MOLECULAR ANALYSIS OF THE AFROTROPICAL BAETIDAE

Jean-Luc Gattolliat[1], Michael T. Monaghan[2,3], Michel Sartori[1], Jean-Marc Elouard[4], Helen Barber-James[5,6], Pascale Derleth[1], Olivier Glaizot[1], Ferdy de Moor[5,6] and Alfried P. Vogler[2,3]

[1] Musée cantonal de Zoologie, CH 1014 Lausanne, Switzerland.
[2] Department of Entomology, The Natural History Museum, London SW7 5BD, UK
[3] Department of Biological Sciences, Imperial College, Silwood Park, Ascott, Berkshire, UK
[4] Institut de Recherche pour le Développement, 34060 Montpellier Cedex 1, France
[5] Department of Freshwater Invertebrates, Albany Museum, Grahamstown 6139, South Africa
[6] Department of Entomology, Rhodes University, Grahamstown 6139, South Africa

Abstract

Recent work on the Afrotropical Baetidae has resulted in a number of important taxonomic changes: several polyphyletic genera have been split and more than 30 new Afrotropical genera have been established. In order to test their phylogenetic relevance and to clarify the suprageneric relationships, we reconstructed the first comprehensive molecular phylogeny of the Afrotropical Baetidae. We sequenced a total of *ca.* 2300 bp from nuclear (18S) and mitochondrial (12S and 16S) gene regions from 65 species belonging to 26 genera. We used three different approaches of phylogeny reconstruction: direct optimization, maximum parsimony and maximum likelihood. The molecular reconstruction indicates the Afrotropical Baetidae require a global revision at a generic as well as suprageneric level. Only four of the 12 genera were monophyletic when represented by more than one species in the analysis. Historically, two conflicting concepts of the suprageneric classification of Afrotropical Baetidae were proposed. One was based on the gathering of sister genera into complexes and the other on the division of the family into a restricted number of subfamilies. According to our reconstruction, neither is completely satisfactory: the major complexes of genera present in Africa are either paraphyletic or polyphyletic and the division of the Afrotropical Baetidae into two subfamilies is probably too simplified.

Key words: Ephemeroptera; Baetidae; Africa; Madagascar; molecular phylogeny; systematics.

Introduction

Taxonomic knowledge of the Afrotropical mayfly fauna has increased substantially in the past two decades. Descriptions of new taxa and several revisions at the supraspecific level have led to a considerable increase in the number of described species and genera (Elouard 2001). This is particularly true for the Baetidae.

Prior to 1980, most African Baetidae species were assigned to Palaearctic genera. South Africa is a good illustration of the case: its fauna has been relatively well known since the middle of the twentieth century, but almost all the species were attributed to a restricted number of European genera, based primarily on imaginal similarities (McCafferty and de Moor 1995). Waltz and McCafferty (1987) were the first to create new African genera to accommodate species that present important modifications in larval morphology. Thereafter, Gilles (Gillies 1985, 1988, 1990, 1991, Gillies and Elouard 1990, Gillies et al. 1990, Gillies and Wuillot 1997), Elouard (Elouard et al. 1990), Wuillot (Wuillot and Gillies 1993a, 1994) and Lugo-Ortiz and McCafferty (1996a, 1996b, 1996c, 1997a, 1997b, 1997c, 1997d, 1997e, 1997f, 1997g, 1998a, 1998b, 1998c, 1999, Lugo-Ortiz and de Moor 2000) greatly contributed to a clarification of the systematics of the Afrotropical Baetidae. In 1990, Gillies transferred all the species attributed to the Palaearctic genus *Centroptilum* to the new genus *Afroptilum*. Gillies' concept of *Afroptilum* was rapidly shown to be polyphyletic and numerous genera were created to accommodate the different species groups (Wuillot and Gillies 1993a, 1993b, 1994, Lugo-Ortiz and McCafferty 1996a, 1996b, 1996c, 1997a, 1997d, 1998a, Barber-James and McCafferty 1997, McCafferty et al. 1997, Lugo-Ortiz et al. 2001). Species previously assigned to *Baetis, Acentrella* and *Pseudocloeon* were reassigned to other genera by Lugo-Ortiz and McCafferty (1997c, 1997f, 1998b) and several new genera were described. In total, more than 30 new Afrotropical genera have been established in the last 25 years.

The reported low diversity of Baetidae in most areas of Africa results, at least partially, from a lack of data and comprehensive analysis of material collected by systematists. It is clear that when intensive sampling is performed over large geographical areas, numerous new taxa are discovered and distribution ranges are greatly extended. For example, West Africa was well investigated by the ORSTOM team during the 1970s to 1980s leading to a large increase in taxonomic knowledge of the mayfly fauna. Previous to this project, the West African Baetidae fauna was virtually unknown. Based on the collected material, about 20 new species and seven new genera of Baetidae were described; many of them were endemic.

A similar progression occurred within the Malagasy Baetidae fauna. Until the mid 1990s, very few taxa were known from Madagascar. Between 1997 and 2004, more than 50 new species and eight new genera were described as part of an aquatic biodiversity project led by ORSTOM (presently IRD (Institut de Recherche pour le Développement, France) (Elouard et al. 2003). One of the results of the increased knowledge of the Malagasy fauna was a strengthened recognition of the importance of affinities between African and Malagasy faunas. This implied that the old concept of an endemic Malagasy fauna evolving separately since the breakup of Gondwanaland is only one of several factors likely to explain the present composition, and that other factors such as colonization and dispersal must be taken into account (Gattolliat and Sartori 2003).

For the present study, we used nuclear and mitochondrial gene sequences from 65 taxa to reconstruct the first comprehensive molecular phylogeny of the Afrotropical Baetidae (Monaghan et al. 2005). The aim of this reconstruction is 1) to clarify the suprageneric relationships, 2) to test the phylogenetic relevance of the recent taxonomic work and 3) to investigate the relative roles of dispersal and vicariance in forming the present-day Malagasy fauna. Only the first two points are developed in this paper; the unravelling of the origin of the Malagasy fauna will be discussed elsewhere (Monaghan et al. 2005).

Material and Methods

Sixty-five species, belonging to 26 different genera of Baetidae, were sequenced for this study (18 of the 24 Malagasy genera and 18 of the 42 Subsaharan genera). Of the 26 included genera, seven were collected only in Madagascar, 10 genera were collected both in Madagascar and Africa, six were collected only in Africa and three were collected in Afrotropical areas and Europe (Switzerland) and/or Asia (Borneo and New Guinea). For the outgroups, four species from the family Tricorythidae (from Madagascar and Africa) were sequenced.

Samples were collected in May–June 2003 in Madagascar and South Africa by the authors and other specimens were taken from collections of the Museum of Zoology in Lausanne or given by different collectors (see acknowledgements). DNA was extracted and four gene fragments (mitochondrial 12S and 16S and two fragments of nuclear 18S, total length 1444–1449 bp) were amplified using methods detailed in Monaghan et al. (submitted). To avoid errors due to contamination or mislabelling, additional individuals of 38 ingroup species were sequenced. All specimens were given a unique number for the study and extracted DNA is stored at the Natural History Museum, London in the frozen collection database (BMNH#s 704056-704136 and 704632-704678).

For phylogeny reconstruction, three different approaches were used: direct optimization as implemented in POY v. 3.0.11 (Gladstein and Wheeler 1999); maximum parsimony with PAUP* v. 4.0b10 (Swofford 2002) and maximum likelihood under a GTR+I+G model (selected in Modeltest 3.06, Posada and Crandall 1998) using PhyML (Guindon and Gascuel 2003). For details of the analysis, see Monaghan et al. (2005).

To test the support of individual clades, Bremer support was calculated using a heuristic procedure implemented in POY on the output tree obtained by direct optimisation and data were bootstrapped (1000 replicates) with PAUP*.

Results

Both direct optimization and parsimony approaches produced a single tree (Fig. 1). The two reconstructions differed only by a single node within the *Cloeon* lineage (Fig. 1 Clade B). Seven well-supported clades (Fig. 1 Clades A, B, C, D, E, F, G) were identified. The relationships between these seven clades were difficult to

Figure 1. Phylogenetic reconstruction of Afrotropical Baetidae based on the single resulting tree from direct optimization of 12S, 16S and two 18S rRNA gene regions using POY. The vertical bar indicates the alternate placement of Cloeon sp. 2 using parsimony reconstruction. Values above branches indicate Bremer Support and values below branches indicate parsimony bootstrap percentage (if above 50%). Letters A–G indicate well-supported lineages (see text). Branches are thickened for Malagasy species. Material from other origins: BO = Borneo; CH = Switzerland; EA = East Africa; NG = New Guinea; SA = South Africa; SE = Seychelles.

establish because of poor support of the deeper nodes. The same seven clades resulted from the likelihood reconstruction, with only minor changes occurring within the different clades (Fig. 2). The likelihood topology differs from the direct optimisation and parsimony tree at deeper nodes, notably by separating Baetidae into two basal sister groups (Fig. 2). Only four genera appeared monophyletic; eight were para- or polyphyletic. The monophyly of the 14 others could not be tested as only one species per genus was included in the analysis (Table 1).

Discussion

Historically, two conflicting concepts of the suprageneric classification of the Afrotropical Baetidae have been proposed. Gillies (1991) considered the African Baetidae to consist of two subfamilies: Baetinae and Cloeoninae (sensus Kazlauskas 1972). Lugo-Ortiz and McCafferty gathered genera in five complexes (*Baetis* complex (Lugo-Ortiz and McCafferty 1998b), *Bugilliesia* complex (Lugo-Ortiz and McCafferty 1996a), *Centroptiloides* complex (Lugo-Ortiz and McCafferty 1998a), *Cloeodes* complex (Lugo-Ortiz and McCafferty 1998c) *Indobaetis* complex (Lugo-Ortiz and de Moor 2000)). Many other genera were not attributed to a complex, as no formal and global reconstruction was undertaken.

Subfamilies. The Baetinae are characterized at the imaginal stage by a forewing with paired intercalaries, and at the larval stage by the absence of setae between the prostheca and mola of the right mandible and a single row of denticles on the tarsal claws. The Baetinae subfamily corresponds to the clade A of our reconstruction (Figs. 1 and 2). The Cloeoninae are characterized at the imaginal stage by a single intercalary in the forewing and, at the larval stage, by a row of setae between the prostheca and mola of the right mandible and by two rows of denticles on the tarsal claws. This subfamily comprises clades B to H (Figs. 1 and 2). Clade A is recovered in all three reconstructions, showing that the Baetinae subfamily represents a monophyletic group. The monophyly of the Cloeoninae cannot be rejected by the reconstructions stemming from direct optimisation and parsimony because of the poor support for most of the basal nodes; however, the reconstructions seem to indicate that the Cloeoninae is paraphyletic. In contrast, the likelihood reconstruction groups clades B to H in a monophyletic lineage. Therefore, the ML reconstruction supports the division of the African Baetidae in two subfamilies made by Gillies (1991), while DO and MP do not allow us to reject or confirm Gillies' classification. Additional sampling and sequencing is required to resolve more fully the basal nodes of the phylogeny and determine the higher-level relationships within and among the subfamilies.

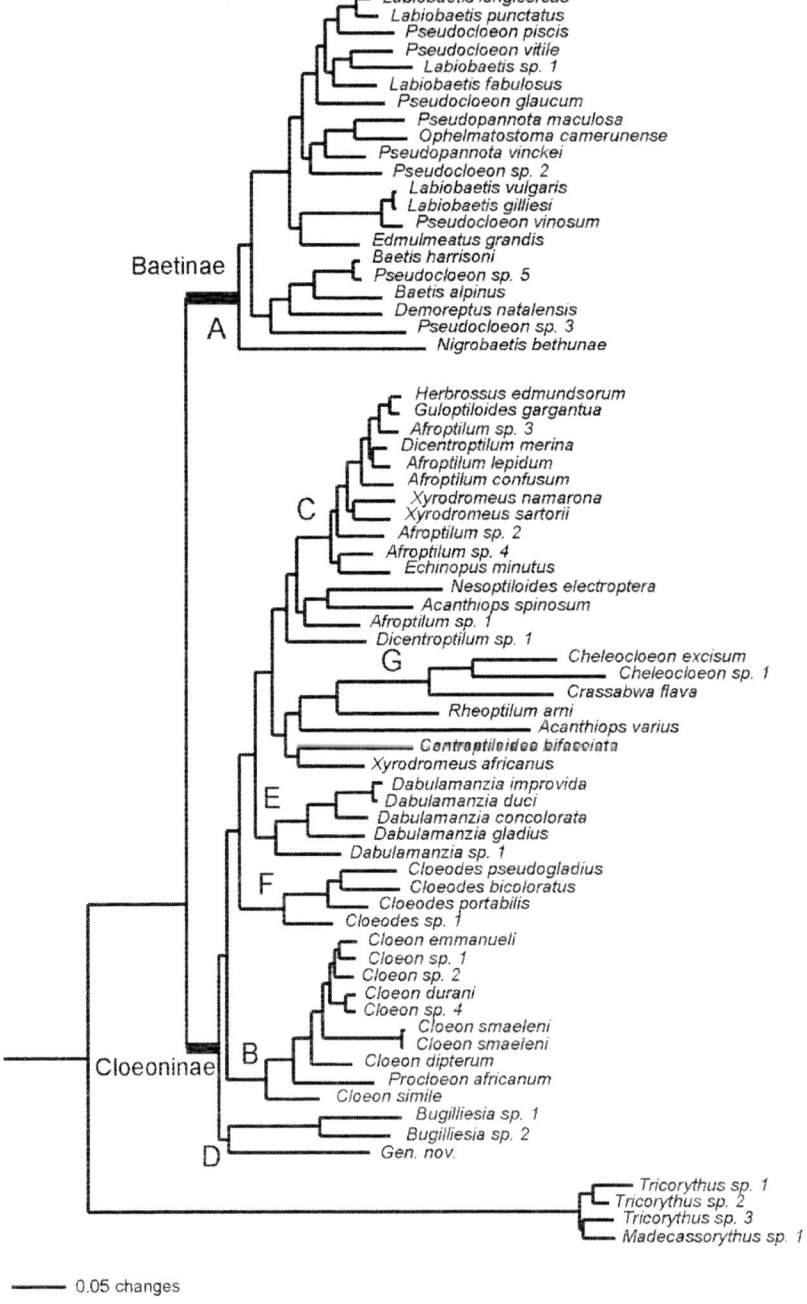

Figure 2. Phylogenetic reconstruction of Afrotropical Baetidae based on the single resulting tree from maximum likelihood reconstruction of 12S, 16S and two 18S rRNA gene regions. Branches are thickened for subfamilies.

Table 1. Distribution and status of the Afrotropical Baetidae genera included in our molecular analysis.

	Distribution	Origin of sequenced material	Status of the genus
Acanthiops	Madagascar + Africa	Madagascar + Africa	polyphyletic
Afroptilum	Madagascar + Africa	Madagascar + Africa	polyphyletic
Baetis	Africa + Palearctic	Africa + Palearctic	polyphyletic
Bugilliesia	Madagascar + Africa	Madagascar + Africa	monophyletic
Centroptiloides	Africa	Africa	nontestable
Cheleocloeon	Madagascar + Africa	Madagascar + Africa	monophyletic
Cloeodes	Pantropical	Madagascar + Africa	monophyletic
Cloeon	World wide except Neotropics	Madagascar + Africa + Palearctic	polyphyletic
Crassabwa	Africa	Africa	nontestable
Dabulamanzia	Madagascar + Africa	Madagascar + Africa	monophyletic
Delouardus	Madagascar	Madagascar	nontestable
Demoreptus	Africa	Africa	nontestable
Dicentroptilum	Madagascar + Africa	Madagascar + Africa	polyphyletic
Echinopus	Madagascar	Madagascar	nontestable
Edmulmeatus	Madagascar	Madagascar	nontestable
Gen nov	Madagascar	Madagascar	nontestable
Guloptiloides	Madagascar	Madagascar	nontestable
Herbrossus	Madagascar	Madagascar	nontestable
Labiobaetis / Pseudocloeon	World wide except Neotropics	Madagascar + Africa + Palearctic	polyphyletic
Nesoptiloides	Madagascar	Madagascar	nontestable
Nigrobaetis	Madagascar + Africa + Palearctic	Africa	nontestable
Ophelmatostoma	Africa	Africa	nontestable
Procloeon	Africa + Palearctic + Neoarctic	Africa	nontestable
Pseudopannota	Madagascar + Africa	Madagascar + Africa	polyphyletic
Rheoptilum	Madagascar	Madagascar	nontestable
Xyrodromeus	Madagascar + Africa	Madagascar + Africa	polyphyletic

Genus Complexes.   We were able to test the monophyly of three complexes recognised by Lugo-Ortiz and McCafferty (1998a, 1998b, 1998c): the *Baetis* complex, the *Centroptiloides* complex and the *Cloeodes* complex.   All were either paraphyletic or polyphyletic, revealing a number of inconsistencies in the classification as proposed by Lugo-Ortiz and McCafferty.   Some can be easily corrected by the inclusion or exclusion of some taxa, e.g., by including *Crassabwa* in the *Centroptiloides* complex. The monophyly of other complexes seems much more difficult to establish (e.g., *Baetis* complex).   Several characters used for the definition of the complexes are either plesiomorphic or homoplasic. Single characters are often misleading when trying to involve phylogenetic placement and combinations of characters offer more useful criteria for resolving the suprageneric classification.

Genera of the *Baetis* complex share larval characters such as the presence of the femoral villopore and the absence of setae between the prostheca and mola of both mandibles.   In the imaginal stage, *Baetis* complex genera possess forewings with double intercalaries (Waltz and McCafferty 1987).   This complex is globally present where Baetidae are found, except for South America.   In Africa, only the genera *Baetis, Demoreptus, Glossidion, Labiobaetis/Pseudocloeon* and *Tanzaniella* are included in this complex (Lugo-Ortiz and McCafferty 1997c, 1998b, Lugo-Ortiz et al. 1999).   In the molecular reconstruction, this complex is represented by *Baetis, Demoreptus* and *Labiobaetis/Pseudocloeon*.   They all belong to the clade A, but do not constitute a monophyletic group due to the inclusion of other genera such as *Ophelmatostoma, Pseudopannota* and *Edmulmeatus* (Figs. 1 and 2).

The *Centroptiloides* complex contains 15 genera and is the most diversified complex in the Afrotropical area, particularly in Madagascar.   This complex was defined by Lugo-Ortiz and McCafferty (1998a) by larvae possessing two subparallel rows of denticles on the tarsal claws and adults with a single marginal intercalary in the forewings.   In our molecular reconstruction, it encompasses the genera *Acanthiops, Afroptilum, Centroptiloides, Cheleocloeon, Dicentroptilum, Echinopus, Guloptiloides, Herbrossus, Nesoptiloides* and *Xyrodromeus* (Figs. 1 and 2).   Consequently, it includes the whole of clade C (called the Malagasy clade as all the taxa included occur in Madagascar), members of clade G as well as genera with uncertain relationships throughout the tree.   This complex appears paraphyletic in this reconstruction mainly due to the inclusion of the genus *Crassabwa* in the clade G and the presence of a clade of Malagasy endemics (see below).

The *Cloeodes* complex is a grouping of genera of Gondwanan origin possessing a more or less developed subproximal arc of setae on the tibiae.   In the Afrotropical area, this complex encompasses *Cloeodes, Crassabwa, Dabulamanzia, Maliqua* and *Nesydemius* (Lugo-Ortiz and McCafferty 1998c).   This complex is polyphyletic in our reconstruction, consisting of clade E (*Dabulamanzia*), Clade F (*Cloeodes*) and the genus *Crassabwa* (included in the Clade G as a sister group of *Cheleocloeon*) (Figs. 1 and 2).   According to the molecular reconstruction, it is not possible to determine whether *Dabulamanzia* is the sister group of *Cloeodes*, but it is clearly demonstrated that *Crassabwa* is not related to these two genera. Consequently, the *Cloeodes* complex is at least diphyletic.

Generic Polyphyly. The reconstruction allowed us to test the monophyly of 12 genera and only four were found to be monophyletic (Table 1). The remaining 14 genera were only represented by a single species and therefore it was not possible to test their monophyly. Several different reasons explain the polyphyly of the 8 remaining genera. For the Afrotropical genera *Afroptilum*, *Dicentroptilum* and *Xyrodromeus*, it is clear that the African and Malagasy species do not belong to the same clade. The Malagasy species are included in the Malagasy clade (Clade C), while the African species are either sisters to clade C or have unclear relationships. The two Malagasy species of *Xyrodromeus* are sister species, but were not closely related to *X. africanus*. The inclusion of the Malagasy species in this African genus is mainly due to convergences observed in the mouthparts, especially the modified mandibles adapted for scraping epilithic algae. Several unrelated Malagasy genera present this kind of adaptation in one or more species; thus it seems to be a relatively common characteristic that has evolved independently in several different Malagasy lineages (Gattolliat and Sartori 2000, Gattolliat 2001a, 2001b, 2002).The inclusion by Lugo-Ortiz and McCafferty (1998a) of Malagasy *Dicentroptilum merina* in the African genus *Dicentroptilum* required a revision of the generic diagnosis with the deletion of several characters from the original description. The imaginal stage of *D. merina* was not known at that time and is still undescribed. The observation of imagos from material reared by the LRSAE team (Laboratoire de Recherche sur les Systèmes Aquatiques et leur Environnement, Antananarivo, Madagascar) clearly indicate that *D. merina* differs from African species; the Malagasy species does not possess the specialised characters of the type species such as hindwing with three longitudinal veins and two widely separated costal spurs (Wuillot and Gillies 1994). Consequently, it is necessary to erect new genera to include the Malagasy species of *Xyrodromeus* and *Dicentroptilum*.

The phylogeny of *Afroptilum* species seems more complicated than that of the Madagascar-Africa dichotomy observed above. For this study, we provisionally assigned undescribed or unnamed species to this genus. The Malagasy clade probably still includes several undescribed genera that have a low degree of adaptation to unusual feeding behaviour or environmental pressure. Nevertheless, the African *Afroptilum* is not the sister group of any Malagasy species. As the genus *Afroptilum* has been used for lumping different taxa *incertae cedis,* a complete revision of this genus is necessary.

With five genera (*Guloptiloides, Herbrossus* and *Nesoptiloides* from Madagascar; *Centroptiloides* and *Barnumus* from South Africa), carnivorous Baetidae are quite abundant and diverse in the Afrotropical area (Gattolliat and Sartori 2001). Two of them (*Herbrossus* and *Guloptiloides*) are sister-taxa and members of the Malagasy clade, but the two others included in the molecular reconstruction (*Nesoptiloides* and *Centroptiloides*) are not closely related (Figs. 1 and 2). It implies that this unusual adaptation has appeared at least twice.

Labiobaetis/Pseudocloeon, even in its most restricted definition, remains polyphyletic. It is clear from the data that the Afrotropical taxa belong to two different lineages, both of which include taxa shared with southeastern Asia. This wide distribution suggests strong dispersal abilities in evolutionary time. Moreover, other non-Afrotropical species of *Labiobaetis/Pseudocloeon* were included in another group of taxa among the clade A. This shows very clearly that this genus greatly needs a complete revision that can only be made by including taxa from all the different biogeographic areas. It also means that the *Pseudocloeon* concept is not sufficiently defined (Lugo-Ortiz et al. 1999) and that new morphological characters need to be added. Because of the number of taxa and the wide geographical range involved, this revision constitutes a great challenge. Adults of the type species of *Pseudocloeon* need to be reexamined and nymphs correlated.

The eight species of *Cloeon* do not constitute a monophyletic group because of the inclusion a single species assigned to *Procloeon* among them. This conflicts the validity of the genus *Cloeon*. It implies that either the genus *Cloeon* must be restricted to a limited number of species and some of the species assigned to other genera, or the generic characterization of *Procloeon africanum* must be rectified. Gillies (1997) considered that there are no reliable characters for distinguishing adults of *Cloeon* and *Procloeon*. The only difference lies in the degree of development of the gills. According to this character, he attributed the different African species to one of the two genera. These attributes remain rather questionable and the concept of the genus *Procloeon* needs a global revision.

Cloeon smaeleni possesses a wide distribution including the whole Afrotropical area as well as the Arabian Peninsula (Gillies 1985); it was recently collected in Madagascar and is therefore the only Malagasy species of Ephemeroptera which is not endemic to Madagascar (Gattolliat and Rabeantoandro 2002). For this species, material from both Madagascar and South Africa was sequenced. The low molecular divergence confirms the specimens from Madagascar and South Africa belong to the same species (Fig. 2).

We conclude that our molecular reconstruction has greatly helped to clarify the Afrotropical Baetidae classification by revealing both the accuracy and inaccuracy of recent systematics research and by highlighting the groups in which taxonomic revisions are necessary. It appears the grouping of genera in different complexes is not satisfactory and the division of the African Baetidae in two subfamilies is probably too simplified. At the generic level despite recent improvements, many genera still require revision and several new genera must be erected, especially for Malagasy species previously assigned to African genera.

Acknowledgments

We thank those who helped us during our field trips to Madagascar and South Africa: Joseph Rakotonarivo, Gabrielle Randria and Francois Jarrige at IRD (previously ORSTOM) in Antananarivo; staff of the Kruger National Park, in particular Hendrik Sithole, Bruce Leslie, Velly Ndlouy, Sipho Mokgalaka and Thomas Ndoy; Mick Angliss from the Limpopo Province Department of Environmental Affairs and Stephan Foord from the University of Venda. The Directorate of Museums and Heritage, eastern Cape, is thanked for making research facilities available to its staff and colleagues. We want to thank the Museum of Zoology in Lausanne, the U.K. Biotechnology and Biological Sciences Research Council and the Swiss National Science Foundation (fellowship Nr. 68592) for funding this research. We are also indebted to Mike Dobson for providing specimens from East Africa, Justin Gerlach for providing specimens from Seychelles, and Katayo Sagata for providing specimens from New Guinea.

Literature Cited

Barber-James, H. M., and W. P. McCafferty. 1997. Review and a new species of the African genus *Acanthiops* (Ephemeroptera: Baetidae). Annales de Limnologie **33**:85–92.

Elouard, J.-M. 2001. Knowledge of the African-Malagasy mayflies. Pages 13–20 *in* E. Dominguez, editor. Trends in Research in Ephemeroptera and Plecoptera. Kluwer Academic/Plenum Publishers, New York.

Elouard, J.-M., J. L. Gattolliat, and M. Sartori. 2003. Ephemeroptera, mayflies. Pages 639–645 *in* S. M. Goodman, and J. P. Benstead, editors. The Natural History of Madagascar. University of Chicago Press, Chicago.

Gattolliat, J.-L. 2001. The genus *Cloeodes* (Ephemeroptera, Baetidae) in Madagascar. Revue Suisse de Zoologie **108**:387–402.

Gattolliat, J.-L. 2001. *Rheoptilum*: a new genus of two–tailed Baetidae (Ephemeroptera) from Madagascar. Aquatic Insects **23**:67–81.

Gattolliat, J.-L. 2002. Two new genera of Baetidae (Ephemeroptera; Insecta) from Madagascar. Aquatic Insects **24**:143–159.

Gattolliat, J.-L. 2003. The genera *Demoulinia* Gillies and *Potamocloeon* Gillies (Ephemeroptera: Baetidae) in Madagascar. Zootaxa **184**:1–18.

Gattolliat, J.-L., and S. Z. Rabeantoandro. 2002. The genus *Cloeon* (Ephemeroptera, Baetidae) in Madagascar. Mitteilungen der Schweizerischen Entomologischen Gesellschaft **74**:195-209 (2001).

Gattolliat, J.-L., and M. Sartori. 2000. Contribution to the systematics of the genus *Dabulamanzia* (Ephemeroptera: Baetidae) in Madagascar. Revue Suisse de Zoologie **107**:561–577.

Gattolliat, J.-L., and M. Sartori. 2001. Predaceous Baetidae in Madagascar: an uncommon and unsuspected high diversity. Pages 321–330 *in* E. Dominguez, editor. Trends in Research in Ephemeroptera and Plecoptera. Kluwer Academic/Plenum Publishers, New York.

Gattolliat, J.-L., and M. Sartori. 2003. An overview of the Baetidae of Madagascar. Pages 135–144 *in* E. Gaino, editor. Research Update on Ephemeroptera and Plecoptera. University of Perugia, Perugia, Italy

Gillies, M. T. 1985. A preliminary account of the East African species of *Cloeon* Leach and *Rhithrocloeon* gen. n. (Ephemeroptera). Aquatic Insects 7:1–17.

Gillies, M. T. 1988. Description of the nymphs of some afrotropical Baetidae (Ephemeroptera). I. *Cloeon* Leach and *Rhithrocloeon* Gillies. Aquatic Insects 10:49–59.

Gillies, M. T. 1990. A revision of the African species of *Centroptilum* Eaton (Baetidae, Ephemeroptera). Aquatic Insects 12:97–128.

Gillies, M. T. 1991. A diphyletic origin for the two-tailed baetid nymphs occurring in East African stony streams with a description of the new genus and species *Tanzaniella spinosa* gen. sp. nov. Pages 175–187 *in* J. Alba-Tercedor and A. Sanchez-Ortega, editors. Overview and Strategies of Ephemeroptera and Plecoptera. Sandhill Crane Press., Inc., Gainesville.

Gillies, M. T. 1997. A new species of *Procloeon* Bengtsson from the forest zone of West Africa (Ephem., Baetidae). Entomologist's Monthly Magazine 133:247–250.

Gillies, M. T., and J.-M. Elouard. 1990. The mayfly-mussel association, a new example from the River Niger Basin. Pages 289–298 *in* I. C. Campbell, editor. Mayflies and Stoneflies: Life Story and Biology. Kluwer Academic Publishers, Dordrecht.

Gillies, M. T., J.-M. Elouard, and J. Wuillot. 1990. Ephemeroptera from West Africa: the genus *Ophelmatostoma* (Baetidae). Revue d'Hydrobiologie Tropicale 23 (2):115–120.

Gillies, M. T., and J. Wuillot. 1997. *Platycloeon*, a new genus of Baetidae (Ephemeroptera) from East Africa. Aquatic Insects 19:185–189.

Gladstein, D., and W. Wheeler. 1999. POY. Phylogeny Reconstruction via Optimization of DNA data. Program and documentation. American Museum of Natural History, New York.

Guindon, S., and O. Gascuel. 2003. A simple, fast, and accurate algorithm to estimate large phylogenies by maximum likelihood. *Systematic Biology* 52:696–704.

Lugo-Ortiz, C. R., H. M. Barber-James, W. P. McCafferty, and F. C. de Moor. 2001. A nonparaphyletic classification of the Afrotropical genus *Acanthiops* Waltz & McCafferty (Ephemeroptera: Baetidae). African Entomology 9:1–15.

Lugo-Ortiz, C. R., and F. C. de Moor. 2000. *Nigrobaetis* Novikova & Kluge (Ephemeroptera: Baetidae): first record and new species from southern Africa, with reassignment of one northern African species. African Entomology 8:69–73.

Lugo-Ortiz, C. R., and W. P. McCafferty. 1996. The *Bugilliesia* complex of African Baetidae (Ephemeroptera). Transactions of the American Entomological Society 122:175–197.

Lugo-Ortiz, C. R., and W. P. McCafferty. 1996. The composition of *Dabulamanzia*, a new genus of Afrotropical Baetidae (Ephemeropera), with descriptions of two new species. Bulletin de la Société d'Histoire Naturelle de Toulouse **132**:7–13.

Lugo-Ortiz, C. R., and W. P. McCafferty. 1996. *Crassabwa:* a new genus of small minnow mayflies (Ephemeroptera: Baetidae) from Africa. Annales de Limnologie **32**:235–240.

Lugo-Ortiz, C. R., and W. P. McCafferty. 1997. Contribution to the systematics of the genus *Cheleocloeon* (Ephemeroptera: Baetidae). Entomological News **108**:283–289.

Lugo-Ortiz, C. R., and W. P. McCafferty. 1997. *Edmulmeatus grandis:* an extraordinary new genus and species of Baetidae (Insecta: Ephemeroptera) from Madagascar. Annales de Limnologie **33**:191–195.

Lugo-Ortiz, C. R., and W. P. McCafferty. 1997. *Labiobaetis* Novikova & Kluge (Ephemeroptera: Baetidae) from the Afrotropical region. African Entomology **5**:241–260.

Lugo-Ortiz, C. R., and W. P. McCafferty. 1997. *Maliqua*: a new genus of Baetidae (Ephemeroptera) for a species previously assigned to *Afroptilum*. Entomological News **108**:367–371.

Lugo-Ortiz, C. R., and W. P. McCafferty. 1997. New Afrotropical genus of Baetidae (Insecta: Ephemeroptera) with bladelike mandibles. Bulletin de la Société d'Histoire Naturelle de Toulouse **133**:41–46.

Lugo-Ortiz, C. R., and W. P. McCafferty. 1997. A new genus and redescriptions for African species previously placed in *Acentrella* (Ephemeroptera: Baetidae). Proceedings of the Entomological Society of Washington **99**:429–439.

Lugo-Ortiz, C. R., and W. P. McCafferty. 1997. New species and first reports of the genera *Cheleocloeon, Dabulamanzia*, and *Mutelocloeon* (Insecta: Ephemeroptera: Baetidae) from Madagascar. Bulletin de la Société d'Histoire Naturelle de Toulouse **133**:47–53.

Lugo-Ortiz, C. R., and W. P. McCafferty. 1998. The *Centroptiloides* Complex of Afrotropical small minnow mayflies (Ephemeroptera: Baetidae). Annals of the Entomological Society of America **91**:1–26.

Lugo-Ortiz, C. R., and W. P. McCafferty. 1998. A new *Baetis*-complex genus (Ephemeroptera: Baetidae) from the Afrotropical region. African Entomology **6**:297–301.

Lugo-Ortiz, C. R., and W. P. McCafferty. 1998. Phylogeny and biogeography of *Nesydemius* n. gen., and related Afrotropical genera (Insecta: Ephemeroptera: Baetidae). Bulletin de la Société d'Histoire Naturelle de Toulouse **134**:7–12.

Lugo-Ortiz, C. R., and W. P. McCafferty. 1999. *Delouardus*, a new *Centroptiloides* complex genus from Madagascar and its relationship with *Cheleocloeon* Wuillot & Gillies (Ephemeroptera: Baetidae). African Entomology **7**:63–66.

Lugo-Ortiz, C. R., W. P. McCafferty, and R. D. Waltz. 1999. Definition and reorganization of the genus *Pseudocloeon* (Ephemeroptera: Baetidae) with new species descriptions and combinations. Transactions of the American Entomological Society **125**:1–37.

McCafferty, W. P., and F. C. de Moor. 1995. South African Ephemeroptera: problems and priorities. Pages 463–476 *in* L. D. Corkum and J. J. H. Ciborowski, editors. Current directions in research on Ephemeroptera. Canadian Scholars' Press, Toronto.

McCafferty, W. P., C. R. Lugo-Ortiz, and H. M. Barber-James. 1997. *Micksiops*, a new genus of small minnow mayflies (Ephemeroptera: Baetidae) from Africa. Entomological News **108**:363–366.

Monaghan, M. T., J.-L. Gattolliat, M. Sartori, J.-M. Elouard, H. Barber-James, P. Derleth, O. Glaizot, F. de Moor, and A. P. Vogler. 2005. Trans-oceanic and endemic origins of the small minnow Mayflies (Ephemeroptera, Baetidae) of Madagascar. Proceedings of the Royal Society B: Biological Sciences **272**(1574): 1829–1836 doi: 10.1098/rspb.2005.3139.

Posada, D., and K. A. Crandall. 1998. MODELTEST: testing the model of DNA substitution. Bioinformatics **14**:817–818.

Waltz, R. D., and W. P. McCafferty. 1987. New genera of Baetidae (Ephemeroptera) from Africa. Proceedings of the Entomological Society of Washington **89**:95–99.

Waltz, R. D., and W. P. McCafferty. 1987. Systematics of *Pseudocloeon, Acentrella, Baetiella,* and *Liebebiella*, new genus (Ephemeroptera: Baetidae). Journal of New York Entomological Society **95**:553–568.

Wuillot, J., and M. T. Gillies. 1993. *Cheleocloeon*, a new genus of Baetidae (Ephemeroptera) from West Africa. Revue d'Hydrobiologie tropicale **26**:213–217.

Wuillot, J., and M. T. Gillies. 1993. New species of *Afroptilum* (Baetidae, Ephemeroptera) from West Africa. Revue d'Hydrobiologie tropicale **26**:269–277.

Wuillot, J., and M. T. Gillies. 1994. *Dicentroptilum*, a new genus of Mayflies (Baetidae, Ephemeroptera) from Africa. Aquatic Insects **16**:133–140.

PHYLOGENETIC RELATIONSHIPS OF THE AUSTRALIAN LEPTOPHLEBIIDAE

K.J. Finlay[1] and Y.J. Bae, [2]

[1] *Department of Biological Sciences, Monash University, Clayton, Victoria 3800, Australia*
[2] *Department of Biology, Seoul Women's University, Seoul 139-774, Korea*

Abstract

The phylogeny of Australian Leptophlebiidae has not been studied in detail although previous research has indicated close relationships with the Neotropical fauna. Phylogenetic relationships of Australian Leptophlebiidae were examined using a cladistic analysis of 34 morphological characters and 21 genera. Character polarity was assessed by out-group comparison and then analysed with NONA (version 2). The three most parsimonious trees produced were used to construct a strict consensus tree. Relationships among the Australian fauna specify some monophyletic groups and unresolved terminals. This study elucidates some previously unknown relationships among the Australian fauna and indicates that the currently recognised genera of Leptophlebiidae in Australia require further definition. Comparison with hypotheses previously proposed for Australian Leptophlebiidae demonstrate partial agreement, recognizing '*Meridialaris*' and '*Atalonella*' clades (*sensu* Pescador and Peters 1980a) and support for a sister relationship between the burrowing genera *Jappa* and *Ulmerophlebia*. Cladistic characters are tabulated and discussed with illustrations. Examined taxa, materials and comprehensive bibliographic sources are provided.

Key words: phylogeny, Australia, Leptophlebiidae, Ephemeroptera, mayflies, cladistics.

Introduction

The Leptophlebiidae, considered to be one of the most diverse mayfly families (Peters 1988), are distributed worldwide with over 100 genera, predominantly found in the Southern Hemisphere (Hubbard 1990). The monophyly of the Leptophlebiidae, within the superfamily Leptophlebioidea, is now well established (Landa and Soldán 1985, McCafferty 1991). However, phylogenetic relationships within the family have not been considered in total, due to the large number of species involved, but rather have been studied in smaller groups. These groups entail biogeographical regions such as the Southern and Eastern Hemisphere (Tsui and Peters 1975, Peters and Edmunds 1970); South America (Pescador and Peters 1980a), New Zealand (Towns and Peters 1979, 1996), New Caledonia (Peters et al. 1978; Peters and Peters 1979, 1981a, 1981b; Peters et al. 1990, 1994), Africa (Peters and Edmunds 1964) and

233

Madagascar (Peters and Edmunds 1984) or genus level relations, e.g., *Meridialaris* and *Massartellopsis* (Pescador and Peters 1987), *Miroculis* (Savage and Peters 1983), *Nousia* (Pescador and Peters 1985), *Penaphlebia* (Pescador and Peters 1991), *Thraulus* (Grant 1985) and *Ulmeritus* (Domínguez 1995).

Only two studies of the evolutionary relationships of Australian genera have previously been undertaken. As part of the analysis of the cool-adapted Leptophlebiid fauna of South America, Pescador and Peters (1980a) inferred affinities between South American and Australian taxa. Later, Christidis (2001) performed an analysis of the Australian Leptophlebiidae, focussing on one evolutionary branch described by Pescador and Peters (1980a), which included several Australian representatives.

Herein we report the relationships between the Australian Leptophlebiid genera and compare and contrast the relationships of the Australian Leptophlebiid genera with hypotheses already proposed, namely the grouping of the Australian genera into five distinct evolutionary lineages based primarily on morphological similarity.

Methods

We examined representative species of the 19 known Australian genera, along with the South American subgenus *Nousia (Nousia)*, for phylogenetic analysis. Taxa (n=21) and characters (n=34) were compiled (Appendix A). NONA version 2.0 (Goloboff 1993) was used to construct a cladogram using the tree bisection-reconnection command (mult*). A strict consensus tree (nelsen command) was constructed from the resulting most parsimonious trees. WinClada version 0.9.99i (beta) (Nixon 1999) was used to redraw the tree with the characters and character states mapped.

We borrowed type species material from the Museum of Victoria (MV), the Australian National Insect Collection (ANIC), the Australian Museum in Sydney (AM), the Queensland Department of Primary Industries (QDPI), the Natural History Museum, London (NHM), the Swedish Museum of Natural History (SMNH) and the Florida Agricultural and Mechanical University collection (FAMU). Additional specimens were borrowed from the private collections of I.C. Campbell (Campbell collection) and P.J. Suter (Suter collection). We also had specimens from our personal collections (Finlay and Bae collections).

Morphological characters and character states (Appendix B) were determined from examination of the specimens and from the species descriptions in the literature. Key sources of information for the ingroup genera included: *Atalophlebia* (Tillyard 1933; Suter 1986); *Atalomicria* (Campbell and Peters 1993), *Austrophlebioides* (Campbell and Suter 1988, Parnong and Campbell 1997), *Bibulmena* (Dean 1987), *Garinjuga* (Campbell and Suter 1988), *Jappa* (Skedros and Polhemus 1986, Bae and Finlay 2003, Bae et al. 2004), *Kalbaybaria* (Campbell 1993), *Kaninga* (Dean 2000), *Kirrara* (Campbell and Peters 1986), *Koorrnonga* (Campbell and Suter 1988), *Loamaggalangta* (Dean et al. 1999), *Neboissophlebia* (Dean 1988), *Nousia* (Pescador and Peters 1980a; Campbell and Suter 1988), *Nyungara* (Dean 1987), *Thraulophlebia* (Demoulin 1955a, Campbell and Suter 1988), *Thraulus* (Grant 1985, Suter 1992), *Tillyardophlebia* (Dean 1997) and *Ulmerophlebia* (Suter 1986, Bae et al. 2004). In addition, the taxa *Nousia* (*Australonousia*) and *Thraulophlebia* were recently revised (Finlay 2002) leading to the synonymisation of Thraulophlebia with *Koorrnonga* over which it has priority (Finlay, Suter and Campbell, unpublished data) and the establishment of two new genera: 'New Genus A' and 'New Genus B' (Finlay, unpublished data).

In agreement with Edmunds and Allen (1966), Riek (1973) and Pescador, and Peters (1980a), larval characters were found to be more taxonomically informative than those of the imago, hence the disproportionately low number of adult characters. Further, wing venation, one of the major characteristics of the adult, is known to be subject to significant parallel evolution (Edmunds 1972).

Character polarities were assessed across all available outgroups (Watrous and Wheeler 1981). The nearest outgroup was the Leptophlebiinae, comprising eight genera (*Paraleptophlebia, Leptophlebia, Habroleptoides, Habrophlebia, Calliarcys, Habrophlebioides, Dipterphlebioides, Gillesia*) considered a primitive furcation in the evolution of the Eastern Hemisphere Leptophlebiidae (Peters and Edmunds 1970) and sufficiently different to warrant the establishment of a new subfamily (Peters 1980). Within this subfamily the plesiotypic *Paraleptophlebia* and *Leptophlebia* provided particularly valuable cladistic information.

The latest revisions of the higher classification of mayflies (McCafferty 1991, McCafferty 2002), encompassing the work of Landa and Soldán (1985), provide the next nearest outgroups within the Infraorder Lanceolata: that of the sister group Ephemeroidea (Polymitarcyidae, Euthyplociidae, Potamanthidae, Ephemeridae, Palingeniidae) followed by the superfamily Caenoidea (Ephemerellidae, Tricorythidae and Caenidae). Although the superfamily Behningoidea (containing the single family Behningiidae) is considered more closely related to Leptophlebioidea (McCafferty 1991) than the previous two superfamilies, its use as an outgroup is limited due to its highly distinctive and specialized nature in both adult and nymphal forms (Edmunds 1959).

Key literary sources of information for the outgroup taxa are as follows: Leptophlebiinae - *Paraleptophlebia, Leptophlebia* (Burks 1953, Peters and Edmunds 1970), *Habroleptoides, Habrophlebia, Calliarcys, Habrophlebioides, Dipterphlebioides* (Peters and Edmunds 1970), *Gillesia* (Gillies 1951, Peters and Edmunds 1970), Ephemeroidea - Ephemeridae: *Aethephemera* (McCafferty 1971a, McCafferty 1973), *Afromera* (Demoulin 1955b, McCafferty and Gillies 1979, Elouard 1986a), *Eatonica* (McCafferty 1971b, Elouard 1986b, Elouard et al. 1998), *Ephemera* (McCafferty 1973, McCafferty 1975, Hubbard 1982, Hubbard 1983; Balasubramanian et al. 1991, Kang and Yang 1994, Bae 1995, Ishiwata 1996), *Hexagenia* (Spieth 1941, McCafferty 1975, Keltner and McCafferty 1986), *Ichthybotus* (Eaton 1899), *Litobrancha* (Lestage 1939, McCafferty 1975), Euthyplociidae: *Afroplocia* (Demoulin 1952a), *Campylocia* (Demoulin 1952a,

Pereira and Da Silva 1990), *Euthyplocia* (Lestage 1918, Lestage 1939, Demoulin 1952a), *Exeuthyplocia* (Lestage 1918, Lestage 1939, Gillies 1980), *Proboscidoplocia* (Demoulin 1966), *Polyplocia* (Demoulin 1952a), *Mesoplocia* (Demoulin 1952a); Palingeniidae: *Cheirogenesia* (McCafferty and Edmunds 1976, Sartori and Elouard 1999), *Chankagenesia* (Demoulin 1952b), *Palingenia* (Sartori 1992), *Pentagenia* (Lestage 1918, McCafferty 1972, McCafferty 1975, Keltner and McCafferty 1986); Polymitarcyidae: *Campsurus* (Eaton 1868-69, McCafferty 1975), *Ephoron* (Lestage 1918, Spieth 1933, Demoulin 1952a, McCafferty 1975, Ishiwata 1996), *Tortopus* (Needham and Murphy 1924, McCafferty 1975, McCafferty and Bloodgood 1989, Lugo-Ortiz and McCafferty 1996), *Povilla* (Lestage 1918; Lestage 1939; Hubbard 1984), Potamanthidae (Bae and McCafferty 1991), *Anthopotamus* (McCafferty and Bae 1990), *Neopotamanthus* (Wu and You 1986), *Potamanthodes* (You 1984, You and Su 1987), *Potamanthus* (Uéno 1928, McCafferty 1975, Elpers and Tomka 1994, Kang and Yang 1994, Vuori 1999), *Rhoenanthus* (Soldán and Putz 2000) and *Stygifloris* (Bae et al. 1990). See Appendix C for a detailed full list of material examined.

Results

Relationships of the Australian Leptophlebiidae are shown in the strict consensus tree (length = 62; $ci = 0.66$; $ri=0.82$, Fig. 1) constructed from the three most parsimonious trees initially produced. Based on the cladogram, the Australian Leptophlebiidae are basally delineated from the outgroup by the following synapomorphies: the presence of square dorsal eye facets in the male imago, the presence of hypopharynx lateral processes and prognathous mouthparts in the nymphs.

Although relationships between some groups of the ingroup taxa have not been fully resolved, certain clades are relatively well supported. *Jappa* and *Ulmerophlebia* form a monophyletic clade united by the following synapomorphies: body and gills fringed with fine setae (Figs. 2–5); abdominal terga with prominent setae on lateral margins and one prominent median denticle on the labrum (Figs. 6 and 7). *Jappa*, however, is clearly separated from *Ulmerophlebia* by the presence of autopomorphic frontal horns.

Atalomicria and *Atalophlebia* are also monophyletic, grouped together by the possession of medium, as opposed to large or contiguous, eye size (*sensu* Bae and McCafferty 1991). There is strong support for a clade, which includes three genera: *Kirrara, Austrophlebioides* and *Tillyardophlebia*. These are united by eight synapomorphies, including two convergences, most of which are multistate and show what are considered to be the most highly derived conditions. Thus, the labrum is much wider than the clypeus, the lateral margins of labrum are angular and the lateral margins of clypeus slightly diverge towards the anterior (Figs. 8–10), the mandibles have angular outer margins (Figs. 11–13), the maxillae galea-lacinae is broad at the apex (Figs. 14–16) and the labium submentum has no lateral setae.

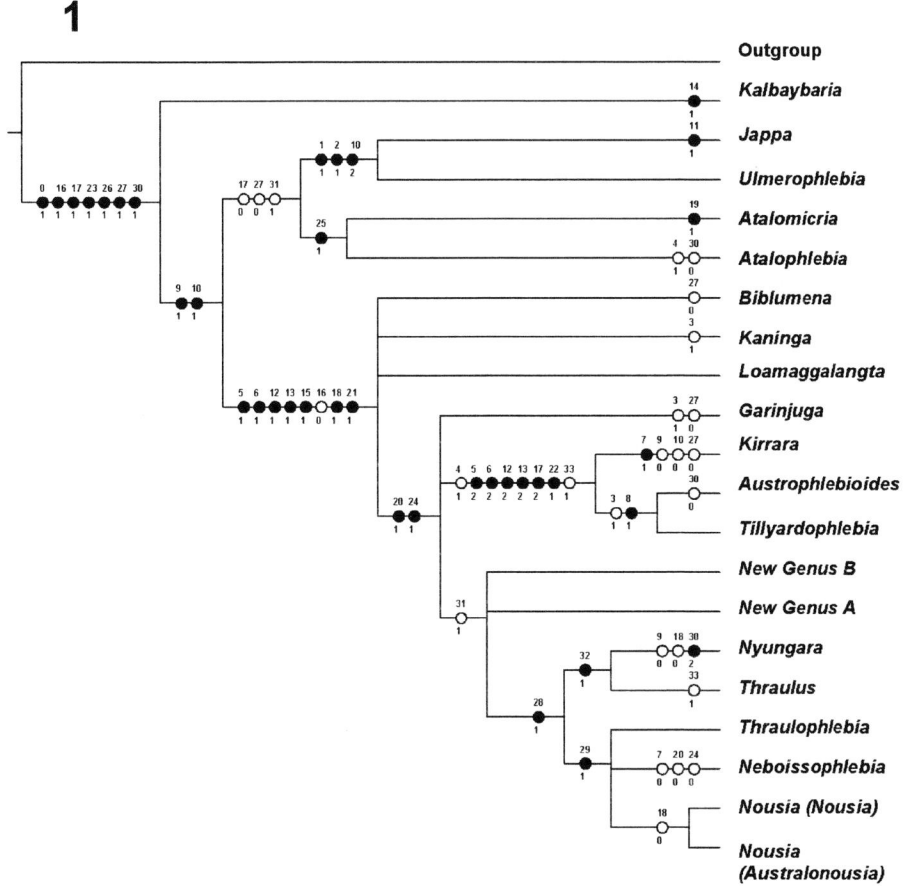

Figure 1. Strict consensus tree (length = 62; ci = 0.66; ri=0.82) of the relationships of the Australian Leptophlebiidae.

Finally, a reduction in the number of crossveins in the forewing costal space defines a clade containing the taxa *Nousia, Nyungara, Thraulus, Thraulophlebia* and *Neboissophlebia* (Figs. 17–22). In the present study, *Neboissophlebia* forms a tricotomy with *Thraulophlebia* and *Nousia*. These taxa, in turn, share a sister group relationship with the *Nyungara-Thraulus* clade. *Nousia, Thraulophlebia* and *Neboissophlebia* share the condition of an absence of crossveins in the male forewing costal space (Figs. 17, 18, 21 and 22). *Nyungara* and *Thraulus* are also grouped by the possession of a well-developed midlength costal projection of the hind wing (Figs. 23 and 24).

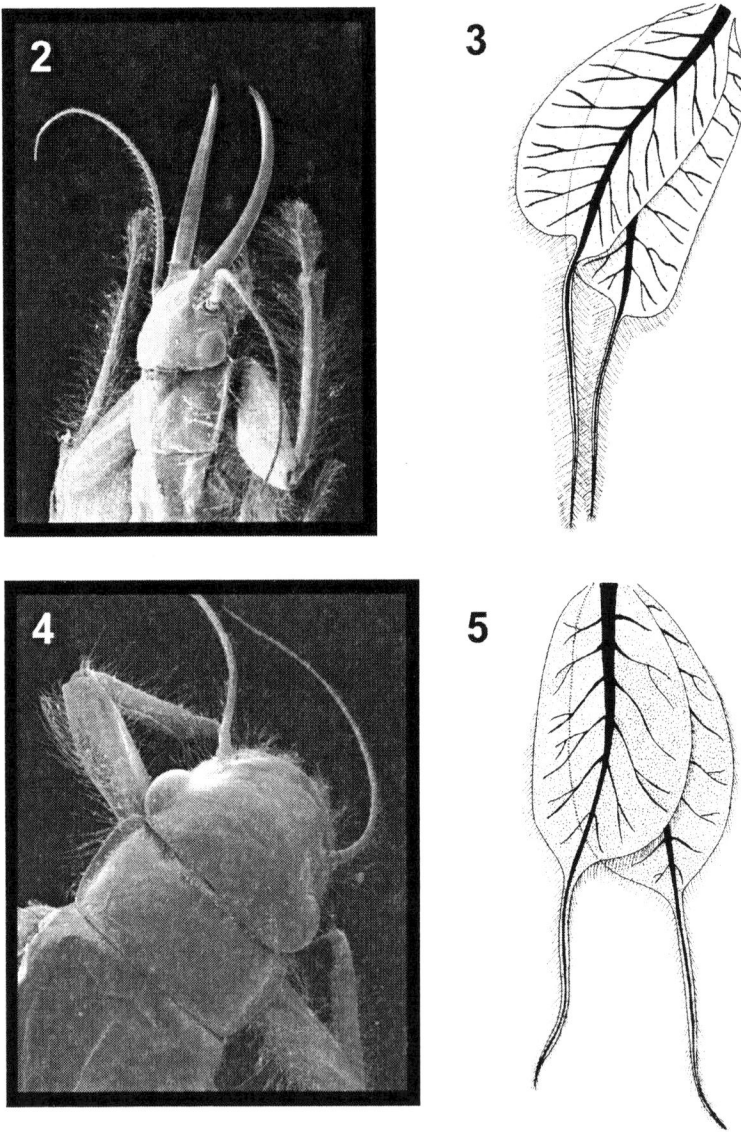

Figures 2–5. *Jappa* (Figs. 2 and 3) and *Ulmerophlebia* (Figs. 4 and 5) have bodies and gills fringed with fine setae.

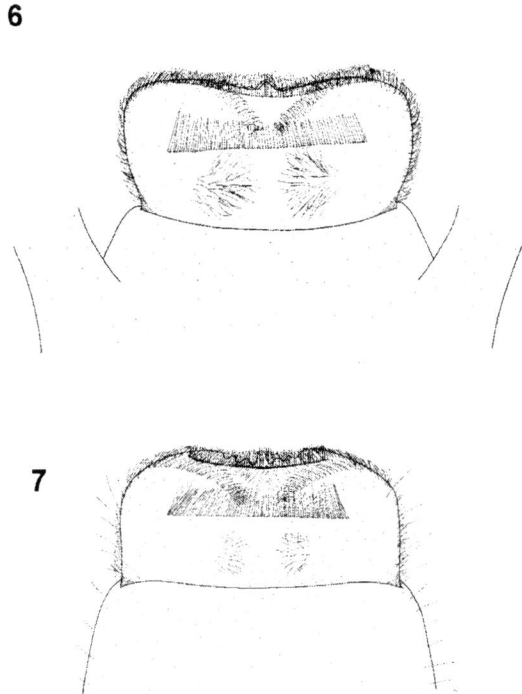

Figures 6 and 7. *Jappa* (Fig. 6) and *Ulmerophlebia* (Fig. 7) Labrum showing prominent median denticle.

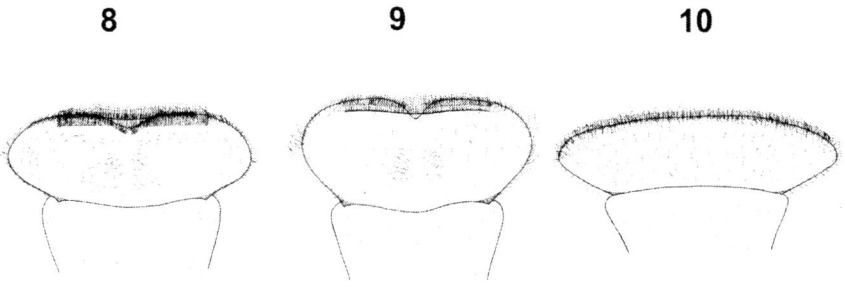

Figures 8–10. Labrum and clypeus morphology of *Austrophlebioide*s (Fig. 8), *Kirrara* (Fig. 10) and *Tillyardophlebia* (Fig. 12).

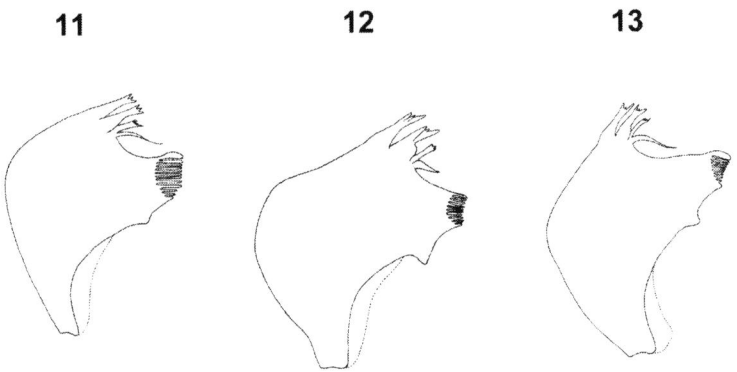

Figures 11–13. Mandible morphology of *Austrophlebioides* (Fig. 11), *Kirrara* (Fig. 12) and *Tillyardophlebia* (Fig. 13), mandible.

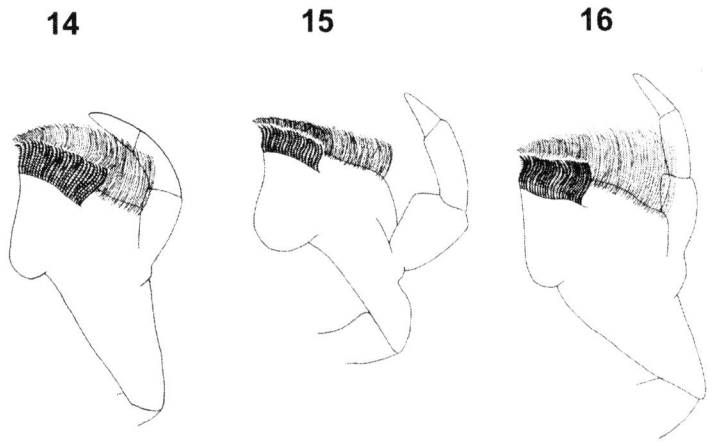

Figures 14–16. Maxilla morphology of *Austrophlebioides* (Fig. 14), *Kirrara* (Fig. 15) and *Tillyardophlebia* (Fig. 16).

17

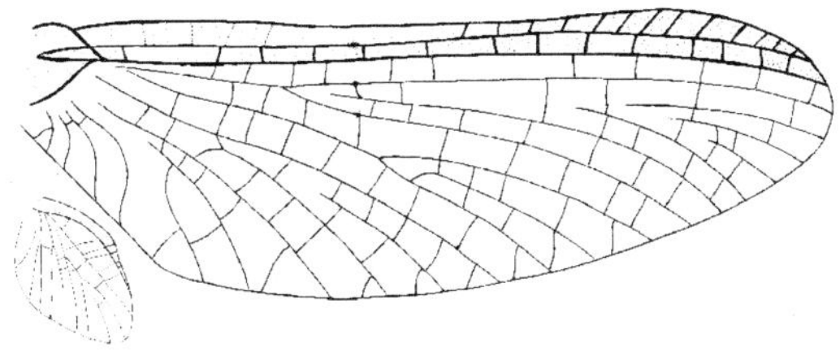

18

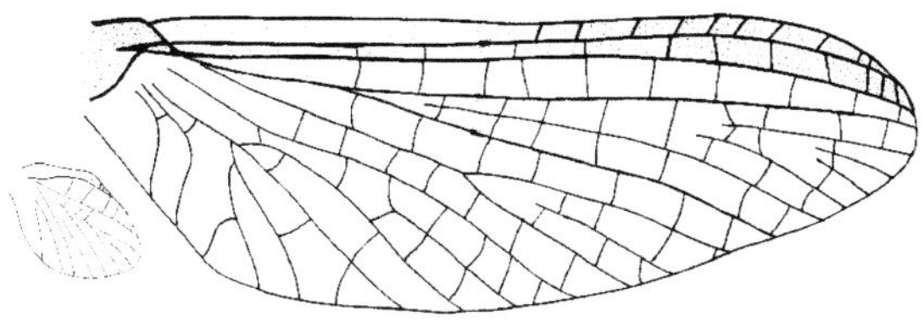

19

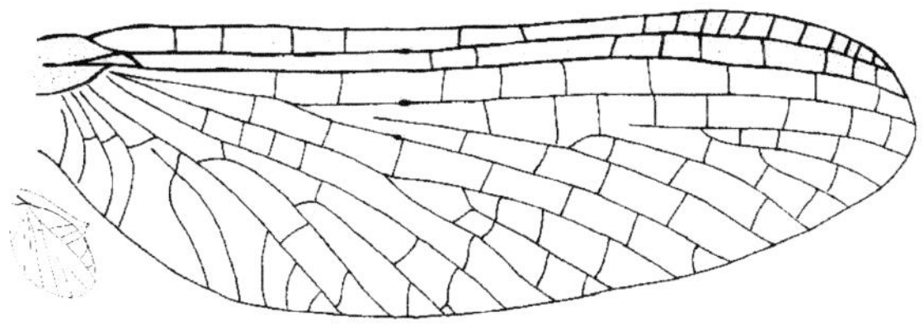

Figures 17–19. Male imago forewings of *Nousia (Nousia)* (Fig 17), *Nousia (Australonousia)* (Fig 18), and *Nyungara* (Fig. 19). Note: reduction in the number of crossveins in the costal space, in particular the proximal half.

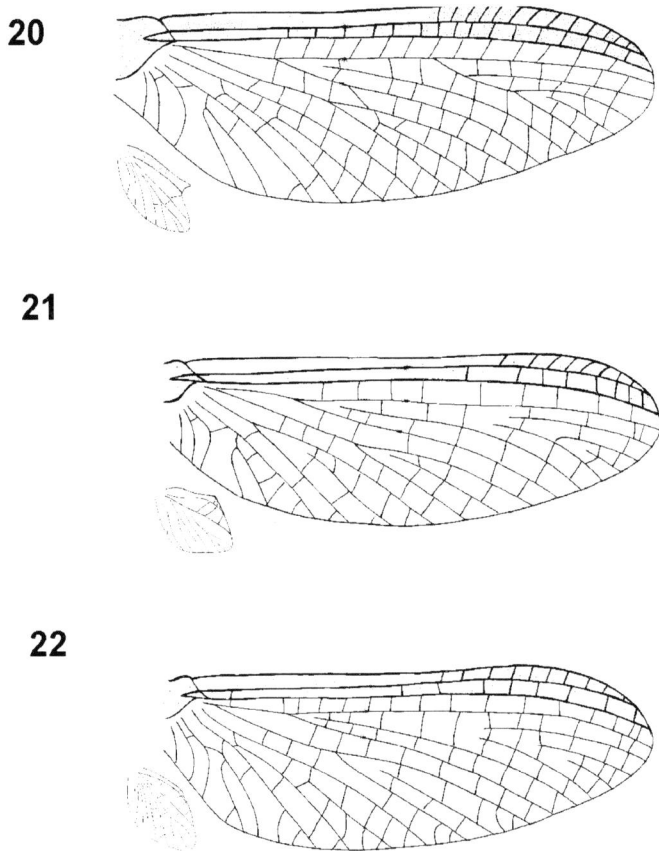

Figures 20–22. Male imago forewings of *Thraulus* (Fig. 20), *Thraulophlebia* (Fig. 21), and *Neboissophlebia* (Fig. 22). Note: reduction in the number of crossveins in the costal space, in particular the proximal half.

23   **24**

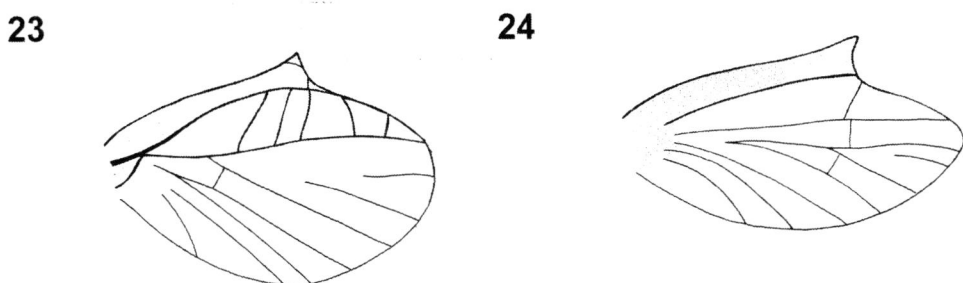

Figures 23–24. Male imago hindwings of Nyungara (Fig. 23) and Thraulus (Fig. 24) showing costal projection.

Discussion

Relationships of the Australian Leptophlebiidae. The characters defining the basal clade of the ingroup (square facets, lateral processes on the hypopharynx, head position) are included in the familial definition of Leptophlebiidae as outlined by Peters and Edmunds (1964) and Towns and Peters (1996). But as these characters are present amongst the Atalophlebiinae worldwide, an analysis including all Gondwanan fauna would be required to investigate the potential monophyly of the Australian group.

Square facets are characteristic of all but one of the Atalophlebiinae taxa (Peters and Gillies 1995); a condition thought to have evolved to catch a greater proportion of the ultraviolet light available (Horridge et al. 1982). The presence of lateral processes on the hypopharynx generally applies to Atalophlebiinae worldwide and is present in all the Australian taxa (Figs. 25–43, however, it may be secondarily lost in some species (Peters and Edmunds 1970). The evolutionary function of this structure is unknown although one could presume it is related to feeding behaviour.

Finally, head position in the Leptophlebiidae varies widely and can be somewhat ambiguous such as the described semiprognathous of Peters and Edmunds (1964). Despite this, the mouthpart position tends to be relatively stable. For example, the position of the head of both *Jappa* and *Ulmerophlebia* can sometimes be oriented partly downwards, although the mouthparts of each are most commonly directed forward. The habit of the Atalophlebiinae as sprawlers, swimmers and clingers (Edmunds and Waltz 1996) translates into a streamlined body capable of maintaining a position of least resistance on the streambed. The Leptophlebiinae tend to be more laterally rather than dorso-ventrally flattened, often with prognathous mouthparts (Peters and Edmunds 1970). The Ephemeroidea are very robust and laterally flattened to accommodate their burrowing behavior and their mouthparts are generally hypognathous for filtering and gathering detritus.

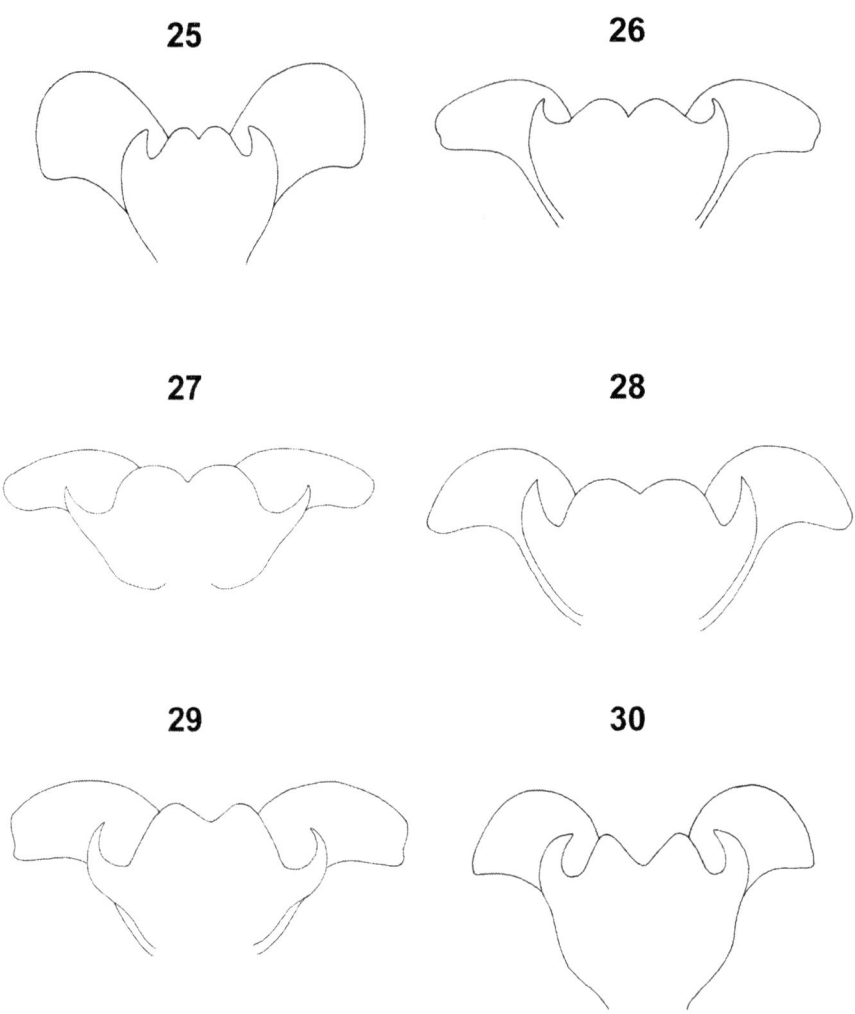

Figures 25–43. Hypopharynx morphology of Australian genera: *Atalomicria* (Fig. 25), *Atalophlebia* (Fig. 26), *Austrophlebioides* (Fig. 27), *Bibulmena* (Fig. 28), *Garinjuga* (Fig. 29), *Jappa* (Fig. 30).

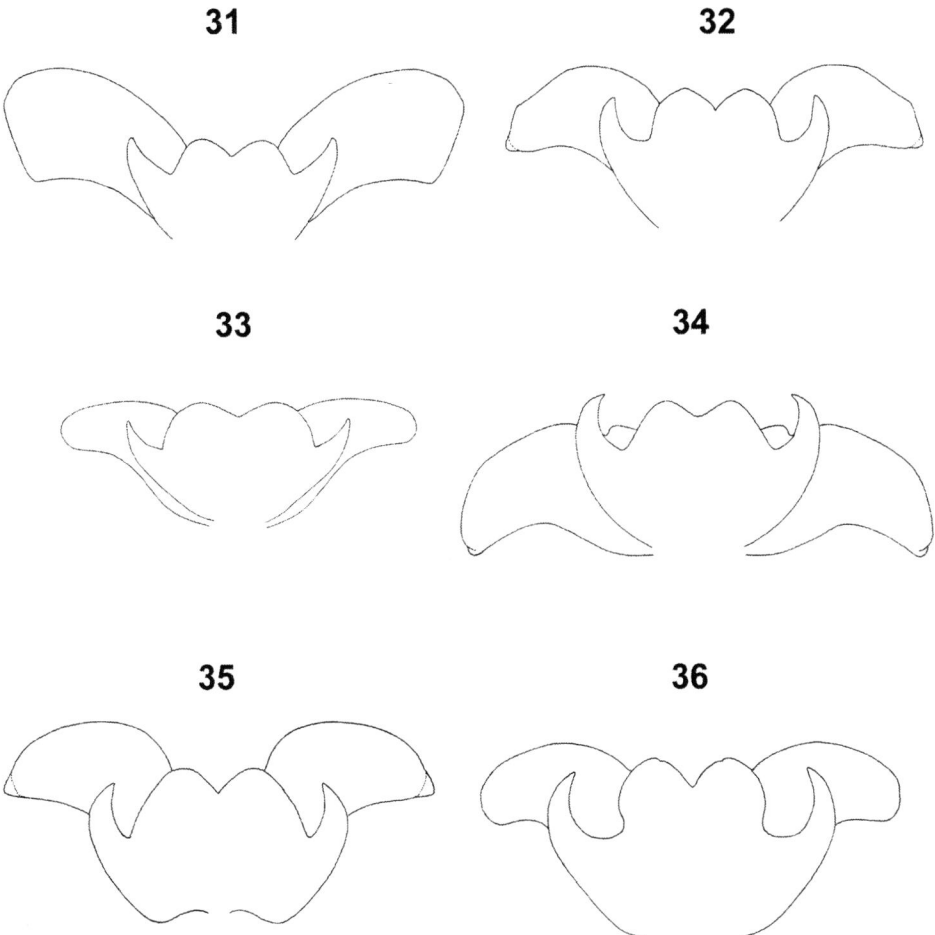

Figures 31–36. Hypopharynx morphology of Australian genera: *Kalbaybaria* (Fig. 31), *Kaninga* (Fig. 32), *Kirrara* (Fig. 33), *Loamaggalangta* (Fig. 34), *Neboissophlebia* (Fig. 35), *Nousia (Australonousia)* (Fig. 36).

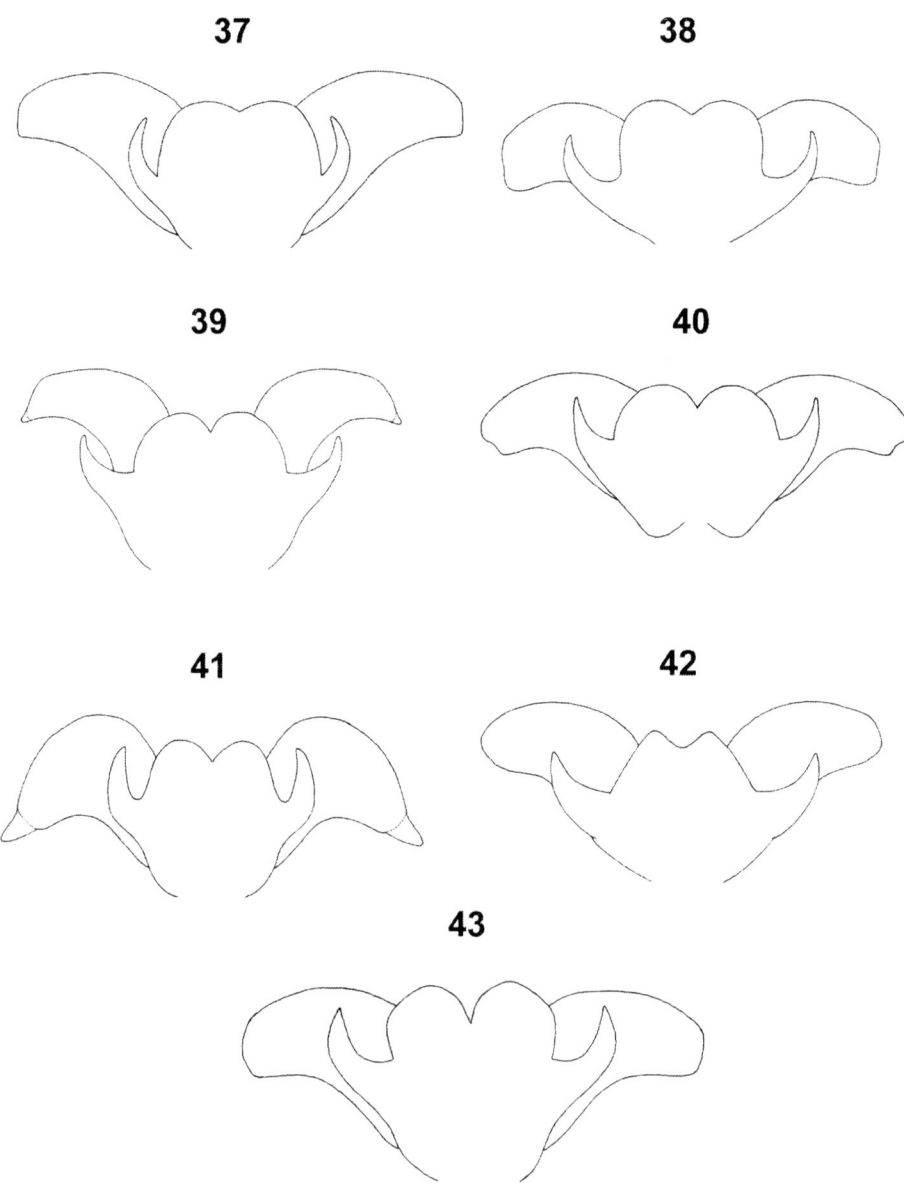

Figures 37–43. Hypopharynx morphology of Australian genera: *Nyungara* (Fig. 37), *Thraulophlebia* (Fig. 38), *Thraulus* (Fig. 39), *Tillyardophlebia* (Fig. 40), *Ulmerophlebia* (Fig. 41), New Genus A (Fig. 42), and New Genus B (Fig. 43)

The monotypic *Kalbaybaria* is separated from the rest of the ingroup by the autapomorphy of the mandible extending anteriorly into an enlarged flattened process (Fig. 44), a feature not homologous with the horns of *Japp*, which are derived from the head capsule. Certain Leptophlebiinae (e.g., *Paraleptophlebia*) also possess tusks derived from the outer incisor of the mandible (Needham et al. 1935, Bae and McCafferty 1995), but more work is required to establish the possible homology of *Kalbaybaria* and Ephemeroidea mandibular tusks. The distinct morphology and habitat of the tropical genus *Kalbaybaria* found only in Far North Queensland may indicate a highly evolved condition despite the obscurity of its adaptive and historical origins. There are some indications that tusk robustness and setation of the Ephemeroidea is more related to burrowing and filter feeding (Bae and McCafferty 1995) as opposed to the spatulate processes of *Kalbaybaria*, which, it is suggested, is used to navigate through leaf packs (Campbell 1993).

The sister relationship and similar characteristics of *Jappa* and *Ulmerophlebia* is presumed to be due to their similar burrowing habits. Morphological differences between the two taxa are not readily apparent apart from the highly visible frontal horns. Suggestions have been made that the two taxa are congeneric (Riek 1970, Suter 1986) due to the significant similarities other than the frontal horn, but this is currently being refuted (Bae et al. 2004).

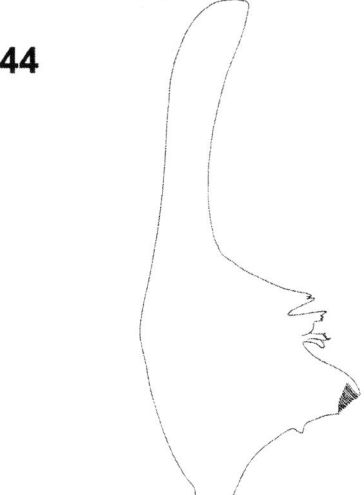

44

Figure 44. Mandible morphology of *Kalbaybaria* showing enlarged flattened process.

The grouping of *Atalomicria* and *Atalophlebia* is based on their comparatively small eye size. This character has previously been noted as an apomorphy in the Potamanthidae (Bae and McCafferty 1991) but is considered liable to convergence. Interestingly, certain members of the supposed more primitive subfamily, Leptophlebiinae (*Paraleptophlebia, Leptophlebia*) also possess large eyes, whereas, some members of Ephemeroidea do not.

The function of the elongate maxillary palps (Fig. 45) of *Atalomicria* is unknown. Although certain ephemerid mayflies (e.g., *Ephemera*) also possess elongated maxillary palps, the morphologies of the two are quite distinct. In *Atalomicria*, segments one and two make up the majority of length, whereas, in *Ephemera* all segments are equally elongate.

45

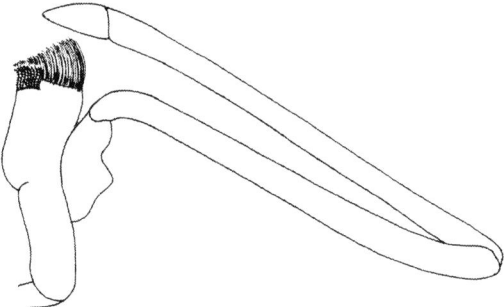

Figure 45. Maxilla morphology of *Atalomicria* showing elongate maxillary palps.

Undoubtably, the most well supported clade is the *Kirrara, Austrophlebioides* and *Tillyardophlebia* trichotomy grouped by eight synapomorphies mostly relating to the nymphal mouthparts as shown above (see Figs. 8–16). Some of these characters are not independent; for example, the diverging lateral margins of the clypeus may be required to accommodate a labrum that is significantly wider than its clypeus. Pescador and Peters (1980a) have labeled a number of these characters as phenoclines, where the character states have evolved from the most primitive state through a series of intermediary stages towards the most derived condition. These character states cannot always be clearly delineated as multistate characters across all genera, but the above genera seem to possess the most highly derived character states which can be easily polarized from all the other taxa.

Austrophlebioides and *Tillyardophlebia* share an apomorphic U-shaped labrum cleft or "hood" (see Figs. 8 and 9). An unnamed Genus 'Z' from southeastern

Australia (Dean 1999) is not well known but also has a wide labrum and distinct V-shaped cleft with many similar characteristics to *Austrophlebioides* and may prove to be closely related.

Kirrara is distinguished from the trichotomy by the possession of a labrum, which is triangular in cross section with quite a wide anterior margin in apical view (Fig. 46). The previously named 'Genus T' (Dean and Suter 1996), which has recently been referred to *Kirrara* as an unnamed species (Christidis 2001), also possesses an apically expanded labrum where the frontal setae have been modified to form a suction disc. *Kirrara* is very similar to the highly derived *Deleatidium* of New Zealand (Towns and Peters 1996) and *Lepegenia* of New Calendonia (Peters et al. 1978), which have laterally and apically expanded labrum and gills forming a suction disc on the venter of the abdomen. However, these structures are not present in all *Deleatidium*. These are considered adaptations to extremely rapid flow that enables the species to cling tightly to the substrate (Towns and Peters 1996). 'Genus T', *Lepegenia* and several species of *Deleatidium*, have all been found in association with vertical rock faces of waterfalls. Other distinguishable characters of *Kirrara* include the loss of the secondary hair fringe and anteromedian denticles of the labrum, which appear to have evolved in an earlier clade in the Australian Leptophlebiidae.

46

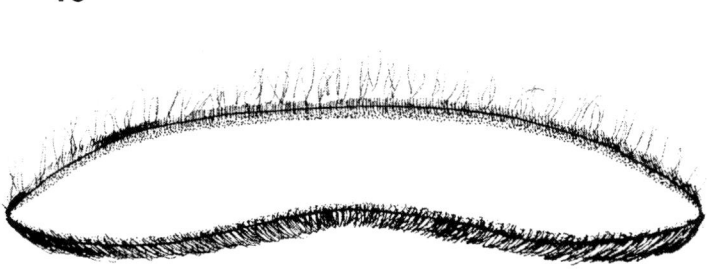

Figure 46. Apical view of *Kirrara procera* labrum; apical view.

A reduced number of crossveins in the forewing costal space defines a clade comprising the taxa *Neboissophlebia, Thraulophlebia, Nousia, Nyungara* and *Thraulus* (see Figs, 17–22); a state that could be related to the relatively small size of the imagos where the cross support between the costal and subcostal spaces is not needed. In *Nousia, Thraulophlebia and Neboissophlebia* (see Figs. 17, 21 and 22), a maximum reduction of crossveins has occurred. They all share the condition of the absence of crossveins in the male forewing costal space. The evolutionary significance of this character state is unknown (W. L. Peters, *pers. comm.*) but as with a reduction in the number of costal crossveins in the male forewing. This character may also be considered as the second step in the apomorphic state of the costal crossvein reduction but was analysed separately in this study because it could also be related to the relatively smaller size only in the male imagos. Ecological investigation will be required to determine the significance of this character state, which is not possessed by females of the same species.

Despite the considerable morphological variation between *Thraulophlebia* and *Nousia* which warrants their separate generic status (Finlay 2002), there appear to be few phylogenetically informative characters separating the two taxa in this clade. The inclusion of the full complement of Gondwanan fauna in an analysis may elucidate more evolutionary information for these taxa in the future. Similarly, there is little phylogenetic information supporting the subgeneric separation of *Nousia*, but as the latest revision of Australian *Nousia* is preliminary, largely due to lack of material (Finlay, unpublished data) the subgenera should remain separate based on the consistent nymphal character states outlined by Campbell and Suter (1988).

Nyungara and *Thraulus* are also grouped by the possession of a well-developed midlength costal projection of the hind wing (see Figs. 23 and 24). Although this character is also possessed by the Leptophlebiinae in some cases (e.g., *Habrophlebiodes*, *Gillesia*), it is not present in the Ephemeroidea, which usually possess an apical costal projection. The Gondwanan status of *Thraulus* is unclear, being primarily distributed throughout the Oriental and Ethiopian regions and Europe (Grant 1985) and known only in Australia from the Northern Territory, north Queensland and north Western Australia. Evidence suggests there are two centers of origin for Leptophlebiidae: one with Holarctic-Oriental distribution and the other Gondwanan (Peters and Edmunds 1970, Edmunds 1972), and the distribution of this genus suggests affinities with the former. However, only a more taxonomically inclusive analysis could address the question of its geographic origins. Further, our knowledge of *Thraulus* in Australia is rudimentary. Two morphospecies were described from the Alligator River region of Northern Territory (Suter 1992) but were subsequently considered conspecific, being referred to as *Thraulus* sp. 'AV1' (Dean 1999). From the specimens examined, it is clear that specimens of *Thraulus* sp. 'AV1' are the same as *Thraulus* sp. A. from Magela Creek of the Northern Territory (Marchant 1982). Only two other unnamed species from Queensland are recognized (Dean 1999) but are based only on nymphs. Adult material is essential to further elucidate the total contingent of species in Australia.

Garinjuga possesses strongly diverging veins ICu_1 and ICu_2 in the male forewing (Fig. 47) that are considered plesiomorphic state and tend to be associated with more triangular shaped forewings, where the junction of the anal and distal margin forms an approximate 90-degree angle. This tends to expand the length of the wing hind margin and therefore separates ICu_1 and ICu_2 as the wing margin is approached. This character is generally distributed amongst the larger sized taxa such as *Jappa*, *Ulmerophlebia*, *Atalomicria*, *Atalophlebia* and *Kirrara*, which are in possession of larger and more robust forewings.

47

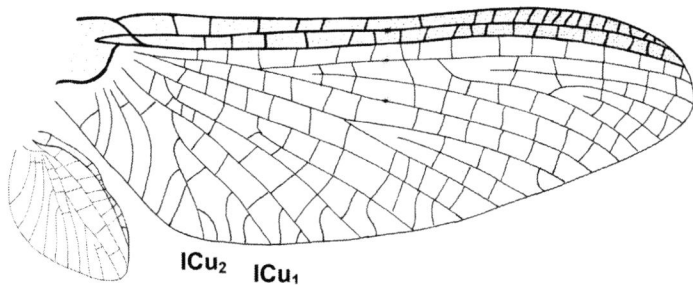

Figure 47. Male imago forewing of *Garinjuga*.

The position of the remaining taxa: *Bibulmena*, *Kaninga* and *Loamaggalangta*, 'New Genus A' and 'New Genus B' are also poorly resolved within the cladogram. *Kaninga* has only recently been established as a genus (Dean 2000) and its single species is confined to southwest Western Australia. It has been considered similar to *Bibulmena* (Dean 2000) and *Loamaggalangta* (Dean et al. 1999), but this is defined only by symplesiomorphies such as smooth tarsal claws in the nymph (Figs. 48 and 49) and the labial glossae turned over (Figs. 50 and 51). *Bibulmena* and *Kaninga* are found only in southwest Western Australia, an indication that the origins of these species may differ from their east Australian counterparts. *Loamaggalangta* has been found solely in Tasmania where it occurs only at water depths of greater than five metres clinging to submerged objects (Dean et al. 1999), which may account for some of its more unusual features such as extremely elongate leg lengths of the nymph including a long curved smooth tarsal claw (see Fig. 49), which may be used to grip tightly to the substrate. 'New Genus A' and 'New Genus B' are also unresolved in the cladogram but do not appear closely related to *Nousia (Australonousia)*, the genus to which they were previously assigned (Dean 1999).

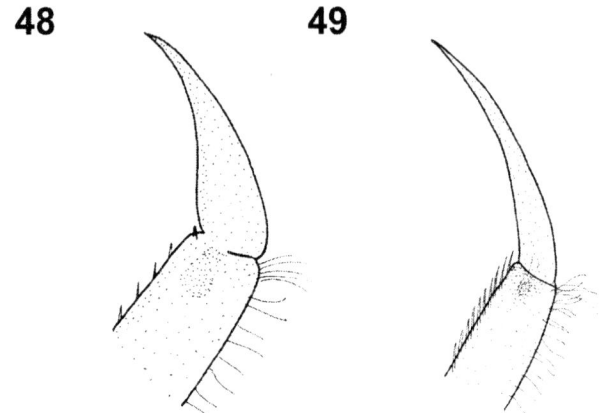

Figures 48 and 49. Nymph tarsal claw morphology of *Bibulmena* (Fig. 48) and *Loamaggalangta* (Fig. 49).

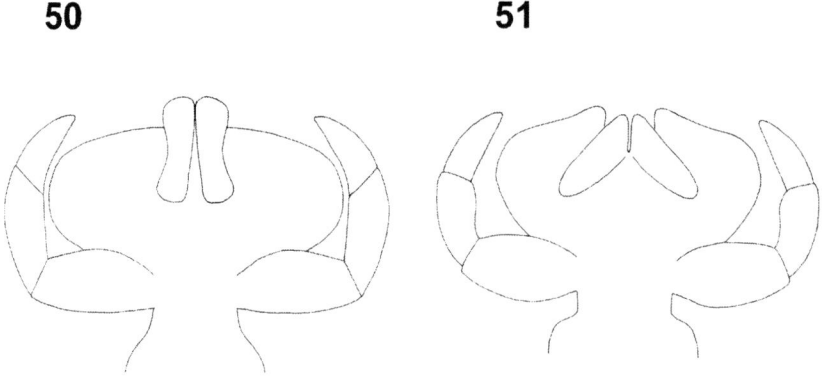

Figures 50 and 51. Labium morphology of *Bibulmena* (Fig. 50) and *Loamaggalangta* (Fig. 51); dorsal view.

Comparisons with previous studies of Australian Leptophlebiidae. The phylogenetic analysis of the major groups of Leptophlebiidae was first attempted by Peters and Edmunds (1970). They proposed a phylogeny for the Eastern Hemisphere genera of Leptophlebiidae including all Palaearctic, Ethiopian, and Oriental regions but excluding Australia and New Zealand as they considered these taxa to have closer affinities with the Neotropical fauna. This hypothesis had been inferred from a previous revision of the Ethiopian Leptophlebiidae (Peters and Edmunds 1964). A similar hypothesis was first intimated by Brundin (1966) while studying the biogeography of chironomid midges and coincides with the generally accepted view of the sequential break up of Gondwana (Norton and Sclater 1979). Amongst the Leptophlebiidae, however, this has not been cladistically tested.

The proposed close affinity between Gondwanan fauna was examined in a study of the cool-adapted taxa of South America referring to cool mountain waters and regular freezing episodes (Pescador and Peters 1980a). This small but significant paper provided a basis for comparison for future phylogenetic work on the Leptophlebiidae. Five lineages, one of which was described subsequently (Pescador and Peters 1980b), were established based on an analysis of the cool-adapted Leptophlebiidae of southern South America (Fig. 52). Each lineage, although named arbitrarily after a prominent South American genus, represented a group of taxa, the names being for convenience only. The lineage at the base of the tree (*'Hapsiphlebia'*) was considered the most primitive and each successive branch more advanced until the *'Meridialaris'* lineage.

52

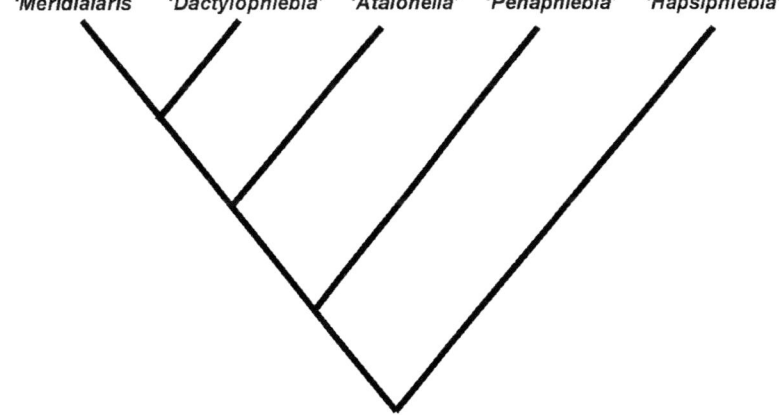

'Meridialaris'   'Dactylophlebia'   'Atalonella'   'Penaphlebia'   'Hapsiphlebia'

Figure 52. Proposed relationships for South American and related Southern Hemisphere fauna (reproduced from Pescador and Peters 1980a with additions from Pescador and Peters 1980b).

Although other Southern Hemisphere fauna were not included as part of the study, cursory examination of the Leptophlebiid fauna from Australia, New Zealand, New Caledonia, Madagascar and Africa inferred affinities with some of these lineages. Thus, with respect to the Australian genera, it was determined that the genera *Jappa, Ulmerophlebia, Atalomicria* and *Atalophlebia* belonged to the *'Hapsiphlebia'* lineage, *Nousia* belonged to the *'Atalonella'* lineage and *Austrophlebioides* belonged to the *'Meridialaris'* lineage. Based on this, it was hypothesized that the fauna of South Americas more closely related to Australia rather than New Zealand (Pescador and Peters 1980a). This is exemplified, in part, by the genus *Nousia* that Australia and New Zealand share (Hubbard 1990). Examination of the phylogenetic relationships of the entire Ephemeropteran Gondwanan fauna may lend further support to this.

Since the publication of Pescador and Peters (1980a), authors of newly erected or redescribed Australian genera have assigned placement of these taxa into the various lineages based on morphological similarities. Australian genera currently thought to belong to each of the various lineages are outlined in Table 1.

Table 1. Australian genera currently thought to belong to the five lineages outlined by Pescador and Peters (1980a). References are given for those genera added since the publication of the original paper.

Lineage	Genus	Reference
1. *Hapsiphlebia*	*Atalophlebia*	
	Atalomicria	
	Jappa	
	Kalbaybaria	Campbell 1993
	Ulmerophlebia	
2. *Penaphlebia*	*Garinjuga*	Campbell and Suter 1988
	Bibulmena?	Dean 1987
3. *Atalonella*	*Atalonella = Nousia*	Pescador and Peters 1985
	Nyungara	Dean 1987
4. *Dactylophlebia*	No Australian representatives	
5. *Meridialaris*	*Austrophlebioides*	
	Kirrara	Campbell and Peters 1986
	Tillyardophlebia	Dean 1997

There is an inherent difficulty in comparing this cladistic analysis with the largely phyletic methods of Pescador and Peters (1980a) and where taxa in each successive lineage were compared only with taxa in the lineages above. For comparison on an equal footing, therefore, data in the original paper was used to perform a cladistic based analysis. The lineages were used as terminal taxa (n= 6), characters (n=23) were sequentially numbered directly from Table 1 and character polarities remained unchanged. The resultant matrix (Appendix D) was used to perform an analysis using the method outlined above. Only one tree was produced (Fig. 53) (tree length of 23, ci = 1.0, ri = 1.0).

53

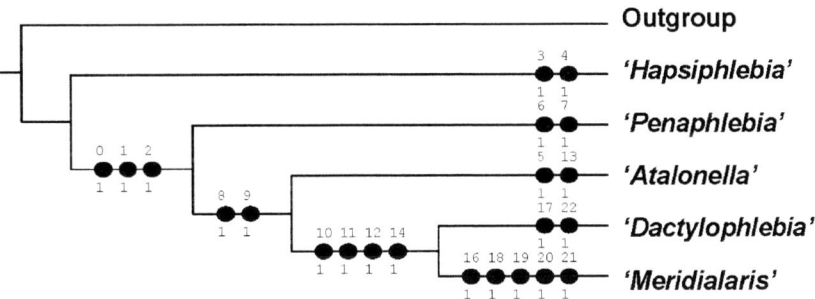

Figure 53. Cladistic based analysis of Pescador and Peters (1980a) data.

With this as a basis of comparison, there is only partial agreement between the relationships of the Australian taxa and those determined for Southern Hemisphere fauna as analyzed by Pescador and Peters (1980a). Of the four lineages considered to have representatives in Australia, only two are well supported in the current analysis; that of *'Meridialaris'* and *'Atalonella'*. The *'Hapsiphlebia'* taxa in the current analysis are basal to the rest of the ingroup as in the Pescador and Peters (1980a). There is also a question mark regarding the monophyly of the *'Penaphlebia'* lineage in Australia as *Garinjuga* is the only Australian representative.

The *'Meridialaris'* lineage is considered to include *Austrophlebioides* (Pescador and Peters 1980a), *Tillyardophlebia* (Dean 1987) and *Kirrara* (Campbell and Peters 1986) and there is strong support for such a grouping in this study. Of the six synapomorphies defining the lineage in the present study, four mirrored those of Pescador and Peters (1980a). The remaining two (lateral margins of the labrum angular and lateral margins of the clypeus divergent) identified a clade containing members of both the *'Meridialaris'* and *'Dactyophlebia'* lineages, but as there are apparently no members of the Australian *'Dactyophlebia'* fauna, these two characters states became synapomorphies for Australian members of *'Meridialaris'*. Therefore in an inclusive Gondwanan phylogeny, these characters could not be used to distinguish the *'Meridialaris'* lineage as currently defined here. Pescador and Peters (1980a) also determined a greater number of subapical pectinate setae on the maxillae galea-lacinae as synapomorphic for the lineage. In the present study, however, the number of setae was somewhat subjective, varied greatly between genera and was found to be homoplasic.

The Christidis (2001) phylogeny can not be directly compared.. Her study aimed to test the monophyly only of the *'Meridialaris'* taxa in Australia (*Austrophlebioides, Tillyardophlebia* and *Kirrara*). To this end, she studied members of the Australian *'Meridialaris'* fauna at a specific level and included many undescribed species ('Northern Kirrara', 'WT species 1 & 2', 'Paluma', 'Henrietta' and 'Daintree') whose generic affinities were postulated. Note: two of these morphospecies have been recently assigned to *Austrophlebioides* (Christidis and Dean 2005). Other Australian genera were included at the generic level to investigate higher level relationships. Support was found for a monophyletic clade of the Australian representatives of the *'Meridialaris'* lineage based on the absence of hairs on the lateral margins of the submentum and the ninth sternum of the female entire or only slightly cleft. An entire ninth sternum of the female imago serves as a synapomorphy for the Australian *'Meridialaris'* tricotomy, although it is homoplasic in a cladistic analysis of the phylogeny of Pescador and Peters (1980a).

According to the cladistic analysis of the Pescador and Peters data, the *'Atalonella'* lineage is defined by two synapomorphies: the labrum wider than the clypeus and the antero-median emargination of the labrum with broad subapical denticles. The latter character was found to be ambiguous, especially with respect to the definition of 'broad', and as it could not be objectively defined, so was not included in the phylogenetic analysis. There was some question of the affinities of the genus *Neboissophlebia*, thought to be related to the *'Hapsiphlebia'* genera, Jappa and *Ulmerophlebia, and* the *'Atalonella'* fauna (Dean 1988). We conclude that *Neboissophlebia* is more closely allied with the *'Atalonella'* fauna.

Christidis (1991) also found support for the *'Atalonella'* lineage where *Nyungara, Nousia (Australonousia)* and *Thraulophlebia* (as *Koorrnonga*) formed a monophyletic group, although there are no synapomorphies to support this. Notably, this did not include *Neboissophlebia* whose relationship was unresolved. Neither Pescador and Peters (1980a) nor Christidis (1991) included *Thraulus* in their analyses.

The cladistic analysis of the Pescador and Peters data defines the *'Penaphlebia'* lineage according to the presence of pectinate setae on the inner margin of the second maxillary palp and prominent spines/tassel-like setae on the posterior margins of the abdominal terga. Both conditions are homoplasic in the present analysis. *Garinjuga* is purported to belong to the *'Penaphlebia'* lineage based on the following character states (Campbell and Suter 1988): rounded labrum lateral margins, smoothly curved outer margin of mandibles, the relatively smooth right mandible inner margin of the outer incisor, inner margin of labial palp segment 3 with peg-like spines and inner margin of maxillary palp segment 2 with pectinate spines. Again for all characters included in the present analysis, no synapomorphies were found to support these hypotheses. It has been argued that *Bibulmena* is intermediate between the *'Hapsiphlebia'* and *'Atalonella'* lineages sharing many characters of both and therefore belongs to *'Penaphlebia'* lineage (Dean 1987), although this is not supported in the present analysis.

Loamaggalangta is considered close to *Bibulmena* (Dean et al. 1999), although it has not been formally 'allocated' a lineage *sensu* Pescador and Peters (1980a). The position of both genera remains unresolved in the present study. The evolutionary relationships of *Bibulmena, Loamaggalangta* (as Genus 'K') and *Garinjuga* among the Australian Leptophlebiidae are also unresolved in the Christidis (1991) phylogeny.

Conclusions

The inordinately large number of Leptophlebiid taxa worldwide (Hubbard 1990) has so far prevented an analysis of the phylogenetic relationships of the family as a whole, although assessments have been made at regional spatial scales. We followed this approach of analyzing regional fauna separately. Although there are still abundant unsolved phylogenies in the Australian Leptophlebiidae, parallel conclusions are reached between the results of this study and the findings of Pescador and Peters (1980a) and Christidis (1991) especially in relation to close relationships between *Austrophlebioides, Tillyardophlebia*, and *Kirrara* (the *'Meridialaris'* fauna); *Nousia, Nyungara* and *Thraulophlebia* (the *'Atalonella'* fauna), and *Jappa* and *Ulmerophlebia* (part of the *'Hapsiphlebia'* lineage). The clarification of some of the phylogenetic relationships between the Australian fauna, although preliminary, is valuable as a basis for inclusion of Australian Leptophlebiibae into a future phylogenetic analysis of the Gondwanan fauna and ultimately a phylogeny of the Leptophlebiidae worldwide.

Acknowledgments

We thank Richard Marchant and Peter Lillywhite (Museum of Victoria, Melbourne), Max Moulds (Australian Museum, Sydney), Ngaire Phillips (Queensland Department of Primary Industries, Brisbane), the Collections Manager (Australian National Insect Collection Canberra), David T. Goodger (Natural History Museum, London, U.K.), Kjell Arne Johanson (Naturhistoriska Riksmuseet, Stockholm, Sweden), Bill and Jan Peters (Florida Agricultural and Mechanical University, Florida, USA), Ian Campbell, Fred Govedich and Dennis O'Dowd (Monash University), John Dean (Victoria EPA), Phil Suter (LaTrobe University) and Nigel Ainsworth (DPI Victoria) for specimen loans, assistance in the field and useful comments. Part of this research was conducted while YJB is on sabbatical at Monash University (2000–2001) supported by the Korea Research Foundation Grant (KRF-2001-013-G00007).

Literature Cited

Bae, Y. J. 1995. *Ephemera separigata*, a new species of Ephemeridae (Insecta: Ephemeroptera) from Korea. The Korean Journal of Systematic Zoology **11**(2):159–166.

Bae, Y. J., and K. J. Finlay. 2003. A new species of an Australian burrowing mayfly (Leptophlebiidae, Ephemeroptera). Pan-Pacific Entomologist **79**(2):90–99.

Bae, Y. J., and W. P. McCafferty. 1991. Phylogenetic systematics of the Potamanthidae (Ephemeroptera). Transactions of the American Entomological Society **117**(3–4):1–143.

Bae, Y. J., and W. P McCafferty. 1995. Ephemeroptera tusks and their evolution. Pages 377–405 *in* Proceedings of the Seventh International Conference on Ephemeroptera - Current Directions in Research on Ephemeroptera. Orono, Maine, Canadian Scholars Press.

Bae, Y. J., K. J. Finlay, and I. C. Campbell. 2004. Taxonomic status of the Australian mayfly genera *Jappa* and *Ulmerophlebia* (Ephemeroptera: Leptophlebiidae). Entomological News **115**(1):1–10.

Bae, Y. J., W. P. McCafferty, and G. F. Edmunds Jr. 1990. *Stygifloris*, a new genus of mayflies (Ephemeroptera: Potamanthidae) from Southeast Asia. Annals of the Entomological Society of America **83**(5):887–891.

Balasubramanian, C., K. Venkataraman, and K. G. Sivaramakrishnan. 1991. Life stages of a south Indian burrowing mayfly, *Ephemera (Aethephemera) nadinae* McCafferty and Edmunds 1973 (Ephemeroptera: Ephemeridae). Aquatic Insects **13**(4):223–228.

Brundin, L. 1966. Transantarctic relationships and their significance, as evidenced by chironomid midges (with a monograph of the subfamily Posonominae and Aphroteniinae and the Austral Heptaginae). Kungliga Svenska Vetenskapsakademiens Handlingar **Series 4, 11**(1):1–472.

Burks, B. D. 1953. The mayflies, or Ephemeroptera, of Illinois. Bulletin of the Natural History Survey Division **26**(1):1–216.

Campbell, I. C. 1993. A new genus and species of Leptophlebiid mayfly (Ephemeroptera: Leptophlebiidae: Atalophlebiinae) from Tropical Australia. Aquatic Insects **15**(3):159–167.

Campbell, I. C., and W. L. Peters. 1986. Redefinition of *Kirrara* Harker with a redescription of *Kirrara procera* Harker (Ephemeroptera: Leptophlebiidae: Atalophlebiinae). Aquatic Insects **8**(2):71–81.

Campbell, I. C., and W. L. Peters. 1993. A revision of the Australian Ephemeroptera genus *Atalomicria* Harker (Leptophlebiidae: Atalophlebiidae). Aquatic Insects **15**(2):89–107.

Campbell, I. C., and P. J. Suter. 1988. Three new genera, a new subgenus and a new species of Leptophlebiidae (Ephemeroptera) from Australia. Journal of the Australian Entomological Society **27**:259–273.

Christidis, F. 2001. A cladistic analysis of *Austrophlebioides* and related genera (Leptophlebiidae: Atalophlebiinae). Pages 305–312 *in* Proceedings of the Ninth International Conference on Ephemeroptera and the Thirteenth International Symposium on Plecoptera - Trends in Research in Ephemeroptera and Plecoptera. Tucumán, Argentina, Kluwer Academic Publishers.

Christidis, F., and J. C. Dean. 2005. Three new species of *Austrophlebiodes* Campbell and Suter (Ephemeroptera: Leptophlebiidae: Atalophlebiinae) from the Wet Tropics bioregion of northeastern Victoria. Australian Journal of Entomology **44**:132–143.

Dean, J. C. 1987. Two new genera of Leptophlebiidae (Insecta:Ephemeroptera) from southwestern Australia. Memoirs of the Museum of Victoria **48**(2):91–100.

Dean, J. C. 1988. Description of a new genus of Leptophlebiid mayfly from Australia (Ephemeroptera: Leptophlebiidae: Atalophlebiinae). Proceedings of the Royal Society of Victoria **100**:39–45.

Dean, J. C. 1999. Preliminary keys for the identification of Australian mayfly nymphs of the family Leptophlebiidae. Albury, Co-operative Research Centre for Freshwater Ecology and Murray Darling Freshwater Research Centre.

Dean, J. C. 2000. Descriptions of new Leptophlebiidae (Insecta: Ephemeroptera) from Australia. II. *Kaninga*, a new monotypic genus from southwestern Australia. Records of the Western Australian Museum **20**:87–94.

Dean, J. C., G. N .R. Forteath, and A. W. Osborn. 1999. *Loamaggalangta pedderensis* gen. & sp. nov.: A new mayfly from Tasmania (Ephemeroptera: Leptophlebiidae: Ataloplebiinae). Australian Journal of Entomology **38**:72–76.

Dean, J. C., and P. J. Suter. 1996. Mayfly nymphs of Australia. Albury, Co-operative Research Centre for Freshwater Ecology and Murray Darling Freshwater Research Centre.

Demoulin, G. 1952a. Contribution a l'étude des Ephoronidae Euthypolciinae (Insectes Éphéméroptères). Bulletin d'Institut Royal des Sciences Naturelles de Belgique **28**(45):1–22.

Demoulin, G. 1952b. Sur deux Palingeniidae (Insectes Éphéméroptères) mal connus. Bulletin d'Institut Royal des Sciences Naturelles de Belgique **28**(33):1–11.

Demoulin, G. 1955a. Les Brachycercidae Australiens. Le genre *Tasmanocoenis* Lestage. Bulletin de l'Institut Royal des Sciences naturelles de Belgique **31**(10):1–7.

Demoulin, G. 1955b. Éphéméroptères Nouveaux ou Rares du Chili. Bulletin de l'Institut Royal des Sciences naturelles de Belgique **73**:1–30.

Demoulin, G. 1966. Contribution a l'étude des Euthyplociidae (Ephemeroptera) IV un nouveau genre de Madagascar. Annals Société Entomologie de France **11**(4):941–949.

Domínguez, E. 1995. Cladistic analysis of the *Ulmeritus-Ulmeritoides* group (Ephemeroptera, Leptophlebiidae) with descriptions of five new species of *Ulmeritoides*. Journal of New York Entomological Society **103**(1):15–38.

Eaton, A. E. 1868–69. An outline of a re-arrangement of the genera of Ephemeridae. The Entomologist's Monthly Magazine **5**:82–95.

Eaton, A. E. 1899. An annotated list of the Ephmeridae of New Zealand. Transactions of the Entomological Society of London Part III:285–293, 1 pl.

Edmunds, G. F. Jr. 1959. Subgeneric groups within the mayfly genus Ephemerella (Ephemeroptera: Ephemerellidae). Annals of the Entomological Society of America **52**:543–547.

Edmunds, G. F. Jr. 1972. Biogeography and evolution of Ephemeroptera. Annual Review of Entomology **17**:21–42.

Edmunds, G. F. Jr., and R. K. Allen. 1966. The significance of the nymphal stages in the study of Ephemeroptera. Annals of the Entomological Society of America **59**(2):300–303.

Edmunds, G. F. Jr., and R. D. Waltz. 1996. Ephemeroptera. Pages 126–163 *in* R. W. Merritt, and K. W. Cummins (editors). An Introduction to the Aquatic Insects of North America. Kendall Hunt Publishing, Dubuque, Iowa.

Elouard, J.-M. 1986a. Éphémères d'Afrique de l'Ouest: le genre *Afromera* (Éphéméridae). Revue d'Hydrobiologie tropicale **19**(3–4):169–176.

Elouard, J.-M. 1986b. Éphémères d'Afrique de l'Ouest: le genre *Eatonica* (Éphéméridae). Revue d'Hydrobiologie tropicale **19**(2):87–92.

Elouard, J.-M., R. Oliarinony, and M. Sartori. 1998. Biodiversité aquatique de Madagascar. 9. Le genre *Eatonica* Navàs (Ephemeroptera, Ephemeridae). Bulletin de la Société Entomologique Suisse **71**:1–9.

Elpers, C., and I. Tomka. 1994. Mouthparts of the predaceous larvae of the Behningiidae (Insecta: Ephemeroptera). Archiv für Hydrobiologie / Supplementum 99 **4**:381–413.

Finlay, K. J. 2002 .Taxonomy, distribution patterns and phylogeny of the Australian Leptophlebiidae (Ephemeroptera), Department of Biological Sciences, Monash University, Melbourne, 2 vols.

Gillies, M. T. 1951. Further notes on Ephemeroptera from India and southeast Asia. Proceedings of the Royal Entomological Society of London *(B)* **20**(11–12):121–130.

Gillies, M. T. 1980. The African Euthyplociidae (Ephemeroptera) (Exeuthyplociinae subfam.n.). Aquatic Insects **1**(4):217–224.

Goloboff, P. A. 1993. NONA. Tucumán, Argentina.

Grant, P. 1985. Systematic revision of the *Thraulus* group genera (Ephemeroptera: Leptophlebiidae: Atalophlebiinae) from the Eastern Hemisphere. Department of Biological Sciences. Tallahasee, Florida, Florida State University - College of Arts and Sciences:326pp.

Horridge, G. A., L. Marcelja, and R. Jahnke. 1982. Light guides in the dorsal eye of the male mayfly. Proceedings of the Royal Society of London B **216**:137–150.

Hubbard, M. D. 1982. Two new species of *Ephemera* from South India (Ephemeroptera: Ephemeridae). Pacific Insects **24**(2):192–195.

Hubbard, M. D. 1983. Ephemeroptera of Sri Lanka: Ephemeridae. Systematic Entomology **8**(4):383–392.

Hubbard, M. D. 1984. A revision of the genus *Povilla* (Ephemeroptera: Polymitarcyidae). Aquatic Insects **6**(1):17–35.

Hubbard, M. D. 1990. Mayflies of the World - A Catalog of the Family and Genus group Taxa (Insecta: Ephemeroptera). Flora and Fauna Handbook No. 8. Gainsville, Sandhill Crane Press Inc.

Ishiwata, S.-I. 1996. A study of the genus *Ephoron* from Japan (Ephemeroptera: Polymitarcyidae). The Canadian Entomologist **128**:551–572.

Kang, S.-C., and C.-T. Yang. 1994. Ephemeroidea of Taiwan (Ephemeroptera). Chinese Journal of Entomology **14**:391–399.

Keltner, J., and W. P. McCafferty. 1986. Functional morphology of burrowing in the mayflies *Hexagenia limbata* and *Pentagenia vittigera*. Zoological Journal of the Linnean Society **87**:139–162.

Landa, V., and T. Soldán. 1985. Phylogeny and higher classification of the order Ephemeroptera: a discussion from the comparative anatomical point of view. Praha, Publishing House of the Czeckoslovak Academy of Sciences.

Lestage, J. A. 1918. Les Ephémères d'Afrique (Notes critiques sur les espèces connues). Revue Zoologique Africaine **6**:65–114.

Lestage, J. A. 1939. Contribution a l'etude des Ephéméroptères XXIII Les Polymitarcidae de la faune africaine et description d'un genre nouveau du Natal. Bulletin et Annales de la Société Royale Entomologique de Belgique **79**:135–138.

Lugo-Ortiz, C. R., and W. P. McCafferty. 1996. Central American *Tortopus* (Ephemeroptera: Polymitarcyidae): a unique new species and new country records. Entomological News **107**(1):23–27.

Marchant, R. 1982. The macroinvertebrates of Magela Creek, Northern Territory, Supervising Scientist for the Alligator Rivers Region, 40 pp.

McCafferty, W. P. 1971a. New genus of mayflies from Eastern North America (Ephemeroptera: Ephemeridae). New York Entomological Society **79**:45–51.

McCafferty, W. P. 1971b. New burrowing mayflies from Africa (Ephemeroptera: Ephemeridae). Journal of the Entomological Society of South Africa **34**(1):57.

McCafferty, W. P. 1972. Pentageniidae: a new family of Ephemeroidea (Ephemeroptera). Journal of the Georgia Entomological Society **7**(1):51–56.

McCafferty, W. P. 1973. Subgeneric classification of *Ephemera* (Ephemeroptera: Ephemeridae). The Pan-Pacific Entomologist **49**:300–307.

McCafferty, W. P. 1975. The burrowing mayflies (Ephemeroptera: Ephemeroidea) of the United States. Transactions of the American Entomological Society **101**:447–504.

McCafferty, W. P. 1991. Towards a phylogenetic classification of the Ephemeroptera (Insecta): a commentary on systematics. Annals of the Entomological Society of America **84**(4):343–360.

McCafferty, W. P. 2002. http://www.mayflycentral.com. Higher classification of the mayflies of North America, Department of Entomology, Purdue University, West Lafayette, Indiana.

McCafferty, W. P., and Y. J. Bae. 1990. *Anthopotamus*, a new genus for North American species previously known as Potamanthus (Ephemeroptera: Potamanthidae). Entomological News **101**(4):200–202.

McCafferty, W. P., and D. W. Bloodgood. 1989. The female and male coupling apparatus in *Tortopus* mayflies. Aquatic Insects **11**(3):141–146.

McCafferty, W. P., and G. F. Edmunds, Jr. 1976. The larvae of the Madagascar genus *Cheirogenesia* Demoulin (Ephemeroptera: Palingeniidae). Systematic Entomology **1**:189–194.

McCafferty, W. P., and M. T. Gillies. 1979. The African Ephemeridae (Ephemeroptera). Aquatic Insects **3**:169–178.

Needham, J. G., and H. E.Murphy. 1924. Neotropical Mayflies. Bulletin of the Lloyd Library of Botany, Pharmacy and Materia Medica - Entomological Series **4**:1–79, 13 pl.

Needham, J. G., J. R. Traver, and Y.-C. Hsu. 1935. The Biology of Mayflies. New York, Comstock.

Nixon, K. C. 1999. Winclada (BETA) ver. 0.9.9 Published by the author, Ithaca, New York.

Norton, I. O., and J. G. Sclater. 1979. A model for the evolution of the Indian Ocean and the breakup of Gondwanaland. Journal of Geophysical Research **84**:6803–6830.

Parnrong, S., and I. C. Campbell. 1997. Two new species of *Austrophlebioides* Campbell and Suter (Ephemeroptera: Leptophlebiidae) from Australia, with notes on the genus. Australian Journal of Entomology **37**:121–127.

Pereira, S. M., and E. R. Da Silva. 1990. Noca espécie de *Campylocia* Needham & Murphy, 1924 com notas biológicas (Ephemeroptera: Euthyplociidae). Boletim do Museu Nacional Nova Série Rio de Janeiro (Brasil) **336**:1–12.

Pescador, M. L., and W. L. Peters. 1980a. Phylogenetic relationships and zoogeography of cool-adapted Leptophlebiidae (Ephemeroptera) in southern South America. Pages 43–56 *in* Proceedings of the Third International Conference on Ephemeroptera - Advances in Ephemeroptera Biology. Winnipeg, Canada, Plenum Press, New York.

Pescador, M. L., and W. L. Peters. 1980b. Two new genera of cool-adapted Leptophlebiidae (Ephemeroptera) from southern South America. Annals of the Entomological Society of America **73**(3):332–338.

Pescador, M. L., and W. L. Peters. 1985. Biosystematics of the genus *Nousia* from southern South America (Ephemeroptera: Leptophlebiidae, Atalophlebiinae). Journal of the Kansas Entomological Society **58**(1):91–123.

Pescador, M. L., and W. L. Peters. 1987. Revision of the genera *Meridialaris* and *Massartellopsis*. Transactions of the American Entomological Society **112**:147–189.

Pescador, M. L., and W. L. Peters. 1991 Biosystematics of the genus *Penaphlebia* (Ephemeroptera: Leptophlebiidae: Atalophlebiinae) from South America. Transactions of the American Entomological Society 117(1):1–38.

Peters, W. L. 1980. Phylogeny of the Leptophlebiidae: an introduction. *In* Proceedings of the Third International Conference on Ephemeroptera - Advances in Ephemeroptera Biology. Winnipeg, Canada, Plenum Press, New York.

Peters, W. L. 1988. Origins of the North American Ephemeroptera fauna, especially the Leptophlebiidae. Memoirs of the Entomological Society of Canada 144:13–24.

Peters, W. L., and G. F. Edmunds, Jr. 1964. A revision of the generic classification of the Ethiopian Leptophlebiidae. Transactions of the Royal Entomological Society of London 116(10):225–253.

Peters, W. L., and G. F. Edmunds, Jr. 1970. Revision of the generic classification of the Eastern Hemisphere Leptophlebiidae. Pacific Insects 12(1):157–240.

Peters, W. L., and G. F. Edmunds, Jr. 1984. A redescription and phylogenetic relationships of *Nesophlebia* (Ephmeroptera, Leptophlebiidae, Atalophlebiinae). Pages 27–35 *in* Proceedings of the Fourth International Conference on Ephemeroptera, Bechyne, Czechoslovakia, Czechoslavak Academy of Science.

Peters, W. L., and J. G. Peters. 1979. The Leptophlebiidae of New Caledonia (Ephemeroptera). Part 2 - Systematics. Cahiers ORSTOM, Série Hydrologie XIII(1–2):61–82.

Peters, W. L., and J. G. Peters. 1981a. The Leptophlebiidae: Atalophlebiinae of New Caledonia (Ephemeroptera). Part 3 - Systematics. Revue d'Hydrobiologie tropicale 14(3):233–243.

Peters, W. L., and J. G. Peters. 1981b. The Leptophlebiidae: Atalophlebiinae of New Caledonia (Ephemeroptera). Part 4 - Systematics. Revue d'Hydrobiologie tropicale 14(3):245–252.

Peters, W. L., and M. T. Gillies. 1995. Square facets in a hexagonal world. Pages 371–375 *in* L. D. Corkum, and J. J. H Ciborowski, editors. Current Directions in Research on Ephemeroptera. Canadian Scholars Press, Toronto.

Peters, W. L., J. G. Peters, and G. F. Edmunds Jr. 1978. The Leptophlebiidae of New Caledonia (Ephemeroptera). Part 1. Introduction and Systematics. Cahiers ORSTOM, Série Hydrologie 12(2):97–117.

Peters, W. L., J. G. Peters, and G. F. Edmunds Jr. 1990. The Leptophlebiidae: Atalophlebiinae of New Caledonia (Ephemeroptera). Part 5 - Systematics. Revue d'Hydrobiologie tropicale 23(2):121–140.

Peters, W. L., J. G. Peters, and G. F. Edmunds Jr. 1994. The Leptophlebiidae: Atalophlebiinae of New Caledonia (Ephemeroptera). Part 6 - Systematics. Revue d' Hydrobiologie tropicale 27(2):97–105.

Riek, E. F. 1970. Ephemeroptera. Insects of Australia. Melbourne University Press, Melbourne.

Riek, E. F. 1973. The classification of Ephemeroptera. Pages 160–178 *in* W. L. Peters, and J. G. Peters, editors. Proceedings of the First International Conference on Ephemeroptera, Florida, E. J. Brill., Leiden.

Sartori, M. 1992. Mayflies from Israel (Insecta: Ephemeroptera) I. Heptageniidae, Ephemerellidae, Leptophlebiidae and Palingeniidae. Revue Suisse de Zoologie **99**(4):835–858.

Sartori, M., and J.-M. Elouard. 1999. Biodiversité aquatique de Madagascar 30: le genre *Cheirogenesia* Demoulin 1952 (Ephemeroptera: Palingeniidae). Revue Suisse de Zoologie **106**(2):325–337.

Savage, H. M., and W. L. Peters. 1983. Systematics of *Miroculis* and related genera from Northern South America (Ephemeroptera: Leptophlebiidae). Transactions of the American Entomological Society **108**:491–600.

Skedros, D. G., and D. A. Polhemus. 1986. Two new species of *Jappa* from Australia (Ephemeroptera: Leptophlebiidae). Pan-Pacific Entomologist **62**(4):311–315.

Soldán, T., and M. Putz. 2000. The larva of *Rhoenanthus distafurcus* Bae et McCafferty (Ephemeroptera: Potamanthidae) with notes on distribution and biology. *Aquatic Insects* **22**(1):9–17.

Spieth, H. T. 1933. The phylogeny of some mayfly genera. Journal of the New York Entomological Society **41**(1,2,3):55–86, 327–391.

Spieth, H. T. 1941. Taxonomic studies on the Ephemeroptera. II. The genus *Hexagenia*. The American Midland Naturalist **26**(2):233–280.

Suter, P. J. 1986. The Ephemeroptera (Mayflies) of South Australia. Records of the South Australian Museum **19**(17):339–397.

Suter, P. J. 1992. Taxonomic key to the Ephemeroptera (Mayflies) of the Alligator Rivers Region, Northern Territory.

Tillyard, R. J. 1933. The trout-food insects of Tasmania. Part 1 - A study of the genotype of the mayfly genus *Atalophlebia* and its life history. *Papers and Proceedings of the Royal Society of Tasmania: 1–16, pl.1–2.*

Towns, D. R., and W. L. Peters. 1979. Phylogenetic relationships of the Leptophlebiidae of New Zealand. Pages 56–69 *in* J. F. Flannagan, and K. E. Marshall, editors. Proceedings of the Third International Conference on Ephemeroptera - Advances in Ephemeroptera Biology. Winnipeg, Canada, Plenum Press, New York.

Towns, D. R., and W. L. Peters. 1996. Fauna of New Zealand - Ko te Aitanga Pepeke o Aotearoa, Leptophlebiidae (Insecta: Ephemeroptera). Canterbury, Manaaki Wheena Press.

Tsui, P. T. P., and W. L. Peters. 1975. The comparative morphology and phylogeny of certain Gondwanian Leptophlebiidae based on the thorax, tentorium, and abdominal terga (Ephemeroptera). Transactions of the American Entomological Society **101**:505–595.

Uéno, M. 1928. Some Japanese mayfly nymphs. Memoirs of the College of Science, Kyoto Imperial University (B) **4**(no. 1, article 2.):1–63.

Vuori, K.-M. 1999. *Potamanthus luteus* L. (Ephemeroptera, Ephemeridae) found for the first time in Finland: notes on the morphology and habitats of the nymphs. Entomologica Fennica **10**:171–174.

Watrous, L. E., and Q. D. Wheeler. 1981. The out-group comparison method of character analysis. Systematic Zoologist. **30**(1):1–11.

Wu, X.-Y., and D.-S. You. 1986. A new genus and species of Potamanthidae from China (Ephemeroptera). Acta Zootaxonomica Sinica **11**(4):405–408.

You, D.-S. 1984. A revision of genus *Potamanthodes* with a description of two new species (Ephemeroptera: Potamanthidae). Pages 101–107 *in* V. Landa, T. Soldán, and M. Tonner, editors. Proceedings of the Fourth International Conference on Ephemeroptera, Bechyne, Czechoslovakia, Czechoslavak Academy of Science.

You, D.-S., and C.-R. Su. 1987. Descriptions of the nymphs of *Choropterps nanjingensis, Potamanthodes fujianensis* and *Isonychia kiangsiensis* (Ephemeroptera: Leptophlebiidae, Potamanthidae, Siphlonuridae). Acta Zootaxonomica Sinica **12**(3):332–336.

Appendix A

Data matrix of phylogenetic character states for the Australian genera of Leptophlebiidae.

| | 11111 | 11111 | 22222 | 22222 | 3333 | | |
	01234	56789	01234	56789	01234	56789	0123
Outgroup	00000	00000	00000	00000	00000	00000	0000
Atalomicria	10000	00001	10000	01001	00010	11000	1100
Atalophlebia	10001	00001	10000	01000	00010	11000	0100
Austrophlebioides	10011	22011	10220	10210	11111	01100	0001
Bibulumena	10000	11001	10110	10110	01010	01000	1000
Garinjuga	10010	11001	10110	10110	11011	01000	1000
New Genus A	10000	11001	10110	10110	11011	01100	1100
Jappa	11100	00001	21000	01000	00010	01000	1100
Kalbaybaria	10000	00000	00001	011-0	00010	01100	1000
Kaninga	10010	11001	10110	10110	01010	01100	1000
Kirrara	10001	22100	00220	-0210	11111	01000	1001
Loamaggalangta	10000	11001	10110	10110	01010	01100	1000
New Genus B	10000	11001	10110	10110	11011	01100	1100
Neboissophlebia	10000	11001	10110	10010	01010	01111	1100
Nousia (Nousia)	10000	11001	10110	10100	11011	01111	1100
Nousia (Australonousia)	10000	11001	10110	10100	11011	01111	1100
Nyungara	10000	11000	10110	10100	11011	01110	2110
Thraulophlebia	10000	11001	10110	10110	11011	01111	1100
Thraulus	10000	11001	10110	10110	11011	0111-	1111
Tillyardophlebia	10011	22011	10220	10210	11111	01100	1001
Ulmerophlebia	11100	00001	20000	01000	00010	01000	1100

Appendix B

Morphological characters and character states used in the cladistic analysis of Australian Leptophlebiidae.

P= plesiomorphic, A = apomorphic.

Larval Characters

Body
- 0. Mouthparts: 0, hypognathous (P); 1, prognathous (A).
- 1. Body and gills: fringed with fine setae: 0, no (P); 1, yes (A).
- 2. Setation or spination on lateral margins of abdominal terga: 0, bare or minute setae (P); 1 prominent spines (A).
- 3. Spination on posterior margins of abdominal terga: 0, none or small (P); 1, prominent (A).
- 4. Posterolateral spines on abdomen: 0, on segments 4, 5, 6, 7 or 8 to 9 (P); 1, on segments 8 to 9 (A).

Head and mouthparts
- 5. Labrum width: 0, narrower than clypeus (P); 1, subequal to slightly wider than clypeus (A); 2, significantly wider than clypeus (A).
- 6. Labrum lateral margins: 0, parallel (P); 1, rounded (A) ; 2 angular (A
- 7. Labrum cross section: 0, oval (P); 1, parallel (A).
- 8. Labrum 'hood': 0, absent (P); 1, present (A).
- 9. Labrum secondary hair fringe: 0, absent (P); 1, present (A).
- 10. Labrum denticles on anteromedian emargination: 0, absent (P); 1, present (A).
- 11. Frontal horns: 0, absent (P); 1, present (A).
- 12. clypeus lateral margins: 0, converging towards anterior (P); 1, parallel (A); 2, diverging towards anterior (A).
- 13. Mandible outer margins: 0, relatively straight (P); 1, smoothly curved (A); 2, angular (A).
- 14. Mandible enlarged process: 0, absent (P); 1, present (A).
- 15. Mandible prosthecal tuft: 0, wide (P); 1, flap-like, setae on apical and lateral margins (A); 2, spine-like; hairs on lateral margins only (A).
- 16. Right mandible, outer incisor, inner margin: 0, smooth or spinose (P); 1, denticulate (A).
- 17. Maxillae galea-lacinae width-length ratio (W/L ratio): 0, narrow at apex, W/L <1 (P); 1, relatively broad at apex, W/L approx <1 (A); 2, very broad at apex: W/L > 1 (A).
- 18. Average number of pectinate setae on maxillae: 0, less than or equal to 15 (P); 1, > 17 (A).

19. Maxillary palp morphology: 0, extending just beyond galea-lacinae (P); 1, greatly elongate (A).
20. Labium glossae: 0, turned over ventrally (P); 1, straight, upright (A).
21. Spines on inner margins of labial terminal palps: 0, prominent (P); 1, minor or absent (A). 22. Labium submentum lateral setae: 0, present (P); 1, absent (A).
22. Hypopharynx lateral process: 0, absent (P); 1, present (A).
23. Tarsal claws dentition: 0, smooth or with minor processes (P); 1, denticulate (A).

Imago Characters

Head

24. Size of male imago dorsal eye lobes. ES=B/D (Bae 1991): 0, eye size large , ≤ 0.2 or contiguous (P); 1, eye size medium, ≥ 0.3 (A).
25. Male dorsal eye facet shape: 0, hexagonal (P); 1, square (A).

Wings

26. Position of forewing veins ICu1 and ICu2 approaching wing margin: 0, strongly diverging (P); 1, parallel to slightly diverging (A).
27. Average number of crossveins in the costal space: 0, approx 20 or more (P); 1, approx 10-15 (A).
28. Costal crossveins present in proximal half of male forewings: 0, yes (P); 1, reduced number or absent (A).
29. Hindwing length relative to forewing: 0, hindwing large -approx 1/3th the length of forewing (P); 1, hindwing smaller - approx 1/4 to 1/5th the length of forewing (A); 2, hindwing very small -approx 1/10th the length of forewing (A).
30. Length of hindwing subcostal vein relative to hindwing: 0, more than or equal to 0.9 (P); 2, less than or equal to 0.85 (A).
31. Hindwing midlength costal projection: 0, absent (P); 1, present (A).

Abdomen

32. Sternum nine of female: 0, cleft (P); 1, entire (A).

Appendix C

Material examined:

LEPTOPHLEBIIDAE: ATALOPHLEBIINAE
Atalomicria Harker
Atalomicria banjdjalama Campbell and Peters: holotype, male imago; paratypes, female imago, male and female nymphs; slide material; Booloumba Creek, Connondale Ranges, QLD, 29 xi 89, coll. I.C. Campbell (MV);

Atalomicria bifasciata Campbell and Peters: holotype, male imago; paratypes, female imago, subimagos, nymphs; slide material; Booloumba Creek, Connondale Ranges, QLD, 5 xii 89, coll. I.C. Campbell (MV);

Atalomicria dalagara Campbell and Peters: holotype, male imago; paratypes, nymphs; slide material; Booloumba Creek, Connondale Ranges, QLD, 29 xi 89, coll. I.C. Campbell (MV).

Atalomicria chessmani Campbell and Peters: holotype, male imago; paratypes, male and female imagos, nymphs; slide material; Ferntree Creek, 10km S of Goongerah, VIC, 18 i 1990, coll. I.C.Campbell (MV);

Atalomicria sexfasciata (Ulmer): holotype, male imago; paratype, female imago; Cedar Creek, QLD, Dr Mjöbergs Swedish Expedition to Australia 1910-1913 (SMNH); slide material; Yuccabine Creek, Kirrama Rd, QLD, 27 ix 87, coll. I.C. Campbell (MV);

Atalomicria uncinata (Ulmer): holotype, male imago; Cedar Creek, QLD, Dr Mjöbergs Swedish Expedition to Australia 1910-1913 (SMNH);

Atalomicria sp. 'AV1' (Dean 1999): nymphs and imagos, NSW, QLD, VIC (MV); nymphs, NSW, VIC (Finlay collection); slide material, QLD (Campbell collection).

Atalophlebia Eaton
Atalophlebia albiterminata Tillyard: nymphs and imagos, reared, VIC (Finlay collection);

Atalophlebia australis (Walker): nymphs and imagos, reared, VIC, NSW, NT, TAS (MV) Finlay collection, Campbell collection); slide material, VIC (MV);

Atalophlebia sp. 'AV2' , 'AV4' , 'AV5' , 'AV7', 'AV9', 'AV13', 'AV15' ,'AV21' (Dean 1999) nymphs and imagos, reared, VIC, TAS, NSW (Finlay collection);

Atalophlebia spp.: nymphs and imagos, reared, VIC, NSW, TAS (Finlay collection).

Austrophlebioides Campbell and Suter
Austrophlebioides booloumbi Parnrong and Campbell: holotype, male imago & paratypes, male and female nymphs; Booloumba Creek, Connondale Ranges, QLD, 5 xii 89, coll. I.C. Campbell (MV);

Austrophlebioides marchanti Parnrong and Campbell: holotype, male imago & paratypes, female imago, female and male nymphs; Loch River, Noojee, VIC, i

1995, coll. S. Parnrong (MV); nymphs, imagos, slide material, NSW, VIC (Finlay collection);

Austrophlebioides pusillus (Harker): nymphs, imagos, NSW, VIC (MV); nymphs, imagos, slide material, NSW, VIC (Finlay collection); slide material (Campbell collection);

Austrophlebioides sp. 'AV2' (Dean 1999): nymphs, imagos, NSW, VIC (Finlay collection);

Austrophlebioides spp: nymphs, imagos, VIC (Finlay collection).

Bibulmena Dean

Bibulmena kadjina Dean: holotype, male imago, reared, North Dandalup River, North Dandalup, WA, 01 vi 1984, coll. S. Bunn (MV); paratypes, male and female imago, slide material, Foster Brook, North Dandalup, 1983, coll. S. Bunn (MV); paratype, nymph, slide material, Wungong Brook, Jarrahdale, WA, 2 xii 1981, coll. S. Bunn (MV).

Garinjuga Campbell and Suter

Garinjuga maryannae Campbell and Suter: holotype, male imago & paratypes, subimagos, female and male nymphs; Thredbo River, Summit Rd at entrance to Kosciusko National Park, NSW, 19 x 1995, coll. I.C. Campbell (ANIC); nymphs, imagos, slide material, NSW, VIC (Finlay collection); nymphs, imagos, slide material, NSW (Campbell collection);

Garinjuga sp. 'AV1' (Dean 1999): nymphs, imagos, NSW, TAS, VIC (MV); nymphs, imagos, slide material, NSW, VIC (Finlay collection);

Garinjuga spp.: nymphs, imagos, slide material, NSW, VIC (Finlay collection).

Jappa Harker

Jappa edmundsi Skedros and Polhemus: holotype and paratypes, nymphs, Hutchinson Creek, Cape Tribulation Rd, N of Daintree Landing, 17 viii 1983, coll. D.A. & T.J. Polhemus (ANIC); slide material, QLD (Campbell collection);

Jappa kutera Harker: nymphs, imagos, slide material, NSW, NT, QLD; nymphs, NSW, VIC (Finlay collection); slide material, VIC (Suter collection);

Jappa serrata Skedros and Polhemus: holotype, nymph, Hutchinson Creek, Cape Tribulation Rd, N of Daintree Landing, QLD, 17 viii 1983, coll. D.A. & T.J. Polhemus (ANIC); nymphs, imagos, slide material, QLD (Campbell collection);

Jappa campbelli Bae and Finlay: holotype, Licola, Wellington R., 3km upstream from Alpine National Pk entrance, VIC, 146° 37' E, 37° 34' S, 6 Jan 2002 (emerged 13 Jan 2002); paratypes, same collection data, 2 male (T-17940, 17941) and 2 female (T-17942, 17943) imagos (reared, with larval exuviae), 10 larvae, 6 male (T-17944-17949) and 4 female (T-17950-17953) larvae (MV), coll. KJ. Finlay, Y.J. Bae & N. Ainsworth (MV);

Jappa sp. 'AV3' (Dean 1999): nymphs, imagos, VIC (Finlay collection);

Jappa spp.: nymphs, imagos, VIC (Campbell collection); slide material, QLD (Campbell collection).

Kalbaybaria Campbell

Kalbaybaria doantrangae Campbell: holotype, male imago; paratypes, male and female imago, nymphs; slide material, Romeo Creek, near Helenvale, QLD, 5 v 1988, coll. I.C. Campbell (ANIC); slide material, Palmer River, QLD, 20 vi 1971, coll. E.F. Riek (ANIC); slide material, Millstream Falls, W of Ravenshoe, QLD, 25 vi 1971, coll. E.F. Riek (ANIC)

Kaninga **Dean**

Kaninga gwabbalitcha Dean: nymphs, slide material, WA (MV);
Genus Q sp. 'AV1' (Dean 1999): 'holotype', male imago; 'paratypes', male imagos, Carey Brook, Staircase Road, WA, 15 vii 1989 coll. I. Growns (MV); nymphs, subimagos, WA (MV).

Kirrara **Harker**

Kirrara procera Harker: nymphs, imagos, slide material, VIC, NSW (MV); nymphs, imagos, slide material, VIC (Finlay collection);
Kirrara sp. 'AV1' (Dean 1999): nymphs, slide material, QLD (Campbell collection).

Loamaggalangta **Dean, Forteath & Osborn**

Loamaggalangta pedderensis Dean, Forteath & Osborn: holotype, male imago; paratypes, male imagos, subimagos; Lake Pedder, Trappes Bay, TAS, ii 1997, coll. N. Forteath (ANIC); nymphs, slide material, TAS (MV);
Genus K sp. 'AV1'& 'AV2' (Dean 1999): nymphs, NSW, VIC, QLD (MV); nymphs, TAS (Finlay collection).

Neboissophlebia **Dean**

Neboissophlebia hamulata Dean: holotype, male imago; paratypes, male imago, female imagos, subimagos; slide material, Tarago River, 7km W Neerim, VIC, 1 iii 1972, coll. A. Neboiss (MV); nymphs, imagos, NSW, VIC (MV); nymphs, imagos, NSW, VIC (Finlay collection);
Neboissophlebia occidentalis Dean: paratypes, male imagos, female imagos, slide material, Harvey River, 15km E Harvey, WA, 21 xi 1978, coll. A. Neboiss (MV).

Nousia **Navás,** subgenus *Nousia* **Navás**

Nousia delicata Navás: male imagos, Santiago Province, CHILE, xi 1972, coll. M.L. Pescador & G. Barria; nymphs, Río Caren, Hacienda Illapel, Coquimbo Province, CHILE, 17 xi 1972, coll. M.L. Pescador (FAMU); nymphs, Río Caren, Hacienda Illapel, Coquimbo Province, CHILE, 18 xi 1972, coll. M.L. Pescador & G. Barria (FAMU);
Nousia grandis (Demoulin): nymphs; female imago reared, male subimago reared; El Coigual, Curico Province, CHILE, i 1964, coll. L. Peña (FAMU);
Nousia minor (Demoulin): male imago, Valdivia, CHILE, 19 x 1957, coll. J. Illies; nymphs, Río Piquiquen, El Manzano, 35km W Angol, Malleco Province, CHILE, xii 1972, coll. M.L. Pescador (FAMU).

Nousia **Navás,** subgenus *Australonousia* **Campbell and Suter**

Nousia (Australonousia) fusca (Ulmer) comb. nov.; *Atalophlebia fusca* Ulmer: holotype, Cedar Creek, QLD, III 1910-1913, male imago, paralectotypes, same

collection data male and female subimago, coll. Mjöberg Expedition (SMNH);
Atalophlebia brunnea (Tillyard): holotype, South Esk River, Clarendon, TAS, 9
III 1933, female imago; paratype, same collection data, male imago, coll by E.
Scott, R.J. Tillyard collection (NHM);
Nousia (Australonousia) fuscula (Tillyard).;   *Atalophlebia fuscula* (Tillyard),
River Shannon, TAS, male imago, 27 I 1933, coll. R.J. Tillyard (NHM).
Nousia (Australonousia) nigeli sp. nov.(Finlay, unpublished data): holotype,
Taggerty River, Lady Talbot Drive, outside Marysville, VIC, 17 X 1988,
145°46′28″ 37°30′20 ″, male imago; paratypes, same collection data, 17 X 1988,
male imago Donnelly Creek, Donnelly Weir Rd, VIC, 6 IV 1997, 145°32′03″
37°37′30″, female imago, coll. K.J. Finlay (MV).

Nyungara **Dean**
Nyungara bunni Dean: holotype, male imago reared; paratype, nymph; slide
material, Foster Brook, North Dandalup, WA, 22 ix 1983, coll. S. Bunn (MV);
paratypes, male imago, female imago, nymphs, slide material, Waterfall Gully,
Jarrahdale, WA, 1981-1982, coll. S. Bunn (MV); imagos, nymphs, slide material,
WA (MV)

Thraulophlebia **Demoulin**
Thraulophlebia lucida (Ulmer) comb. nov.: *Atalophlebia lucida* Ulmer:
holotype, Cedar Creek, QLD, III 1910-1913, coll. Mjöberg's expedition, male
imago; paratype, same collection data, male imago, head and genitalia detached
(SMNH); *Nousia pilosa* Suter: holotype, Wannon River, Grampians, VIC, 25 XI
1997, male imago; paratypes, same collection data, male imagos, female and
male nymphs; Hitchcock Drain, SA, 25 XI 1997, male imago, male nymph, coll.
D.N and P.J. Suter (MV);
Thraulophlebia inconspicua (Eaton) comb. nov.: Brownhill Creek, SA, 3 III
1976, 150m, 138°38′ 34°59′, male and female imago, coll. unknown; Bull Creek,
"'The Cliff", SA, 31 X 1989, nymphs, Deep Creek Tributary, Castambul, SA, 04
III 1977, 138°45′ 34°52′, nymphs, coll. J.E Bishop and A. Wells; Finniss River,
"Riverdale" & Meadows Creek, Fingerboard corner, SA, 31 X 1989, nymphs,
coll. P. Suter and S. Sheerlock; North East River, near Carnarvan, Kangaroo
Island, SA, 19 XI 1977, 136°59′ 35°56′, male imagos, coll. J.E. Bishop; North
Pava River, SA, 19 VIII 1983, nymphs, coll. P. Suter; North Pava River,
Tanunda, SA, 21 X 1991, 136°59′ 35°56′, nymphs, coll. P. Waller; Rocky River,
Flinders Chase National Park, Kangaroo Island SA, 13 XII 1976 & 18 XII 1976,
, 136°44′ 35°56′, nymphs, coll. Bill Williams; Scott Creek, SA, 9 XI 1994,
nymphs, coll. P. Goonan & C. Madden; South West River, Brigadoon, Kangaroo
Island, SA, 19 XI 1977, 136°50′ 35°52′, male imago, coll. J.E. Bishop; Spring
Creek, SA, 11 X 1995, nymphs, coll. Monitoring River Health Initiative; Stunsail
Broom River, Kangaroo Island, SA, 12 XII 1976, 137°00′ 35°59′, nymphs, coll.
Bill Williams; Sturt River, Bedford Park, SA, 27 XI 1976, 138°33′ 35°02′,

nymphs, coll. J.E. Bishop; Sturt River, upstream Minno Creek, SA, 26 X 1994, 138°38′ 35°02′, nymphs, coll. Monitoring River Health Initiative; Sturt River, Coramandel Valley SA, 29 IV 1976, 138°57′ 35°03′, male and female imago, coll. P. Suter; Tookayerta Creek, 22 X 1984, nymphs, coll. P. Suter; Unnamed Creek, Parawa Rd, near Yankalilla, 2 XI 1978, 138°21′ 35°28′, coll. J.E. Bishop and A. Wells (Suter collection);

Thraulophlebia parva (Harker) comb. nov.: *Atalophlebia parva.* Harker: holotype, Serpentine River, Point Lookout, NSW, X 1948, male imago, coll. J. Harker; paratypes, same collection data, male imago, subimago and nymph, coll. J. Harker (AM).

Thraulus Eaton
Thraulus sp. 'A' sp. nov.(informal description, Grant 1985): 'holotype', male imago, Drysdale River, WA, viii 1975, coll. I.F. Common and M.D. Upton (ANIC); 'paratypes', male imagos, Nourlangie Creek, 6km W Cahill, NT, 18 xi 1972, coll. J.C. Cardale (ANIC); 'paratypes', male subimagos, Cooper Creek, 19km SE of Mt Borrdaile, NT, vi 1973, coll. J.C. Cardale (ANIC);
Thraulus sp. 'AV1' 'AV2' 'AV3' (Dean 1999): imagos, nymphs, NT, QLD (MV);
Thraulus spp.: nymphs, QLD (DPIQ).

Tillyardophlebia Dean
Tillyardophlebia alpina Dean: imagos, nymphs, NSW (MV);
Tillyardophlebia rufosa Dean: holotype, male imago; paratypes, male and female imagos, nymphs; slide material, Badger Creek downstream weir, VIC, 23 ii 1984, coll. J. Dean (MV); imagos, nymphs, NSW, VIC (MV); imagos, nymphs, NSW, VIC (Finlay collection);
Tillyardophlebia spp.: imagos, nymphs, NSW, VIC (Finlay collection).

Ulmerophlebia Demoulin
Ulmerophlebia pippina Suter: nymphs, subimago, VIC (Finlay collection);
Ulmerophlebia sp. 'AV2' (Dean 1999): nymphs, imagos, NSW, VIC (Finlay collection); nymphs, imagos, NSW, VIC (Finlay collection);
Ulmerophlebia sp. 'AV6' -*mjöbergi*? (Dean 1999): nymphs, imagos, QLD, (MV)
Ulmerophlebia spp.: nymphs, imagos, VIC (Finlay collection).

'New Genus A' gen.nov.
'New Genus A' *wiltkorringae* comb. nov; *Nousia (Australonousia) wiltkorringae* *F*inlay: holotype, male imago, Cement Creek, Mt Donna Buang Rd, outside Warburton, VIC, 05 III 1998, 145°42′20″ 37 °42′48″, coll. K.J. Finlay (MV). Paratypes; male and female imagos, subimagos and nymphs, same collection data, 1 XII 1976, EPH 1587, EPH 1588, EPH 1602; 23 XI 1978, EPH 1589, EPH 1590, EPH 1591, EPH 1592, EPH 1594, EPH 1596, EPH 1597, EPH 1598, EPH 1599; 3 III 1980, EPH 1593, coll. J. Dean (MV); 05 III 1998, 29 XI 1998, 23 I 1999, , 20 III 1999, coll. K. Finlay (MV).
'New Genus B' gen. nov.

'New Genus B' *kala* comb. nov., *Atalophlebia kala* (Harker): holotype, Lake Albina, Mount Kosciusko, NSW, 2 II 1929, male imago, coll. R.J. Tillyard (NHM);

'New Genus B' *adamus* comb. nov., holotype, Frying Pan Raceline tributary, Telmark St, Falls Creek, VIC, 5 II 1999, 1560m, 147°16′50″ 36°52′00″, coll. K.J. Finlay; paratypes, same collection data, male imagos, nymphs; McKay Creek tributary (waterfall), Mt McKay, Alpine National Park, VIC, 5 II 1999, 1700m, 147°15′20″ 36°52′19″, female imago; Tanjil River east branch headwaters, Mt Baw Baw Alpine Village, VIC, I6 II 1999, 1440m, 146°15′45″ 36°50′25″, male imago; coll. K.J. Finlay (MV).

LEPTOPHLEBIIDAE: LEPTOPHLEBIINAE

Gillesia hindustanica (Gillies): imagos, Assam, INDIA (FAMU);

Leptophlebia cupida (Say): nymphs, Indiana, USA (Bae collection);

Leptophlebia sp.: nymphs, Missouri, USA (Bae collection);

Paraleptophlebia bicornuta (McDonnough): nymphs, Idaho, USA (Bae collection);

Paraleptophlebia chocolata Imanishi: nymphs, imago, reared, Kyonggi-do, KOREA (Bae collection);

Paraleptophlebia packi (Needham): nymph, imago, reared, Utah, USA (Bae collection);

Paraleptophlebia sp.: nymphs, Wyoming, USA (ANIC).

EPHEMEROIDEA

Ephemera danica Müller: nymphs, near Silkeborg, DENMARK (Campbell collection);

Ephemera simulans Walker: nymphs, Wyoming, USA (ANIC);

Euthyplocia spp.: nymphs, San Martin Province, PERU (ANIC);

Ichthybotus hudsoni McLachlan: nymphs, Taupo, NEW ZEALAND (ANIC);

Povilla adusta Navás: nymphs, Lake Kivu, BELGIAN CONGO (ANIC);

Rhoenanthopsis spp.: nymphs, Chiangmai Province, THAILAND (ANIC).

Appendix D

Data matrix of Pescador and Peters (1980a)

Outgroup	00000	00000	00000	00000	000
Hapsiphlebia	00011	-----	-----	-----	---
Penaphlebia	11100	01100	-----	-----	---
Atalonella	11100	10011	00010	-----	---
Dactyophlebia	11100	00011	11101	10100	001
Meridialaris	11100	00011	11101	11011	110

TAXONOMY OF *EPEORUS FRISONI* (BURKS) AND A KEY TO NEW ENGLAND SPECIES OF *EPEORUS*

Steven K. Burian[1], Beth I. Swartz[2] and Philip C. Wick[2]

[1] *Department of Biology, Southern Connecticut State University, New Haven, Connecticut 06516 USA*
[2] *Maine Department of Inland Fisheries and Wildlife, Wildlife Division, Bangor, Maine 04401-5654 USA*

Abstract

Among all of the species of *Epeorus* recorded from New England (i.e., Connecticut, Massachusetts, Maine, New Hampshire, Rhode Island and Vermont), *E. frisoni* (Burks) has been the least studied. Since Burks published the original description of *E. frisoni* in 1946, no other specimens have been found. Thus our knowledge of this species is based on a single specimen (the holotype) and a vague type locality label ("Mt. Katahdin, Roaring Brooks"...Maine). Until recently, further study at the type locality was not possible because of regulations protecting the wilderness nature of the area. However, because of efforts to assess the status of rare or presumably endangered aquatic organisms in Maine by the Maine Department of Inland Fisheries and Wildlife, access to the type locality was granted to search for *E. frisoni*.

Sampling during the summer of 2003 provided the first new specimens of *E. frisoni* since 1939 (when the holotype was collected). Larvae were reared and all life stages were associated. Study of these specimens and the holotype showed *E. frisoni* was distinct from its presumed sister species *E. fragilis* (Morgan) and *E. pleuralis* (Banks). Comparative study of the new *E. frisoni* material has led to the development of new species level keys for male imagos and well developed male larvae. Currently, female imagos and some female larvae can not be reliably determined.

Key words: Ephemeroptera; Heptageniidae; *Epeorus*, *E. frisoni*; taxonomy; key; New England species.

Introduction

The focus of this study is what has been called New England's only endemic mayfly, *Epeorus frisoni* (Burks). The original description was based on a single male imago collected in 1939 by T. H. Frison from the "Roaring Brooks" area east of Mt. Katahdin, Baxter State Park, Maine. Over the past 57 years, no other specimens identifiable as *E. frisoni* have been found. This has led to the suggestion that perhaps *E. frisoni* was a variant of one of the other common species of *Epeorus* or that it may have gone extinct. Part of the problem in searching for *E. frisoni* has been in knowing what habitats to search. Certainly flowing water, but not knowing precisely what microhabitat variables are important, makes looking for an apparently rare species even worse. Therefore, to really understand this species and begin to answer some of the questions concerning its biology and conservation status, it was necessary to start at the type locality.

As part of a larger program to assess rare and presumably endangered aquatic organism in Maine, the Maine Department of Inland Fisheries and Wildlife (MDIFW) in conjunction with officials governing Baxter State Park developed a plan to systematically sample streams in and around the area. The holotype of *E. frisoni* was collected with the following goals: first, verify the existence of *Epeorus frisoni* at or about the presumed type locality; second, rear larvae to associate adult and larval stages; third, collect habitat data necessary to facilitate future searches for *E. frisoni* outside of Baxter State Park; and finally, provide new material for comparative study to develop new keys for the species of *Epeorus* that occur in New England.

Methods

The study area was located in Baxter State Park, Maine, USA and focused on the area of Roaring Brook upstream from Roaring Brook campground and its tributaries along the trail to Chimney Pond. Because Baxter State Park is protected by special regulations to maintain the wilderness nature of the park, sampling was done to minimize effects to aquatic habitats and aquatic insects not the focus of this study. Benthic samples were taken using a 1-m kick net at three sites along Roaring Brook, one site on Pamola Brook and one site on Blacksmith Brook. Larvae were sorted alive and those at or near the final instar were retained for rearing. All rearing was done on site. Aerial nets were used to sweep for adults and a battery-powered blacklight was used at the base camp on Roaring Brook. All specimens were preserved in 95% ETOH. Water temperatures were recorded at each site from June – September 2003 and at the Pamola Brook site dissolved oxygen, pH, specific conductance, velocity and depth were recorded for the 26 August sampling. In addition, substrate composition and changes in channel morphology were qualitatively assessed at each sampling time.

New material of *E. frisoni* was comparatively studied using the holotype of *E. frisoni* and specimens of *E. fragilis*, *E. pleuralis* and *E. vitreus* from New England and adjacent areas. All specimens were observed for morphological characters and body and appendage coloration under stereoscopic and compound light microscopy (up to 1000x magnification. All measurements were made using a calibrated ocular micrometer (nearest 0.01 mm). Larval exuviae were measured with exuviae held as flat as possible without causing excessive distortion. Means and SDs were calculated for all continuous data; medians were estimated by interpolation. Lengths of the foreleg segments of male imagos were compared to the lengths of the fore tibiae and expressed as ratios. Ranges of ratios are given with each ranges median value. Eggs were dissected in 95% ethanol and slide mounted in CMCP® --10. Eggs were removed from the lower oviducts to minimize differences related to maturation. Egg chorion features were observed under phase contrast microscopy (400 and 1000X). Length measurements were made on 10 eggs chosen haphazardly using an ocular grid.

Material Studied

Holotype of *Epeorus frisoni*: USA, MAINE: Piscataquis Co., Mt. Katahdin, Roaring Brooks, 26.VIII. 1939, T.H. Frison, 1M (INHS), Collection # 16253 and Slide # 20562; Baxter State Park, Pamola Brook [45° 55.263'N/068° 52.570'W], 15.VII.2003, B. Swartz, 1M (NEL); same, 29.VII.2003, P.C. Wick, 4M,1F (NEL); same, 18.VIII.2003, 1SF, 1LEX, 1L (NEL); same, 26.VIII.2003, S.K. Burian and P.C. Wick, 2ML (NEL); Baxter State Park, Blacksmith Brook [45° 55.268'N/068° 52.690'W], 24.IX.2003, P.C. Wick, 1ML (NEL); same, 9.X.2003, P.C. Wick, 1M (NEL); trib. to Abol Stream, Baxter State Park, elevation 1144' (349 m) [45° 51. 6833'N/068° 57.6000'W], 12.VII.1982, A.C. Graham, 1ML (SWRC);VERMONT: Bennington Co., South Fork of Goodman Brook, tributary of the West Branch of the Batten Kill, Elev. 1420' (433m) [43° 17.7833'N/073° 07.1833'W], 14.VII.1982, D.I. Rebuck, 1M, 1F, 2LEX (SWRC).

Other Material Studied: In addition, specimens of *E. fragilis*, *E. pleuralis*, and *E. vitreus* were studied from the following areas:

E. fragilis: USA: MAINE: Piscataquis Co., Nesowadnehunk Stream, Baxter State Park, elevation 1050' (320 m) [45° 54.0333'N\069° 2.3666'W], 28. VI. 1988, D.I. Rebuck and D.H. Funk, 2M (SWRC); same, 14. VII. 1982, A.C.Graham, 2L (SWRC); same, Little Abol Stream (abol trib. #1), Baxter State Park, 27. IX. 1981, D.H. Funk and A.C.Graham, 4L (SWRC); PENNSYLVANIA: Eire Co., Fourmile Creek, trib. site #1, lower section, 6. VI. 1990, E.C.Mastellar, 1M, 1F (NEL); VERMONT: Bennington Co., Gilbert Brook below Daley Brook on West Branch of Batten Kill (River), 0.5 mi. SW of Dorset, elevation 1100' (325 m) [43° 14.4833'N/073° 6.4833'W], 20. VII. 1982, D.I. Rebuck, 58L (SWRC); VIRGINIA: Rappahannock Co., Bearwallow Creek (above bridge), 2.2 mi. W Jct. of CR630 and Hwy. 522 on CR630, elevation 920' (280 m) [38° 47.1166'N/078° 8.9000'W], 18. V. 1980, C.E. Dunn, 1M (SWRC) CANADA: NEW BRUNSWICK: Northumberland Co., Catamaran Brook, 27. VI. 1995, M.Dobrin , 1M (NEL); same, 28.VI.1995, M. Dobrin and D. Giberson, 1M (NEL); QUEBEC: Saguenay Co., Riviere aux Loups Marlins, above Rt. 138, elevation 050' (15 m) [50° 16.7333'N/065° 43.2000'W], 28. VII. 1982, J.A.G., 7L (SWRC).

E. pleuralis: USA: CONNECTICUT: New Haven Co., Bethany, 2. V. 1983, V.A. Nelson, 5M (NEL); Seymour, from swarm over pool on Molsick Rd., 18. IV. 1998, S.K. Burian, 4M (NEL); same, 23. IV. 1998, S.K. Burian, 1F (NEL); MAINE: Franklin Co., Carrabassett River, dwnstr. of 2nd bridge at Sugarloaf Ski Area, 15. VI. 1985, S.K. Burian, 12M, 3F (NEL); same, 21. VI. 1985, S.K. Burian, 1M, 1F (NEL);

small stream flowing through Gondola Village, Sugarloaf Ski Area, 21. VI. 1985, S.K. Burian, 2M, 2F (NEL); Penobscot Co., Crystal Brook, 3.6 mi. N of Patten at Rt. 11, elevation 670' (204 m) [46° 3.2166'N/068° 26.6000'W], 26. V. 1982, A.C. Graham and M.K. Butcher, 7L (SWRC); Piscataquis Co., Nesowadnehunk Stream, Baxter State Park, elevation 1050' (320 m) [45° 54.0333'N/069° 2.3666'W], 2. VI. 1982, A.C. Graham and M.K. Butcher, 1L (SWRC); NEW YORK: Delaware Co., East Fork of Delaware River, 0.8mi. SW of Downsville, elevation 1090' (332 m) [42° 4.3166'N/075° 0.4166'W], 27. IV. 1982, J.W.P. and D.H. Funk, 4M, 2F, 6LEX (SWRC); same, 25. IV. 1982, J.W.P. and D.H. Funk, 1M, 1F, 2LEX (SWRC); same, 19. IV. 1983, P.D., J.D., and D.I. Rebuck, 20L (SWRC); PENNSYLVANIA: Chester Co., Bog Hollow Creek, West Branch of Brandywine Creek, 1.9mi. SE of Mortonville, elevation 330' (100 m) [39° 55.4166'N\075° 45.4500'W], 11. IV. 1978, D.H. Funk, 22L (SWRC); same, 12. IV. 1978, D.H. Funk, 1M, 1LEX (SWRC); same, 14. IV. 1978, D.H. Funk, 1M, 1LEX (SWRC); same, 15. IV. 1978, D.H. Funk, 1M, 1LEX (SWRC); Sesquehanna Co., trib. of Partners Creek, 1.9mi. SW of Harford on Rt. 944, elevation 1360' (414 m) [41° 45.4500'N/075° 44.7666'W], 29. IV. 1980, D.T. Mulvey, 1M, 1LEX (SWRC); same, 15. IV. 1982, D.I. Rebuck, 16L (SWRC); trib. to Wylusing Creek, 3.0mi. W of Montrose, elevation 1220' (372 m) [41° 49.1833'N/075° 56.0000'W], 16. IV. 1980, D.T. Mulvey, 1M (SWRC); same, 26. IV. 1980, D.T. Mulvey, 1M, 1LEX (SWRC); same, 29. IV. 1980, D.T. Mulvey, 1M, 1LEX (SWRC); trib. to Meshoppen Creek, 2.0 mi. E of Dimmock on Rd. T508 at Jct. of Rd. T518, elevation 1060' (323 m) [41° 44.2166'N/075° 51.6000'W], 21. II. 1980, D.H. Funk and P.J. Dodds, 6L (SWRC); same, 27. V. 1980, D.T. Mulvey, IM, 1F (SWRC); same, 4. VI. 1980, D.T. Mulvey, 1M (SWRC); VIRGINA: Bedford Co., unnamed trib. of Sheep Creek, 0.65 mi. NW of Jct. of CR680 and CR614 on CR680, elevation 1250' (381 m) [37° 24.9666'N/079° 38.7666'W], 20. IV. 1980, P.J. Dodds, 1M, 1LEX (SWRC); Sheep Creek, 0.7mi. N of Jct. CR614 and CR680 on CR614, elevation 1240' (378 m) [37° 25.3500'N/079° 38.4000'W], 27. III. 1980, D.H. Funk and P.J. Dodds, 1SM (SWRC); same, 7. IV. 1980, D.H. Funk and P.J. Dodds, 1M (SWRC); Rappahannock Co., Bearwallow Creek (above bridge), 2.2 mi. W of Jct. CR630 and Hwy. 522 on CR630, elevation 920' (280 m) [38° 47.1166'N/078° 8.9000'W], 17. IV. 1980, C.E. Dunn, 5M (SWRC); same, 24. IV. 1980, C.E. Dunn, 2F, 2LEX (SWRC); same, 8. IV. 1981, M.B. Griffith and D.I. Rebuck, 5L (SWRC); same, 8. V. 1981, P.J. Dodds, 12M, 4F (SWRC); Hittles Mill Stream at Jct. of CR630 and CR638, elevation 740' (225 m) [38° 47.6333'N/078° 7.0666'W], 12. IV. 1981, D.I. Rebuck, 2M, 1LEX (SWRC); same, 13. IV. 1981, D.I. Rebuck, 1M, 1LEX (SWRC);CANADA: QUEBEC: Saquenay Co., Ruisseau du Cran Carre, above Rt. 138, elevation 160' (49 m) [50° 17.6000'N/065° 55.5000'W], 13. VI. 1982, A.G., 3M, 2LEX (SWRC); same, 14. VI. 1982, J.A.G., 5L (SWRC); same, 2. III. 1982, J.A.G., 2M, 2LEX (SWRC).

E. vitreus: USA: CONNECTICUT: Litchfield Co., Blackberry River, dwnstr. of Rt. 44 bridge, Norfolk [42° 0.4000'N/073° 14.0833'W], 8. VI. 1997, S.K. Burian, 6L

(NEL); Housatonic River, fly fishing area of Housatonic Meadows State Park, 29. II. 2004, L.M. Rojas, 2L (NEL); MAINE: Hancock Co., Whitten Parritt Stream, T7 SD, Rt. 1, 21. VI. 2004, S.K. Burian, 1F, 1LEX (NEL); Piscataquis Co., Nesowadnehunk Stream, Baxter State Park, elevation 1050' (320 m) [45° 54.0333'N/069° 2.3666'W], 20. VI. 1988, D.H. Funk and D.I. Rebuck, 1M, 1F (SWRC); same, 6. VIII. 1988, A.C. Graham, 2L (SWRC); Washington Co., Narraguagus River, Little Falls National Marine Fisheries Field Station site at Little Falls north of Cherryfield, off of Rt. 193, 23. VI. 2004, S.K. Burian, 12L (NEL); MINNESOTA: Lake Co., West Branch of Splitrock River at Rt. 3 [47° 14.5300'N/091° 28.9116'W], 29. V. 1999, Sk.Burian, 6L (NEL); NORTH CAROLINA: Madison Co., Walnut Creek, Marshall, 6. VI. 1983, S.K. Burian, 4L (NEL); NEW HAMPSHIRE: Hills Co., Souhegan River, 2.5mi. N of Greenville, 11. Vi. 1998, J. Burger and D. Chandler, 1L (NEL); NEW YORK: Delaware Co., East Creek, trib. of Delaware River, 0.8 mi. SW of Downsville, elevation 1090' (332 m) [42° 4.3166'N/075° 0.4166'W], 13. VII. 1982, D.H. Funk and J.W.P., 10L (SWRC); PENNSYLVANIA: Chester Co., East Fork of East Branch of White Clay Creek, 0.6 mi. W of London Grove, 15. V. 1980, D.H. Funk, 1M, 1LEX (SWRC); same, 0.8 mi. WSW of London Grove, elevation 345' (105 m) [39° 51.7833'N/075° 47.1166'W], 14. V. 1981, A.C. Graham, 5L (SWRC); Wet Lab at Stroud Water Research Center, London Grove, elevation 325' (99 m) [39° 51.5333'N/075° 47.0333'W], 27. V. 1979, D.H. Funk, 1M (SWRC); same, 30. V. 1987, D.H. Funk, 1M (SWRC); VERMONT: Bennington Co., West Branch of Batten Kill (river), 2.0 mi. S of South Dorset on Rt. 30, elevation 879' (268 m) [43° 13.1666'N/073° 4.3000'W], 4. VI. 1980, A.C. Graham, 1M, 1LEX (SWRC); CANADA: QUEBEC: Saguenay Co., Riviere Matamec below Beaver Creek, elevation 060' (18 m) [50° 18.3500'N/065° 56.1833'W], 28. VII. 1981, J.A.G., 5L (SWRC); Riviere Pigou above Rt. 138, elevation 075'(23m) [50° 16.9500'N/065° 38.5166'W],8. IX. 1982, D.H. Funk, 2L (SWRC).

Specimen Abbreviations: M= Male imago; F= Female imago; SM= Subimago male; SF= Subimago female; L= Larva; LEX= Larval exuviae. Numbers of specimens studied precede abbreviations. Institution Abbreviations: NEL= Northeast Ephemeroptera Laboratory, at the Department of Biology, Southern Connecticut State University, New Haven, CT USA; SWRC= Stroud Water Research Center, Avondale, PA USA. Deposition of all materials is with the institutions indicated with each record.

Systematic Account and Key

Epeorus frisoni (Burks)

Iron frisoni Burks 1946: 608; Edmunds et al. 1976: 193; Burian and Gibbs 1991: 33.

Male imago (Figs. 5–7 in alcohol):   Body Length: 6.83–9.42 mm (8.40±0.80, mean±SD, n=8); Forewing Length: 7.42–9.83 mm (9.02±0.79, n=8); Forewing Width: 2.71–3.93 mm (3.39±0.41, n=7); Caudal Filaments: 15.00–26.00 mm (23.0±3.57, n=7)

Range of ratios and median values of the foreleg segments referenced to the tibia: Femur = 0.44–0.86, Med.= 0.77; Tibia = 1.00; Tarsus 1 = 0.28–0.39, Med. = 0.31; Tarsus 2 = 0.31–0.38, Med. = 0.34; Tarsus 3 = 0.29–0.37, Med. = 0.32; Tarsus 4 = 0.19–0.27, Med. = 0.22; Tarsus 5 = 0.09–0.11, Med. = 0.10.

Head: Pale with bases of ocelli black.   Antennae scape and pedicle light brown, flagellum slightly darker. Eyes light grayish brown to kaki colored meeting dorsally.

Thorax: Pronotum mostly pale brown with deep posterior medial notch and dark brown along posterior edge.   Mesonotum with dark brown on infrascutellem, medium brown along the medioscutum and on posterior scutal protuberances. Pleural areas mostly pale yellowish brown with heavier sclerotized edges darker amber brown. Metanotum colored similar to mesonotum.

Wings: Forewing and hind wing as in Figs. 1 and 2.   All veins pale.   No anastomosing of cross veins in stigma.   Stigmatic area of wing membrane slightly clouded.

Legs: Mostly pale. Femora with a distinct dark brown medial spot on dorsal surface. Apex of femora slightly shaded brown.   Joints of tarsal segments with dark edge, tarsal claws blunt/sharp with dark brown shading in bases of claws.

Abdomen: Segments 2–7 translucent with subcuticular tracheols dark and visible dorsally. Extremely faint brown shading along posterior margins of terga (Fig. 3). Sterna 2–7 completely pale.   Segments 8–10 opaque cream colored.   Genitalia modified pleuralis-type with a dorsal forked process that merges with the inner apical margins of each pene forming a hook-shaped process (Fig. 5).

Caudal Filaments: Surfaces of annuli with many small setae. Filaments light brown near base becoming paler toward tips.

Female Imago (Figs. 8 and 9 in alcohol): Body Length: 7.08–8.58mm (7.77±0.62, n=4); Forewing Length: 8.58–10.00 mm (9.48±0.62, n=4); Forewing Width:3.25–3.58 mm (3.48±0.16, n=4); Caudal Filaments: 12.50–16.00 mm (14.25±2.47, n=2). Head similar in color to male. Eyes smaller, darker and widely separated. Antennae similar to male in size and color. Thorax similar in color and morphology to male. Legs similar in coloration to male. Abdominal terga mostly pale, but in contrast to male have a light reddish brown area anteriorly on segments 2–10. Posterior margins of abdominal terga with distinct dark brown edge. Subcuticular tracheols dark and visible dorsally on terga. Caudal filaments with similar setae and color as in male.

Eggs (Fig. 4): Maximum Length: 0.165–0.180 mm (0.172±0.007, n=10); Maximum Width: 0.110–0.130 mm (0.112±0.008, n=10). Eggs oval shaped without any distinct chorionic sculpturing detectible at 1000x phase. Each egg with multiple micropylar devices (Mpd), Mpd positioned around the middle of each egg and somewhat equally spaced out. Mpd as in Fig. 4.

Larva (Figs. 8–17, 19, 22, 24 in alcohol): Body Length [all sexes]: 8.17–10.66 mm (9.16±1.18, n=4); Body Length ♂: 8.17–9.55 mm (8.86±0.97, n=2); Body Length ♀: 8.25–10.66 mm (9.45±1.70, n=2); Head Capsule Width ♂: 2.06–2.40 mm (2.23±0.14, n=3); Head Capsule Width ♀: 2.24 (n=1); Ratio of Head Capsule Width to Distance between Antennae: 2.36–2.73, Med.=2.47, n=7; Head and Thorax Length ♂: 2.96–3.63 mm (3.36±0.35, n=3); Head and Thorax Length ♀: 3.64–3.84 mm (3.74±0.14, n=2).

Head: Broad with anterior portion expanded (Fig. 8). Long hair-like setae present along anterior margin. Coloration as in Fig. 8, distinctive v-shaped mark present on anterior portion. Outer edges of head capsule smoothly curving posteriorly. Outer posterior edges of eyes only slightly extend beyond edge of head capsule (Figs. 8 and 22). Antennae with small brown spot near apical edge of last 10–15 annuli.

Mouthparts: Mouthparts as in Figs. 9–12, 16. Dorsal and ventral aspects of labrum as in Fig. 9. Labrum with shallow anterior median emargination and 2 pairs of setae medially and 1 pair of longer setae lateral to medial group of setae. Lateral edges with several long thick setae. Hypopharynx as in Fig. 10. Left and right mandibles as in Figs. 11 and 12. Right mandible with tuft of setae on anterior margin and row of long setae below molar surface. Left mandible with scattered long setae below molar surface. Incisors of both mandibles with 6–8 teeth long inner edge. Maxillae as in Fig. 13. Galea-lacinia with massive tridentate tip with apical spine tip hooked upward. Dorsal surface with secondary row of about 29 long setae set back from brush of setae along inner margin. Dorsal and ventral aspects of labium as in Fig. 16. Dorsal surfaces of glossa and paraglossa with brush of setae. Labial palps with dense row of long simple setae and large stout pectinate setae on ventral surface of segment 3.

Thorax: Pronotum with broad transverse median ridge margined with setae. Coloration as in Fig. 8 with dark subcuticular tracheols visible. Mesonotum as in Figs. 8 and 22. Pigmented areas strongly contrast with pale regions. Profile of margins of mesonotum from base of wing pads to anterior edge of pronotum not deeply concave (Fig. 22).

Legs: Legs as in Figs. 8, 14, 15. Femora with basal row of 5–6 large flat setae that can extend to anterior margin (Fig. 14). Dorsal surface of femora with scattered smaller flat setae over middle portion with dark median mark (Fig. 14). Posterior margins of femora with row of long blade-like setae. Tibiae with dorsal row of hair-like setae oriented perpendicular relative to those on the posterior margins of femora. Tarsi with scattered setae. Tarsal claws with 4 small subapical denticles ventrally (Fig. 15).

Abdomen: Abdomen as in Fig. 8. Terga 2–10 with medial longitudinal patch of setae. Terga 2–9 with small denticles along the posterior margin separated medially by area lacking denticles, but marked by a brown spot. Posterolateral projects on segments 2–9 short (Fig. 24).

Gills: Gills as in Figs. 8 and 19. Anterior portion of all gills extend beneath the abdomen and except for gill 1 overlap. Anterior portion of gill 1 elongate and broadly rounded (Fig. 19). Elongate anterior portion reaches to about midline of body ventrally, but does not touch or overlap the opposite gill. Medial sclerotized strip of gill 1 angled at about 90° (Fig. 19).

Caudal Filaments: Filaments as in Fig. 8. Two long filaments present and annuli lack long setae, but a few small setae may be present on basal annuli.

Because larva of all the Nearctic species of *Epeorus* have not been studied in detail it is not possible to provide a definitive diagnosis. Therefore, this diagnosis is now restricted to the group of species known for New England.

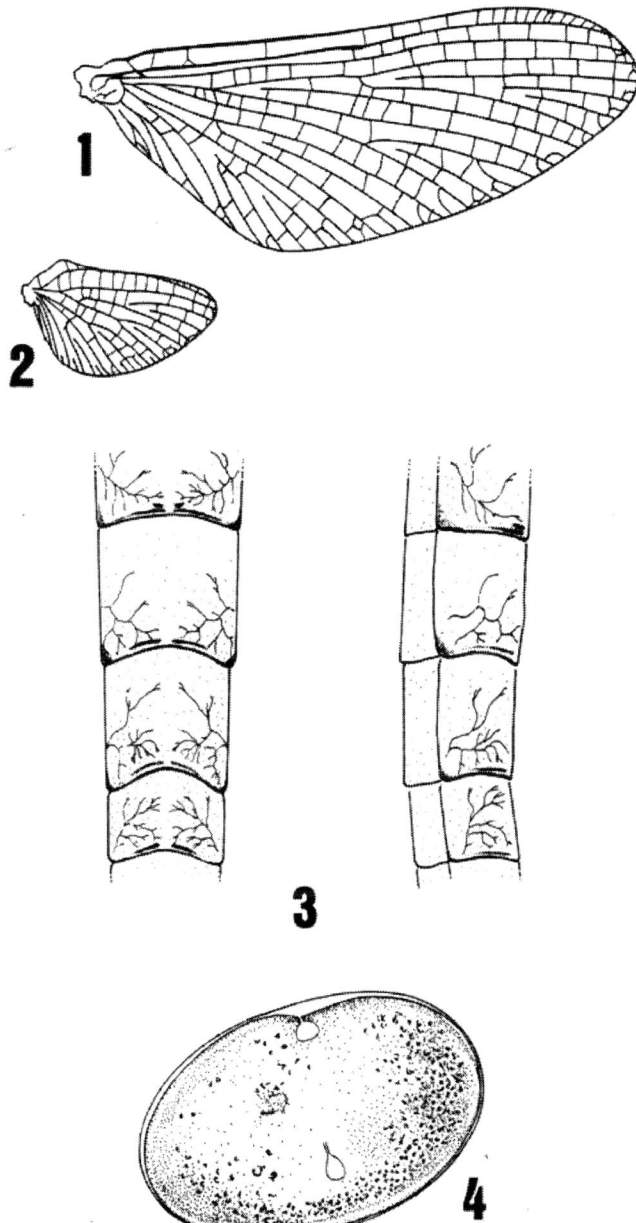

Figures 1–4. Imago of *Epeorus frisoni* 1—Forewing. 2—Hind wing. 3—Abdominal terga 3–6 of male imago (dorsal and left lateral view). 4—Egg showing micropylar opening.

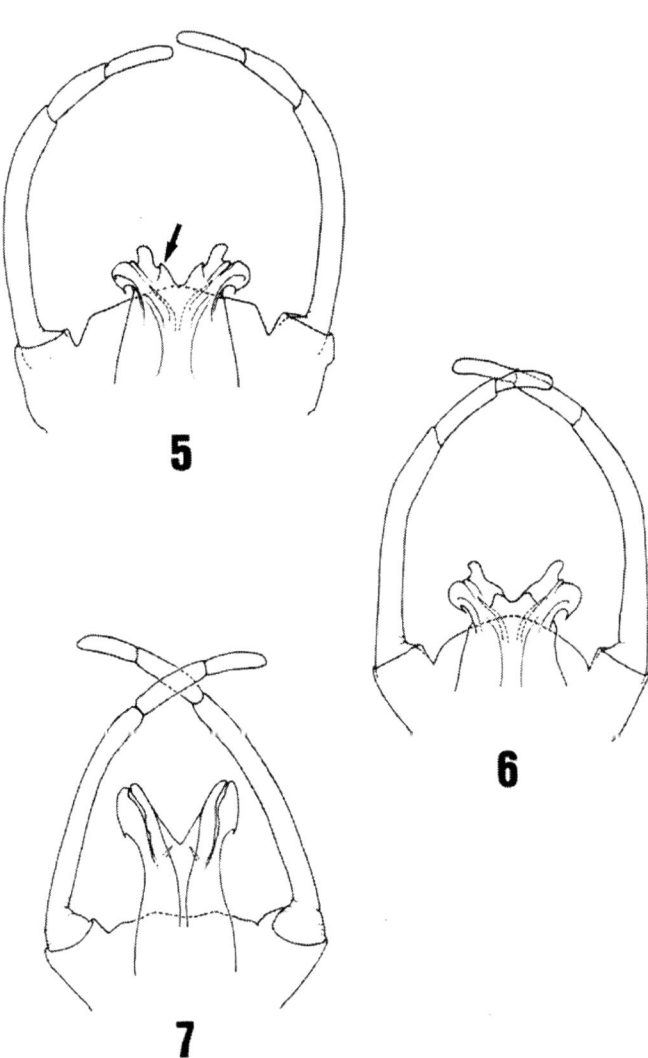

Figures 5–7. Male genitalia (ventral view). 5—*Epeorus frisoni* (arrow indicates hook-like process on inner apical margin of pene). 6—*E. fragilis*. 7—*E. vitreus*.

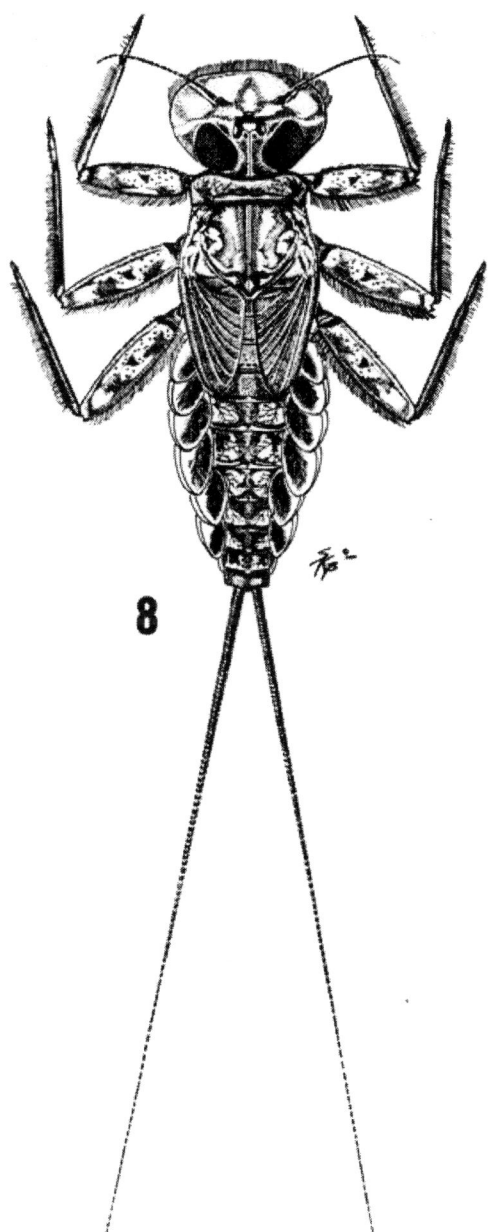

Figure 8. Dorsal view of final instar male larva of *Epeorus frisoni*.

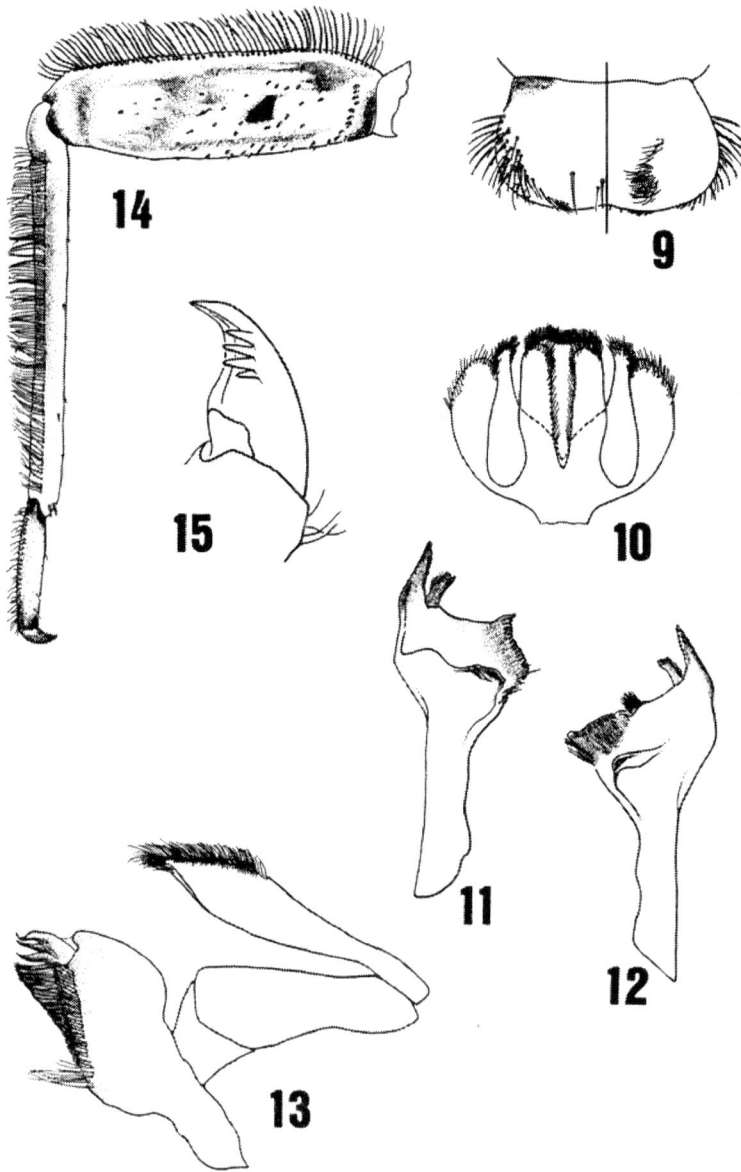

Figures 9–15. Final instar of *Epeorus frisoni*. 9—Labrum with dorsal surface on left and ventral on right. 10— Hypopharynx (dorsal view). 11—Left mandible (dorsal view). 12—Right mandible (dorsal view). 13—Right maxilla (dorsal view). 14—Right foreleg. 15—Right foreclaw (ventral view showing denticles).

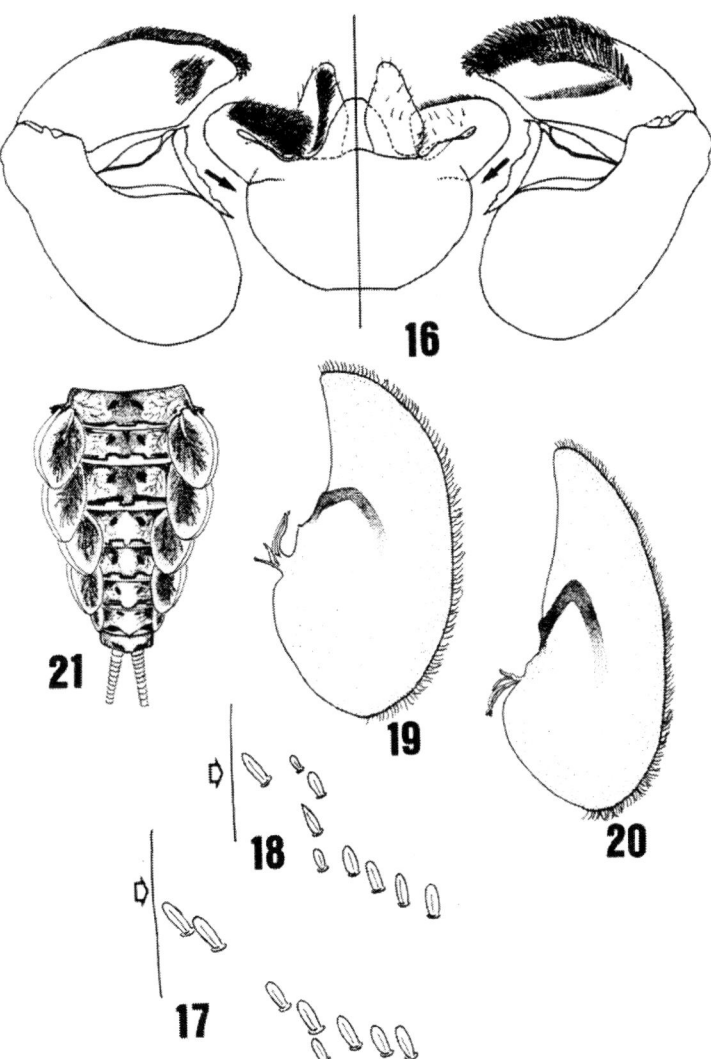

Figures 16—21. *Epeorus frisoni, E. pleuralis,* and *E. fragilis* larva. 16—Labium of *E. frisoni* (dorsal surface on left and ventral on right).
17—Row of setae on dorsal surface of forefemora of *E. frisoni* (near joint with coax). 18—Row of setae on dorsal surface of forefemora of *E. fragilis* (near joint with coax). 19—Gill 1 of *E. frisoni* (dorsal view). 20—Gill 1 of *E. fragilis* (dorsal view). 21—Abdominal terga 3-10 of *E. pleuralis* (dorsal view).

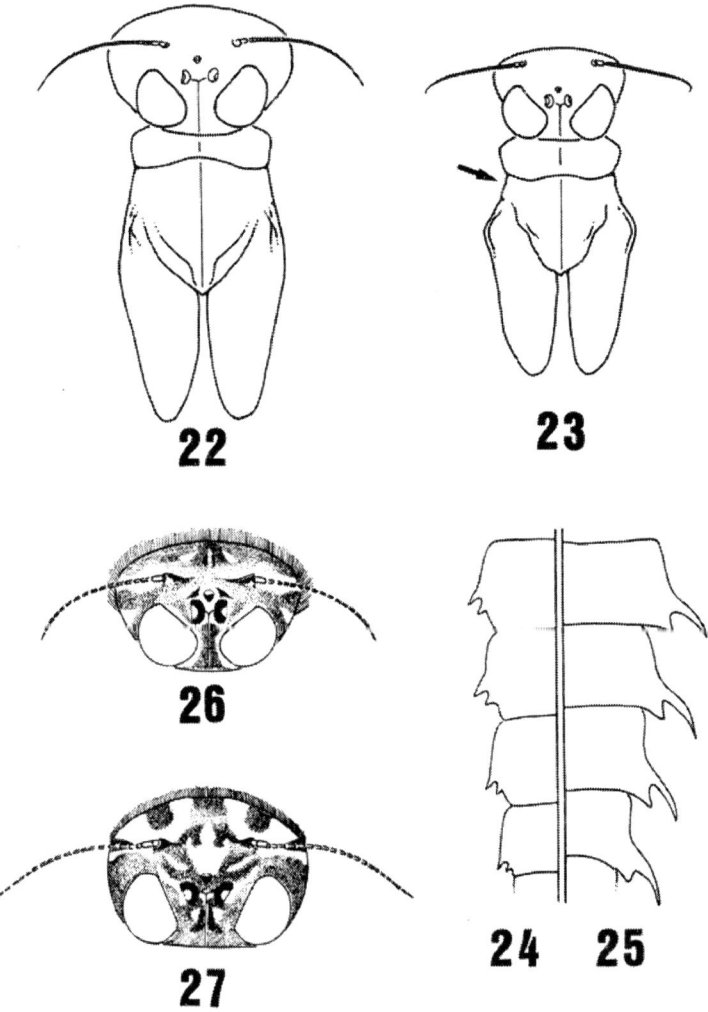

Figures 22–25. *Epeorus frisoni*, *E. fragilis*, and *E. vitreus* larva.
22— Head and thorax outline of *E. frisoni* (dorsal view). 23—Head and thorax outline of *E. fragilis* (arrow indicates deeply concave area of profile). 24—Posterolateral projections of abdominal sterna 4–7 of *E. frisoni*. 25—Posterolateral projections of abdominal sterna 4–7 of *E. vitreus*. 26—Head capsule of final instar male *E. pleuralis* (dorsal view). 27—Head capsule of final instar male *E. vitreus* (dorsal view).

Diagnosis

In New England, the early emergence of *E. pleuralis* and distinctive morphology of *E. vitreus* allow these species to be easily separated from *E. frisoni*. However, separating larvae of *E. frisoni* from those of *E. fragilis* is more difficult. Several diagnostic characters observed during this study are subtle and with the discovery of a population in VT, suggest the need of additional study. However, the consistent characters of male near final or final instar larvae are: shallowly concave profile of side of the thorax from the base of the wing pads to the anterior edge of the pronotum; ratio of the width of the head capsule to the distance between the antennae bases, 2.36 – 2.73, Med=2.47; head capsule with smoothly curving outer edges; posterior outer edges of compound eyes only slightly extending beyond outer edge of head capsule; anterior portion of gill 1 elongate, but broadly rounded with a sclerotized strip angled at about 90°; medial dark spot on femora not part of a streak or band, spot more vertical in position and sometimes appearing coma-shaped or triangular; and row or 5-6 large flat setae at base of forefemora. The coloration of the femora of *E. fragilis* varies from that of *E. frisoni* in that the medial spot often exists as part of a streak or band and is usually oval in shape.

Male imagos of *E frisoni* can be easily separated from *E. pleuralis* and *E. vitreus* not only by emergence time, but also body size, color and type of genitalia. Male imagos of *E. frisoni*, despite the overlap of emergence time and sympatry with *E. fragilis,* can be easily separated from those of *E. fragilis* by: larger body size (8.17—9.55 mm); pale color of head, thorax and especially abdomen; unique genitalia type with medial apical hook-shaped projections; much lighter grayish brown color of compound eyes after death.

Key to Mature Male Larvae of *Epeorus* of New England

1   Posterolateral projections of abdominal terga long and curving (Fig. 25); abdominal gills not greatly extended under abdomen and anterior portion of gill 1 not elongate; head capsule with outer margin flat curved and distinctive color pattern (Fig. 27) ....*E. vitreus.*

1'   Posterolateral projects of abdominal terga short (Fig. 24); abdominal gills extended under the abdomen and anterior portion of gill 1 elongate so that the apices about meet or exceed the midline of the abdomen; outer margin of head capsule smoothly rounded (Figs. 22, 23, 26)..2

2   Abdominal terga 3–7 with small paired medial dark spots (sometimes these are faint) (Fig. 21); body length ≥ 9.85 mm; anterior portion of head capsule with indistinct v-shaped mark (Fig. 26); final instar larvae collected from May to early June in most areas ....*E. pleuralis.*

2'   Abdominal terga 3–7 without paired dark medial spots (Fig. 8); body length < 9.55 mm; anterior portion of head capsule with distinct or moderately

distinct v-shaped mark (Fig. 8); final instar larvae collected from early July
to early October 3

3   Head capsule outer margin with abrupt transition near outer anterior
        corners of compound eyes; ratio of head capsule width to distance between
        antennae 2.11–2.45, Med.=2.21; posterior edges of compound eyes hide
        much of posterolateral edge of head capsule (Fig. 23); profile of thorax
        deeply concave (Fig. 23); body length ≤ 6.66 mm; anterior portion of gill 1
        narrow with sclerotized strip approximating a 45° (Fig. 20); forefemora
        with 3–4 large flat setae in a row that may extend to anterior edge of
        femora (Fig. 18) ....*E. fragilis.*

3'   Head capsule oval in shape; outer margin smoothly curving posteriorly; ratio
        of head capsule width to distance between antennae 2.36–2.73, Med.=2.47;
        outer edge of head capsule visible at posterior edges of compound eyes
        where eyes extend slightly beyond edge (Figs. 8 and 22); profile of thorax
        shallowly concave (Fig. 22); body length 8.17–9.55 mm; anterior portion of
        gill 1 broad with sclerotized strip approximating a 90° (Fig. 19); forefemora
        with 5--6 large flat setae in a row that may extend to anterior edge of femora
        (Fig. 17) ....*E. frisoni.*

Key to Male Imagos of *Epeorus* of New England

1   Genitalia as in Fig. 7; abdominal terga mostly pale with median black marks
        centered on posterior margins of terga 3–5 and 7–9 (rarely abdomen mostly
        brown) ....*E. vitreus*

1'   Genitalia as in Figs. 5 and 6; abdominal terga either brown or pale without
        dark medial marks 2

2   Genitalia as in Fig. 5 with dorsal process meeting inner apical margins of
        penes forming small hook-like projections; abdominal terga pale white with
        only faint yellowish shading along posterior third of terga, small areas of
        brown along posterior margins and posterolateral areas as in Fig. 3;
        compound eyes grayish or kaki colored after death ....*E. frisoni*

2'   Genitalia as in Fig. 6 without hook-like projections; abdominal terga either
        mostly medium brown or pale yellowish brown occasionally pale yellowish
        white; compound eyes after death dark brown to blackish brown 3

3   Body length > 9.00 mm; abdominal terga mostly medium brown with
        occasional pale areas laterally (rarely with posterior margins of tergites dark
        brown); emergence May to early June in most areas ....*E. pleuralis*

3'   Body length < 6.80 mm; abdominal terga pale yellowish brown to pale
        yellowish white; abdominal terga with faint dark brown on posterior
        margins; emergence late June–early August in most areas ....*E. fragilis*

Discussion

Results of this study have greatly increased our understanding of the *Epeorus* of New England and has provided several new directions for future study. After studying the New England subset of the eastern Nearctic species of *Epeorus*, it is now clear that there is much unexplored variation within and among species. Many of the characters studied for larvae and adults of *E. frisoni* are subtle and some are within the range of variation for *E. fragilis* and *E. pleuralis*. Specimens of *E. frisoni* from Vermont, although consistent in adult characters with specimens from the type locality in Maine, had larvae that were more variable compared to those from Maine. This suggests that at least some larval characters presented here as "diagnostic" may need to be modified as more material becomes available for study.

Direct study of the habitats of the type locality has provided much valuable data. We now know the types of clean, cold, higher elevation streams to search for additional populations of *E. frisoni*. Efforts are now under way by some of us to identify these types of habitats in Maine using GIS technologies and these will be the next areas searched.

Lastly, with a much better understanding of the structure of adults and larvae, it is possible to reexamine previously collected material and search for overlooked or misidentified specimens. For example, study of a specimen from New York reported as *E. frisoni* (Jacobus and McCafferty 2001) has showed it to be a misidentified specimen of *E. pleuralis*. Whereas, study of material from the SWRC previously determined to be *E. fragilis* has resulted in one new record for *E. frisoni* for Vermont. With the discovery of *E. frisoni* in Vermont, it is clear the species is not a narrow-endemic. At least one of us believes that a thorough search of suitable higher elevation, cold streams in the White Mountains of New Hampshire, the Adirondacks of New York and the Gaspe area of Quebec should result in the discovery of several new records.

Conclusions

Epeorus frisoni (Burks) is a valid species and is distinctive from its presumed sister species. It is still present at the type locality and there is now no immediate evidence to suggest that its population is not stable or in decline at that site. The habitat association of *E. frisoni* with cold (even in summer) continuously flowing, high gradient, minimally disturbed streams at or above 1200' (~365 m) could be used to identify some of the best high elevation lotic habitats for conservation purposes. Variation among the New England subset of the eastern Nearctic *Epeorus* strongly points to the need for a full revision of this genus.

Acknowledgments

We are especially grateful to the Governing Board for Baxter State Park for granting permission to conduct this study in the park, the Illinois Natural History Survey for providing access to the holotype of *E. frisoni*, and David Funk of the Stroud Water Research Center for providing a large series of *Epeorus* from eastern North America. We also thank Luke Jacobus of the Department of Entomology, Purdue University for providing the specimen from New York.

Literature Cited

Burian, S. K., and K. E. Gibbs. 1991. The Mayflies of Maine: An Annotated Faunal List. Maine Agricultural Experiment Station Technical Bulletin **142**:1–109.

Burks, B. D. 1946. New heptagenine mayflies. Annals of the Entomological Society of America **39**:607–615.

Edmunds, G. F., Jr., S. L. Jensen, and L. Berner. 1976. The Mayflies of North and Central America. University of Minnesota Press, Minneapolis, Minnesota. 330 pp.

Jacobus, L. M., and W. P. McCafferty. 2001. The mayfly fauna of New York State (Insecta: Ephemeroptera). Journal of New York Entomological Society **109**:47–80

SPECIES OF *MIROCULIS* FROM THE SERRANÍA DE CHIRIBIQUETE IN COLOMBIA

J. G. Peters[1], E. Domínguez[2] and A. Currea Dereser[3]

[1] *Entomology, Florida A&M University, Tallahassee, Florida 32307 USA*
[2] *CONICET-Facultad de Ciencias Naturales, Universidad Nacional de Tucumán, Tucumán, Argentina*
[3] *Universidad de Los Andes, Bogotá, Colombia*

Abstract

Miroculis (Miroculis) chiribiquete new species is described from male and female imagos from streams in the region of Serranía de Chiribiquete National Park, Caquetá Department, Colombia. The new species is distinguished from other species of *Miroculis* by hyaline wings in both sexes, the presence of prominent basal hooks on the penes of the male and an expanded opening on sternum 7 of the female. The subgenus *Miroculis* s.s. is redefined to include all species with dorsally-directed, stalked eyes and 5–20 facets in the longest row of the dorsal portion. The female imagos and nymphs of *Miroculis (Miroculis) nebulosus*, a new subgeneric combination, are described for the first time, and the description of the male imago is expanded. *M. nebulosus* is a new record for Colombia. Ecological data and notes on swarming of *M. chiribiquete* new species are included.
Key words: Ephemeroptera; Leptophlebiidae; *Miroculis;* Colombia; Chiribiquete.

Introduction

The genus *Miroculis* was established by Edmunds (1963) for the species *Miroculis rossi* Edmunds from Peru. Eleven more species were later described by Savage and Peters (2003) and Savage (1987b) from Brazil, Colombia and Venezuela. Additional records extend the genus into Argentina in the South (ED), Ecuador in the West (ED) and Trinidad in the North (Savage 1987a). The genus is distinguished by several apomorphies, most notably the tent-shaped costal projection of the hind wing and the long-spike like projection on the male foreclaw in the imago, and in the nymph by the combination of a long distal filament on the gill, ventral glossae and a thick row of pectinate setae on the maxilla (Savage and Peters 1983). Males of many species have stalked eyes, and Savage and Peters (1983) with later modifications by Savage (1987b), divided the genus into subgenera based on the male eyes: *Miroculis* s.s. (long stalks with length greater than width, 10 facets or less in longest row of dorsal portion), *Yaruma* (short wide stalks), *Ommaethus* (without stalks), and *Atroari* (without stalks or with short stalks and more than 11 facets in longest row of dorsal portion). Recent collections near the Parque Nacional Natural Serranía de Chiribiquete in the Amazonian region of Colombia have added to our knowledge of *Miroculis* and revealed a new species that we describe here. Specimens of *Miroculis nebulosus* Savage were also collected, making possible the description of the female imago and nymph.

The Serranía del Chiribiquete is marked by sandstone ridges and a layer of igneous rock relating the region to the ancient Guyanan and Brazilian shields (Zloty and Pritchard 2001). The area receives 3 to 4 thousand mm rainfall each year (Goulding et al. 2003). The study stream of the junior author (AC-D), "La Piscina," is located on the right bank of the Mesay river southwest of the Puerto Abeja Biological Station at the southeastern border of Chiribiquete National Park. It is a small blackwater stream, approximately 2.5-m wide and at most 1.5-m deep, subject to wide variations in depth between dry and rainy season and to smaller oscillations from local rain. During the dry season, sandy river banks are visible. From July to December 2001 water temperature ranged between 20°C and 26°C. In July, 1996, three streams within 15–30 minutes walking distance south of Puerto Abeja were sampled by G. Pritchard and J. Zloty, their Stream #3 being equivalent to "La Piscina" (Zloty and Pritchard personal communication). Stream #1, closest to the station, was 1–2 m wide with dirty brown water, and Stream #2 (15 min walk uphill from #1) was very small with many pools (Zloty, pers. comm.), acidic (pH 5.3) and chemically thin (5.4 mg/l TDS) (Zloty and Pritchard 2001).

Miroculis (Miroculis) chiribiquete new species

(Figs. 1–22)

Male imago (in alcohol). Length of body 5.7–6.4 mm, forewing 5.0–6.1 mm. Head brown, darker posteriorly and around base of eyes, pale between ocelli (Figs. 5 and 7). Antennal scape light brown, pedicel brown, flagellum pale. Dorsal portion of eyes on stalk, with 11–20 facets in longest row (16 on holotype) (Figs. 5–7). Pronotum light brown medially with blackish posterior margin, each lateral third lobed posteriorly and surrounded by blackish brown band; mesonotum brown, a little lighter dorsally in some specimens, with median longitudinal suture occasionally brown to pale, anteronotal transverse impression darker, scutellum blackish brown laterally in some specimens; prosternum brown, meso- and metasterna brown except lateral protuberances of furcasternum darker brown to blackish brown. Wings (Figs. 1–3): longitudinal veins light brown, C, Sc and R_1 of forewing and bases of C, Sc and R of hind wing a little darker, crossveins yellowish brown; bulla prominent on Sc and major branches of Rs; membrane of fore and hind wings pale tan to hyaline. Legs: ratios of segments of foreleg to tibia (tibia 2.2–2.9 mm): femur 0.52–0.59, tarsal segments 0.03–0.04, 0.29–0.32, 0.21–0.26, 0.14–0.19, 0.06–0.07; coxae and trochanters brown; femora of prothoracic legs light brown with blackish brown band subapically and with lateral blackish streak on inner surface, tibiae lighter brown and

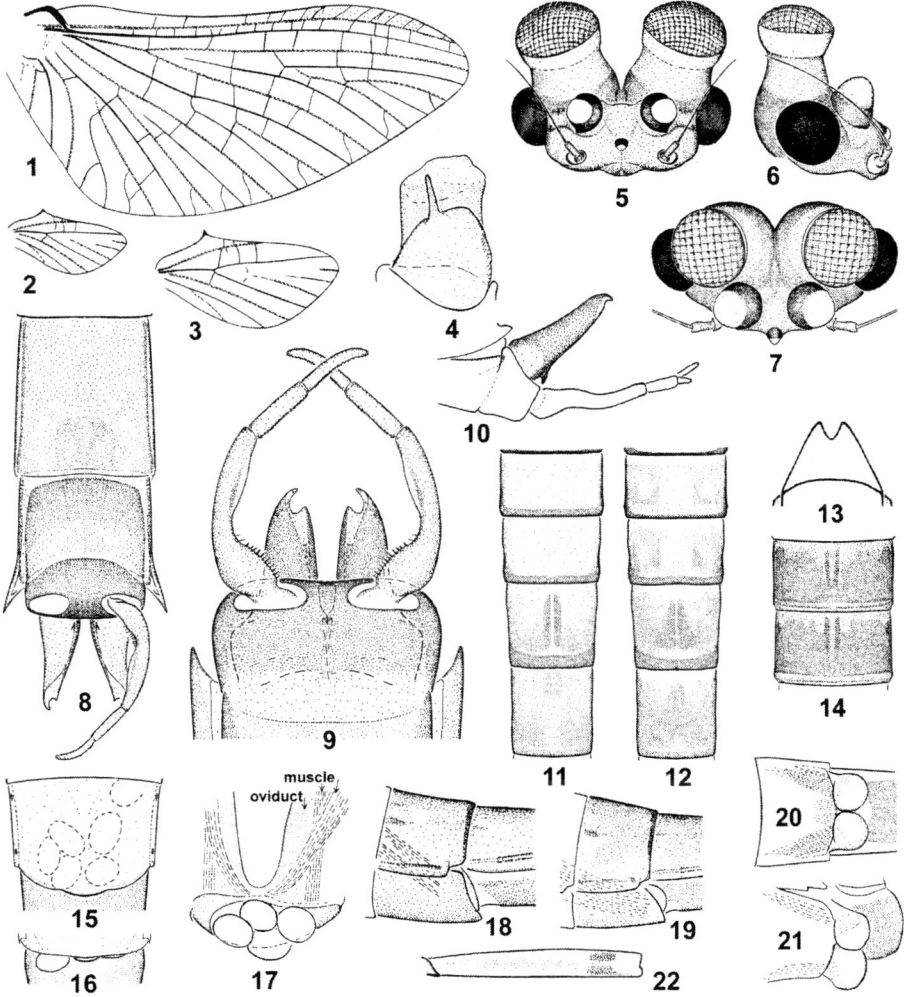

Figures 1–22. *Miroculis chiribiquete* new species. Figs. 1–12, male imago: 1-3, forewing, hind wing and hind wing enlarged; 4, foreclaw; 5, frontal view of head; 6–7, lateral and dorsal views of head (December 2001); 8, sterna 8–9 and genitalia (free); 9, genitalia (ventral, slide mounted); 10, genitalia (lateral); 11–12, terga 4–7 (12, December 2001). Figs. 13–21, female imago: 13, apex of sternum 9; 14, terga 5–6 (December 2001); 15–16, 20–21, sternum 7 variations (21, sublateral view of specimen in Fig. 20); 17, internal aspect of apex of sternum 7 showing broad opening with eggs; 18–9, abdominal segment 7, lateral, variations; 22, female mesothoracic

femur. (Figures prepared from material collected in July 1996 unless indicated; concave veins stippled in wing figures.)

tarsal segments fading to yellowish brown; meso- and metathoracic legs light brown to yellow-brown, femora with dark blackish mark on apical sixth of femur and medial blackish streak on inner surface as in Fig. 22, tibiae with subapical blackish band, tarsal segments pale with apical margins distinct; foreclaw as in Fig. 4. Abdomen: base color of all terga a translucent golden brown with posterior blackish bands on all terga or reduced on tergum 8; tergum 1 dark brown posteriorly to dark brown entirely, terga 2–3 lighter brown posteriorly to brown over most of terga, terga 1–3 with anteromedian dark line; terga 4–5 pale medially with darker posterolateral markings (Fig. 11), to terga 4–5 with small submedian markings (Fig. 12); tergum 6 with heavy median blackish brown longitudinal lines which may form a weak inverted "V"; tergum 7 similar but markings less distinct (Figs. 11 and 12); terga 8–9 brown; tergum 10 brown with median line and heavy posterior band; spiracles weakly indicated, tracheae hyaline to margined in dark gray; sterna translucent golden brown, sometimes with narrow posterior blackish bands, sternum 9 brown anteriorly, paler toward styliger plate (Fig. 8). Genitalia (Figs. 8–10): length of styliger plate about half width, dark brown to blackish brown apically; forceps and penes as in Figs. 8–10, basal segment of forceps brown, apical segments paler; penes brown with hook-like spines near base. Cerci approximately 3 times length of body, terminal filament a little shorter; all caudal filaments light brown, fading apically.

Female imago (in alcohol). Length of body 4.0–6.2 mm, forewing 4.3–6.4 mm. Markings of head, thorax, and legs as in male, with bands on femora (Fig. 22) and tibiae usually darker. Abdominal segments brownish with lateral tracheae black to indistinct, muscles to egg guide visible through integument on 6 and 7 (Figs. 18–19, 21); each tergum with narrow posterior blackish band; terga 1–6 with heavy blackish brown wash dorsally, wash paler laterally and anterolaterally (Figs. 14, 18–19), terga 7–9 with similar pattern but a lighter brown, tergum 10 a richer brown with median brown marking; submedian longitudinal black bars on terga 5–6, bars and wash heaviest on tergum 6 (Fig. 14), bars sometimes present but less distinct on terga 4 and 7; sterna 1–6 light brown, sternum 7 brown, sternum 8 darker brown, sternum 9 dark brown posteriorly. Sternum 7 with broad egg guide (genital extension) (Figs. 15–21); sternum 9 with apical cleft (Fig. 13). Caudal filaments broken off and missing.

Material. Holotype male imago: COLOMBIA: Caquetá: Colombian Amazonia, Stream #1, Trib. of Rio Mesay, approximately 1 km S of Puerto Abeja, NW of Araracuara, 2-VII-1996, G. Pritchard & J. Zloty. Allotype female imago, same data as holotype. Paratypes: 5 male and 5 female imagos, same data as holotype, 2-VII-1996 and 5-VII-1996; 6 male imagos, same data as holotype except Stream #3, 3-VII-1996; 3 male and 5 female imagos, Parque National Natural de Chiribiquete, Quebrada La Piscina (0° 04' 16" N; 72° 26' 48" W), cuenca Rio Mesay, Estacion

Puerto Abeja, trampa de luz, 7-XII-2001, A Currea-Dereser. Holotype, allotype and one male paratype deposited at the Instituto Alexander von Humboldt, Villa de Leyva, Colombia; paratypes at Florida A&M University, Tallahassee; the Fundacíon Istituto Miguel Lillo, Tucumán, Argentina; and Museo de Entomología, Universidad del Valle, Cali, Colombia.

Diagnosis. Miroculis (Miroculis) chiribiquete n. sp. may be distinguished from all species of *Miroculis* by the combination of hyaline wings in both sexes and the following characters: in the male by 1) spines present at base of the penes (Figs. 8–10), and 2) eyes on stalks (length of stalk ≥ than width) with 11–20 facets in the longest row (Figs. 5–7); and in the female by 1) sternum 7 with broad egg guide (genital extension) so some eggs extruding laterally (Figs. 15–19), 2) muscles to genital extension visible through integument on abdominal segments 6–7 (Figs. 18–19, 20–21), and 3) submedian longitudinal black bars heaviest on tergum 6 (Fig. 14).

Discussion. There is variation on markings of the leg. The apical bands on the femora of all legs are always present and distinct, but the inner medial markings and the bands at the apex of the tibiae are faded or absent and never conspicuously marked. The base of the male genital ducts is visible on sternum 8 in some of the specimens (Fig. 8). Although the site was studied for six months, only specimens from July and December can be located, representing the beginning and end of the study period for AC-D. Specimens from July are larger (wing length of males 5.5–6.1 mm, of females 6.0–6.4 mm) than specimens collected in the December dry season (wing length of males 5.0–5.1 mm, of females 4.3–5.2 mm) and color markings are generally darker and more extensive in December (Figs. 12 and 14), although color variations occur among specimens from both collecting dates. The number of facets in the dorsal portion of the eye is greater in July [stream #1: 16–18 (Fig. 5), stream #3 (La Piscina): 18–20) than in December [La Piscina: 11–12 (Figs. 6 and 7)]. Illustrations from earlier in the season at La Piscina show 15 facets (AC-D).

Biology. Miroculis chiribiquete n. sp. is the only species of the genus with hook-like spines at the base of the penes (Figs. 8–10). It is also the only species where female oviducts appear expanded or pulled (compared against *M. nebulosus* Savage, *M. fittkaui* Savage and Peters and *M.* sp. nr. *mourei*). Detailed observations of mating behavior of *Miroculis* sp. nr. *mourei* have shown strong male/male competition for females and secondary swarm behavior as males approach mated females at the oviposition site (ED, unpublished data). In most examined females of *M. chiribiquete* n. sp., a few eggs remained in lateral positions, but in one specimen with no eggs, the oviducts were completely everted (Figs. 20 and 21). We suggest that the basal hooks on *M. chiribiquete* n. sp. are derived to pull female oviducts and prevent other males from fertilizing eggs.

Mating flights were observed every other day at La Piscina from 16 July to 10 December, 2001, and adults were present the entire season. However, peak activity occurred during the last two weeks of August with a smaller peak in early November. The flight took place from 0800 to 1000 hours, with highest densities early in the swarm. The swarm was composed of 60 to 100 individuals, flying at 50 to 70 cm above the water. Males flew in large companies over the main part of the stream where water was swiftest. Observations by Zloty in early July were similar: "1–2 m above the water in the middle of the river, with swarms in excess of 100 individuals; on this occasion, swarming occurred in full sunlight starting about 1100 following morning rain." More specimens, mostly female, were found in the water near the sandy river banks, apparently close to the area of oviposition (AC-D) or concentrated on the surface in one small area near the bank (Zloty). Very few copulas were observed (5 or 6) during the whole period. More females than males were collected, but females were more accessible in funnel traps set in the sandy margins. Most males could only be collected at light. (Males collected by Zloty were collected with a long-handled net.)

Subgenera of Miroculis. Most species of *Miroculis* are described from short series from isolated localities, and the range of individual variation is unknown. *Miroculis chiribiquete* n. sp., described from a longer series from one locality, although from different collectors in different years, displays seasonal variation in coloration, size and the number of facets in the male eyes. For this reason, the number of facets can no longer be used to distinguish subgenera, and it is necessary to redefine the limits of the subgenus *Miroculis* to include all species with dorsally-directed, stalked eyes in the male (presently known to have 5–20 facets in the longest row of the dorsal portion). Except for the number of facets, this follows the key of Savage and Peters (1983). The species *Miroculis (Atroari) nebulosus* is transferred to the subgenus *Miroculis.* Couplet number 1 in the key to subgenera (Savage and Peter 1983) is modified to read as follows:

1   Upper portion of eyes on narrow, dorsally-directed stalks (Figs. 5–7, 26 and Figs. 44–46 in Savage and Peters 1983, Figs. 14 and 15 in Savage 1987b), 5-20 facets in longest row of dorsal portion, eyes separated on meson of head by a distance 0.5–1.5 width of an upper portion .... *Miroculis (Miroculis)*
1'   Upper portion of eyes not on stalks (Figs. 49–54 in Savage and Peters 1983), or upper portion on short, wide stalks (Figs. 47 and 48 in Savage and Peters 1983) separated on meson of head by a distance 0.1–0.2 maximum width of an upper portion; 27–40 medium sized facets in longest row ....
     .... other subgenera of *Miroculis* as keyed in Savage and Peters (1983)

Miroculis (Miroculis) nebulosus Savage, 1987, new subgeneric combination

(Figs. 23–39)

Miroculis (Atroari) *nebulosus* Savage, 1987b:104.

Male imago (in alcohol). As described by Savage (1987b), except body length of male to 6.4 mm, forewing length to 5.1 mm, foreleg length to 5.4 mm, wing color banding heavier and more distinct (Figs. 23 and 24), distance between dorsal portion of compound eyes equal to width of dorsal portion and dorsal portion almost circular (Fig. 26) without evident median projection. Abdomen: tergum 1 brown, terga 2-8 light brown, translucent, with sublateral darker marks as in Fig. 27, marks indistinct on terga 3-4; posterior dark brown bands on terga 2–5, variable on terga 6–8; terga 9–10 uniformly brown, but a little darker medially and posteriorly; spiracles black, tracheae outlined in gray; sterna light brown, same intensity as base color of terga. Genitalia (Fig. 25) as described by Savage (1987b) except apex of styliger plate rounded.

Female imago (in alcohol). Body length 5.2 mm, forewing length 4.4 mm. Head with blackish brown wash, heaviest anteriorly, pale posterior to ocelli and between eyes. Antennal scape pale, pedicel blackish brown, flagellum light brown. Pronotum yellow-brown with blackish median carina and posterior band, each lateral third lobed posteriorly and surrounded by dark brown band with heavier black band medially; mesonotum brown; thoracic sterna brown except a little lighter medially. Wings (Figs. 28 and 29): longitudinal and cross veins brown, with small brownish band around most crossveins, bands heaviest anteriorly (Fig. 28), base and apex of wing also tinged with brown; hind wing veins brown in basal half, fading to hyaline apically, membrane with small brownish apical band (Fig. 29). Coxae brown, trochanters a darker brown, remainder of legs yellow-brown faded to pale on tarsal segments; heavy medial and apical blackish bands on femora of all legs, bands on forefemora heavier than illustrated for mesothoracic femur (Fig. 34), tibiae with a subapical black band. Abdominal terga a light brown with heavy blackish brown wash dorsally (Fig. 32), submedian longitudinal marks and blackish wash anterolaterally on terga 2–6 (Fig. 33), wash a little lighter on terga 7–8; terga 9 and 10 dark posteriorly (Fig. 32); all terga with narrow posterior blackish bands; spiracles black, tracheae margined with black (Fig. 33); sterna light brown, color darker posteriorly on sterna 7–9. Sternum 7 illustrated in Fig. 31 with eggs apparently extruded from single medial opening; sternum 9 with apical cleft typical for genus (Fig. 30). Caudal filaments broken off and missing (from male and nymph, probably with dark rings at base of annulations).

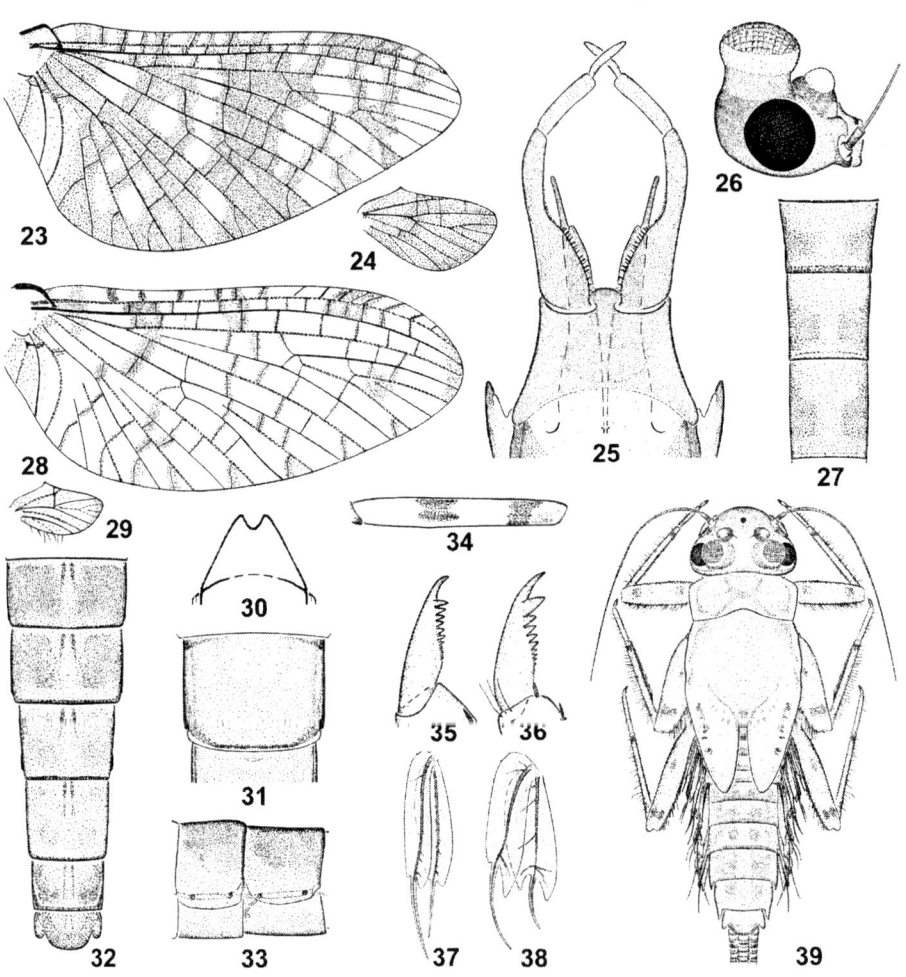

Figures 23–39. *Miroculis nebulosus* Savage. Figs. 23–27, male imago: 23–24, forewing and hind wing; 25, ventral view of genitalia; 26, lateral view of head; 27, terga 5–7. Figs. 28–34, female imago: 28–29, forewing and hind wing; 30, apex of sternum 9; 31, sternum 7; 32, terga 5–10; 33, lateral view of abdominal segments 5–6; 34, mesothoracic femur. Figs. 35–39, mature nymph: 35, foreclaw; 36, mesothoracic claw; 37–38, gill 4 (variations); 39, full nymph, male. (Concave veins stippled in wing figures.)

Mature nymph (in alcohol). Head brown with pale area between eyes in both sexes; mouthparts characteristic for genus (Savage and Peter 1983); antennae pale except pedicel dark brown. Pronotum light yellowish brown, with adult markings visible as in Fig. 39; mesonotum light brown; developing wings in wing pads with numerous mottles; thoracic sterna pale yellowish, except prosternum with darker margins. Legs pale yellowish brown with heavy posterior blackish brown band on femora and tibiae and a second band just basal to middle of femora, this band heavier on inner surface of leg; prothoracic and mesothoracic claw illustrated in Figs. 35 and 36 (see discussion). In males, abdominal terga 1-10 with posterior blackish brown band; terga 1–2 brown medially; terga 3–4 pale medially with lateral brownish markings, terga 5–8 with heavier sublateral and lateral blackish brown marking, marks reduced on terga 9–10 (Fig. 39); sterna pale yellowish brown. In females, a uniform blackish brown wash over terga 1–10, but lighter medially on terga 8–10; sterna yellowish brown. Gills well developed, with additional tracheae forking from median trachea on some gills and outer lobe of dorsal portion as in Figs. 37 and 38; gill membrane light gray. Posterolateral spines on terga 3 or 4–9, progressively larger from terga 5 or 6–9. Caudal filaments with dark bands at annulations.

Material. 4 male imagos, 1 female imago, 4 nymphs. COLOMBIA: Caquetá: Colombian Amazonia, Stream #2 (0° 04' 44" N, 72° 26' 50" W), Trib. Rio Mesay, S of Puerto Abeja and upstream from Stream #1, 2-4-VII-1996, G. Pritchard & J. Zloty.

Diagnosis. The male is keyed in Savage (1987b). In the only key to females (Savage and Peters 1983), *M. nebulosus* is closest to *M. fittkaui* from which it is distinguished by the heavy wing markings (Figs. 28 and 29), heavy medial and apical blackish bands on the femora (Fig. 34), and markings on abdominal tergum 6 which repeat those of preceding terga (Fig. 32). Nymphs will key to *M. colombiensis* Savage and Peters (or to *M. amazonicus* Savage and Peters) from which male nymphs can be distinguished by the developing male eye stalks (Fig. 39). Female nymphs may possibly be distinguished by the continuation of developing dark adult coloration on terga 8-9, but this and possible claw characters may be variable. Although not mentioned in Savage and Peters (1983), the paratype nymphs of *M. colombiensis* show the same developing wing coloration as illustrated for *M. nebulosus* (Fig. 39).

Discussion. *Miroculis nebulosus* was described from Cerra de la Neblina, Venezuela, by Savage (1987b) from male imagos. The presence of this species at Chiribiquete, Colombia, represents a range extension for the species and a new record for Colombia. There are only minor differences between these males and those described by Savage (1987b): the apex of the styliger plate is rounded instead of shallowly notched (Fig. 25) and the wing coloration is more intense (Figs. 23 and 24). The color pattern of the Chiribiquete males matches a variation described for *M. nebulosus* (Savage 1987b) where lateral bands and submedial bands appear to merge (Fig. 27). Eyes are within the variation described by Savage (1987b), although more circular. In the male wing illustrated (Fig. 23), vein MP_2 attaches to vein CuA, but this atypical condition is not found on the other wing of this specimen or on the other males.

The nymphal color pattern is that of imagos, as all nymphs examined were about to emerge. The broad gills with heavy branched tracheation resemble *M. colombiensis*; however, we examined the paratype male nymph of *M. colombiensis*, and its eyes do not show developing stalks and have at least 20 facets in the longest row (eye damaged when mouthparts were dissected, so precise number not known). The illustrated foreclaw (Fig. 35) shows an apical denticle only a little larger than the preceding denticle, but other specimens of *M. nebulosus* have an apical denticle much larger than the preceding denticles resembling that illustrated for the mesothoracic leg (Fig. 36). Paratypes of *M. colombiensis* also have a larger apical denticle on the mesothoracic leg. Because all specimens of both species are teneral adults, the denticle on the foreclaw may be worn in some specimens. Based on the present material, only the developing male eyes will clearly separate *M. nebulosus* from *M. colombiensis*.

Biology. The habitat (Stream #2) was very small with trickling water and many small pools filled with leaf litter. Nymphs were collected from rocks and leaf packs in the pools, and adults were collected from rocks at the margins.

Acknowledgments

We sincerely thank J. Zloty (University of Calgary, Alberta, Canada) for specimens of *Miroculis chiribiquete* n. sp. and *M. nebulosus* and for information on the habitat. We also thank M. del C. Zuñiga (Universidad del Valle, Cali, Colombia) for specimens of *M. chiribiquete.* We especially thank H. M. Savage (Centers for Disease Control, Fort Collins, Colorado) for advice on the status of subgenera, helpful suggestions and review of the manuscript.

Literature Cited

Currea Dereser, A. 2002. Estudio de adultos de Ephemeroptera en el Parque Nacional Natural Chiribiquete (Caquetá-Colombia). Tesis de Licenciatura. Universidad de Los Andes, Bogotá, Colombia.

Edmunds, G. F., Jr. 1963. A new genus and species of mayfly from Peru (Ephemeroptera: Leptophlebiidae). Pan-Pacific Entomologist **39**:34–36.

Goulding, M., R. Barthem, and E. Ferreira. 2003. Atlas of the Amazon. Washington, Smithsonian Books.

Savage, H. M. 1987a. Biogeographic classification of the Neotropical Leptophlebiidae (Ephemeroptera) based upon geological centers of ancestral origin and ecology. Studies on Neotropical Fauna and Environment **22**:199–222.

Savage, H. M. 1987b. Two new species of *Miroculis* from Cerro de la Neblina, Venezuela with new distribution records for *Miroculis fittkaui* and *Microphlebia surinamensis* (Ephemeroptera: Leptophlebiidae). Aquatic Insects **9**:97–108.

Savage, H. M., and W. L. Peters. 2003. Systematics of *Miroculis* and related genera from Northern South America (Ephemeroptera: Leptophlebiidae). Transactions of the American Entomological Society, 2002, **108**:491–600.

Zloty, J., and G. Pritchard. 2001. *Cora chiribiquete* spec. nov., a new damselfly species from Colombia (Zygoptera: Polythoridae), Odonatologica **30**:233–254

THE ANTENNAL SENSILLA OF THE NYMPH OF *EPHEMERA DANICA*

Manuela Rebora and Elda Gaino

Dipartimento di Biologia Animale ed Ecologia, Università degli Studi di Perugia, Via Elce di Sotto, 06123 Perugia, Italy

Abstract

In the present study, a first ultrastructural investigation under Scanning and Transmission Electron Microscopy (SEM, TEM) on the sensilla located on the flagellum of the nymphal antennae of *E. danica* is presented. Each antenna consists of scape, pedicel and a flagellum of 26–27 segments. The flagellar sensilla are mainly located on the dorsal side of the flagellum and they are represented by long trichoid sensilla and shorter basiconic sensilla. The internal structure of these sensilla reveals that the long trichoid sensilla are mechanoreceptors while the shorter basiconic sensilla are uniporous sensilla with a chemo-mechanical function. These chemo-mechanosensory sensilla seem to be very common in mayflies. The presence of these sensilla is discussed in relation with the behavioral ecology of this burrowing insect.

Key words: Sensilla; aquatic insects; mechanoreceptors; chemo- mechano-receptors.

Introduction

Ephemera danica is a burrowing mayfly commonly present in freshwater streams and characterized by a semivoltine life-cycle. At present, the knowledge on the sensory structures of the nymph of this genus and, in particular, on its sensilla, is limited to an ultrastructural investigation by Schmidt (1974), who described the mononematic scolopidia located on the pedicel of the nymph of *Ephemera* sp. The author considered this structure as homologous of the Johnston's organ of other insects, thereby performing a mechanosensory function allowing insects to sense the movement of the flagellum over the pedicel.

The present study represents the first ultrastructural investigation under scanning and transmission electron microscopy (SEM, TEM) on the sensilla located on the flagellum of the nymphal antennae of *E. danica*.

Methods

Mature nymphs of *Ephemera danica* Müller, 1764 were collected in the Nera River (Perugia, Umbria Region, Italy) in spring 2003. The antennae of the nymphs were dissected under a stereomicroscope. The antennal flagella were fixed in 2.5% glutaraldehyde buffered in cacodylate, pH 7.2, for 12 h, repeatedly rinsed in the samebuffer and postfixed in 1% osmium tetroxide for 1 h. For scanning electron microscopy (SEM) observations, the specimens were dehydrated by using ethanol gradients, followed by critical-point drying in a CO_2 Pabisch CPD apparatus. Specimens were mounted on stubs with silver conducting paint, sputter-coated with gold-palladium in a Balzers Union Evaporator, observed and photographed in a Philips EM XL30 of the Electron Microscopy Centre of the University of Perugia.

For transmission electron microscopy (TEM) analysis, selected material was dehydrated in the graded ethanol series and embedded in Epon-Araldite mixture resin. Thin sections, cut on a Reichert ultramicrotome, were collected on formvar-coated copper grids and stained with uranyl acetate and lead citrate. The thin sections were examined with Philips EM 400 transmission electron microscope of the Electron Microscopy Centre of the University of Perugia.

Results

The antenna are held horizontally over the mandibles (Fig. 1), pointing anteriorly to the head. Each antenna consists of scape, pedicel and a flagellum of 26–27 segments. The flagellar sensilla are mainly located on the dorsal side of the antenna (Fig. 2), whose cuticle is irregularly spiny (inset of Fig. 2). The ventral side of the antenna is almost hairless (Fig. 3) and shows a smooth cuticle (inset of Fig. 3). The dorsal side shows in its proximal portion numerous setae located in the middle of each segment (Fig. 2). These setae are represented by two lateral groups of long (500–800 μm long, 7–8 μm wide) trichoid sensilla and one central group of shorter (150–200 μm long, 6–7 μm wide) basiconic sensilla (Fig. 4). Each lateral group of trichoid sensilla shows 6 hairs while the central group of basiconic sensilla is composed of 4–6 hairs. The sensilla tend to decrease in number towards the antennal distal region where only few groups of 2–4 basiconic sensilla are located (Fig. 5).

The very long and thin trichoid sensilla are mechanoreceptors: indeed they emerge from a well developed socket (Fig. 6), showing in section an elastic joint membrane connecting the hair to the socket and a socket septum supporting the tubular body (Fig. 7). The basiconic sensilla emerge from a socket (Fig. 6) and show an apical pore (Fig. 8). In section, the socket shows an elastic joint membrane connecting the hair to the socket and a socket septum supporting the tubular body (Fig. 9). The shaft has two longitudinally separated lumina as shown in longitudinal (Fig. 10) and in cross sections (Fig. 11): the inner lumen containing two dendrites extending along the shaft up to the apical pore and an outer lumen (Figs. 10 and 11). The three dendrites innervating the sensillum (one ending at the base of the hair with a tubular body and two extending along the hair up to the terminal pore) are surrounded by a dendritic sheath (Fig. 12). This morphology is consistent with a chemo-mechanosensory (gustatory) function.

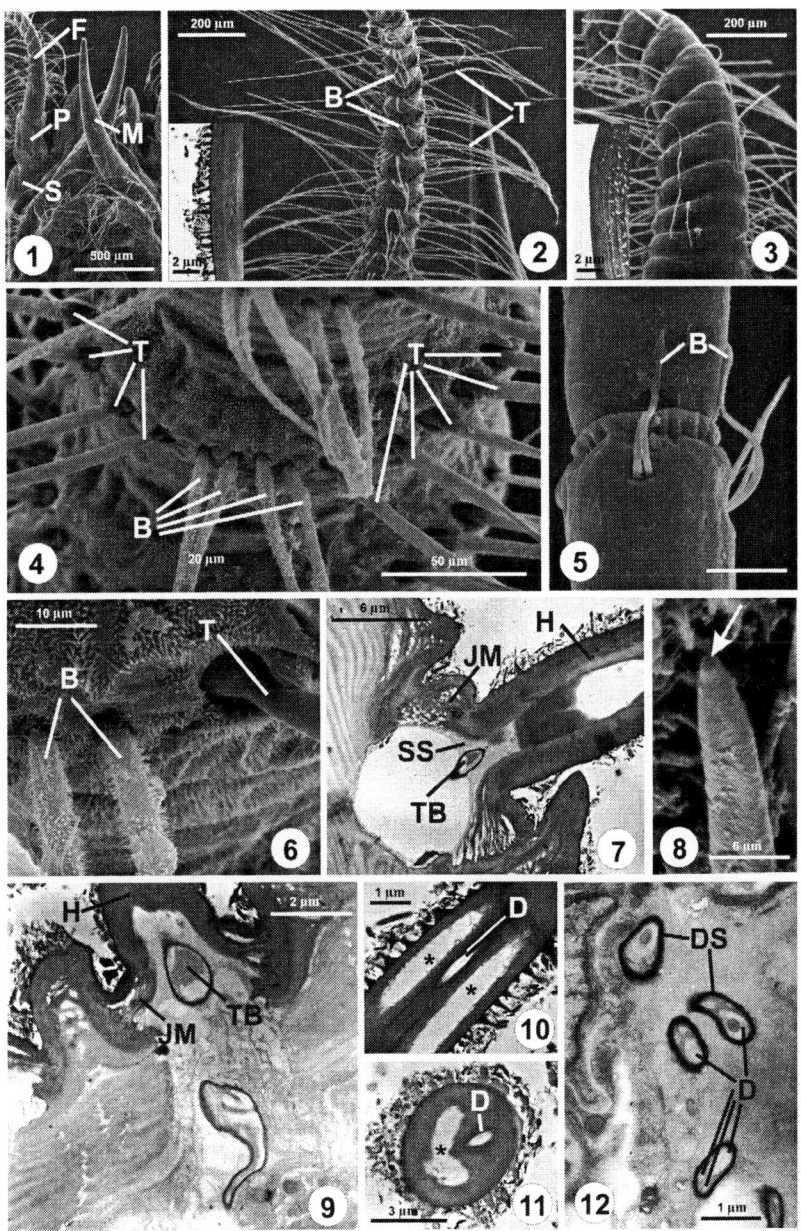

Figures 1–12. Flagellar sensilla of the nymph of *E. danica* under SEM (Figs. 1–6, 8) and TEM (insets of Fig. 2 and 3, Figs. 7, 9–12): Fig. 1. Ventral view of the mandibles (M) and antenna with scape (S), pedicel (P) and flagellum (F); Fig. 2.

Dorsal side of the flagellum in its proximal portion. Note the numerous sensilla (long trichoid sensilla, T and short basiconic sensilla, B) located in the middle of each flagellar segment. Inset shows the irregularly spiny appearance of the cuticle; Fig. 3. Ventral side of the flagellum in its proximal portion. Inset shows the smooth surface of the cuticle; Fig. 4. Detail of Fig. 2 showing two lateral groups of long trichoid sensilla (T) and one central group of shorter basiconic sensilla (B); Fig. 5. Distal portion of the flagellum. Note the low number of basiconic sensilla (B) in this area; Fig. 6. Long trichoid sensilla (T) and basiconic sensilla (B) emerging from a well developed socket. Note the spiny appearance of the cuticle; Fig. 7. Transversal section of a mechanoreceptor at the level of the socket. Note the elastic joint membrane (JM) connecting the hair (H) to the socket and the socket septum (SS) supporting the tubular body (TB); Fig. 8. Apical portion of a basiconic sensillum showing the apical pore (arrow); Fig. 9. Transversal section of a basiconic sensillum at the level of the socket. Note the elastic joint membrane (JM) connecting the hair (H) to the socket and the socket septum (SS) supporting the tubular body (TB); Fig. 10. Longitudinal section of a basiconic sensillum at the level of the shaft. Note the two longitudinally separated lumina. The inner lumen contains two dendrites (D) extending along the shaft; asterisks point out the outer lumen; Fig. 11. Transversal section of a basiconic sensillum at the level of the shaft. Note the two longitudinally separated lumina. The inner lumen contains two dendrites (D) extending along the shaft; asterisk points out the outer lumen; Fig. 12. Three dendrites (D) innervating the basiconic sensillum wrapped by the dendritic sheath (DS).

Discussion

This ultrastructural investigation showed that all the hairs located on the dorsal antennal surface of the nymph of *E. danica* are sensilla. Most of them are groups of very long mechanoreceptors among which groups of shorter uniporous basiconic sensilla with chemo-mechanosensory (gustatory) function are located. The different distribution of sensilla on the two sides of the antenna (the gathering of sensilla on the dorsal side and their absence on the ventral side of the antenna) is in agreement with the behavioural ecology of this species. Indeed, *E. danica* nymphs live in very depressed "U" shaped burrows in sandy sediment (Ladle and Radke, 1990). The burrows are generally closed by coarse sediment anteriorly and the nymph, with upward thrusting movements of the head, creates and maintains a sediment-free space beneath the head and body. This sediment-free space is important for the water current around the insect. Ladle and Radke (1990) stress the importance of the antennae "thickly clothed in long, stiff hairs and held horizontally" in preventing coarse particle of sediment from falling into the space anterior to and ventral to the head. From our investigation, it emerges that the longest hairs are mechanoreceptors probably involved in sensing the presence and dimensions of the sediment over the head. The relevance of tactile stimuli on the dorsal surface of the body in the burrowing of *Ephemera* sp. nymphs was already reported by Grandi (1960) on the base of some behavioural observations.

Also, the spiny pattern of the antennal dorsal cuticle may be correlated with the distribution of the sediment around the nymphal body: the spiny pattern could protect the cuticle from the abrasive effect of the sand grains during the burrowing of the insect. This assumption is confirmed by the smooth surface of the antennal ventral side, which is not in contact with the sediment.

The uniporous chemo-mechanosensory basiconic sensilla show an internal structure very similar to that of the so-called "flat-tipped sensillum", first described in the larvae of *Baetis rhodani* (Gaino and Rebora 1996, 1997, 1998, 1999a) and then observed on their antennae (Gaino and Rebora 1998), legs, tergites and cerci (Gaino and Rebora 1999b), maxillary and labial palps (Gaino and Rebora 2003). A sensillum similar to the flat-tipped sensillum has been also observed on the nymphal antennae of the heptageniid *Rhithrogena* and *Ecdyonurus* (Gaino and Rebora 1996). The bulk of data indicate that this sensillum, having both a mechanical and a chemical sensory function, is fairly common in Ephemeroptera. Behavioral studies on Ephemeroptera proved *Baetis tricaudatus* nymphs use chemical and hydrodynamic stimuli to detect Plecopteran predators (Peckarsky 1980, Peckarsky and Penton 1989). These two types of stimuli seem to interact together, as hypothesized by Ode and Wissinger (1993) in a leptophlebiid. On the basis of previous investigations, it seems acceptable that this sensillum, scattered on the whole insect body, can support a "general gustatory function", useful not only to detect predators but also to locate feeding resources. The position of these gustatory sensilla on the antennae of *E. danica* is particularly relevant in consideration of the peculiar feeding of this mayfly. Indeed, suspended particles that the nymph is going to ingest are drawn by the water current through the "plug" of sediment just above the antennae (Ladle and Radke 1990).

Literature Cited

Gaino, E., and M. Rebora. 1996. Fine structure of flat-tipped antennal sensilla in three species of mayflies (Ephemeroptera). Invertebrate Biology **115**:145–149.

Gaino, E., and M. Rebora. 1997. Antennal cuticular sensilla in some mayflies (Ephemeroptera). Pages 317–325 *in* P. Landolt P., and M. Sartori, editors. Ephemeroptera and Plecoptera: Biology-Ecology-Systematics. Fribourg, MTL.

Gaino, E., and M. Rebora. 1998. Ultrastructure of the antennal sensilla of the mayfly *Baetis rhodani* (Pictet) (Ephemeroptera: Baetidae). International Journal of Insect Morphology and Embryology **27**:143–149.

Gaino, E., and M. Rebora. 1999a. Flat-tipped sensillum in Baetidae (Ephemeroptera): a microcharacter for taxonomic and phylogenetic considerations. Invertebrate Biology **118**:68–74.

Gaino, E., and M. Rebora. 1999b. Larval antennal sensilla in water-living insects. Microscopy Research and Technique **47**:440–457.

Gaino, E., and M. Rebora M. 2003. The sensilla on the labial and maxillary palps of the nymph of *Baetis rhodani* (Ephemeroptera, Baetidae). Pages 445–451 *in* E. Gaino, editor. Research Update on Ephemeroptera and Plecoptera. University of Perugia, Perugia, Italy.

Grandi, M. 1960. Ephemeroidea. Fauna d'Italia. Vol. Edizioni Calderini Bologna.

Ladle, M., and R. Radke R. 1990. Burrowing and feeding behaviour of the larva of *Ephemera danica* Müller (Ephemeroptera: Ephemeridae). Entomologist's Gazette **41**:113–118.

Ode, P., and S. Wissinger. 1993. Interaction between chemical and tactile cues in mayfly detection of stoneflies. Freshwater Biology **30**:351–357.

Peckarsky, B., 1980. Predatory-prey interactions between stoneflies and mayflies: behavioral observations. Ecology **61**(4):932–943.

Peckarsky, B., and M. Penton. 1989. Early warning lowers risk of stonefly predation for a vulnerable mayfly. Oikos **54**:301–309.

Schmidt, K. 1974. Die Mechanorezeptoren im Pedicellus der Eintagsfliegen (Insecta: Ephemeroptera). Zeitschrift fuer Morphologie der Tiere **78**:193–220.

THE TRICORYTHIDAE OF THE ORIENTAL REGION

Pavel Sroka[1] and Tomáš Soldán[2]

[1] *Biological Faculty, the University of South Bohemia, Branišovská 31, 370 05 České Budějovice, Czech Republic*
[2] *Institute of Entomology, Branišovská 31, 370 05 České Budějovice, Czech Republic*

Abstract

Based on detailed taxonomic revision of predominantly larval material of the family Tricorythidae (Ephemeroptera) so far available from the Oriental Region, a new genus, *Sparsorythus* gen. n., is established to include six new species: *S. bifurcatus* sp. n. (larva, imago male and female), *S. dongnai* sp. n. (larva, imago male and female), *S. gracilis* sp. n. (larva), *S. grandis* sp. n. (larva), and *S. ceylonicus* sp. n. (larva), and *S. multilabeculatus* sp. n. (imago male), respective differential diagnoses are presented. *S. jacobsoni* (Ulmer 1913) comb. n. is transferred from the genus *Tricorythus*, now supposed to cover only a part of Afrotropic species of this family. Further five species are described but left unnamed since the larval stage is still unknown. The egg stage (a single polar cap and usually hexagonal exochorionic structures) is described for the first time, relationships of *Sparsorythus* gen. n. to all other genera of the family and their composition are discussed with regard to classical extent of knowledge and rather confusing data in the past. Available data on biology of this new genus are summarized and its distribution with regard to historical biogeography id briefly discussed.

Key words: Tricorythidae; Oriental region; *Sparsorythus* gen. n; new species; taxonomy; biogeography.

Introduction

Eaton (1868) established the genus *Tricorythus* on the basis of *Caenis varicauda* Pictet, 1843–1845 described in adult stage from the Upper Egypt. The same author (Eaton 1884) mentioned adults of two species, *T. varicauda* and *T. discolor* ranging the genus to his Section 7 of Third Series of Group II. of the Genera. Moreover, Eaton (1884: Table 15, Fig. 25) figured the wing of a species called *Tricorythus* (Malay sp.), but he did not describe or mention this material in his text. Ulmer (1913) noted the venation figured by Eaton (1884) corresponds to his new species (*T. jacobsoni*), described after adult material from Java. Later, he revised the adult description (Ulmer 1925) and published a description of larval stage assigned just to this species (Ulmer 1940).

The genus has been classified in the Caenidae and Ephemerellidae and, finally, in the separate family with apparently polyphyletic components by Lestage (1942) to include also the present families Leptohyphidae, Ephemerythidae and Machadorythidae defined later (Edmunds and Traver 1954, Edmunds et al. 1963, Landa and Soldán 1985, McCafferty and Wang 2000).

Three additional genera, namely *Dicercomyzon, Tricorythurus* and *Neurocaenis*, had been established from the continental Afrotropic region.

The genus *Dicercomyzon* Demoulin, 1954 constitutes the monotypic subfamily Dicercomyzinae Edmunds and Traver, 1954.

The genus *Tricorythurus* Lestage, 1942 (type species *Tricorythus latus* Ulmer, 1916, by monotypy, type locality the Congo River, Zaire, Kingshasa, other locality, Bahr-el-Djebel) seems to be disputable. According to Kimmins (1960), the genus *Tricorythurus* was erected on the basis of possibly 3-segmented forceps (Ulmer 1916, Lestage 1942), female and larvae were unknown. Type species was reexamined by Demoulin (1954). He found the forceps to be actually 2-segmented and also other characters are congeneric with the genus *Tricorythus*. Consequently, he synonymized *Tricorythurus* Lestage 1942 with *Tricorythus* Eaton, 1868 (cf. also Demoulin 1970). But some authors, e.g., Hubbard (1990) maintain the opinion on *Tricorythurus* to be a valid taxon, the subgenus of the genus *Tricorythus* in this case.

The genus *Neurocaenis* Navás, 1936 (type species *N. fuscata* Navás, 1936: by original designation, type locality Zaire, Beni, male and larva unknown) was originally defined mainly on the basis of minor differences in the arrangement of cross veins (Navás 1936; Demoulin 1954, 1970) to include, besides type species, 5 species originally described in the genus *Tricorythus*. The only Oriental species known so far, *T. jacobsoni* Ulmer (1913) was also transferred to the genus *Neurocaenis* (Demoulin 1954). Later Demoulin (1970) admitted subgeneric status of this genus. Demoulin (1970) also summarized all Afrotropic taxa of the Tricorythidae: There are 11 species (6 in *Tricorythus* and 5 in *Neurocaenis*) described mostly according to adult stage with the exception of *T. reticulatus* Barnard, 1932 and *T. discolor* (Burmeister 1938), the larval characters of which are mentioned by Barnard (1932) and Crass (1947).

Generic status of *Neurocaenis* was followed by Edmunds and Traver (1954), Edmunds et al. (1963), Hubbard and Peters (1978), Hubbard and Pescador (1978), Hubbard (1990), the latter author also mentioned the occurrence of this genus in Madagascar most probably on the basis of Demoulin's (1958) record on unidentified species of *Neurocaenis* from Madagascar. Although the synomyny of *Tricorythus-Neurocaenis* has been intuitively supposed for a long time (cf., e.g., Soldán 1983, 1991, Sivaramakrishnan and Venkataraman 1987), the respective formal taxonomic act was conducted by Oliarinony et al. (1998). These authors, besides discussing the problems of cross veins character value, also described nine new species of *Tricorythus* from Madagascar.

Recently, new genera *Madecassorythus, Spinirythus* and *Ranorythus* were established by Elouard and Oliarinomy (1997), Oliarinomy and Elouard (1998a), and Oliarinomy and Elouard (1998b).

After description of *Tricorythus jacobsoni* (Ulmer 1913) only sparse literature data exist about occurrence of Tricorythids in the Oriental region. Hubbard and Peters (1978) and Hubbard and Pescador (1978) mentioned the genus *Neurocaenis* (not identified species) from Srí Lanka. Under the generic name *Tricorythus*, the Oriental Tricorythidae were mentioned by Soldán (1983, 1991) from southern part of Vietnam, Sivaramakrishnan and Venkataraman (1987) from southern India (Madras State), and McCafferty and Wang (2000) from Indonesia.

Based on a relatively extensive material from the Oriental region, the principal objectives of this study are, as follows: (i) to determine the proper generic identity of *Tricorythus jacobsoni* Ulmer; (ii) to compare all the other specimens available from the Oriental region with this species and find possible differences in both adult and larval arrangement of morphological characters and (iii) to discuss in detail the relationships of the Oriental species to other Tricorithidae genera (and subfamilies).

Systematic Part

Sparsorythus gen. n.

Mature larva (in alcohol):
Head – Apparently wider than long. Antennae longer than head length. Scape and pedicle well differentiated, pedicle about twice longer than scape.
Labrum – Oval, about twice as wide as long. A single row of medialy diminishing bristles at the anterior margin. Uniformly scattered bristles on dorsal surface. Two submarginal groups of shorter dense tiny bristles on the ventral side of labrum.
Hypopharynx – Lingua rounded, ellipsoidal, with medial incurvation. Lingua longer than superlinguae. Superlinguae triangular, rounded or bluntly pointed at apex, with a row of bristles in distal half of outer margin, diminishing apically.
Mandibles – Outer incisors triangular, with numerous bristles on the ventral side. Apex simple or apically with a pair of short rounded projections. Inner incisors approximately on the same shape and length, with bristles on the vental side and tiny branched setae on the dorsal side. Right prostheca shorter by 1/3 than left one, expanded apically, bifurcated or with several pointed teeth. A group of branched setae longer than prostheca inserted at its base. Left prostheca as long as or slightly shorter than the inner incisor, rod-like, with simple bluntly pointed apex or apex bearing several bluntly pointed projections. A group of branched setae as long as or shorter than prostheca inserted at its base. Outer margin of mandibles with a row of long filtering setae. Short transversal row of setae on the ventral side near mandible base.
Maxillae – Suture of stipes and galeolacinia apparent. Maxillae roughly oblong-shaped or elipsoidal. Apical part of maxilla nearly truncate, the outer apical (galeal) lobe well apparent, produced. Maxilla about by 1/3 longer than wide. Outer margin of maxilla convex without any setation. Anterolateral part of maxilla with a group of long setae. Similar, but smaller setae also on the anterior and medial margin of the

galeolacinia. A regular oblique ventral transversal row of stout setae situated at distal third of galeolacinia. Maxillary palps completely missing. No sclerotised structure at the place of insertion of maxillary palp recognizable.

Labium – Glossae and paraglossae fused into rounded triangular plate with two groups of lateral submarginal setae. The whole plate surrounded with a regular row of setae diminishing apically. Labial palps three-segmented. First segment oblong-shaped, about by 1/3 shorter than the second one, without any setation. Second segment curved, apically bluntly pointed, with a row of stout marginal setae at its outer margin and tiny submarginal setae at its inner margin. Third segment very small, bluntly pointed at apex, without any setation.

Pronotum – Oblong-shaped, as wide as head, about twice longer than wide and about by 1/4 shorter than mesonotum.

Legs – Femora flat, shorter than tibiae. Fore femora with a conspicuous transversal row of flat rounded articulated spines and concave posteromedial margin. Foretibiae with a longitudinal row of spines or bristles near their inner margin. Claws strongly hooked, with two teeth approximately in the middle and one or two subapical spines. Surface of the middle and hind femora covered by spines of various sizes. Posterior margins of the middle and hind femora with spines and setae.

Abdomen – Abdominal segments bearing gills only moderately compressed. Segments VIII, IX and X only slightly longer than segments I–VII. Posterolateral spines of abdominal segments well apparent, as long as 1/3 or 1/4 of segment length. Anlagen of male external genitalia (penis and forceps) well apparent in larvae of the last instar.

Gills – Six or five pairs of gills on abdominal segments II – VII or II – VI. Gills on segments II – VI alike, with rounded or elipsoidal plates and two branched ventral membranous parts with rich filaments. Plates simple, thin, not enforced, with only several tiny and short marginal bristles. Gills on abdominal segment VII strongly reduced (if present). Dorsal plate always missing, ventral membranous part reduced to a single or bifurcated filament.

Caudal filaments – Paracercus always apparently longer than cerci. Sexual dimorphism in arrangement of caudal filaments well visible, cerci and paracercus of males much wider and compressed at base than those in females. Segment of cerci and paracercus without hires and bristles. Spines of different length and shape only round the posterior margin of individual segments.

Imago male (in alcohol): Body smaller and slimmer than female. Head apparently wider than long. Composed eyes large. Antennal pedicle much longer than scape. Pronotum approximately as long as head. Fore wings translucent, colourless or coloured mainly in basal half with dark grey smudges. Costal field with maximally 15 crossweins (if present, mostly badly visible), pterostigma not developed. Posterior margin covered with fine hairs, diminishing distally. Otherwise the venation follows the general tricorythid plan, including the typical "tricorythid fork".

Posterior margin covered with fine hairs, diminishing distally. Hindwings absent. Fore legs with two rounded claws, other pairs with claws dissimilar, one hooked and one rounded. Subgenital plate entire, not divided. Forceps evidently two-segmented. Basal segment of forceps always shorter than distal one. Last segment of forceps provided with numerous attached structures. Penis lobes completely fused, usually forming a rod-like structure, moderately extending basal segment of forceps. The apex of penis rounded, often with apparent medial nick indicating the original separation of mesomeres. Caudal filaments longer than body, without hairs. Paracercus longer than cerci.

Imago female (in alcohol): Body large and robust. Head apparently wider than long. Composed eyes smaller than those in males. Antennal pedicle much longer than scape. Pronotum approximately as long as head. Fore wings translucent, colourless or coloured mainly in basal half with dark grey smudges. Costal field with maximally 15 crossweins (if present, mostly badly visible), pterostigma not developed. Posterior margin covered with fine hairs, diminishing distally. Hindwings absent. Legs slender, long. All pairs with claws dissimilar, one hooked and one rounded. Caudal filaments shorter than body, covered with fine hairs. Paracercus longer than cerci.

Subimago (in alcohol): Similar to imago, with darker wing coloration.

Egg (dissected from mature female, critical point dried, gold-coated, and electronmicrograms taken by scanning microscope Jeol JSM 6300 at 10–15 kV): Generally oval-shaped, always apparently longer than wide, about 150–200 m in length and 70–130 m in width. A single polar cap (type I - noncoiled, single unit cap according to Koss and Edmunds 1974) of about from 1/4 to 1/2 of the egg length always present. Egg pole opposite to the polar cap rounded or bluntly pointed like. Polar cap itself always rounded at apex. Egg surface regularly covered with numerous polygonal (usually hexagonal) exochorionic structures of about 25–30 m in diameter. Micropyle unknown.

Etymology: *Sparsorythus* (m.), from Latin *sparsus* meaning spotted or blotched and *Tricorythus*, related genus. Named after common presence of dark spots or smudges on wings.
Type species: Sparsorythus bifurcatus sp. n.

Species included: *Sparsorythus bifurcatus* sp. n., *S. jacobsoni* (Ulmer 1913) comb. n., *S. dongnai* sp. n., *S. gracilis* sp. n., *S. grandis* sp. n., *S. ceylonicus* sp. n., *S. multilabeculatus* sp. n.

Biology and distribution: Eggs adapted to be attached to substrate (polar caps well developed). Larvae highly rheophilous, at the localities always at places with the highest or very high current velocities preferring stony bottom, passive filtrators, life cycle unknown. Generally known from the Oriental region (Indian subcontinent, South East Asia, Sunda Islands and Philippines).

Sparsorythus bifurcatus sp. n.

(Figs. 1–6, 9–14, 35–40, 52, 67)

Tricorythus sp.1 (partim): Soldán, 1991: 8.
Tricorythus sp. 2 (partim): Soldán, 1991: 8.

Mature larva (in alcohol): General coloration of body pale yellowish with black markings on dorsal side. Uniformly pale yellowish without any markings on ventral side. Body length 4–5 mm. Cerci approximately 1.1 x longer than body, paracercus approximately 1.3 x longer than body. Head apparently wider than long (ratio length : width 1 : 1.4). Eyes black, ocelli grayish. Composed eyes of males considerably larger than those in females (Figs. 5, 6). Distance between composed eyes in males as long as or slightly larger than the eye width. The ratio of distance between composed eyes in females to the eye width 2.2: 1. Hypopharyngeal lingua approximately as wide as long, divided by a short rill in the middle (Fig. 13). Right prostheca (Fig. 14b) notched, triangular, with concave margins and several short pointed teeth, bearing several setae on the inner side. Length of right prostheca about by 1/3 shorter than that of the inner incisor. Left prostheca (Fig. 14a) rod-like, pointed at apex, as long as the inner incisor. Two stout long setae, subequal to prostheca, inserted on its base. Labial plate without a small nick at the middle of anterior margin. Posterior margin of mesonotum overlapping at most the first abdominal segment. Rudimental gill on abdominal segment VII bifurcated, Y-shaped (Fig. 9f). Legs (Fig. 10) relatively robust. Length ratio femur : tibia : tarsus - 1.9 : 2.2 : 1 (fore legs); 2 : 2 : 1 (middle legs); 2.4 : 2.5 : 1 (hind legs). Ratio of femur length : width 1.8 : 1 in all leg pairs. Posterior margin of the middle and hind femora convex with rounded or bluntly pointed setae, irregularly alternating with tiny hairs. Transversal row of setae on fore femora slightly S-shaped. Fore femoral setae rounded at apex and about 4–5 times longer than wide (Fig. 11a). Foretibiae with conspicuous inner submarginal oblique row of setae, narrower and longer than femoral ones (Fig. 11b). Surface of the middle and hind femora sparsely covered by very small spines. Caudal filaments (Fig. 12) with a circles of sparse and small setae on the rounded posterior margins of individual segments. Setae are smaller than one-tenth of the length of segments.

Imago male (in alcohol): Body length 3.5–5.5 mm. Cerci approximately 1.9 times longer than body, paracercus approximately 2.6 times longer than body. Head dark blackish-brown. Prothorax yellowish-brown with black markings. Mesothorax and metathorax brown. Abdomen pale, brownish, with pale black markings. Legs pale brown-yellowish. Cerci pale, yellowish, posterior margins of segments darker, grey. Composed eyes much larger than those in females (Fig. 35). Distance between composed eyes in males slightly shorter than the eye width. Pedicle longer than scape (ratio length of scape : length of pedicle is 1 : 1.8). Penis lobes apparently constricted subapically (Fig. 52). Penis extending the basal segment of forceps and reaching approximately the 1/4 of the second forceps segment. Penis with apparent medial nick indicating the original separation of mesomeres. Venation of the forewing (Fig. 39) relatively variable in number and pattern of cross veins. Forewing basal half dark coloured and distal half translucent. Femora narrower than those in females (Fig. 36). Length ratio femur : tibia : tarsus – 2.5 : 2.5 : 1 (fore legs); 2.9 : 3 : 1 (middle legs); 3.7 : 3.8 : 1 (hind legs).

Imago female (in alcohol): Body length 5–6 mm. Length of cerci approximately 0.92 x body length, length of paracercus approximately 0.96 x body length. Head dark blackish-grey. Prothorax brownish-grey. Mesothorax and metathorax brown. Abdomen and legs brownish-grey. Composed eyes much smaller than those in males (Fig. 35). The ratio of distance between composed eyes to the eye width 2.8 : 1. Scape very small (ratio length of scape : length of pedicle is 1 : 2.3). Venation of the forewing (Fig. 40) relatively variable, similar to male. Forewing dark coloured in their basal half, distal half of wing translucent. Femora wider than those in males (Fig. 36). Length ratio femur : tibia : tarsus - 3 : 3 : 1 (fore legs); 2.9 : 3.2 : 1 (middle legs); 3.5 : 3.4 : 1 (hind legs).

Subimago: Unknown.

Egg (Fig. 67): 175 m long, 74 m wide. Surface with apparent polygonal (mainly hexagonal) structures. Polar cap covers approximately 1/5 of the surface. Egg pole opposite to the polar cap rounded.

Material examined: Holotype: mature larva, Vietnam, Kinh-Dinh River, Nha-Ho, 16. IV. – 4. V. 1982, T. Soldán leg.; paratypes (parts on slides): 7 mature larvae, 12 immature larvae, 132 male imagines and 283 female imagines, same data as holotype, 1 mature larva, Vietnam, Tuan Hai prov., Kinh-Dinh R., Nha-Ho, 2. XI. 1984, T. Soldán leg, 4 mature larvae and 2 female imagines, Vietnam, Dong-Nai prov., Dong-Nai R., Nam Cat Tien res., 6. – 18. XI. 1989, T. Soldán leg.

Holotype and paratypes deposited in the Institute of Entomology, České Budějovice, Czech Republic.

Etymology: Species is named according to a characteristic bifurcation of the last tracheal gill in larvae.

Biology: The Kinh Dinh river at Nha-Ho, about 10 km W of Phan Rang is a large permanent lowland river (150–200 m across), 20–150 cm in mean depth during dry season and about 4–5 m water level fluctuation in the wet season (daily fluctuation of about 10 cm during dry season). The river is regulated in order to supply a system of artificial irrigation and forms a large number of rapids and backwaters. Judging from the primary plant succession on the river bed, this regulation originates from at least 50–70 years ago (Rejmánek, *pers. comm.*). Water is very turbid (transparency at most 15–20 cm), slightly alkaline (pH = 7.2–8.0) and relatively warm (maximal temperatures 24.6–29.8°C by night and day, respectively, in dry season).

Larvae of *S. bifurcatus* sp. n. were collected in various habitats by kicking technique, or occasionally by Surber sampler in April and May 1982 (dry season) and October–November 1984 (wet season).

They evidently prefer gravel bottom riffles or stones from small to medium size (up to 10–15, or 30–40 cm in diameter, respectively). They were never found on pure coarse sand bottom, at mixed sandy and clayey habitats and organic debris or in plat roots or submerged vegetation of *Elodea* sp. and *Polygonum tomentosum*. They were collected only at places with fast to very fast current velocities being never found at habitats with the current lower than 30–40 cm s^{-1}. Most larvae were collected at gravel bottom riffle with more than 60 cm s^{-1}, however some specimens occurred also at places with about 40–50 cm s^{-1} current velocity. They can easily survive fluctuation of current velocity up to more than 100 cm s^{-1} as well as fluctuation of water level (fluctuation observed by 1.5 m during dry season). On the other hand, their survival is apparently limited by gradual drying up of respective habitats.   They were never found in temporary backwaters or pools isolated temporarily for more than one day. Larvae are always solitary to rare at habitats in question, their standing crop never reached more than about 5%, contrary to, e.g., *Rhoenanthus distafurcus* Bae et McCafferty, 1991 (up to 10%), *Potamanthus* (*Potamanthodes*) *formosus* Eaton, 1892 (up to 10%), Baetidae (mostly *Baetis* spp. and *Pseudocloeon*, up to 25%), Leptophlebiidae (mostly *Choroterpes* [*Euthraulus*] and *Choroterpides* sp., up to 15%), Heptageniidae (mostly *Cinignina* spp. up to 10%), and Ephemerellidae and Caenidae (mostly *Ephemerella*, *Torleya* and *Drunella*, and *Caenis* and *Clypeocaenis*, respectively, up to 20%). Life cycle is generally unknown. During dry season, first subimagines emerged about half an hour before sunset, submarine molting occurred shortly after the emergence. Mating flight followed immediately and was finished shortly after the sunset.

Differential diagnosis: Combination of larval characters distinguishing *S. bifurcatus* sp. n. from other species of *Sparsorythus* gen. n. is apparent from Table 1 in Appendix A. Unique characters no. 9, 11 (reach of mesonotum and shape of the last tracheal gill) distinguish it from all other species of the genus. The species has united characters 10, 21, 22 (number of tracheal gills, shape of middle and hind femoral margins) with *S. jacobsoni* comb. n. and 7, 26 (shape of right prostheca, setation on caudal filaments) with *S. dongnai* sp. n. Adults can be compared only with *S. dongnai* sp. n., *S. jacobsoni* comb. n. and *S. multilabeculatus* sp. n. From these, males can be generally distinguished by wing coloration, penis shape and eye size.

Eggs can be distinguished by shape and arrangement of hexagonal structures and relative size of the polar cap.

Sparsorythus jacobsoni comb. n.

Tricorythus jacobsoni Ulmer, 1913: 105, fig. 5, 6.
Tricorythus jacobsoni: Ulmer, 1924: 50, fig. 23, 24.
Tricorythus jacobsoni: Ulmer, 1939: 521, 638, figs. 336–344.

Mature larva (in alcohol): General coloration of body brownish-yellow. Body length 5–6 mm. Cerci approximately 1.5 x longer than body, paracercus approximately 1.6 x longer than body. Head apparently wider than long. Hypopharyngeal lingua approximately as wide as long, without a rill in the middle, with large U-shaped medial incurvation. Superlinguae pointed at apex. Right prostheca notched, with one long curved projection at distal part, bearing several setae on the inner side. Right prostheca about by 1/3 shorter than the inner incisor. Distal part of the left prostheca extended, with several short pointed teeth. Left prostheca about by 1/3 shorter than the inner incisor. Some stout long setae, subequal to prostheca, inserted on its base. Labial plate with a small nick at the middle of anterior margin. Rudimental gill on abdominal segment VII filamentous. Legs relatively robust. Length ratio femur : tibia : tarsus - 2 : 2.7 : 1 (fore legs); 2.6 : 2.8 : 1 (middle legs); 2.2 : 2.6 : 1. (hind legs). Ratio of femur length : width 2 : 1 in all leg pairs. Posterior margins of the middle and hind femora convex with rounded or bluntly pointed setae, irregularly alternating with tiny hairs. Arrangement of setae on the fore femoral dorsal surface irregular. Foretibiae with conspicuous inner submarginal row of setae (shape of this setation unknown). Surface of the middle and hind femora sparsely covered by very small spines. Caudal filaments with circles of sparse setae rounded the posterior margins of individual segments. Setae are smaller than 1/5 of the length of segments.

Imago male (in alcohol): Body length 5–5.5 mm. Cerci approximately 2.5 times longer than body, paracercus approximately 2.6 times longer than body. Head and prothorax dark blackish. Mesothorax yellowish-brown. Metathorax yellowish-brown. Abdomen pale, greyish. Legs pale greyish, femora darker. Cerci pale, greyish, with darker blackish stripes. Penis lobes apparently only slightly constricted subapically. Penis extending the basal segment of forceps and reaching approximately the 1/4 of the second forceps segment. Penis without apparent medial nick indicating the original separation of mesomeres. Venation of the forewing relatively variable in cross veins number and pattern. Cross veins in costal field present, badly visible. Forewing dull, coloured dark grey, with blackish-grey veins.

Imago female (in alcohol): Body length 6 mm. Length of cerci is approximately 0.8 x body length, length of paracercus is approximately 0.9 x body length. General coloration of body dark greyish-yellow with black markings. Prothorax lighter. Ventral side of body greyish-yellow. Venation of the forewing is relatively variable, similar to male. Forewing dull, coloured dark grey, with blackish-grey veins.

Subimago female (in alcohol): Similar to imago, with darker wing coloration.

Egg: Unknown.

Type locality: Wonosobo, Java.

Distribution: Java, Sri Lanka, Sumatra, Philippines.
    The more detailed description of this species in:
    Ulmer G. (1913) Note V. Ephemeriden aus Java, gesammelt von Edw. Jacobson. Notes Leyden mus. 35:102–120. [male described, figured; type: male, Wonosobo, Java].
    Ulmer, G. (1924) Ephemeropteren von den Sunda-Inseln und den Philippinen. Treubia 6:28–91. [male, female described, figured].
    Ulmer G. (1939–1940) Eintagsfliegen (Ephemeroptera) von den Sunda Inseln. Arch. Hydrobiol., Suppl. 16:443–692. [larva described, figured].

Differential diagnosis: Combination of larval characters distinguishing *S. jacobsoni* comb. n. from other species of *Sparsorythus* gen. n. is apparent from Tab.1. Unique characters no. 8, 11, 19 (presence of nick on labium, shape of the last tracheal gill and irregular arrangement of fore femoral setae) distinguish it from all other species of the genus. The species has united characters mainly with *S. grandis* sp. n. (5, 6 – absence of rill on hypopharynx, shape of apex of the left prostheca), and with *S. bifurcatus* sp. n. (10, 21, 22 – number of tracheal gills, shape of middle and hind femoral margins). Adults can be compared only with *S. bifurcatus* sp. n., *S. dongnai* sp. n. and *S. multilabeculatus* sp. n. From these, males can be distinguished by coloration of wings and penis shape.

Sparsorythus dongnai sp. n.

(Figs. 7, 8, 15–19, 41, 42, 51, 53–55, 68)

Mature larva (in alcohol): General coloration of body pale yellowish-brown with black markings on dorsal side similar to *S. bifurcatus* sp. n. Uniformly pale yellowish without any markings on ventral side. Body length 4–5 mm. Cerci are approximately 1.2 times longer than body, paracercus is approximately 1.3 times longer than body. Head apparently wider than long (ratio length : width is 1 : 1.6). Eyes black, ocelli greyish. Composed eyes of males considerably larger than those in females (Figs. 7, 8). Distance between composed eyes in males shorter than the eye width. The ratio of distance between composed eyes in females to the eye width 2.6 : 1. Hypopharyngeal lingua rounded, approximately as wide as long, divided by a short rill in the middle (Fig. 18). Right prostheca (Fig. 19b) notched, triangular, with several long pointed teeth, bearing several setae on the inner side. Length of right prostheca is about 2/3 of the length of inner incisor. Left prostheca (Fig. 19a) rod-like, pointed at apex, slightly shorter than the inner incisor. Several stout long setae, subequal to prostheca, inserted on its base. Labial plate without a small nick at the middle of anterior margin. Posterior margin of mesonotum overlapping at most the fourth abdominal segment in females and the fifth one in males. Rudimental gill on abdominal segment VII absent. Legs (Fig. 15) relatively robust. Length ratio femur : tibia : tarsus – 2.1 : 2.6 : 1 (fore legs); 2.2 : 2.2 : 1 (middle legs); 2.8 : 3 : 1 (hind legs). Ratio of femur length : width 1.9 : 1 (fore legs); 2.1 : 1 (middle and hind legs). Fore femoral setae about 2.5–4 times longer than wide, with blunt apex (Fig. 16a). Fore tibial setae narrower and longer (Fig. 16b). Dorsal surface of the middle and hind femora sparsely covered by very small spines. Middle femoral posterior margins slightly concave, hind femoral ones convex. Individual segments of caudal filaments sparsely rounded at its posterior margins with very small setae. Setae are always smaller than 1/10 of the length of segments (Fig. 17).

Imago male (in alcohol): Body length 5–5.5 mm. Length of cerci approximately 2 x body length, length of paracercus approximately 2.4 x body length. Head pale brown. Prothorax pale yellowish with dark brown markings. Meso- and metathorax pale brown. Abdomen, legs and cerci whitish. Composed eyes are much larger than those in females. Distance between composed eyes in males slightly shorter than the eye width. Pedicle longer than scape (ratio length of scape : length of pedicle is 1 : 2). Penis (Fig. 51) extending to the basal segment of forceps and extending to approximately 1/4 of the second forceps segment. Penis with apparent medial nick indicating the original separation of mesomeres. Wings (Fig. 41) translucent, very slightly coloured in proximal part. Venation of the forewing relatively variable in cross veins number and pattern. Legs see Fig. 53. Length ratio femur : tibia : tarsus – 2.7 : 3 : 1 (fore legs); 2.6 : 2.7 : 1 (middle legs); 3.4 : 3.4 : 1 (hind legs).

Imago female (in alcohol):   Body length 5–6 mm.   Head, prothorax, legs and abdomen brownish-yellow.   Mesothorax and metathorax   brown.   Cerci whitish. Length of cerci is approximately 0.5 x body length, length of paracercus is approximately 0.6 x body length. Composed eyes much smaller than those in males. The ratio of distance between composed eyes to the eye width 2.7 : 1.   Venation of the forewing (Fig. 42) relatively variable, similar to male.   Forewings dark coloured in their basal half, distal half of wing translucent.   Coloration of basal wing part is much darker than in males of this species.   Legs see Fig. 54.   Length ratio femur : tibia : tarsus – 2.6 : 3 : 1 (fore legs); 3 : 3 : 1 (middle legs); 3 : 2.8 : 1 (hind legs).

Egg (Fig.68):   175 m long, 73 m wide.   Surface with exserted polygonal (mainly hexagonal) structures.   Polar cap covers approximately 1/3 of the surface.   Egg pole opposite to the polar cap rounded.

Subimago: Unknown.

Material examined: Holotype: mature larva, Vietnam, Dong-Nai R., Nam Cat Tien res., 6. – 18. xii. 1989, T. Soldán leg.; paratypes (parts on slides): 8 mature larvae, 5 immature larvae, 6 male imagines and 4 female imagines, same data as holotype.

   Holotype   and   paratypes   deposited   in   the   Institute   of   Entomology,   České Budějovice, Czech Republic.

Etymology: Species is named after its type locality – the Dong-Nai River.

Differential diagnosis: Combination of larval characters distinguishing *S. dongnai* sp. n. from other species of *Sparsorythus* gen. n. is apparent from Tab.1.   Unique character no. 9 (reach of mesonotum) distinguish it from all species of the genus. The species has united characters 7, 26 (shape of right prostheca, setation on caudal filaments) with *S. bifurcatus* sp. n.   Adults can be compared only with *S. bifurcatus* sp. n., *S. jacobsoni* comb. n. and *S. multilabeculatus* sp. n.   From these, males can be generally distinguished by wing coloration, penis shape and eye size.   Eggs can be distinguished by shape and arrangement of hexagonal structures and relative size of the polar cap.

Sparsorythus gracilis sp. n.

(Figs. 20–24)

Mature female larva (in alcohol):   Body length 4.9 mm.   Length of cerci is approximately 0.8 x body length, length of paracercus is approximately 0.9 x body length.   General coloration of body surface brownish-yellow.   Gills very pale yellowish.   Ventral side of body pale yellowish.   Head apparently wider than long (ratio length : width is 1 : 1.4). Eyes black, ocelli grayish. The ratio of distance between composed eyes to the eye width   2.6 : 1.   Hypopharyngeal lingua approximately as wide as long, divided by a long rill in the middle. Small nick at the anterior margin of the lingua (Fig. 23). Right prostheca (Fig. 24b) notched, with several pointed teeth, bearing several setae on the inner side.   Length of right prostheca is about 2/3 of the length of the inner incisor. Left prostheca (Fig. 24a) rod-like, bluntly pointed at apex, as long as the inner incisor. Two stout long setae, subequal to prostheca, inserted on its base. Labial plate without a small nick at the middle of anterior margin. Posterior margin of mesonotum overlapping at most the third abdominal segment. Rudimental gill on abdominal segment VII absent. Legs (Fig. 20) with regard to other species of this genus very slim. Length ratio femur : tibia : tarsus − 1.9 : 2.3 : 1 (fore legs); 2 : 2.3 : 1 (middle legs); 2.5 : 2.9 : 1 (hind legs). Ratio of femur length : width 2.5 : 1 (fore legs); 2.6 : 1 (middle legs), 2.9 : 1 (hind legs). Posterior margin of the middle and hind femora slightly concave at its basal half, with rounded or bluntly pointed setae, irregularly alternating with tiny hairs. Transversal row of setae on the fore femora bow-shaped, with a group of chaotically inserted setae near fore femoral posterior margin.   Fore femoral setae about 5–6 times longer than wide, rounded apically (Fig. 21a). Setae on the fore tibiae very thin, spiky (Fig. 21b). Dorsal surface of the middle and hind femora sparsely covered by spines of the small and medium size. Individual segments of caudal filaments rounded at its posterior margin with setae approximately as long as 1/3 of the length of segments. Lateral margins with spiky setae as long as 2/3 of the length of segments. These long spiky setae on both sides of paracercus and only on the inner sides of cerci (Fig. 22).

Male larvae, imagines and subimagines unknown.

Material examined: Holotype: 1 mature larva, India, Madras state, Poona R., Poona, ix. 1962, V. Landa leg.; paratypes (parts on slides): 2 mature larvae, same data as holotype.

   Holotype and paratypes deposited in the Institute of Entomology, České Budějovice, Czech Republic.

Etymology: From Latin *gracilis* meaning slim, the species is named after tenuous body structures (legs, fore tibial setae).

Differential diagnosis: Combination of larval characters distinguishing S. gracilis sp. n. from other species of Sparsorythus gen. n. is apparent from Tab.1.   Unique characters no. 7 (shape of right prostheca), 15, 16, 17 (shape of legs), 20 (bristle like setae on fore tibiae), and 24, 25, 26 (specific setation on caudal filaments) distinguish it from all species of the genus.   The species seems to be well separated from all other species of Sparsorythus gen. n. Most united characters to *S. gracilis* sp. n. can be found in *S. ceylonicus* sp. n. (4, 19, 23 − presence of nick on hypopharynx, arrangement of fore femoral setae, shape of setae on middle and hind femoral surface).

Sparsorythus grandis sp. n.

(Figs. 25–29)

Mature female larva (in alcohol):   Body length 8 mm.   Length of cerci is approximately 0.9 x body length, length of paracercus is approximately 1.1 x body length. Body coloration dark, brownish-black. Gills blackish with lighter margins. Ventral surface brownish. Head apparently wider than long (ratio length : width is 1 : 1.5). Eyes black, ocelli grayish. The ratio of distance between composed eyes in females to the eye width 3 : 1. Hypopharyngeal lingua apparently wider than long, without a rill in the middle (Fig. 28). Right prostheca (Fig. 29b) notched, triangular, with concave margins and many pointed teeths and extremities, bearing several setae on the inner side. Length of right prostheca is about 2/3 of the length of the inner incisor. Distal part of left prostheca (Fig. 29a) extended, with several pointed teeths. Left prostheca approximately as long as the inner incisor. Some stout long setae, subequal to prostheca, inserted on its base. Labial plate without a small nick at the middle of anterior margin. Posterior margin of mesonotum overlapping at most the third abdominal segment. Rudimental gill on abdominal segment VII absent. Legs (Fig. 25) relatively robust. Length ratio femur : tibia : tarsus – 2.4 : 3 : 1 (fore legs); 2.8 : 2.9 : 1 (middle legs); 3.2 : 3.7 : 1 (hind legs). Ratio of femur length : width 2 : 1 (all leg pairs). Medium femora slightly concave at its basal half, hind femora at basal half approximately straight.   Middle and hind femoral margins with rounded or bluntly pointed setae, irregularly alternating with tiny hairs. Transversal row of setae on the fore femoral dorsal surface S-shaped. Fore femoral setae about 4 to 5.5 times longer than wide, rounded apically, with apical part extended (Fig. 26a). Setae on the fore tibiae narrower, long, not extended at its apical part (Fig. 26b). Surface of the middle and hind femora sparsely covered by small setae. Individual segments of caudal filaments thickly rounded at its posterior margins by setae. Setae approximately as long as 1/5 of the length of segments (Fig. 27).

Male larvae, imagines and subimagines unknown.

Material examined: Holotype (parts on slides): mature larva, Indonesia, West Java, Ciwalen riv., NR Puncak (Gudung Gede), ca 1360 m a. s. l., 8. iv. – 3. ix. 1983, P. Sporrer leg.; paratype: 1 mature larva, same data as holotype.

Holotype deposited in the collection of Agriculture and Mechanical University, Tallahassee, Florida, paratype in the Institute of Entomology, České Budějovice, Czech Republic.

Etymology: From Latin *grandis* meaning large, the species is named after its relatively robust body.

Differential diagnosis: Combination of larval characters distinguishing *S. grandis* sp. n. from other species of *Sparsorythus* gen. n. is apparent from Tab.1. Unique characters no. 3, 7 (shape of hypopharyngeal lingua, shape of right prostheca) distinguish it from all species of the genus. Most united characters to *S. grandis* sp. n. can be found in *S. jacobsoni* comb. n.(5, 6 – absence of hypopharyngeal rill, shape of apex of the right prostheca).

Sparsorythus ceylonicus sp. n.

(Figs. 30–34)

Mature male larva (in alcohol): Body length 4–5 mm. Length of cerci is approximately 0.9 x body length, length of paracercus is approximately 1.1 x body length. General coloration of body yellowish-brown. Uniformly pale yellowish on ventral side. Head apparently wider than long (ratio length : width is 1 : 1.4). Eyes black, ocelli grayish. The ratio of distance between composed eyes to the eye width 2.8 : 1. Size of composed eyes comparable with female larvae of other species of the genus. Hypopharyngeal lingua rounded, approximately as long as wide, divided by a short rill in the middle (Fig. 33). Right prostheca (Fig. 34b) bifurcated, with a small number of projections, bearing several setae on the inner side. Length of right prostheca is about 2/3 of the length of the inner incisor. Left prostheca (Fig. 34a) rod-like, pointed apically, subequal to the inner incisor. Some stout long setae, as long as prostheca, inserted on its base. Labial plate without a small nick at the middle of anterior margin. Posterior margin of mesonotum overlapping at most the second abdominal segment. Rudimental gill on abdominal segment VII absent. Legs (Fig. 30) relatively robust. Length ratio femur : tibia : tarsus – 1.7 : 2.2 : 1 (fore legs); 2 : 2 : 1 (middle legs); 2.6 : 1.9 : 1 (hind legs). Ratio of femur length : width 1.6 : 1 (fore legs); 2 : 1 (middle and hind legs). Posterior margins of the middle and hind femora slightly concave at its basal half, with sparse rounded or bluntly pointed setae, irregularly alternating with tiny hairs. Transversal row of setae on the fore femoral dorsal surface bow-shaped. Setae on the fore femora (Fig. 31a) relatively wide near its base, narrowing apically. Apex blunt. Fore femoral setae about 3 to 4.5 times longer than wide. Fore tibial setae (Fig. 31b) narrower and longer. Dorsal surface of the middle and hind femora very sparsely covered by some setae of various sizes, including relatively big ones. Individual segments of caudal filaments sparsely rounded at its posterior margins by small setae. Setae are smaller than 1/3 of the length of segments (Fig. 32).

Female larvae, imagines and subimagines unknown.

Material examined: Holotype (parts on slides): mature larva, Sri Lanka, Ratnapura dist., Kukula Ganga, Waddagala, 17. iv. 1973, Dawis and Rowe leg.

Holotype deposited in the collection of Agriculture and Mechanical University, Tallahassee, Florida.

Etymology: The species is named after the type locality (Sri Lanka, formerly Ceylon).

Differential diagnosis: Combination of larval characters distinguishing *S. ceylonicus* sp. n. from other species of *Sparsorythus* gen. n. is apparent from Tab. 1. United characters to *S. ceylonicus* sp. n. can be found mainly in *S. gracilis* sp. n. (4, 19, 21, 22, 23 – presence of nick on hypopharynx, arrangement of fore femoral setae, shape of middle and hind femoral posterior margins, shape of setation on the middle and hind femoral dorsal surface).

Sparsorythus multilabeculatus sp. n.

(Figs. 43, 49, 50, 56)

Imago male (in alcohol): Body length 3 mm. Cerci approximately 3.3 x longer than body, paracercus approximately 5 x longer than body. Head, prothorax, abdomen, legs and cerci pale greyish-brown. Meso and metathorax brown. Composed eyes (Fig. 49) moderately enlarged (the ratio of distance between composed eyes to the eye width 2.4 : 1.). Pedicle much longer than scape (ratio length of scape : length of pedicle is 1 : 2.6) Penis extending the basal segment of forceps and reach to approximately 1/3 of the second forceps segment. Some small thorn-like structures on the subgenital plate near penis base. Penis (Fig. 50) with apparent medial nick indicating the original separation of mesomeres. Wing (Fig. 43) with typically organised smudges of various intensity. Most dark smudges in fields C, Sc, A, and near MA1 – MA2 fork. Lighter blotches in fields R1 and MP1. Legs see Fig. 56. Length ratio femur : tibia : tarsus – 2.5 : 2.6 : 1 (fore legs); 2.9 : 2.7 : 1 (middle legs); 3.5 : 3.3 : 1 (hind legs).

Larvae, female imagines and subimagines unknown.

Material examined: Holotype: imago male, Vietnam, Dong-Nai Prov., Dong-Nai R., Nam Cat Tien res., 6. – 18. xi. 1989, T. Soldán lgt.; paratypes (parts on slides): 27 male imagines, same data as holotype.

Holotype and paratypes deposited in the Institute of Entomology, České Budějovice, Czech Republic.

Etymology: The species is named after the presence of more dark spots on its wings.

Differential diagnosis: Adults of this species can be compared only with *S. bifurcatus* sp. n., *S. dongnai* sp. n. and *S. jacobsoni* comb. n. From these, males can be distinguished by coloration of wings, penis shape and eye size.

Sparsorythus sp. 1

(Figs. 44, 57, 58, 69)

Imago female (in alcohol): Body length 3.5–4 mm. Head, prothorax and abdomen greyish-brown. Mesothorax, metathorax and legs brownish-yellow. Dorsal side of body and cerci whitish. Forewing (Fig. 44) dark coloured in their basal half, distal half of wing translucent. Legs see Fig. 57. Length ratio femur : tibia : tarsus – 2.8 : 3.1 : 1 (fore legs); 2.5 : 2.6 : 1 (middle legs); 3.3 : 3.7 : 1 (hind legs).

Egg (Fig. 69): 160 m long, 100 m wide. Surface with exserted polygonal (mainly hexagonal) structures. Polar cap covers approximately 1/4 of the surface. Egg pole opposite to the polar cap bluntly pointed like.

Larvae, male imagines and subimagines unknown.

Material examined: Female imagines (parts on slides), Vietnam, Dong-Nai prov., Dong-Nai riv., Nam Cat Tien res., 6. – 18. xii. 1989, T. Soldán leg.
    Ca 100 ex. deposited in the Institute of Entomology, České Budějovice, Czech Republic.

Sparsorythus sp. 2

(Figs. 45, 59, 60, 70)

Imago female (in alcohol): Body length 4 mm. General coloration of body brownish-black. The ratio of distance between composed eyes to the eye width 1 : 2.5. Cerci pale yellowish with darker grey stripes. Wings (Fig. 45) very lightly brownish, mainly in fields C, Sc. Legs see Fig. 59. Length ratio femur : tibia : tarsus – 2.3 : 2.7 : 1 (fore legs); 2.4 : 2.4 : 1 (middle legs); 2.8 : 2.9 : 1 (hind legs).

Egg (Fig. 70): 190 m long, 115 m wide. Surface with slightly exserted polygonal (mainly hexagonal) structures. Polar cap covers approximately 1/2 of the surface. Egg pole opposite to the polar cap bluntly pointed like.

Larvae, male imagines and subimagines unknown.

Material examined: Female imagines (parts on slides), Thailand, Chiengmai prov., Chiengmai, Mae Ping, 9. iv. 1964, W.L. & J.G. Peters leg.

Two ex. deposited in the Institute of Entomology, České Budějovice, Czech Republic.

Sparsorythus sp. 3

(Figs. 46, 61, 62, 71)

Imago female (in alcohol): Body length 5–6 mm. Length of cerci is approximately 0.9 x body length, length of paracercus is approximately 1.1 x body length. Head, prothorax and legs yellowish-grey. Meso and metathorax brownish, abdomen yellowish-brown. Dorsal side of body very pale yellowish. Cerci whitish. Wings (Fig. 46) translucent, whitish. The ratio of distance between composed eyes to the eye width 2.8 : 1. Legs see Fig. 61. Length ratio femur : tibia : tarsus – 2.5 : 2.9 : 1 (fore legs); 2.7 : 2.9 : 1 (middle legs); 3.3 : 3.3 : 1 (hind legs).

Egg (Fig. 71): 190 m long, 125 m wide. Surface with slightly exserted polygonal (mainly hexagonal) structures. Polar cap covers approximately 1/3 of the surface. Egg pole opposite to the polar cap bluntly pointed like.

Larvae, male imagines and subimagines unknown.

Material examined: Female imagines (parts on slides), Sri Lanka, Kandy dist., Kandy peak, Vievo motel, 17. iv. 1973, Dawis & Rowe leg.

8 ex. deposited in the Institute of Entomology, České Budějovice, Czech Republic.

Sparsorythus sp. 4

(Figs. 47, 63, 64, 72)

Imago female (in alcohol): Body length 4–4.6 mm. Length of cerci is approximately 0.8 x body length, length of paracercus is approximately 0.9 x body length. General coloration of body brownish-yellow. Cerci whitish. Wings (Fig. 47) translucent, with very pale brown smudges in proximal part. The ratio of distance between composed eyes to the eye width 3.2 : 1. Legs see Fig. 63. Length ratio femur : tibia : tarsus – 3 : 3.2 : 1 (fore legs); 3.2 : 3 : 1 (middle legs); 4.8 : 4.2 : 1 (hind legs).

Egg (Fig. 72): 185 m long, 115 m wide. Surface almost smooth, with polygonal (mainly hexagonal) structures only very slightly exserted. Polar cap covers approximately 1/4 of the surface. Egg pole opposite to the polar cap rounded.

Larvae, male imagines and subimagines unknown.

Material examined: Female imagines (parts on slides), Philippines, Mindanao, Mt. Apo School, 15 km SW Davao, ca 500 m a. s. l., 22. – 31. x. 1965, D. Davis leg.

Fifteen ex. deposited in the Institute of Entomology, České Budějovice, Czech Republic.

Sparsorythus sp. 5

(Figs. 48, 65, 66, 73)

Imago female (in alcohol): Body length 4–4.5 mm. Length of cerci approximately 0.5 x body length, length of paracercus approximately 0.6 x body length. General coloration of body black. Cerci greyish with darker grey stripes. Wings (Fig. 48) black. The ratio of distance between composed eyes to the eye width 3.3 : 1. Legs see Fig. 65. Length ratio femur : tibia : tarsus – 3.2 : 4 : 1 (fore legs); 3.9 : 4 : 1 (middle legs); 4.5 : 5 : 1 (hind legs).

Egg (Fig. 73): 190 m long, 120 m wide. Surface with only slightly exserted areas between polygonal structures. Polar cap covers approximately 1/4 of the surface. Egg pole opposite to the polar cap rounded.

Larvae, male imagines and subimagines unknown.

Material examined: Female imagines (parts on slides), Indonesia, Sulawesi – Utara, Dumoga-Bone, NP Sungai Tupah, 5. viii. 1985, D. Dudgeon leg.

26 ex. deposited in the Institute of Entomology, České Budějovice, Czech Republic.

Discussion

Systematics. Altogether, six genera of the family Tricorythidae have been described from the Afrotropic (Ethiopian) Region (including Madagascar) so far.

The genus *Dicercomyzon* Demoulin, 1954 (separate subfamily Dicercomyzinae) is undoubtedly well characterized by its prominent apomorphies mainly in the larval stage: highly expanded femora, dorsoventrally flattened body, pro- and mesosternum with a disc of friction hairs, superlinguae of hypopharynx highly developed laterally (Demoulin 1954, 1970, Kimmins 1957, Edmunds and Traver 1954, Edmunds et al. 1963, Landa and Soldán 1985, McCafferty and Wang 2000). In adults, penes are slightly to deeply divided, associated with auxilliary processes (McCafferty and Wang 2000).

Genera *Madecassorythus*, *Spinirythus* and *Ranorythus* represent either the most plesiomorphic lineage within the Tricorythidae (*Madecassorythus* and *Spinirythus*), or intermediary lineage to that represented by afrotropical tricorythids (*Ranorythus*). Genera *Madecassorythus* and *Spinirythus* were separated into the subfamily Madecassorythinae by Oliarinomy and Elouard (1997). The same authors placed genus *Ranorythus* into the separate subfamily Ranorythinae (Oliarinomy and Elouard 1998). This subfamilial classification was discussed by McCafferty and Wang (2000) who point out, e.g., lack of some synapomorphies in *Tricorythus* and *Ranorythus*, even suggesting by default that *Ranorythus* even belongs to the Tricorythinae lineage.

The genus *Spinirythus* differs from *Sparsorythus* gen. n. in imaginal stage (larvae of *Spinirythus* remain unknown) by penis shape and degree of paracercus reduction. *Spinirythus* has entirely separated penial lobes associated with lamellar auxilliary processes. Penis and auxilliary processes are approximately of the same size and shape, paracercus reduced.

Second genus of the subfamily Madecassorythinae, *Madecassorythus*, is characterized also by separated penial lobes with auxilliary processes (penis and auxilliary processes differently formed, penis lobes much longer than gonostyles), paracercus well developed. Both genera of the subfamily Madecassorythinae have big sexual dimorphism in eye size (composed eyes in males much larger than in females).

On the other hand, *Sparsorythus* gen. n. has completely fused penial lobes without auxilliary processes, paracercus well developed. Sexual dimorphism in eye size often pronounced but probably do not represent a truly consistent character (see below). Larvae of *Madecassorythus* have following characters, which separate them from the *Sparsorythus* gen. n.: presence of maxillar palp, absence of regular row of setae on the fore tibiae. In *Madecassorythus* larvae, tracheal gill on the abdominal segment VII always missing. (But in some *Sparsorythus* gen. n. species also present only five pairs of tracheal gills like in *Madecassorythus*).

The genus *Ranorythus* is characterized in imaginal stage (larvae remain unknown) by partly (in proximal part) fused penial lobes, distally separated. Auxilliary processes are absent. Sexual dimorphism in eye size well apparent. Paracercus reduced in males only, in females paracercus normally developed.

Difference between *Sparsorythus* gen. n. and *Tricorythus* is mainly completely reduced maxillar palp in the larval stage of *Sparsorythus* gen. n. Other characters are not consistent throughout all species of this genus. Sexual dimorphism in eye size in *Tricorythus* always absent, in *Sparsorythus* gen. n. present, but probably in a variable degree. In some species, difference in eye size between sexes is really big (*S. bifurcatus* sp. n., *S. dongnai* sp. n.). On the other hand, some species have relatively small eyes of males (*S. ceylonicus* sp. n., *S. multilabeculatus* sp. n.). Unfortunately, there we have not any comparison with females of these species. Tracheal gill on abdominal segment VII in *Tricorythus* absent, in *Sparsorythus* gen. n. sometimes present, but not in all species.

As noted above, *Sparsorythus jacobsoni* comb. n. was originally described by Ulmer (1913) after male adult stages as *Tricorythus jacobsoni*. Later, the more detailed description of male adults was given by the same author (Ulmer, 1925). In this study, female adults from the same type locality as males were also described. Fifteen years later, Ulmer described a larva, which in his opinion belongs to *T. jacobsoni*. Larvae were collected in a different localities than imagines and there is not any evidence, which associates larvae described by Ulmer in 1940 with imagines described by him before. It is a question if this larva really belongs to *S. jacobsoni* comb. n. The nymph figured by Edmunds et al. (1963), collected in the Philippines, and called "*Neurocaenis jacobsoni* Ulmer ?" is evidently a different species than Ulmer's larva, described in 1940 and declared to be a larva of *T. jacobsoni*. In larva, figured by Edmunds et al. (1963), the tracheal gill on abdominal segment VII is absent and also the shape of its legs is different from Ulmer's larva, similar to *S. gracilis* sp. n.

Establishing *Sparsorythus bifurcatus* sp. n. as a type species of *Sparsorythus* gen. n. (instead of *S. jacobsoni* comb. n.) rests upon our better knowledge of *S. bifurcatus* sp. n. In this species, that the described larva really belongs to imago (the same type locality of these with the only tricorythid species founded) is a greater probability. Another occasion is nonavailability of any comparative material of *S. jacobsoni* comb. n.

However, within the "true" African *Tricorythus* are still open questions. Quite recently, Barber-Jones (2004) indicates there are only two species, namely *T. reticulatus* Barnard, 1932 and *T. discolor* (Burmeister, 1938) among the representatives of the genus Tricorythus in South Africa, which really can be classified within this genus. The others might represent an undescribed genus, possibly related to the genus *Ranorythus*.

Oriental Tricorythidae significantly differ from African ones. The most explicit difference is completely reduced maxillar palp in all oriental species. All African Tricorythidae have big, well-developed maxillar palp. Total reducing of maxillar palp is a big morphological change and it was very probably unique event, which happened to one collective ancestor of all present oriental Tricorythidae. Within the genus *Sparsorythus* gen. n. can be found a big variability (in some species present six pairs of tracheal gills, in some only five; in some species is difference in eye size between sexes big and in some species is small). But absence of maxillar palp shows their collective origin.

Genus *Sparsorythus* gen. n. is an advanced group from the Tricorythidae family with many apomorphies (reduced hind wings, completely fused penial lobes without auxilliary processes, missing maxillar palp, reduced or missing gill on abdominal segment VII, big sexual dimorphism in eye size in some species). Diversity of the family Tricorythidae in Asia is very probably much bigger than species described in this study. Establishing of the genus *Sparsorythus* gen. n. like a taxon involving all Asian Tricorythidae is not certainly the final status.

Distribution and Biogeography

The family Tricorythidae appear to have evolved primarily in Gondwanaland. All the subfamilies occur in Africa or Madagascar. Only the genus *Sparsorythus* is found in Sri Lanka and Southeast Asia. These oriental Tricorythidae are geographically isolated from African species for more than 100 mil. years. After seceding of Indian subcontinent and Africa, Indian species developed separately. After connecting of Indian subcontinent and Asia (about 45 mil. years ago), mayflies from India expanded into the rest of the South-east Asia (Edmunds 1979), Jacob (2003).

The genus *Sparsorythus* was found in almost all regions of the South-East Asia (Sri Lanka, India, Thailand, Vietnam, Indonesia and Philippines).

Acknowledgments

We are indebted to Janice G. and the late professor William L. Peters, A & M University, Tallahassee for providing us with some material. We also thank K. Bláhová for technical assistance. This study was partly supported by Reasearch Project No. 1QS500070505 by Grant Agency of the Academy of Sciences.

Literature Cited

Barber-James, H. M. *In press*. A synpopsis of the Afrotropical Tricorythidae. Pages 187–20 *in* F. R. Hauer, J. A. Stanford, and R. L. Newell (editors). International Advances in the Ecology, Zoogeography and Systematics of Mayflies and Stoneflies. University of California Press. Berkeley, California, USA.

Barnard, K. H. 1932. South African May-flies (Ephemeroptera). Transactions of the Royal Society of South Africa **20**:201–259.

Corbet, P. S. 1960. Larvae of certain East African Ephemeroptera. Revue de zoologie et de botanique africaines **61**:119–129.

Crass, R. S. 1947. The May-Flies (Ephemeroptera) of Natal and the Eastern Cape. Annals of the Natal Museum **11**:37–110.

Demoulin, G. 1954. Recherches critiques sur les Ephéméroptères Tricorythidae d'Afrique et d'Asie. Bulletin et Annales de al Societe Royale d'Entomologie de Belgique **90**:264–277.

Demoulin, G. 1954. Description préliminaire d'un type larvaire nouveau d'Ephéméroptères Tricorythidae du Congo Belge. Bulletin de l'Institut Royale des Sciences Naturelles de Belgique **30**(6):1–4.

Demoulin, G. 1957. Le type larvaire probable des Tricorythus Eaton (Ephemeroptera: Tricorythidae). Bulletin de l'Institut Royale des Sciences Naturelles de Belgique **33**(19):1–4.

Demoulin, G. 1958. Un curieuse larve d'Ephéméroptère de l'Angola portugais. Bulletin et Annales de al Societe Royale d'Entomologie de Belgique **95** (7-8):249–252.

Demoulin, G. 1970. Ephemeroptera des faunes éthiopienne et malgache. South African Animal Life, **14**:24–170.

Eaton, A. E. 1868. An outline of a re-arrangement of the genera of Ephemeridae. Entomologist's Monthly Magazine 5:82–91.

Eaton, A. E. 1883–1888. A Revisional Monograph of Recent Ephemeridae or Mayflies. Transactions of the Linnean Society of London, Second Series, Zoology 3:1-352, 65 pl.

Edmunds Jr., G. F., R. K. Allen, and W. L. Peters. 1963. An Annotated Key to the Nymphs of the Families and Subfamilies of Mayflies (Ephemeroptera). Univ. Utah Biol. Ser. **13**(1), pp. 53.

Edmunds Jr., G. F., and J. R. Traver. 1954. An outline of a reclassification of the Ephemeroptera. Proceedings of the Entomological Society of Washington **56**:236–240.

Edmunds Jr., G. F. 1979. Biogeografical relationships of the Oriental and Ethiopian Mayflies. Pages 11–14 *in* K. Pasternak, and R. Sowa, editors. Proceedings of the Second International Conference on Ephemeroptera.. Państwowe Wydawnictwo Naukowe, Warszawa – Kraków, 312 pp.

Elouard, J.-M., and R. Oliarinony. 1997. Biodiversité aquatique de Madagascar. 6 - *Madecassorythus* un nouveau genre de Tricorythidae définissant la nouvelle sous-famille des Madecassorythinae (Ephemeroptera, Pannota). Bulletin de la Société entomologique de France **102**(3):225–232.

Fernando, C. H. 1965. A Guide to the Freshwater Fauna of Ceylon. Supplement 2. Bulletin of the Fisheries Research Station of Ceylon **17**:177–211.

Hubbard, M. D., and W. L. Peters. 1984. Ephemeroptera of Sri Lanka: An introduction to their ecology and biogeography. Pages 257–274 *in* C. H. Fernando, editor. Ecology and Biogeography in Sri Lanka. Dr. W. Junk, The Hague.

Hubbard, M. D. 1990. Mayflies of the World. A Catalog of the Family and Genus Group Taxa (Insecta: Ephemeroptera), Flora and Fauna Handbook No. 8, Sandhill Crane Press, Gainesville, 119 pp.

Jacob, U. 2003. Africa and its Ephemeroptera: Remarks from a biogeographical view. Pages 317–325 *in* E. Gaino, editor. Research Update on Ephemeroptera and Plecoptera. University of Perugia, Perugia, Italy.

Kimmins, D. E. 1960. Notes on East African Ephemeroptera, with descriptions of newBulletin of the British Museum (Natural History), Entomology **9**:337–355.

Koss, R. W., and G. F. Edmunds Jr. 1974. Ephemeroptera eggs and their contribution to phylogenetic studies of the order. Zoological Journal of the Linnean Society **55**(4):267–349.

Lestage, J. A. 1942. Contribution à l'étude des Ephéméroptères. XXV. Notes critiques sur les anciens Caenidiens d'Afrique et sur l'indépendance de l'évolution tricorythido-caenidienne. Bull. Mus. R. Hist. Nat. Belg., XVIII, 48, 20 pp.

McCafferty, W. P., and T.-Q. Wang. 2000. Phylogenetic systematics of the major lineages of Pannote Mayflies (Ephemeroptera: Pannota). Transactions of the American Entomological Society **126**(1):9–101.

Navás, L. 1936. Insectes du Congo Belge, Série IX. Revue de Zoologie et de Botanique Africaines **28**:333–368.

Oliarinony, R., and J.-M. Elouard. 1998b. Biodiversité aquatique de Madagascar. 8 - *Spinirythus* un nouveau genre de Tricorythidae (Ephemeroptera, Pannota). Bulletin de la Société entomologique de France **103**(3):237–244.

Oliarinony, R., and J.-M. Elouard. 1998a. Biodiversité aquatique de Madagascar. 7 - *Ranorythus* un nouveau genre de Tricorythidae définissant la nouvelle sous-famille des Ranorythinae (Ephemeroptera, Pannota). Bulletin de la Société entomologique de France **102**(5):439–447.

Oliarinony, R., J.-M. Elouard, and N. Raberiaca. 1998. Biodiversité aquatique de Madagascar 19 - neuf nouvelles espèces de *Tricorythus* Eaton [ephemeroptera, Pannota, Tricorythidae]. Revue français d'entomologie **20**:73–90.

Sivaramakrishnan, K. G., and K. Venkataraman. 1987. Biosystematic studies of south Indian Leptophlebiidae and Heptageniidae in relation to egg ultrastructure and phylogenetic interpretations. Proceedings of the Indiana Academy of Science (Animal Sciences) **96**:637–646.

Soldán, T. 1983. Two new species of *Clypeocaenis* (Ephemeroptera, Caenidae) with a description of adult stage and biology of the genus. Acta Entomologica Bohemoslovaca **80**:196–205.

Soldán, T. 1991. An annotated list of mayflies (Ephemeroptera) found in the Nam Cat Tien National Park. Pages 4–9 *in* K. Spitzer, J. Lepš, and M. Zacharda, editors. Nam Cat Tien Czechoslovak Vietnamese Expedition, November 1989, Research Report. Institute of Entomology, Czechoslovak Acad. Sci., České. Budějovice, 45 pp.

Soldán, T., and V. Landa. 1991. Two new species of Caenidae (Ephemeroptera) from Srí Lanka. Pages 235–246 *in* J. Alba-Tercedor, and A. Sanchez-Ortega, editors. Overview and Strategies of Ephemeroptera and Plecoptera. The Sandhill Crane Press, Gainesville, Florida.

Soldán, T., and M. Putz. 2000. The larva of *Rhoenanthus distafurcus* Bae et McCafferty (Ephemeroptera, Potamanthidae) with notes on distribution and biology. Aquatic Insects **22**:9–17.

Ulmer, G. (1913) Ephemeriden aus Java, Gesammelt von Edw. Jacobson, Note V., Notes from the Leyden Museum, **35**:102–120.

Ulmer, G. 1916. Ephemeropteren von Aequatorial-Afrika. Archiv für Naturgeschichte **81**(A) 7 (1915):1–19.

Ulmer, G. 1925. Ephemeropteren von den Sunda-Inseln und den Philippinen. Treubia;Recueil de Travaux Zoologiques, Hydrobiologiques et Oceanographiques **6**(1):28–91.

Ulmer, G. 1940. Eintagsfliegen (Ephemeropteren) von den Sunda-Inseln. Archiv für Hydrobiologie **16**:443–692, p. 690.

Appendix A

Table 1. Critical characters distinguishing larvae of the genus *Sparsorythus* gen. n.

Character/Species	*S. bifurcatus* sp. n.	*S. jacobsoni* (Ulmer,1913)	*S. dongnai* sp. n.	*S. gracilis* sp. n.	*S. grandis* sp. n.	*S. ceylonicus* sp. n.
1. Distance between composed eyes in males	as long as or slightly larger than the eye width	unknown	shorter than the eye width	unknown	unknown	much larger than the eye width
2. Ratio of distance between composed eyes in females to eye width	2.2 : 1	unknown	2.6 : 1	2.6 : 1	3 : 1	unknown
3. Shape of hypopharyngeal lingua	as wide as long	as wide as long	as wide as long	as wide as long	wider than long	as wide as long
4. Presence of medial nick on hypopharynx	absent	absent	absent	present	absent	present
5. Presence of medial rill on hypopharynx	present	absent	present	present	absent	present
6. Apex of the left prostheca	pointed	extended, with several teeth	pointed	bluntly pointed	extended, with several teeth	bluntly pointed
7. Ratio right prostheca length : width	2 : 1	2.4 : 1	2 : 1	3.3 : 1	1.8 : 1	2.4 : 1
8. Medial nick on labium	absent	present	absent	absent	absent	absent

Table 1. (*continued*)

Character/Species	*S. bifurcatus* sp. n.	*S. jacobsoni* (Ulmer, 1913)	*S. dongnai* sp. n.	*S. gracilis* sp. n.	*S. grandis* sp. n.	*S. ceylonicus* sp. n.
9. Wing pads reaching to the abdominal segment	I	unknown	IV (females); V (males)	III	III	II
10. Rudimental gill on abdominal segment VII	present	present	absent	absent	absent	absent
11. Shape of rudimental gill on abdominal segment VII	bifurcated	filamentous	-	-	-	-
12. Ratio femur : tibia : tarsus (fore leg)	1.9 : 2.2 : 1	2 : 2.7 : 1	2.1 : 2.6 : 1	1.9 : 2.3 : 1	2.4 : 3 : 1	1.7 : 2.2 : 1
13. Ratio femur : tibia : tarsus (middle leg)	2 : 2 : 1	2.6 : 2.8 : 1	2.2 : 2.2 : 1	2 : 2.3 : 1	2.8 : 2.9 : 1	2 : 2 : 1
14. Ratio femur : tibia : tarsus (hind leg)	2.4 : 2.5 : 1	2.2 : 2.6 : 1	2.8 : 3 : 1	2.5 : 2.9 : 1	3.2 : 3.7 : 1	2.6 : 1.9 : 1
15. Ratio fore femur length : width	1.8 : 1	2 : 1	1.9 : 1	2.5 : 1	2 : 1	1.6 : 1
16. Ratio middle femur length : width	1.8 : 1	2 : 1	2.1 : 1	2.6 : 1	2 : 1	2 : 1
17. Ratio hind femur length : width	1.8 : 1	2 : 1	2.1 : 1	2.9 : 1	2 : 1	2 : 1

Continued on next page

Table 1. (*continued*)

Character/Species	S. bifurcatus sp. n.	S. jacobsoni (Ulmer, 1913)	S. dongnai sp. n.	S. gracilis sp. n.	S. grandis sp. n.	S. ceylonicus sp. n.
18. Ratio fore femoral setae length : width	4 - 5 : 1	unknown	2.5 – 4 :1	4 - 5.5 : 1	3 - 4.5 : 1	5 - 6 : 1
19. Transversal row of setae on the fore femora	S - shaped	irregular	S - shaped	bow-shaped	S - shaped	bow-shaped
20. Shape of fore tibial setae	spatulated	unknown	spatulated	bristle like	spatulated	spatulated
21. Posterior margin of hind femora (basal half)	convex	convex	convex	concave	straight	concave
22. Posterior margin of middle femora(basal half)	convex	convex	concave	concave	concave	concave
23. Length of setae on surface of middle and hind femora	same	same	same	different	same	different
24. Arrangement of cerci and paracercus setae	same	same	same	different	same	same
25. Length of cerci and paracercus setae	same	same	same	different	same	same
26. Maximal length of setae on caudal filaments	1/10 of the segment length	1/5 of the segment length	1/10 of the segment length	2/3 of the segment length	1/5 of the segment length	1/3 of the segment length

Figures

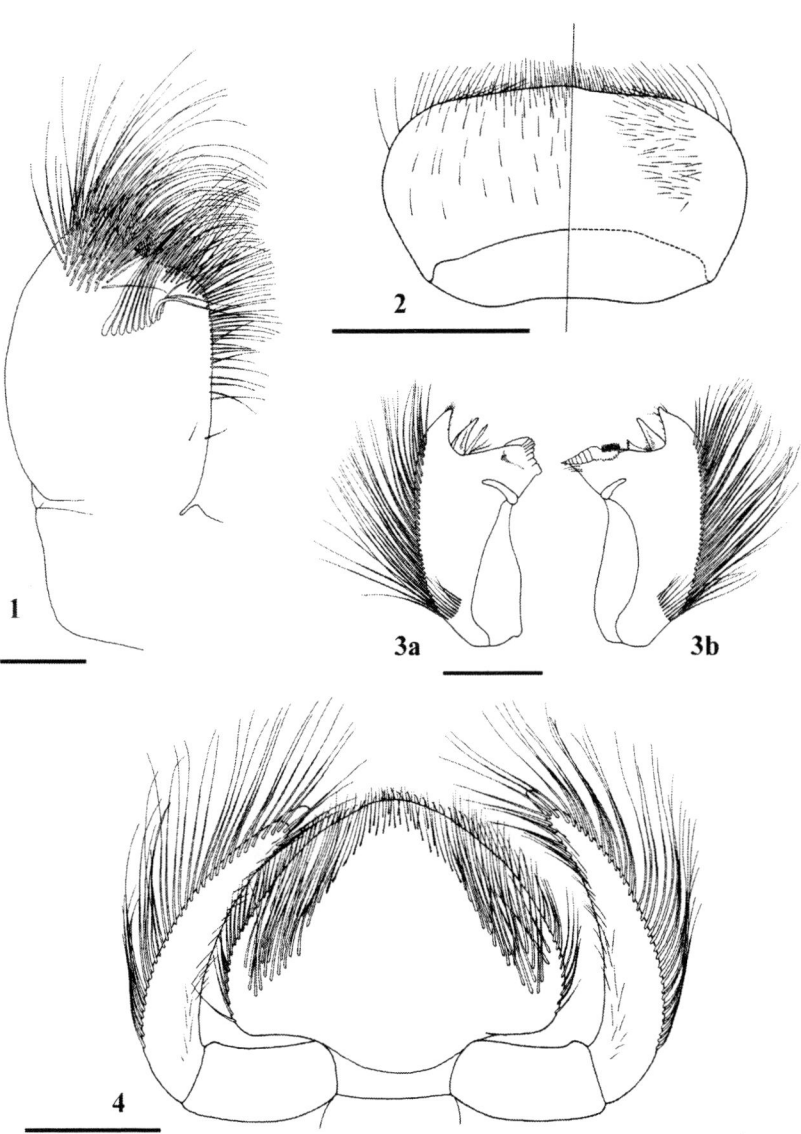

Figures 1–4. *Sparsorythus bifurcatus* sp. n. 1: maxilla (ventral view); 2: labrum (left part dorsal view, right part ventral view); 3a: left mandible (dorsal view); 3b: right mandible (dorsal view); 4: labium (ventral view).
Scale Fig. 1 = 0.1 mm; scale Fig. 2 (= scale Fig. 3–4) = 0.25 mm.

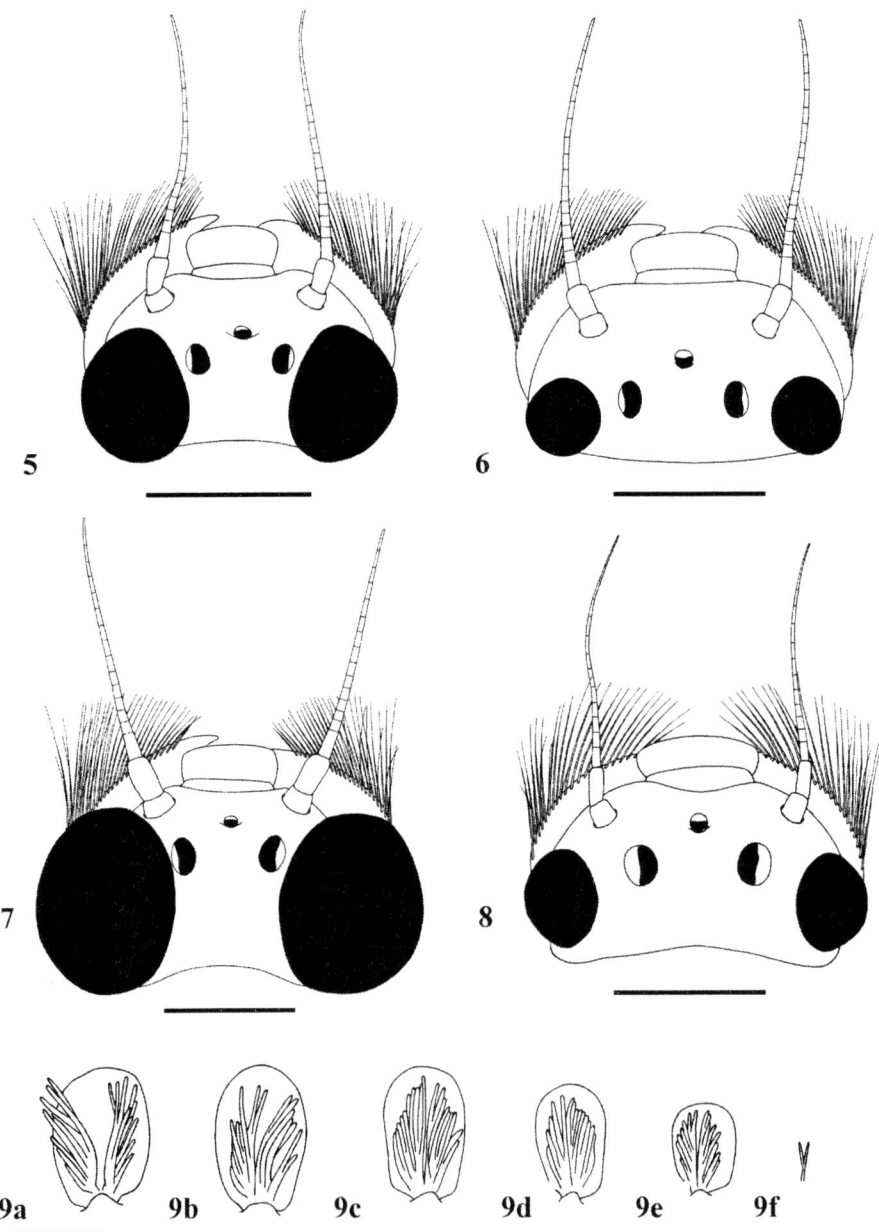

Figures 5–9. *Sparsorythus* spp. 5, 6, 9: *S. bifurcatus* sp. n.; 7, 8: *S. dongnai* sp. n.; 5: head of male larva; 6: head of female larva; 7: head of male larva; 8: head of female larva; 9a–9f: gills on abdominal segments II–VII.
Scale Figs. 5–8 = 0.5 mm; scale Fig. 9 = 0.25 mm.

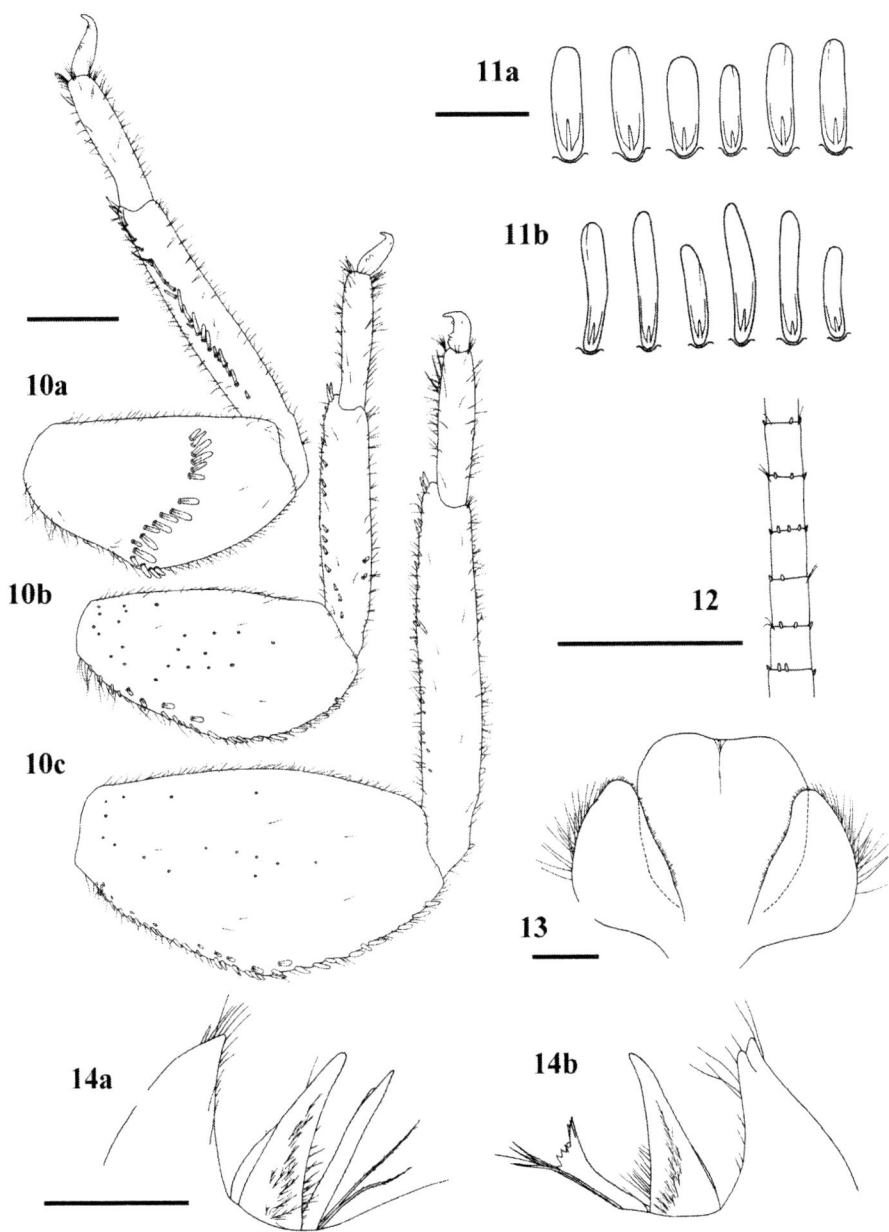

Figures 10–14. *Sparsorythus bifurcatus* sp. n. 10a: fore leg; 10b: middle leg; 10c: hind leg; 11a: fore femoral setae; 11b: fore tibial setae; 12: caudal filament; 13: hypopharynx; 14a: left prostheca; 14b: right prostheca.
Scale Fig. 10 (= scale Fig. 12) = 0.25 mm; scale Fig. 11 (= scale Fig. 13–14) = 0.1 mm.

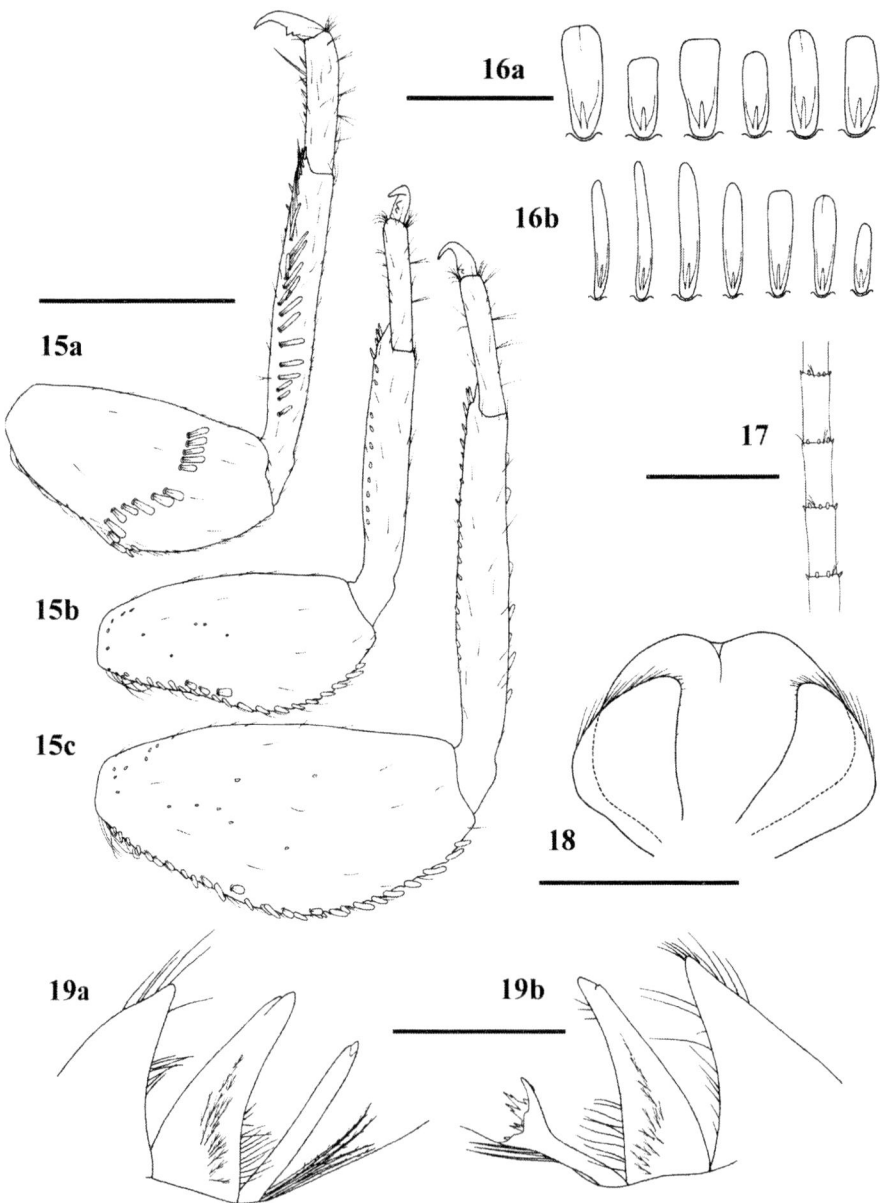

Figures 15–19. *Sparsorythus dongnai* sp. n. 15a: fore leg; 15b: middle leg; 15c: hind leg; 16a: fore femoral setae; 16b fore tibial setae; 17: caudal filament; 18: hypopharynx; 19a: left prostheca; 19b: right prostheca.
Scale Fig. 15 = 0.5 mm; scale Fig. 16 (= scale Fig. 19) = 0.1 mm; scale Fig. 17 (= scale Fig. 18) = 0.25 mm.

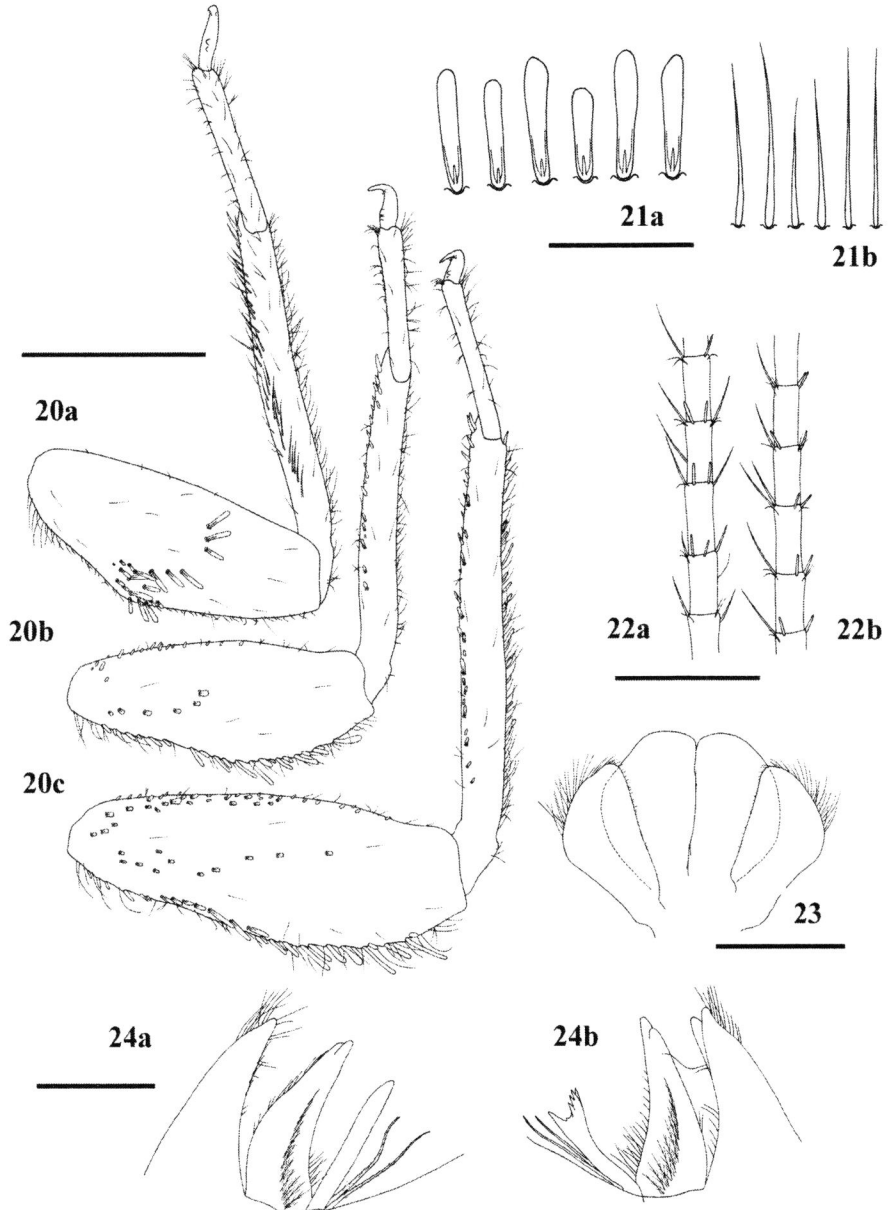

Figures 20–24. *Sparsorythus gracilis* sp. n. 20a: fore leg; 20b: middle leg; 20c: hind
leg; 21a: fore femoral setae; 21b: fore tibial setae; 22a: paracercus; 22b: left
cercus; 23: hypopharynx; 24a: left prostheca; 24b: right prostheca.
Scale Fig. 20 = 0.5 mm; scale Fig. 21 (= scale Fig. 24) = 0.1 mm; scale Fig. 22 (scale
Fig. 23) = 0.25 mm.

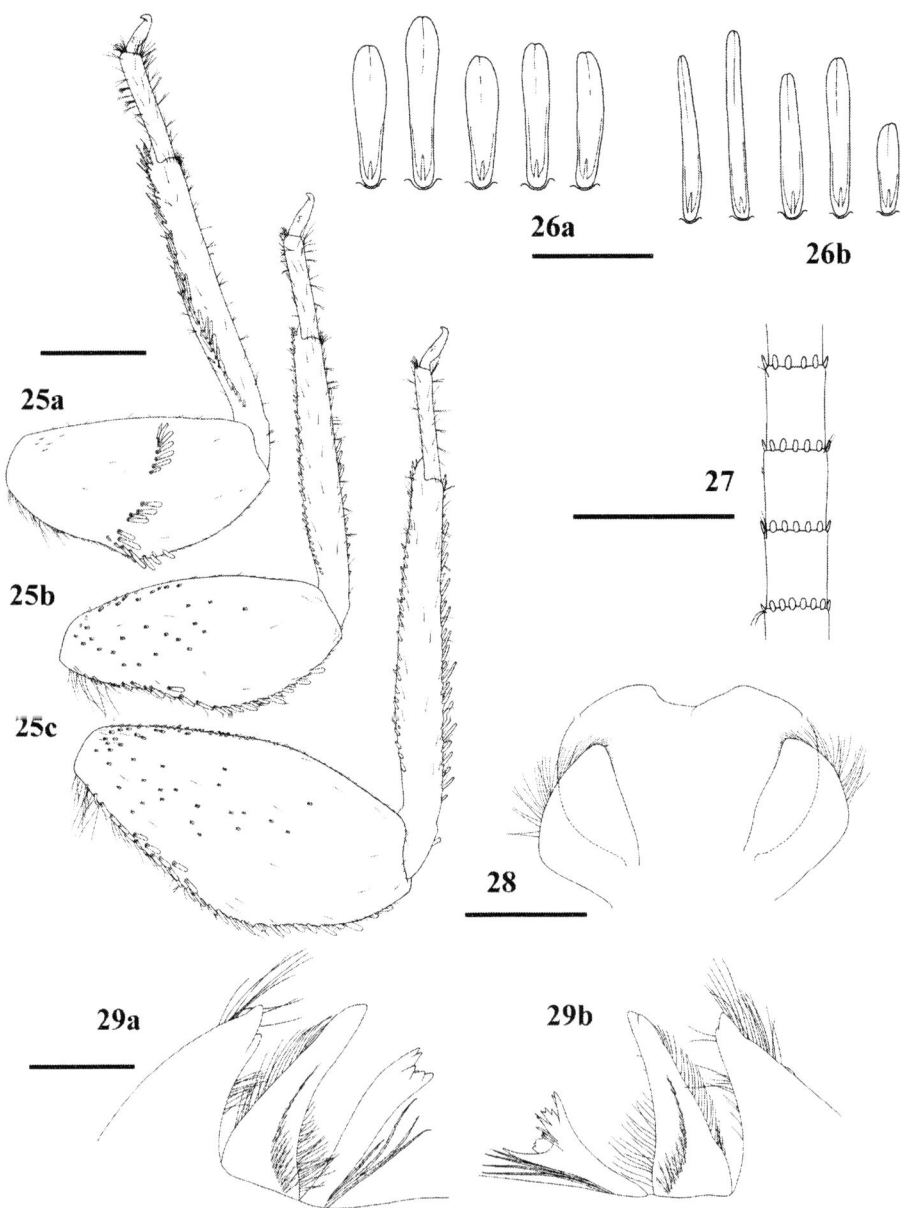

Figures 25–29. *Sparsorythus grandis* sp. n. 25a: fore leg; 25b: middle leg; 25c: hind leg; 26a: fore femoral setae; 26b: fore tibial setae; 27: caudal filament; 28: hypopharynx; 29a: left prostheca; 29b: right prostheca.
Scale Fig. 25 = 0.5 mm; scale Fig. 26 (=scale Fig. 29) = 0.1 mm; scale Fig. 27 (= scale Fig. 28) = 0.25 mm.

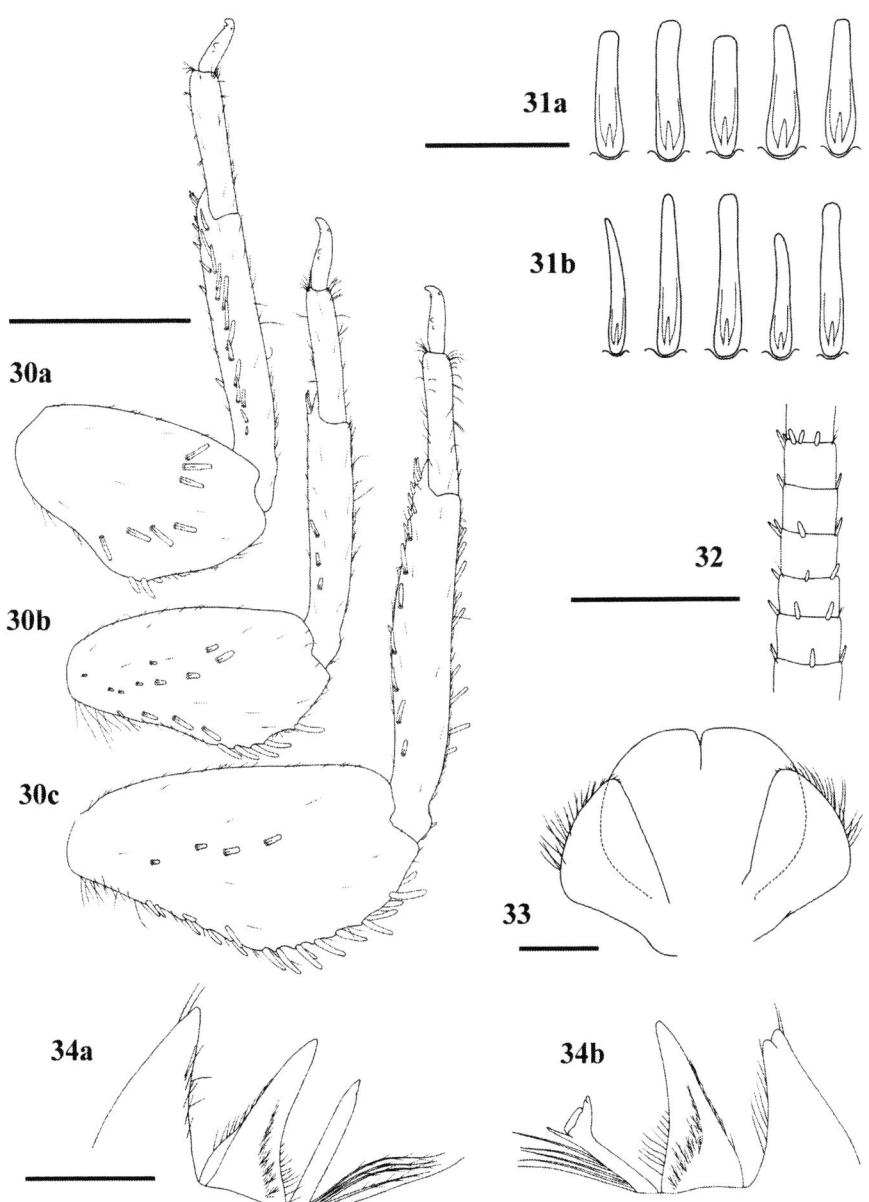

Figures 30–34. *Sparsorythus ceylonicus* sp.n. 30a: fore leg; 30b: middle leg; 30c: hind leg; 31a: fore femoral setae; 31b: fore tibial setae; 32: caudal filament; 33: hypopharynx; 34a: left prostheca; 34b: right prostheca.
Scale Fig. 30 = 0.5 mm; scale Fig. 31 (= scale Fig. 33–34) = 0.1 mm; scale Fig. 32 = 0.25 mm.

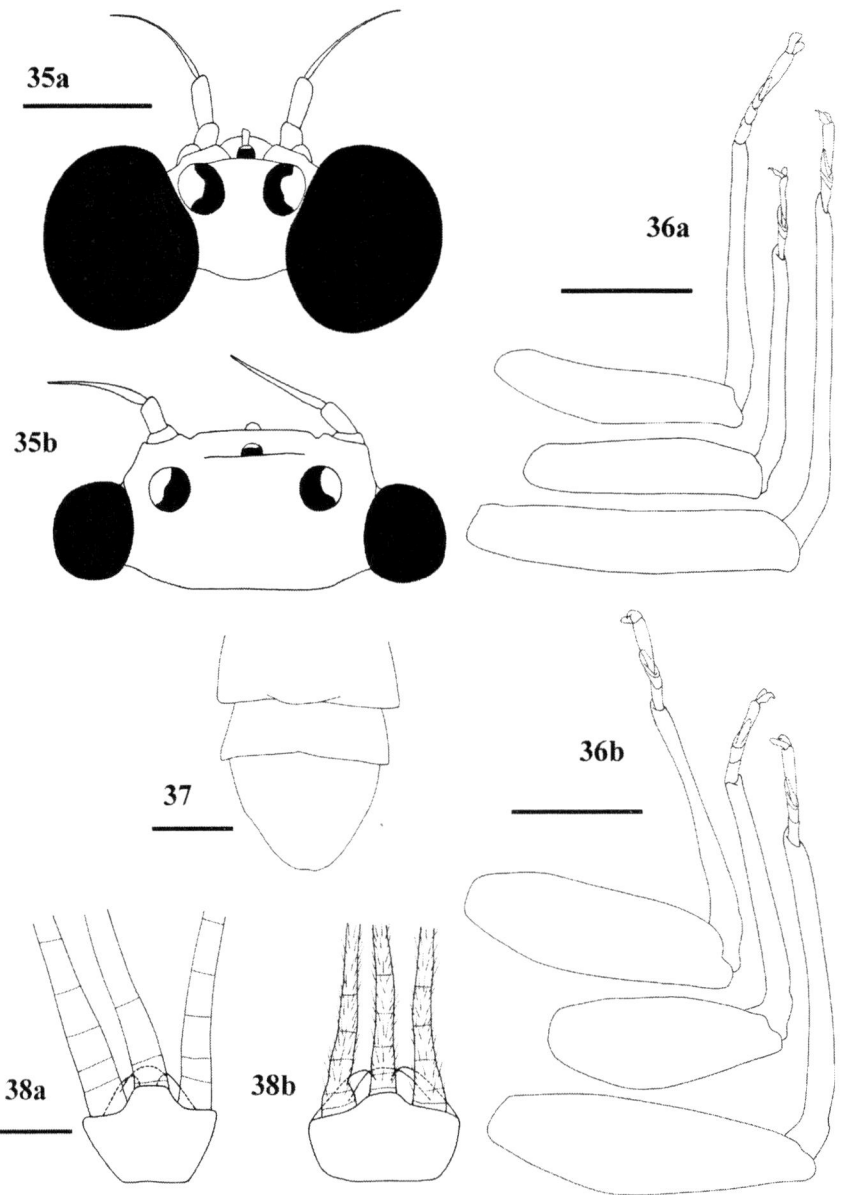

Figures 35–38: *Sparsorythus bifurcatus* sp. n. 35a: head of male imago; 35b: head of female imago; 36a: legs of male imago; 36b: legs of female imago; 37: female subgenital plate; 38a: male caudal filaments; 38b: female caudal filaments.
Scale Fig. 35 = 0.5 mm; scale Fig. 36 (= scale Fig. 37) = 0.5 mm; scale Fig. 38 = 0.25 mm.

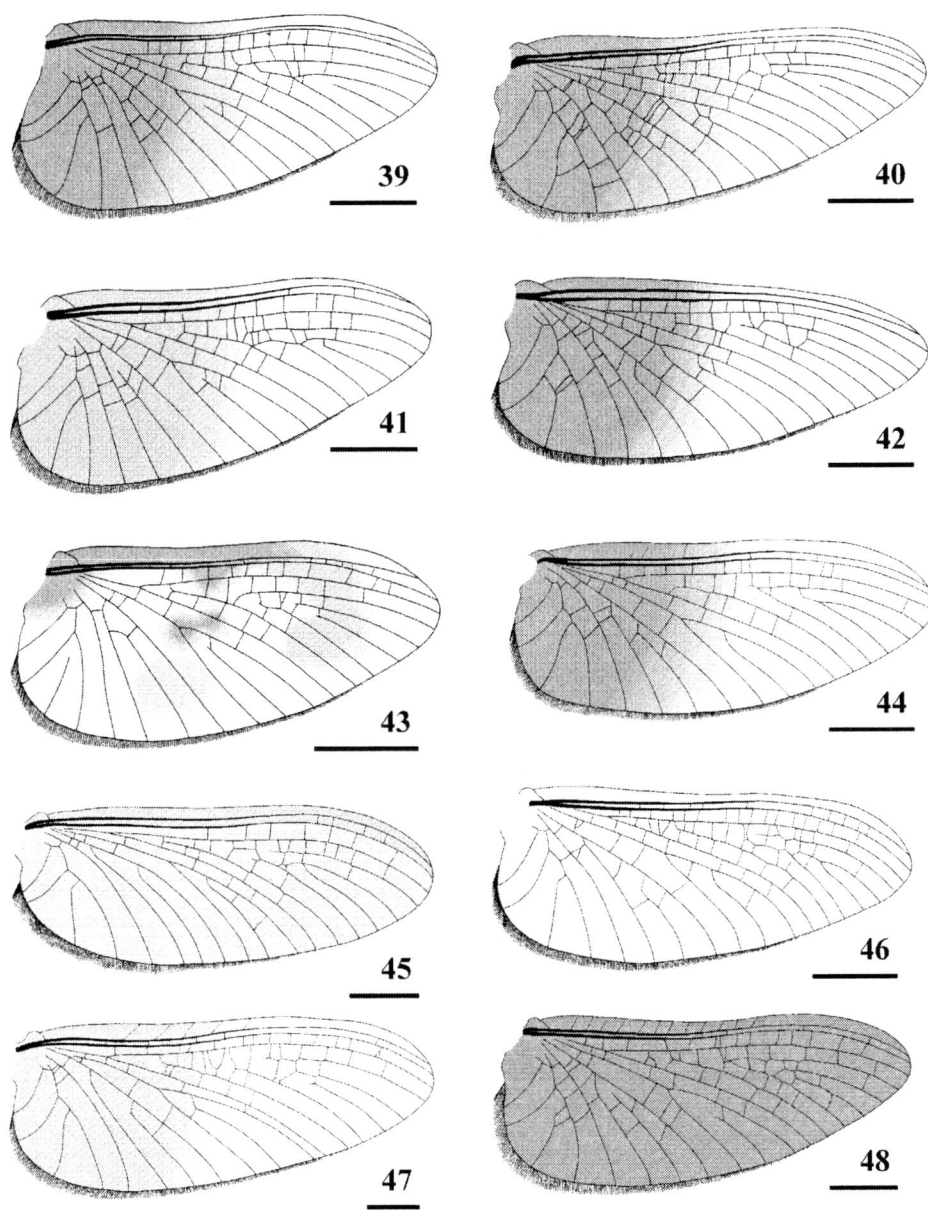

Figures 39–48. *Sparsorythus* spp. 39: *S. bifurcatus* sp. n. (male); 40: *S. bifurcatus* sp. n. (female); 41: *S. dongnai* sp. n. (male); 42: *S. dongnai* sp. n. (female); 43: *S. multilabeculatus* sp.n.; 44: *S.* sp. 1 (female); 45: *S.* sp. 2 (female); 46: *S.* sp. 3 (female); 47: *S.* sp. 4 (female); 48: *S.* sp. 5 (female).
Scale = 1 mm.

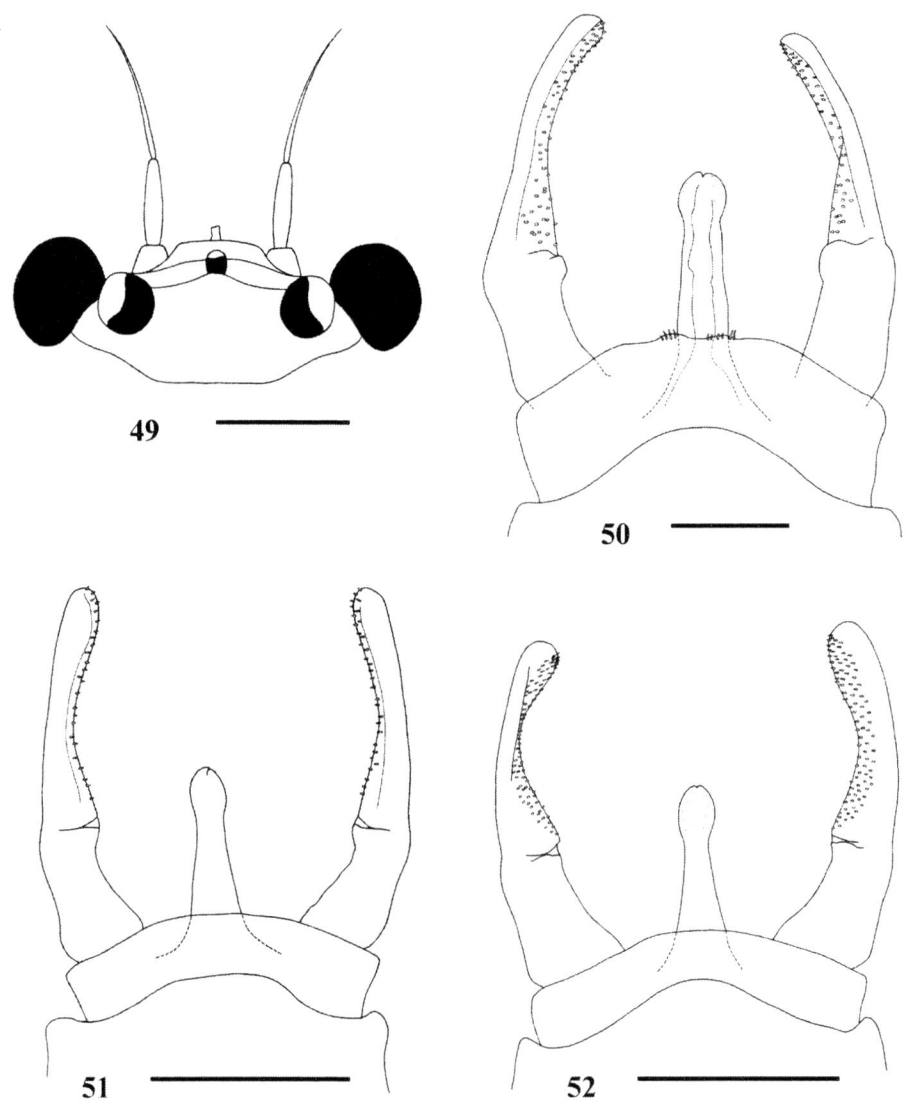

Figures 49–52. *Sparsorythus* spp. 49: *S. multilabeculatus* sp. n., head of male; 50: *S. multilabeculatus* sp. n., penis; 51. *S. dongnai* sp. n., penis; 52: *S. bifurcatus* sp. n., penis.
Scale Fig. 49 (= scale Figs. 51, 52) = 0.25 mm; scale Fig. 50 = 0.1 mm.

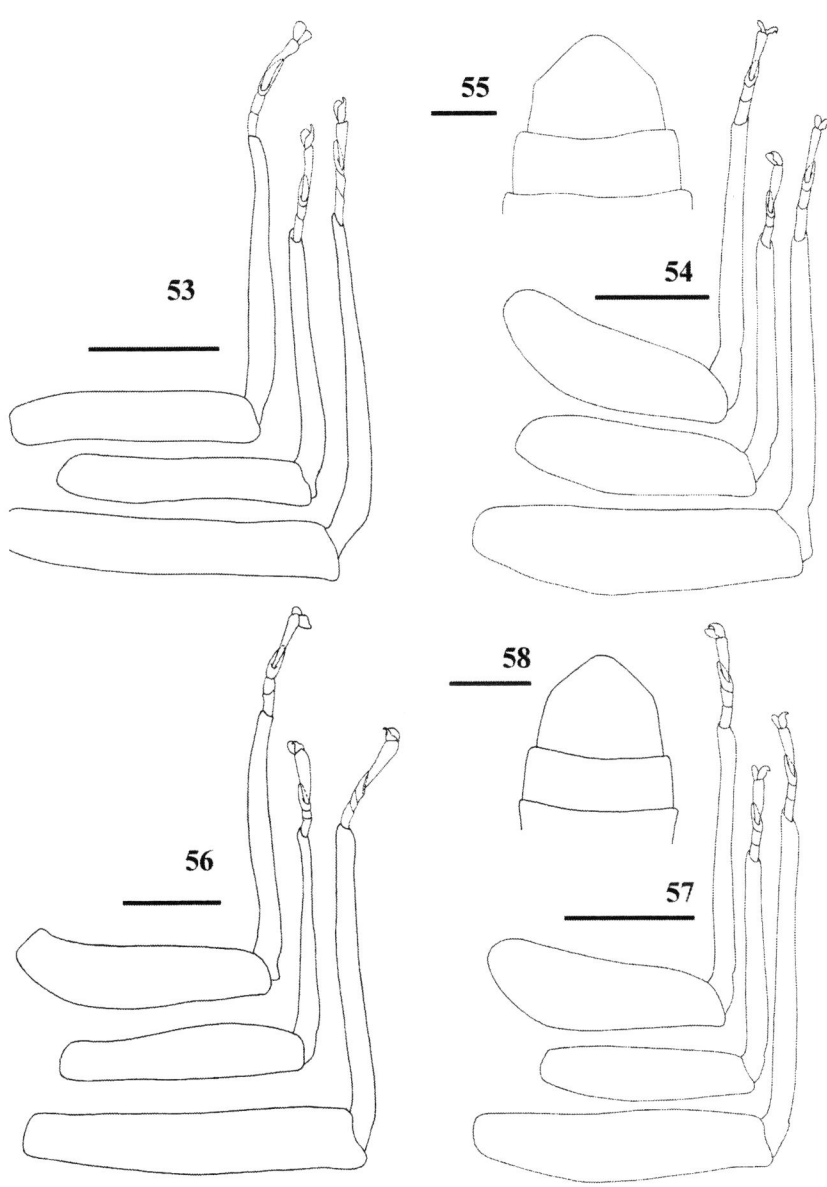

Figures 53–58. *Sparsorythus* spp. 53: *S. dongnai* sp. n., legs of male; 54: *S. dongnai* sp. n., legs of female; 55: *S. dongnai* sp. n., shape of last female sternites; 56: *S. multilabeculatus* sp. n., legs of male; 57: *S.* sp. 1, legs of female; 58: *S.* sp. 1, shape of last female sternites.

Scale Fig. 53 (= scale Fig. 54, 57) = 0.5 mm; scale Fig. 55 (= scale Fig. 56, 58) = 0.25 mm

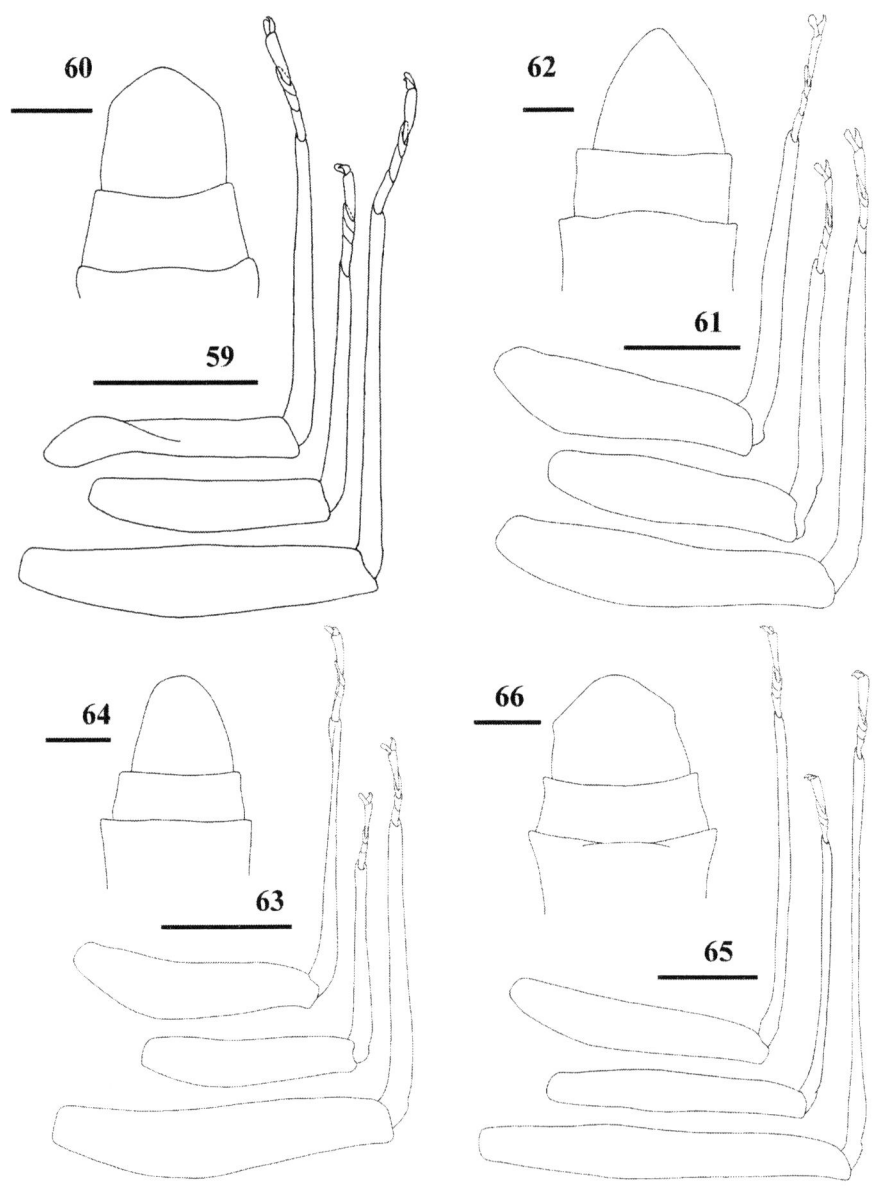

Figures 59–66. *Sparsorythus* spp. 59: *S.* sp. 2, legs of female; 60: *S.* sp. 2, shape of last female sternites; 61: *S.* sp. 3, legs of female; 62: *S.* sp. 3, shape of last female sternites; 63: *S.* sp. 4, legs of female; 64: *S.* sp. 4, shape of last female sternites; 65: *S.* sp. 5, legs of female; 66: *S.* sp. 5, shape of last female sternites.
Scale Fig. 59 (= scale Figs. 61, 63, 65) = 0.5 mm; scale Fig. 60 (= scale Figs. 62, 64, 66) = 0.25 mm.

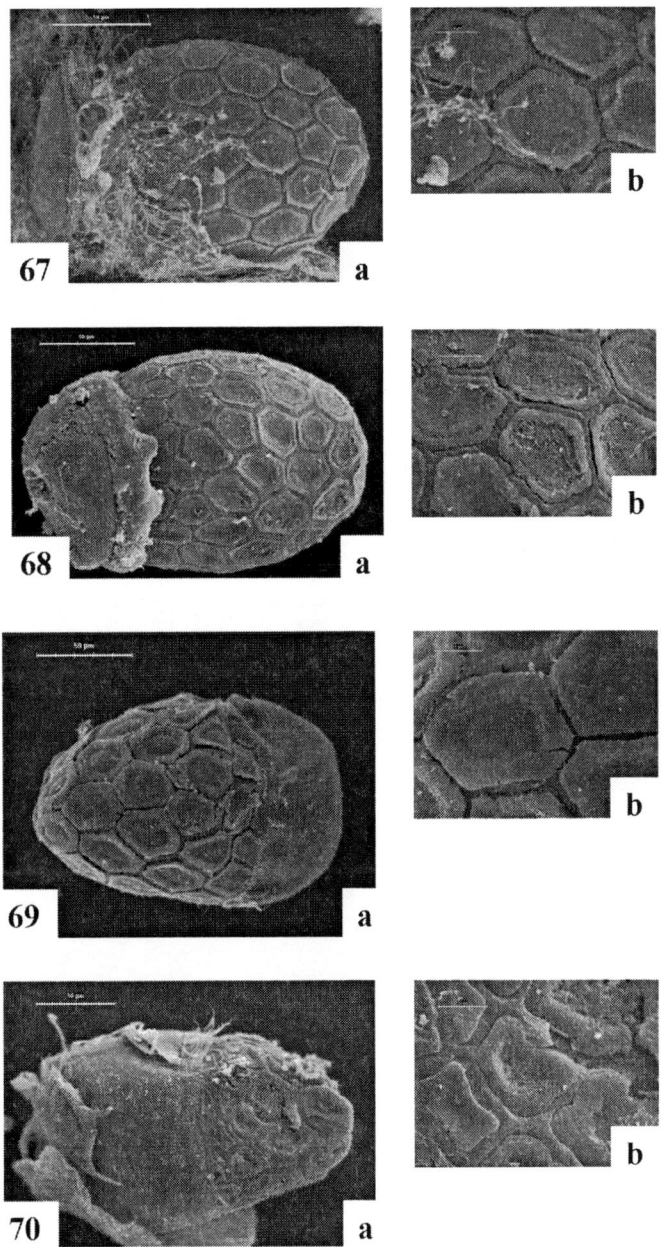

Figures 67–70. *Sparsorythus* spp., eggs (a: entirely egg; b: chorionic surface in detail). 67: *S. bifurcatus* sp. n.; 68: *Sparsorythus dongnai* sp. n.; 69: *Sparsorythus* sp.1; 70: *Sparsorythus* sp. 2.
Scale Figs. 50–53a = 50 m; scale Figs. 50–53b = 10 m.

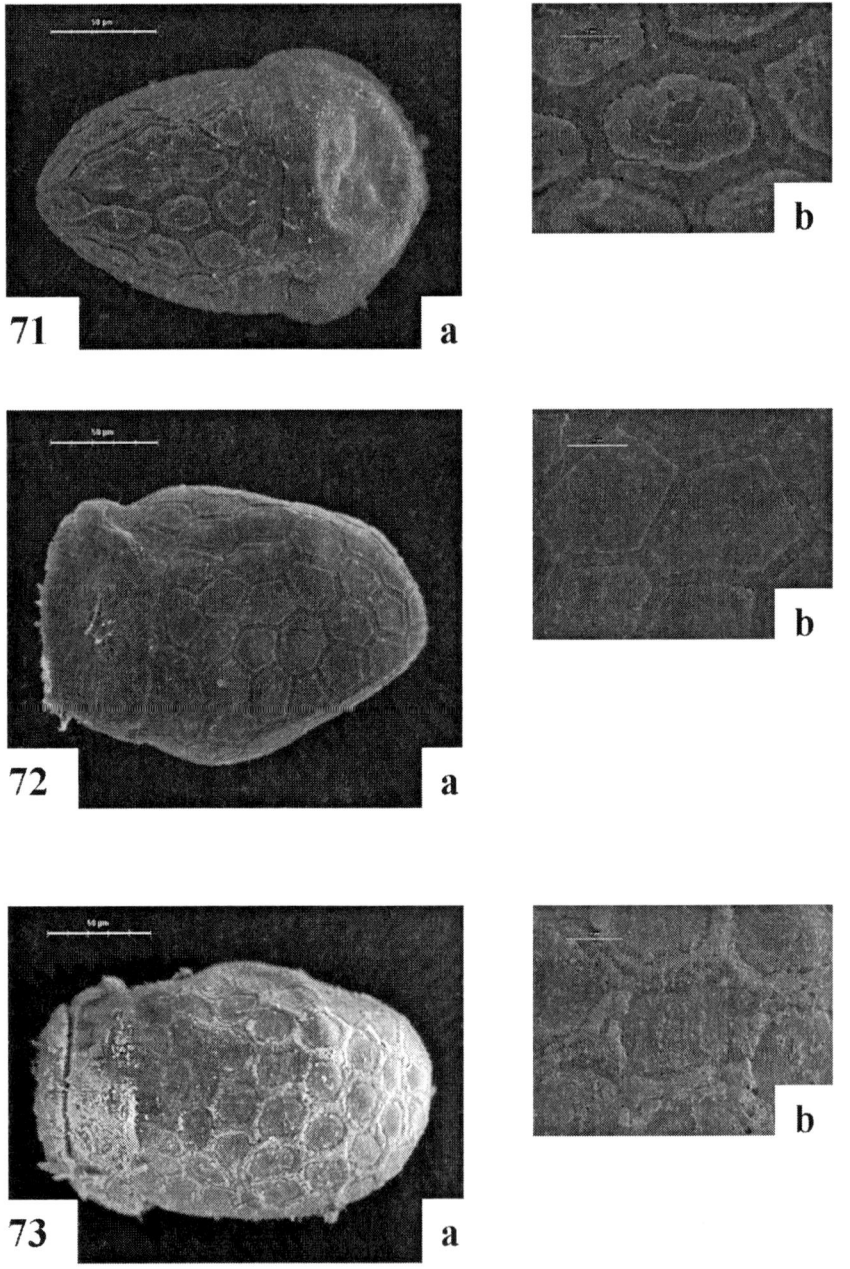

Figures 71–73. Sparsorythus spp., eggs (a: entirely egg; b: chorionic surface in detail). 71: Sparsorythus sp. 3; 72: Sparsorythus sp. 4; 73: Sparsorythus sp. 5.
Scale Figs. 54–56a = 50 m; scale Figs. 54–56b = 10 m.

THE TYPES AND DISTRIBUTION OF SETAE ON THE LARVAL LEGS OF NEOPERLA GENICULATA

Fumiko Tamura[1], Yu Isobe[2] and Tadashi Oishi[1,3]

[1] KYOUSEI Science Center for Life and Nature, Nara Women's University,
Nara 630-8506, Japan
[2] Nara Bunka Women's College, Nara 635-8530, Japan
[3] Graduates School of Humanitis and Sciences, Nara Woman's University,
Nara 630-8506, Japan

Abstract

We investigated the morphology and distribution of setae on each segment of the larval legs in Neoperla geniculata (Perlidae, Plecoptera) using a scanning electron microscope (SEM). Observations were carried out on the ventral and dorsal surfaces of the femur, tibia and the third tarsus of all legs. We found characteristic setae on the ventral side and recognized five types: Spine-like seta, Broad pinnate A, Broad pinnate B, Slender pinnate A and Slender pinnate B. The distribution of the setae showed differences among the segments. The third tarsus was unique in setal types (Broad pinnate A and spine-like seta) and in distribution.

Key words: Setae; the larval leg; SEM, Plecoptera.

Introduction

Many studies have been conducted on the morphology of stonefly nymphs. Stewart et al. (1988) reported that all legs of a larva were similar in coloration and setation in Plecoptera and that the surface of larval legs was covered with many hairs and setae. In *Eccoptura xanthense*, various combinations of silky fringe hairs, fine depressed clothing hairs and stout setae (bristles) of various lengths were observed on the dorsal surface of legs (Stewart et al. 1988).

The setae on the leg surface must function as sensors for physical and chemical stimuli in the aquatic environment (Wiggins 1977). Spinelli and Corallini (2002) did a comparative study of the setae of Hydropsichidae (Trichotera) from Italy. Following examination of larval fore-legs, they speculated that the pinnate setae observed might be linked to net-cleaning behavior. Among three paired-legs, the fore-legs are highly specialized in Trichoptera (Spinelli 2002). On the other hand, three paired legs are not so different in Plecoptera. Legs are flattened ventro-dorsally, especially in a family group Perloidea. We considered the possibility that the ventral side of legs of Perloidea are important for sensing movement or vibration from the substrate. To observe the leg setae of Perloidea more closely, we examined legs using a scanning electron microscope SEM. During our examinations, we found unique pinnate setae on the third tarsus of *Neoperla geniculata*. We observed the setal morphology and distribution in detail on all legs of the species using the SEM, and compared them among the three paired legs.

Materials and Methods

We collected the larvae of *Neoperla geniculata* in a slightly stagnating area of the Kizu River, at Kizu, Kyoto Pref., Japan on 30 Nov. 2002, 20 March 2004 and 12 April 2004. The river is 50 m wide at the site surrounded by rice fields and houses. The larvae collected in November 2002 were fixed in 70% ethanol in the field. Larvae collected in 2004 were kept in several aquaria in the laboratory.

We examined an early instar larva fixed on 30 November 2002 (Table 1 #1) and 5 final-instar larvae fixed after keeping in the laboratory for 24 and 53 days (Table 1 #2, #6). Further information on the specimens is shown in Table 1. The species was identified from the male penis (Uchida 1990). All specimens were fixed in 70% ethanol. Legs were detached and cleaned for 50 sec. (specimen #1, #2, #3) or for 80 sec (#4, #5, #6) using an ultrasonic cleaner (Clean Matic, Shinmeidai·laboratory). Legs were dehydrated in 80~99.9% ethanol series, and then treated with acetone and isopentyl acetate. After drying, using a critical point dryer (Hitachi HCP-2), legs from the larvae were stuck on stubs and coated with gold for 4 minutes (6~8 mA) in an ion coater (Giko IB・3).

We observed one side of all legs of each specimen using the SEM (Hitachi S-2001A): the ventral side of the specimens #1, #2, #4 and #6, and the dorsal side of the specimens #3 and #5. The femur, tibia and the third tarsus were observed for each leg. We recognized five types of setae on the ventral side of the legs and made maps of the setae on the legs with SEM photographs. We measured the setal size and counted the number of teeth of the pinnate setae on the SEM screen.

Table 1. Data of the observed specimens; y: young; f: final instar.

Sample #	instar	Observed surface	Sex	Head width mm	Body length mm	Collection date	Date of fixation [a]
#1	y	Ventral	♀	27	118	30 Nov. 2002	30 Nov. 2002
#2	f	Ventral	♀	29	125	12 Apr. 2004	06 May 2004 (24)
#3	f	Dorsal	♀	29	125	12 Apr. 2004	06 May 2004 (24)
#4	f	Ventral	♀	31	121	20 Mar. 2004	12 May 2004 (53)
#5	f	Dorsal	♀	32	148	20 Mar. 2004	12 May 2004 (53)
#6	f	Ventral	♂	27	110	20 Mar. 2004	12 May 2004 (53)

[a] (days kept in the laboratory)

Results

The Ventral and Dorsal Surfaces of the Leg.   Dense clothing hairs cover the entire dorsal side of the legs (Fig. 1).   Each hair is long and has a small socket and a square tip.   The setae with a large socket are not observed on the dorsal side.   On the ventral side, the clothing hairs are rather sparsely distributed, and many setae can be observed (Fig. 2a).   The clothing hairs on the ventral side are similar to those on the dorsal side, but they are bent in the middle of the hairs.   The features of both sides of the legs are similar to each other.

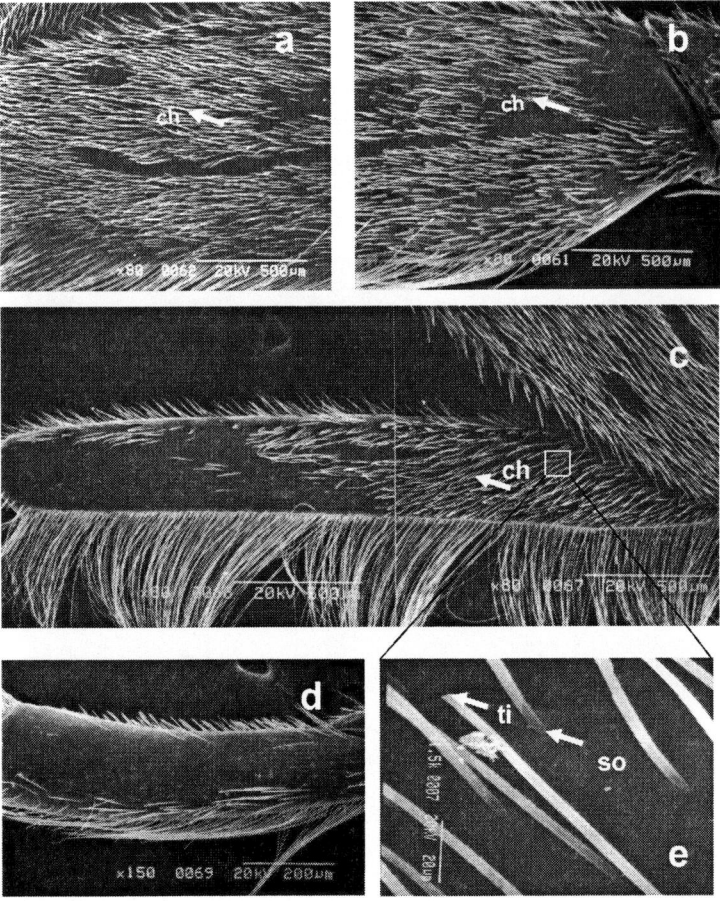

Figure 1. The dorsal surface of the left mid-leg of a final instar lava (# 5). Dense clothing hairs cover whole dorsal side of the leg and have a small socket. a,b: Femur, c: Tibia, d: Third tarsus, e: Ventral hairs (Clothing hairs), ch: Clothing hair, so: Socket, ti: Tip.

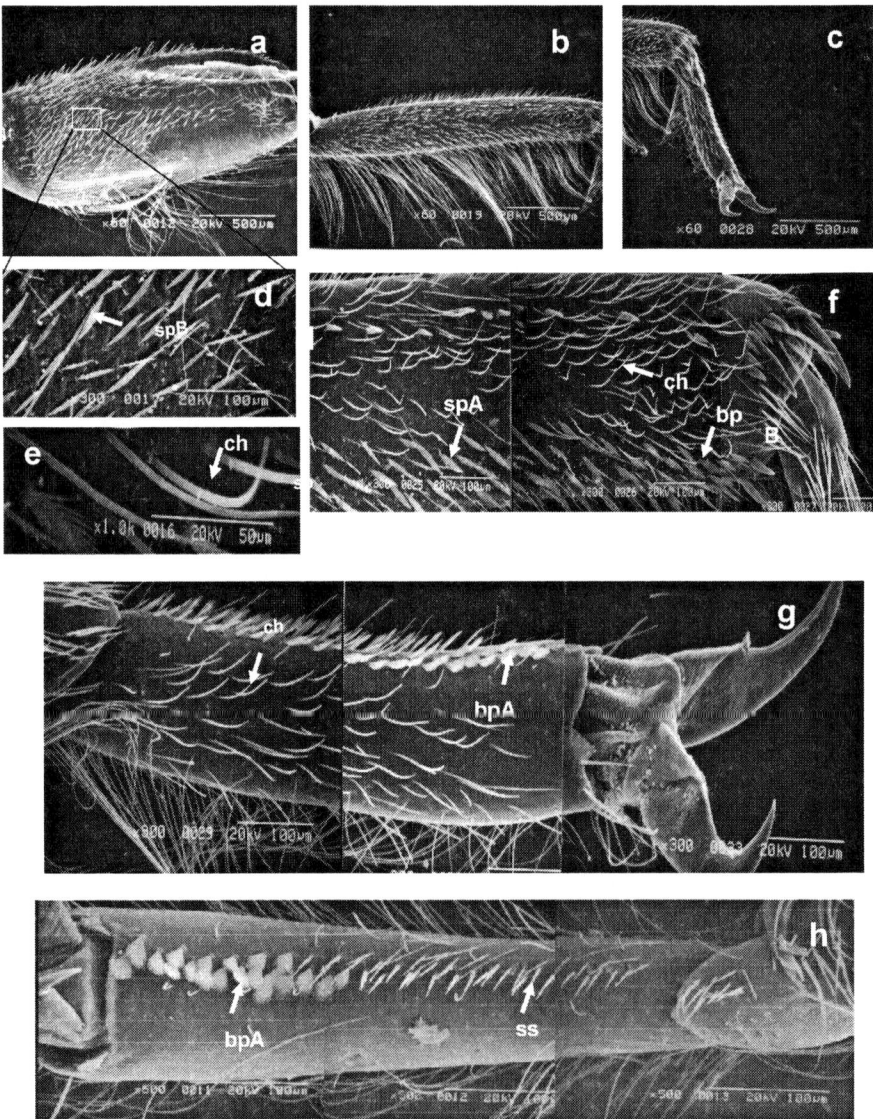

Figure 2. The ventral surface of the left (a-g) and right (h) mid-legs of a young larva
(# 1) The ventral side is cover rather sparsely with clothing hairs and various setae.
The seta has a pointed tip and a large socket. a, d: Femur, b, f: Tibia; c, g, h: Third
tarsus, e: Clothing hairs, ss: Spine-like seta, bpA: Broad pinnate A, bpB: Broad
pinnate B, ch: Clothing hair, spA: Slender pinnate A, spB: Slender pinnate B, so:
Socket, ti: Tip,

Features of Five Types of the Setae on the Ventral Surface. We recognized five types of setae with a large socket on the ventral side: one type of spine-like setae and four types of pinnate setae (Broad pinnate A, Broad pinnate B, Slender pinnate A, Slender pinnate B) (Figs. 3 and 4). The spine-like setae include thick and slender ones that are simple in shape. The pinnate setae have several tips like a leaf. Both early and final instar larvae have similar types of setae. The size and the number of teeth of the setae are shown in Tables 2 and 3. The morphometry of the setae of specimen #6 are presented in Table 4. The thick and slender types of spine–like setae were not significantly different in either length or width.

Three categories of setae (spine-like, broad and slender pinnates) show much differences in length and width. In the broad pinnate setae, types A and B are similar in width but different in length. The broad pinnate A is shorter than B and is similar to the spine–like seta in length. The broad pinnate A shows a tendency to bear more teeth than the broad pinnate B (Table 3). The slender pinnate B is longer than the slender pinnate A, and the longest among all types of setae. The slender type setae bear more teeth than the broad ones. Early- and final-instar larvae show a similar tendency in number of teeth on all types of setae (Table 2), but early-instar (♀# 1) larva bears a slightly shorter setae than final-instar ones (♀ # 2, # 4).

Distribution of Setae. Each segment of the legs bear two types of setae (Table 5). The broad pinnate A and spine-like seta (the shortest ones among the setae) are on the third tarsus. Basal segments have longer setae. In the final-instar larvae, all of fore-, mid- and hind-legs have the same types of setae, while in the early-instar larvae, the broad pinnate B setae are found only on the tibia of hind-legs.

Femora are widely covered with the slender pinnate B and the broad pinnate B in the basal region (Figs. 5 and 6). On the other hand, tibiae are widely covered with the slender pinnate A and with the broad pinnate B, partially in the apical region. Distribution of the setae shows little difference among fore-, mid- and hind-legs, except for the broad pinnate B on the tibiae of the early instar-larvae (Figs. 5 and 6).

The third tarsi bear the spine-like setae and the broad pinnate A in a few lines on the inner surface. The arrangement of two types of the setae is clearly different among fore-, mid- and hind-legs. The broad pinnate A setae are on the apical half of the fore-leg, and the spine-like setae are in the basal region. The pinnate setae occupy almost all regions on the mid-leg, but the spine-like ones are on the hind-leg. It is similar between early- and the final-instar larvae, with some differences in the amount of the setae.

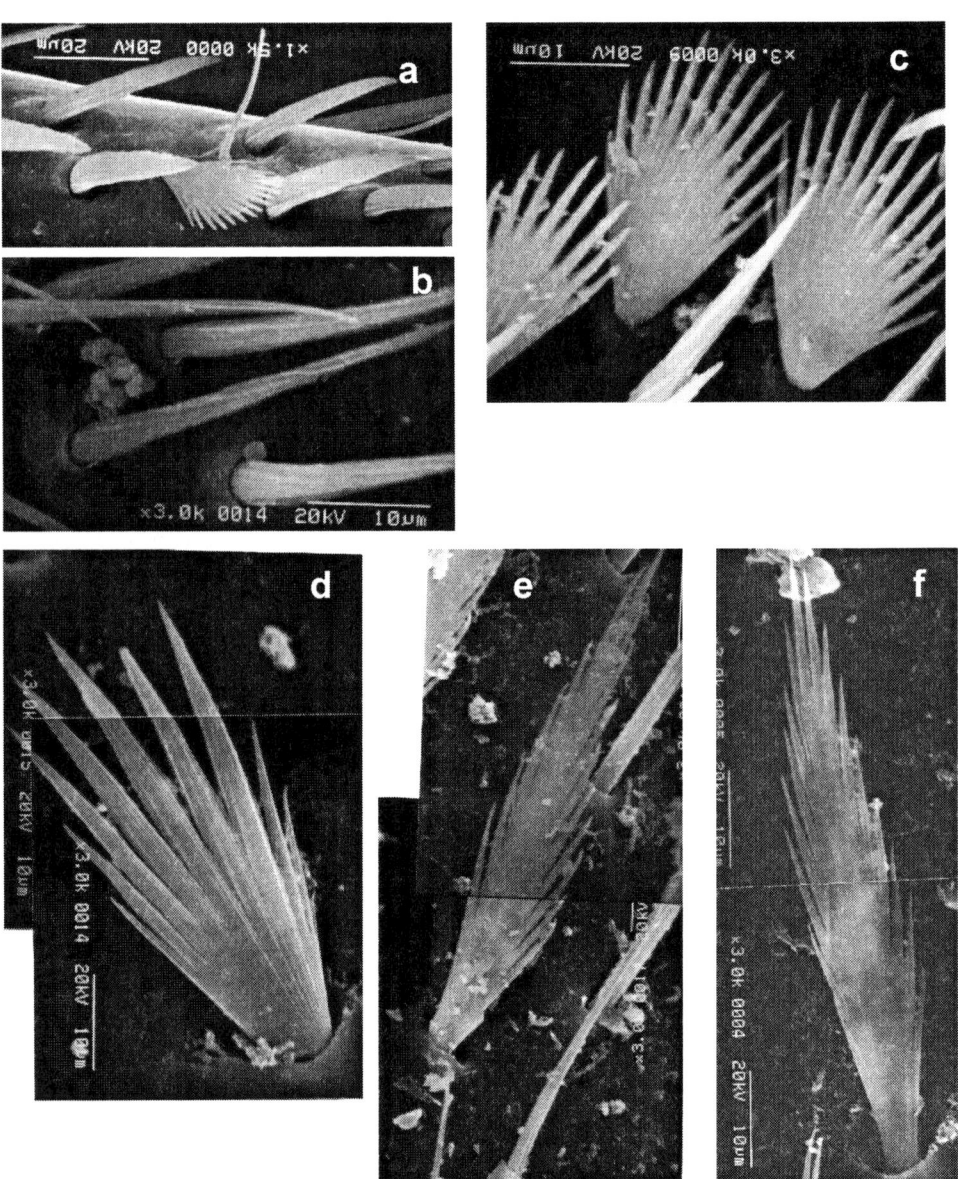

Figure 3. SEM photographs of five types of setae on the ventral side of the left leg in a young larva (sample # 1). We find 5 types in the setae: Spine-like setae (thick and slender) and 4 types of pinnate setae (Broad pinnate A, Broad pinnate B, Slender pinnate A, Slender pinnate B) a: Spine-like seta, thick, b: Spine-like seta, slender, c: Broad pinnate A, d: Broad pinnate B, e: Slender pinnate A, f: Slender pinnate B.

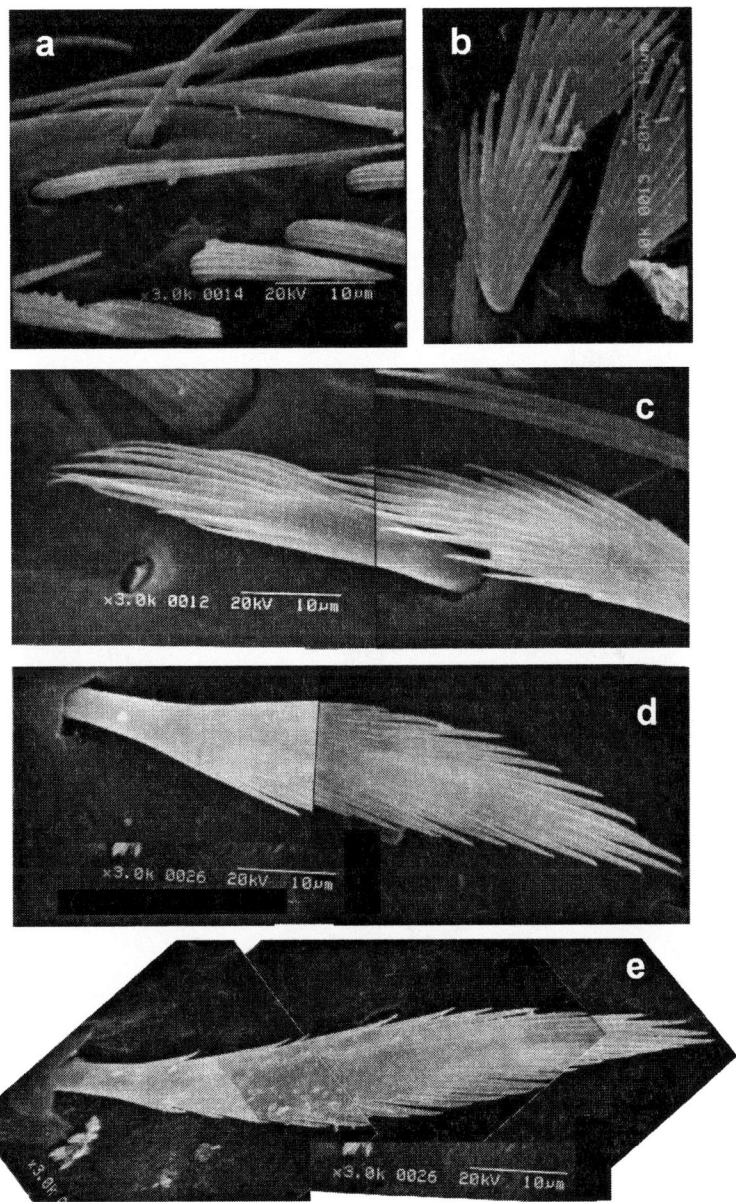

Figure 4. SEM photographs of five types of setae, on the ventral side of left leg in a final instar larva (sample # 6) a: thick (lower side) Spine-like seta, and Slender (upper side) Spine-like seta, b: Broad pinnate A, c: Broad pinnate B, d: Slender pinnate A, e: Slender pinnate B.

Table 2. The size of setae (μm), SD: Standard deviation, N: the number of sample.

Sample	Length (SD)	N	Width (SD)	N
#1				
Seta1 thick	23.3 (-)	1	5.5 (-)	1
Seta1 slender	26.6 (-)	2	3.7 (-)	2
Broad pinnate A	24.6 (-)	2	15.1 (-)	2
Broad pinnate B	50.0 (-)	2	19.6 (-)	2
Slender pinnate A	41.7 (13.95)	5	13.4 (4.24)	5
Slender pinnate B	57.9 (8.81)	7	9.7 (-)	2
#2				
Seta1 thick	28.5 (3.13)	6	4.7 (0.59)	5
Seta1 slender	35.7 (5.47)	5	2.7 (0.82)	5
Broad pinnate A	31.0 (3.03)	5	11.3 (1.41)	5
Broad pinnate B	41.2 (4.82)	7	13.2 (1.77)	3
Slender pinnate A	52.7 (5.03)	8	11.2 (-)	1
Slender pinnate B	62.7 (11.28)	5	9.8 (0.81)	3
#4				
Seta1 thick	34.2 (6.72)	11	5.3 (1.21)	13
Seta1 slender	-		-	
Broad pinnate A	33.3 (4.73)	10	12.7 (3.00)	9
Broad pinnate B	53.7 (8.13)	12	13.3 (1.52)	10
Slender pinnate A	60.3 (9.24)	17	9.7 (1.56)	17
Slender pinnate B	71.0 (12.82)	21	7.7 (1.69)	21
#6				
Seta1 thick	26.8 (4.94)	8	4.0 (1.10)	8
Seta1 slender	25.5 (2.43)	7	1.8 (0.32)	7
Broad pinnate A	23.7 (2.50)	8	12.1 (0.88)	13
Broad pinnate B	46.9 (6.45)	12	12.2 (1.20)	11
Slender pinnate A	53.7 (8.14)	21	8.8 (1.15)	19
Slender pinnate B	62.8 (11.36)	21	8.0 (1.41)	21

Table 3. The number of teeth of the setae (N: sample number).

Sample	Average	(SD)	N	Max	Min
#1					
seta1	-	-	-	-	-
Broad pinnate A	13.9	(1.21)	20	16	12
Broad pinnate B	10.9	(2.57)	9	15	8
Slender pinnate A	17.2	(2.57)	6	19	13
Slender pinnate B	17.0	-	2	20	14
#2					
seta1	-	-	-	-	-
Broad pinnate A	12.9	(1.15)	39	16	10
Broad pinnate B	12.0	(3.31)	7	15	7
Slender pinnate A	16.7	(3.38)	11	23	12
Slender pinnate B	-	-	-	-	-
#4					
seta1					
Broad pinnate A	16.4	(2.69)	15	21	11
Broad pinnate B	12.3	(3.27)	18	18	8
Slender pinnate A	15.7	(1.81)	24	18	12
Slender pinnate B	17.5	(2.57)	24	22	12
#6					
seta1					
Broad pinnate A	14.3	(2.33)	14	18	10
Broad pinnate B	10.8	(1.98)	18	14	8
Slender pinnate A	16.6	(3.91)	23	30	11
Slender pinnate B	19.2	(3.73)	24	26	14

Table 4. Statistical significance in length and width of the specimen # 6 (Student t p<0.05). The same letters means no significance.

Setal type	Length	Width
Spine-like seta (thick)	a	a
Spine-like seta (slender)	a	a
Broad pinnate A	a	b,c
Broad pinnate B	b	b
Slender pinnate A	b	c,d
Slender pinnate B	c	d

Table 5. Region where each type of the setae is located.

	Femur	Tibia	Third Tarsus
Spine-like seta			○
Broad pinnate A			○
Broad pinnate B	○	○	
Slender pinnate A		○	
Slender pinnate B	○		

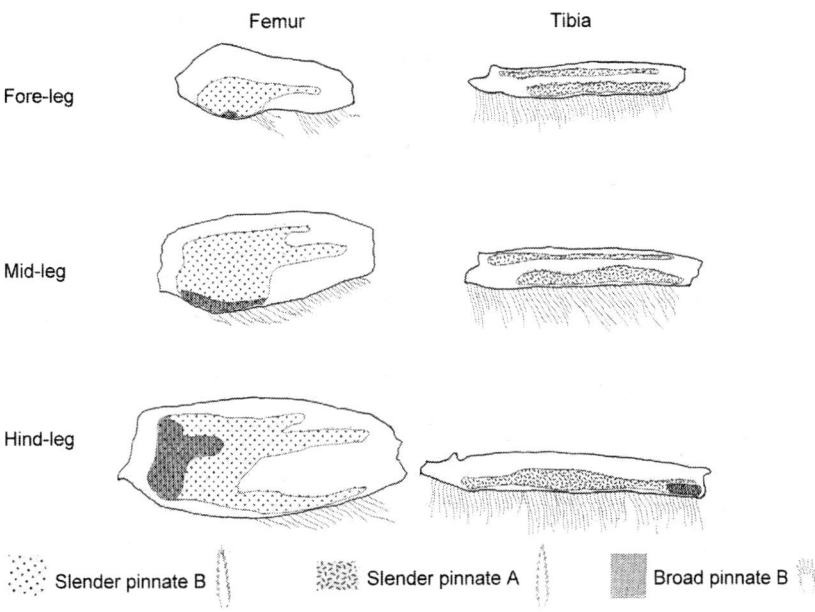

Figure 5. Distributions of the setae on the femora and tibiae, ventral view (young larva).

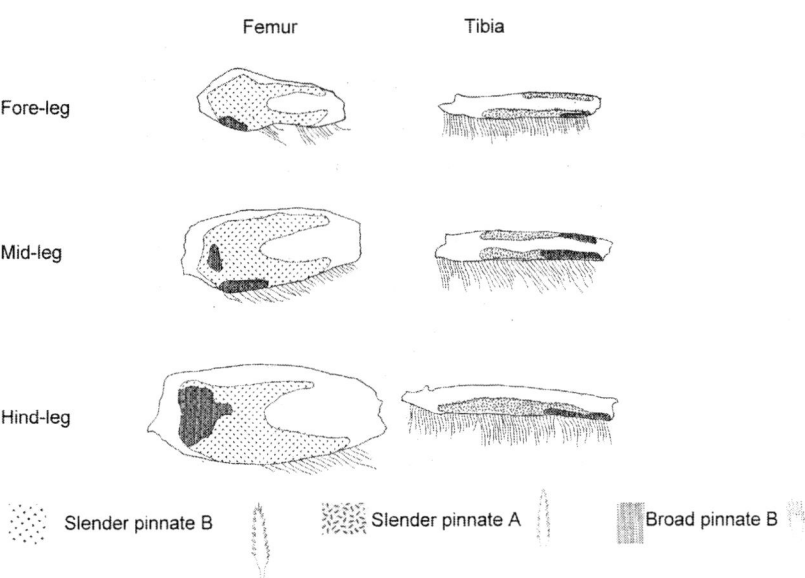

Figure 6. Distributions of the setae on the femora and tibiae, ventral view (final instar larva).

Discussion

We found many setae on the ventral side of the larval legs of *Neoperla geniculata* (Perlidae, Plecoptera), although the dorsal side of the legs bears dense clothing hairs. We recognized five types in the setae. Five types of the setae on the larval legs exhibited similar structures and distributions between the early larvae and the final instar larvae, except for setal density. The setal types were divided into two groups: simple spine-like and pinnate types. We observed the simple spine-like setae, but no pinnate types, on the legs of the other stonefly species (Tamura et. al. unpublished). However, *Neoperla geniculata* bears many pinnate setae.

Each segment of all legs of the species has two types. The basal segment bears longer setae. Femora and tibiae are widely covered with those setae, while the third tarsi bear the setae in a few lines. The setae on the third tarsus, "Broad pinnate A" and "Spine-like seta", have similar length with different width. The arrangments of the two types of the setae on the third tarsus are different among fore-, mid- and hind-legs (Figs. 7 and 8).

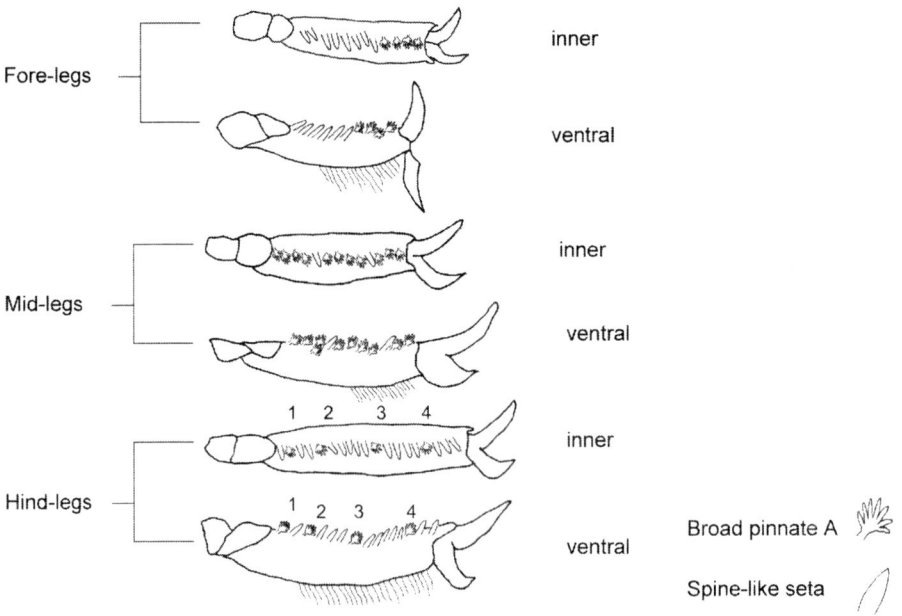

Figure 7. Distributions of the setae on the third tarsi of the left legs, ventral view (young larva).

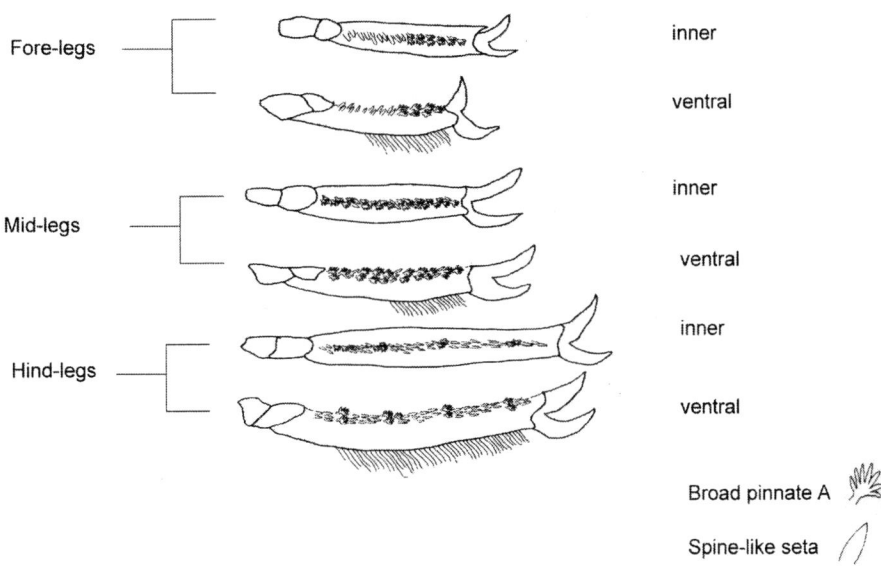

Figure 8. Distributions of the setae on the third tarsi of the left legs, ventral view (final instar larva).

Similar pinnate setae were observed in Trichoptera. The coxa of the fore-leg of net-spinning species, *Cheumatopsyche lepida* and *Hydropsyche pellucidula*, are densely covered with pinnate setae (Spinelli and Corallini 2002). The forelegs are usually used for catching prey, collecting food, and case building in the trichopteran species (Spinelli and Corallini 2002). They speculated that the setae might be linked to the function of net-cleaning. In mayflies, antennal sensillae have been observed by using SEM (Gaino and Rebora 1997), but the authors could not establish a clear relationship between the morphology and the specific function.

In *Neoperla geniculata* (Plecoptera), the pinnate setae are distributed over the ventral surface of all paired legs. It suggests that the setae on the ventral side of legs might function for sensing the stimuli from the substrate (Vaught and Stewart 1974). Furthermore, the third tarsi of the legs are quite characteristic in the distribution of Broad pinnte A and Spine-like setae. Although the general features of the legs in Plecoptera are not so different, the setal distributions are different among fore-, mid- and hind legs. Therefore, we suggest the paired legs may have different functions serving as mechanoreceptors in Plecoptera.

Acknowledgments

We are grateful to Drs. S. Furumi and M. Naruse for their great help on computer technique and others during the study. We also would like to express our sincere thanks to the staffs of the *KYOUSEI* Science Center for Life and Nature and Nara Women's University for their material and spiritual support during this study.

Literature Cited

Gaino, E., and M. Rebora. 1997. Antennal cuticular sensilla in some Mayflies (Ephemeroptera). Pages 317–325 *in* P. Landolt and M. Sartori, editors. Ephemeroptera and Plecoptera, Biology-Ecology-Systematics. MTL-Mauron+Tinguely & Lachat SA, Switzerland.

Spinelli, G., and C. Corallini. W. 2002. Morphology of the first pair of legs of Italian trichopteran larvae: a comparative SEM study. Nova Supplementa Entomologica **15**:29–36.

Stewart, K. W., and B. P. Stark. 1988. Nymphs of North American Plecoptera Genera (Plecoptera). University of North Texas Press.

Uchida, S. 1990. A revision of the Japanese Perlidae (Insect, Plecoptera), with special reference to their phylogeny. Doctoral Thesis, Tokyo Metropolitan University.

Vaught, G. L., and K. W. Stewart. 1974. The life history and ecology of the stonefly *Neoperla clymene* (Newman) (Plecoptera: Perlidae). Annals of the Entomological Society of America **67**:167-178.

Wiggins, G. B. 1977. Larvae of the North American Caddisfly Genera (Trichoptera). University of Toronto Press, Toronto and Buffalo.

MORPHOLOGY, ULTRASTRUCTURE AND FUNCTION OF THE PALMEN BODY AND CONTACT MALLET

T. J. Fink

4493 Pontchartrain Drive, Slidell, Louisiana 70458 USA

Abstract

The Palmen body and the similar contact mallet have been incorrectly and incompletely described since their discovery by Palmén in 1877. The Palmen body of a mature mayfly is made of multiple Palmen body units stacked like unique nested boxes, where the latest formed unit (larger than the previous one of the previous instar) covers and overlaps the previous unit. This overlap gives the illusion of rings, where none really exist. Counts of "rings" are then useful in instar determination.

Two-dimensionally, Palmen body formation and structure/ultrastructure are very similar to tracheae, while in three dimensions Palmen cuticle differs greatly. Each unit has multiple uniformly distributed projections that span the interunit space. Each projection is hollow and open on the outer side, as if poking a finger through dough from the outside in. These projections make counting "rings" of whole Palmen bodies difficult or impossible with compound light microscopy because they scatter the light used to view the rings. Palmen bodies appear to lack the inner epicuticle and granular layer that the tracheal cuticle possess.

Palmen bodies and contact mallets are not innervated and probably have no real function, as evident from body and head tracheal comparisons with Odonata.

Key words: Palmen body; contact mallet; cuticle morphology; cuticle ultrastructure; Ephemeroptera cuticle; Ephemeroptera instars; Ephemeroptera; mayflies.

Introduction

The Palmen body is a cuticular structure that forms the commissure of four head tracheae in mayflies (Ephemeroptera). It is a unique structure known in insects to occur only in the order Ephemeroptera (Landa 1948). The Palmen body apparently is formed at the time of tracheae formation in the second or third instar (Rawlinson 1939, Degrange 1959) and new cuticle is added to the Palmen body during each subsequent instar. The cuticle deposited during different instars appears in the compound light microscope as rings for both whole and sectioned Palmen bodies. The apparent tree-ring like nature of the Palmen body has been infrequently used in determining instars by counting rings (Rawlinson 1939, Degrange 1959, Taylor and Richards 1963, McLean 1970, Benech 1972, Jones 1977, Ruffieux et al. 1996).

The Palmen body was discovered in 1877 by Palmén, and subsequent studies (e.g., Gross 1903, Wodsedalek 1912, Hsu 1933, 1935, Rawlinson 1939, Landa 1948) have briefly described or figured its basic morphology. This structure has not been investigated ultrastructurally and only one study (Wodsedalek 1912) seriously considered its function. These studies did not fully and clearly describe the structure of the Palmen body and major misinterpretations were made concerning basic morphology, formation and function. For example, the Palmen body is shown in the present study not to be composed of rings at all. In 1933, Hsu discovered a Palmen body-like structure at the junction of two transverse tracheae in both the eighth and ninth abdominal segments of a *Stenacron interpunctatum* (Say) larva. Similar structures were subsequently found in other mayflies in the head and abdomen (Rawlinson 1939, Landa 1948, Landa et al. 1980, Landa et al. 1982). Termed contact mallets by Landa (1948), this term has been adopted in the present study. Contact mallets have received much less study than Palmen bodies due to their smaller size.

The purpose of this study is to correctly describe for the first time the correct morphology, ultrastructure and formation of the Palmen body and contact mallet. The apparent function of these cuticular structures was also investigated.

Methods

General Dissections and Compound Light Microscopy. Due to the very small size of these structures, the very finest watchmaker forceps and the finest minuten insect pins mounted in narrow wooden dowels must be used in all dissections. In preserved specimens, the Palmen body was best found by locating the connecting tracheae. First, a portion of the dorsal head cuticle and underlying tissue was carefully removed in the general vicinity of the Palmen body in order to find the connecting tracheae, which served as a guide to the amber colored Palmen body. The Palmen body tracheae were then cut and used as handles to pick up and manipulate the Palmen body in later processing. The head should not be pulled from the body since the connecting tracheae and Palmen body may be pulled out of position or the tracheae will be severed. The Palmen body and its tracheae were much easier to locate in live specimens since these structures were still filled with air and thereby stood out in sharp contrast to the surrounding tissues. This was greatly enhanced by placing live specimens in 100% glycerin which rapidly cleared in several minutes or more obscuring tissues so that the Palmen body and tracheae may be viewed directly through the dorsal head cuticle. Subimaginal and imaginal specimens did not adequately clear in glycerin and were dissected. If the Palmen body was inadvertently separated from its tracheae, then the Palmen body was sucked up into the space between the forceps without being touched.

After dissection, the Palmen body was transferred to a depression slide or regular slide containing a suitable liquid (see below) and any remaining adherent tissues were removed as far as practical. Whole Palmen body slide mounts for light microscopy were obtained by placing the Palmen body directly on the slide in 80% ethanol, letting the excess alcohol evaporate and then adding a drop of 100%

cellusolve (ethylene glycol mono ethyl ether), and allowing the excess to evaporate before adding 95% ethanol and mounting in euparol. Many other specimens were mounted in euparol without cellusolve clearing. Polyvinyl-lactophenol, cellusolve-balsam, ethanol and 100% glycerin "mountants" also gave satisfactory results; none of these yield permanent mounts including the cellusolve-balsam and polyvinyl-lactophenol, which eventually degrade the Palmen body rings. A dorsal or ventral view of the Palmen body was achieved by laying down flat all Palmen body tracheae. Detailed observations and counting of "rings" were made at 1000 X (oil immersion) using bright-field illumination with or without a green filter. Phase contrast light microscopy was not useful at all due to the complicated multiunit structure of the Palmen body and contact mallet (see results/discussion section).

Sections for light microscopy were prepared by dehydrating specimens in a graded series of ethanol, clearing in xylol, embedding in paraffin, sectioning at 4- to 10- m thickness and slide mounting in xylol-balsam. Staining was with hematoxylin and eosin. Some sections were stained with Mallory's triple stain to determine the layers of cuticle (endocuticle, mesocuticle and exocuticle) (Taylor and Richards 1963, Whitten 1972). All observations and counts were made under 1000X (oil immersion) as above.

Scanning Electron Microscopy (SEM). Whole Palmen bodies for scanning electron microscopy were air dried from ethanol, distilled water or amyl acetate, or critical point dried using amyl acetate/carbon dioxide gas, mounted on aluminum tabs with double sided sticky tape, coated with gold palladium and examined in a Cambridge Stereoscan S4-10 scanning electron microscope at Florida State University. Whole Palmen bodies were manipulated by the attached tracheae. However, if views were desired looking into the Palmen body where the tracheae attach, then the tracheae were pulled off while the Palmen body was wet and the Palmen body was then sucked up into the space between a pair of jeweler's forceps and allowed to air dry on a slide or critical point dried. Minuten pins were then used to prod the Palmen body into a cone up (where a trachea attaches, see results) position. Some dried Palmen bodies were rolled on the double-sided sticky tape by pushing with minuten pins. This served to effectively dissect the Palmen body apart.

Paraffin sections examined in the scanning electron microscope were prepared by dissolving off the coverslip in warm xylol, cutting the glass slide into chips (containing chosen sections) small enough to be mounted on aluminum tabs, rinsing in several changes of fresh xylol, air drying, mounting on metal tabs and coating and examination as above.

Transmission Electron Microscopy (TEM). Palmen bodies for TEM were either dissected out of the head or left in the head. The later proved more successful due to ease of handling. Specimens were then placed in phosphate buffered (pH 7.4) primary fix of 2% paraformaldehyde and 2.5% glutaraldehyde for 1 to 2 hours at room temperature, postfixed in phosphate buffered 1% osmium tetroxide for 1 to 2 hours at room temperature, dehydrated in a graded series of acetone, and then embedded in Spurr's low viscosity medium. Thin sections were stained with uranyl acetate and lead citrate and observed with a Phillips 301 transmission electron microscope at Florida State University.

Procedures for Contact Mallets. Procedures used for location, dissection, handling and examination of contact mallets were identical to those used for the Palmen body.

Function of the Palmen Body: Replicating Wodsedalek's Experiments. Wodsedalek's (1912) Palmen body orientation experiments were repeated in the present study on *Maccaffertium exiguum* larvae. An unsuccessful attempt was made to remove the Palmen body as Wodsedalek reports to have done on nymphs of the similar species *Stenacron interpunctatum* (Say) by using two fine needles (minutens) to pierce and cut a small hole in the head cuticle above the Palmen body, cut the tracheal connections and then lift the Palmen body out. Much greater success was attained by a procedure that greatly saved time and amount of handling that appeared to be the main causes of mortality. The nymph was grasped firmly with two pairs of jewelers forceps at the extreme posterior margin of the head, the cuticle was then torn along the coronal suture to a spot just above the Palmen body where a small flap of cuticle was removed, and finally the Palmen body was quickly destroyed by grasping with forceps. Another attempt involved bending the head downward slightly so that forceps could reach in directly to destroy the Palmen body; however, just bending the head resulted in massive bleeding.

Details of the experiments are described in the results and discussion section for greater clarity.

Mayfly Specimens Examined. The Palmen bodies and/or contact mallets of the following mayfly species were examined in this study: *Hexagenia limbata* (Serville), *Hexagenia bilineata* (Say), *Dolania americana* Edmunds and Traver, *Tortopus puella* (Pictet), *Palingenia* longicauda (Olivier), *Siphlonurus spectabilis* Traver, *Baetisca rogersi* Berner, *Neoephemera youngi* Berner, *Callibaetis pretiosus* Banks, *Baetis* sp., *Caenis macafferti* Provonsha, *Caenis diminuta diminuta* Walker, *Caenis* sp., *Maccaffertium exiguum* (Traver), *Maccaffertium carlsoni* (Lewis) and Heptageniidae sp. Specimens were either collected by the author in Florida, Alabama, and Louisiana or were obtained from the mayfly collection at the Department of Entomology, Florida A & M University.

Arthropods Other Than Mayflies Examined for Palmen Body-Like Structures. Arthropods, other than mayflies, examined for the presence of Palmen body-like structures include the following. Several specimens of the insect orders Trichoptera, Plecoptera and Megaloptera (*Corydalus cornutus* Linnaeus) were briefly examined throughout the head and body by the glycerin treatment or were clear enough without glycerin or were dissected.   Five fairly mature Thysanura (Lepismatidae) specimens were carefully examined by the glycerin treatment and dissection (glycerin treatment was only partially effective, especially in the head).   In the Odonata (all specimens were half-mature to mature larvae), two *Enallagma* sp. (Zygoptera, Coenagrionidae), one *Pachydiplax longipennis* (Burmeister) (Anisoptera, Libellulidae) and two *Celithemis* sp. (Libellulidae) specimens were carefully examined throughout the head and body by the glycerin treatment; the head and body of three preserved *Tramea* sp. nymphs (Libellulidae) were carefully examined and dissected; the heads of mature preserved nymphs of one *Aeschna* sp. (Anisoptera, Aeschnidae), one *Progomphus obscurus* (Rambur) and one *Gomphus* sp. (Anisoptera, Gomphidae) were briefly dissected.   Two centipedes (Chilopoda, Scolopendromorpha) and five millipedes (Diplopoda) were examined throughout the head and body by the glycerin treatment and dissection (the glycerin treatment was only partially effective, especially for the millipedes).

Results and Discussion

Palmen Body and Contact Mallet Location, Size and Occurrence in Mayflies. The basic mayfly tracheal system consists of two large lateral longitudinal trunks, one on each side of the body, which run the length of the abdomen and thorax and give rise to branches which supply the gills, caudal filaments, legs, digestive tube, gonads, ventral nerve cord and other areas of the body (Plate 1). Generally in the mesothorax or prothorax each lateral trunk branches (this may occur in the head as shown for *Oligoneuriella rhenana* Imhoff (Landa 1948) into a dorsal and ventral trunk which supply the dorsal and ventral areas of the anterior thorax and the head.

The Palmen body is located in the head at the junction of four tracheal branches, of the head tracheal trunks, just underneath the coronal suture. In most species it is located near the back of the head but never anterior of the anterior-posterior midpoint of the head (Plate 2).

A small structure, the Palmen body of a mature specimen when measured transversely and longitudinally (measurement directions are in reference to axes of the body and are made from a dorsal or ventral view) ranges in size from 317 by 191 m in a very large mayfly species (*Palingenia longicauda* subimago) to 31 by 27 m and 22 by 19 m respectively for two very small species *Caenis diminuta diminuta* imago and in *Caenis macafferti*.

The Palmen body has been found in all mayflies examined for its presence. This has included over 68 species in 15 families (data gathered from the present study and the following: Palmén 1877, Gross 1903, Wodsedalek 1912, Hsu 1933, Hsu 1935, Rawlinson 1939, Landa 1948, Landa et al. 1980, Landa et al. 1982, Degrange 1959, Taylor and Richards 1963, McLean 1970, Benech 1972, Jones 1977, Ruffieux et al. 1996).

Contact mallets have been found in the eighth and ninth abdominal segments of a *Stenacron interpunctatum* nymph (Hsu 1933), in the eighth abdominal segment of

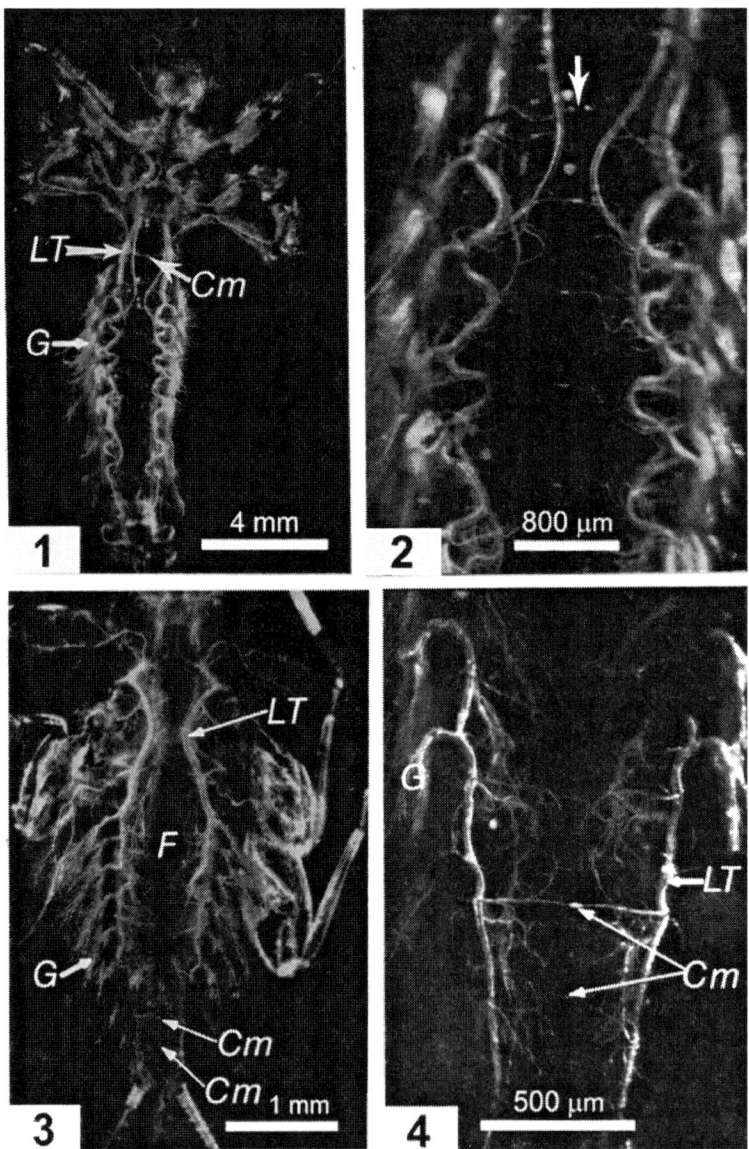

Plate 1: Figure 1—Ventral view of a mature male *Hexagenia* sp. larvae cleared in 100% glycerin. Seven contact mallets are visible just dorsal to the abdominal sternites, where each one forms the anastomosis at the junction of two transverse tracheae in one of the abdominal segments 1–7. Figure 2—Enlargement of Fig. 1. Arrow points to broken contact mallet in segment 2. Figure 3—Ventral view of body of a mature *Maccaffertium exiguum* larvae cleared in 100% glycerin. Two contact mallets visible just dorsal to the sternites of abdominal segments 8 and 9. Figure 4— Enlargement of Fig. 3. (See Appendix A for plate abbreviations.)

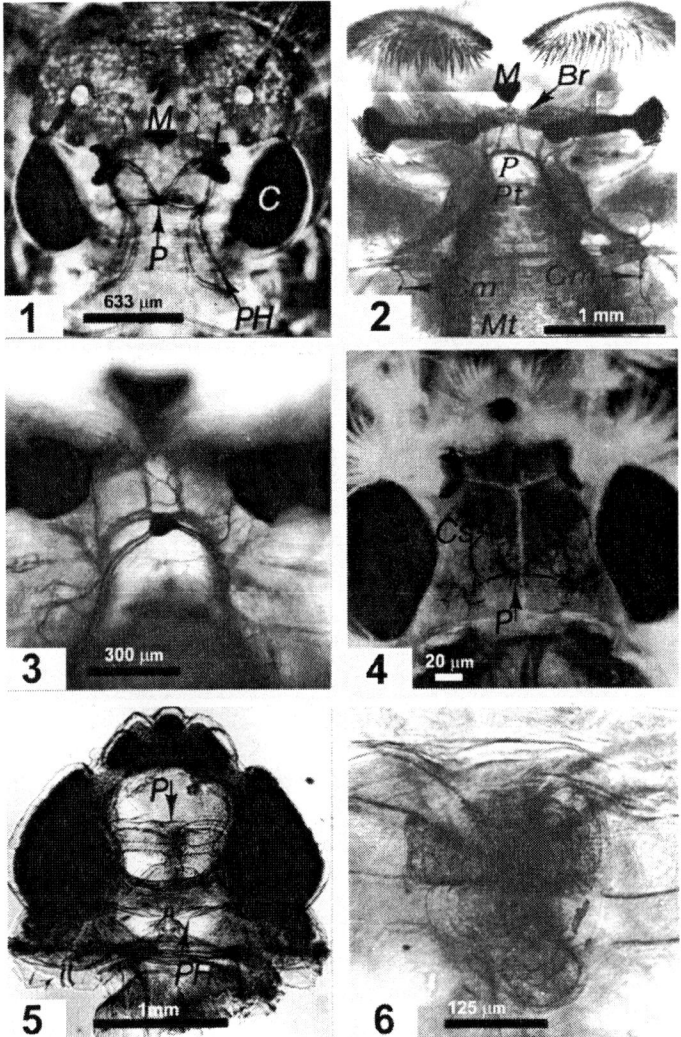

Plate 2: Figure 1—Dorsal view of head of an immature *Maccaffertium exiguum* larvae cleared in 100% glycerin showing location of Palmen body. Figure 2—Dorsal view of head and pro- and mesothorax of an immature female *Dolania americana* larvae cleared in 100% glycerin showing location of Palmen body and contact mallets. Figure 3—Enlargement of Palmen body area of Fig. 2. Figure 4—Dorsal view of head and prothorax of a mature male *Hexagenia* sp. larvae cleared in 100% glycerin showing the location of the Palmen body. Figure 5—Dorsal view of head of a female *Siphlonurus spectabilis* subimago in which much of the dorsal cuticle was removed to reveal the Palmen body and associated tracheae. Figure 6—Enlargement of the Palmen body of Fig. 5. The Palmen body of this species is unique due to the waviness of the rings.

Ecdyonurus venosus (Fabricius) (Rawlinson 1939), in other abdominal segments and in the head of a variety of species (Landa 1948).

In the present study, contact mallets were found in the mesothorax of *Dolania americana* (Plate 2: Fig. 2). In a mature larva of *Hexagenia* sp., contact mallets were found in abdominal segments one through seven just dorsal to the sternites (Plate 1: Figs. 1 and 2). In a mature larva of *Maccaffertium exiguum*, contact mallets were found in abdominal segments 8–9 just dorsal to the sternites (Plate 1: Figs. 3 and 4).

Landa in his 1948 monograph of comparative mayfly tracheation states that contact mallets are found at all tracheal commissures. However, in the present study, contact mallets are absent in the ninth abdominal transverse tracheae in a mature male *Baetis* sp. nymph (Plate 3: Fig. 5), and in a mature male *Callibaetis pretiosus* larva (Plate 3: Fig. 6).

Whereas the Palmen body is the anastomosis or commissure of four tracheae, the contact mallet is the anastomosis of two tracheae and is therefore smaller. The contact mallets seen in Figs. 1–3 of Plate 3 are respectively in size: 53 x 33 m (*Maccaffertium exiguum*), and 59 x 42 m and 77 x 41.5 m (*Hexagenia* sp.), while the respective approximate Palmen body sizes are: 130 x 83 m (*Maccaffertium exiguum*) and 101 x 103 m (or greater, *Hexagenia* sp). Contact mallets in the abdomen are located in the midline just dorsal to the cuticle of the sternites (Plate 1).

Palmen Body and Contact Mallet Structure as Viewed in the Compound Light Microscope. The Palmen body and contact mallet structure have only been described at the light microscope level. This is partly why these structures have been erroneously described by all previous investigators. Examining dorsal/ventral mounts of the Palmen body in Plate 4 from several species appears to indicate a ringed structure, similar to the circuli of a fish scale. The "rings" appear to have fine lines or striations perpendicular to the running axis of the ring. In reality the observed rings are really the peripheral area of underlying Palmen body units formed during earlier instars, and the lines are inner projections of the Palmen body units. This will be become obvious in the next section where Palmen bodies observed under the scanning electron microscope are described. The projections also create the illusion of a very pebbly appearance on the surface of the Palmen body (Plate 3: Fig. 4) when what one is really viewing are the many projections from the many Palmen body units superimposed on one focal plane. The attached tracheae never show such a pebbly surface appearance (Plate 3: Fig. 4).

The contact mallet is made identically to the Palmen body except only two tracheae attach to it. False rings are apparent (Plate 3: Figs. 1–3) and the "pebbly surface" appearance is readily observed (Plate 3: Fig. 3). Landa (1948) has stated that contact mallets form on the tracheal ends prior to the junction of two transverse tracheae. This deduction is probably the result of viewing broken contact mallets (see Plate 1: Fig. 2). Intact contact mallets show a false ring structure over the intact mallet that could not result by fusion of separate halves (Plate 3: Figs. 1–3).

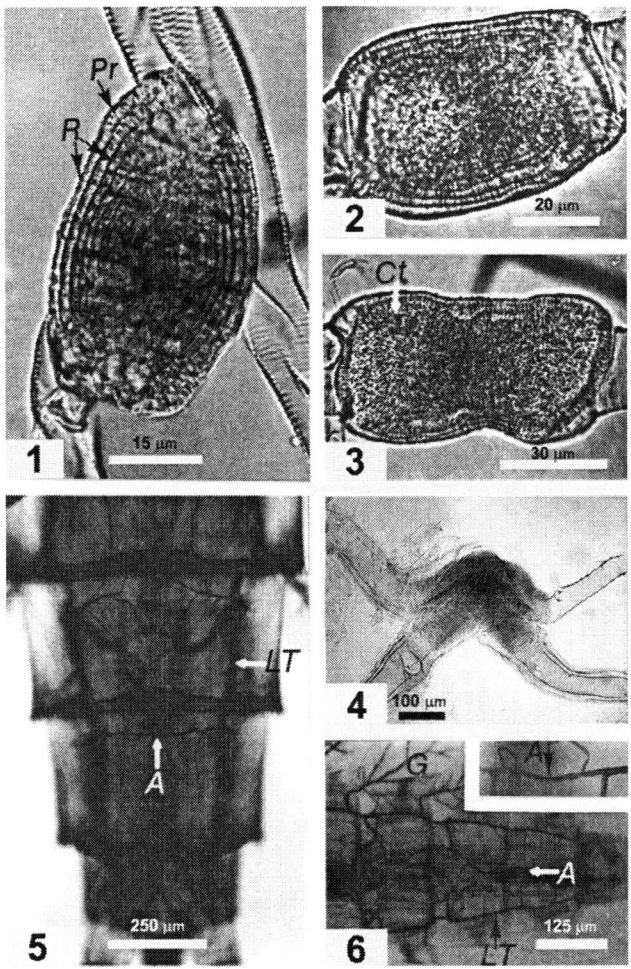

Plate 3: Figure 1—Ventral view of eighth abdominal segment contact mallet from a mature *Maccaffertium exiguum* larvae. Figure 2—Ventral view of fourth abdominal segment contact mallet from a mature male *Hexagenia* sp. larvae. Figure 3—Surface focus on the first abdominal segment contact mallet from larva of Fig. 2. The projections of the underlying Palmen body units superimposed on this focal plane give the illusion of a pebbly surface. Figure 4—Dorsal view of the surface of the Palmen body from a *Palingenia longicauda* female subimago. The surface appears pebbly due to the projections of the underlying Palmen body units. Figure 5— Ventral view of abdominal segments 7–10 of a mature male *Baetis* sp. larvae. No contact mallet is present at the junction of the two transverse tracheae that join together in the anterior ninth segment. Figure 6—Ventral view of abdominal segments 6–10 of a *Callibaetis pretiosus* larvae. No contact mallet is present at the

anastomosis of the two transverse tracheae which join together in the anterior ninth segment.

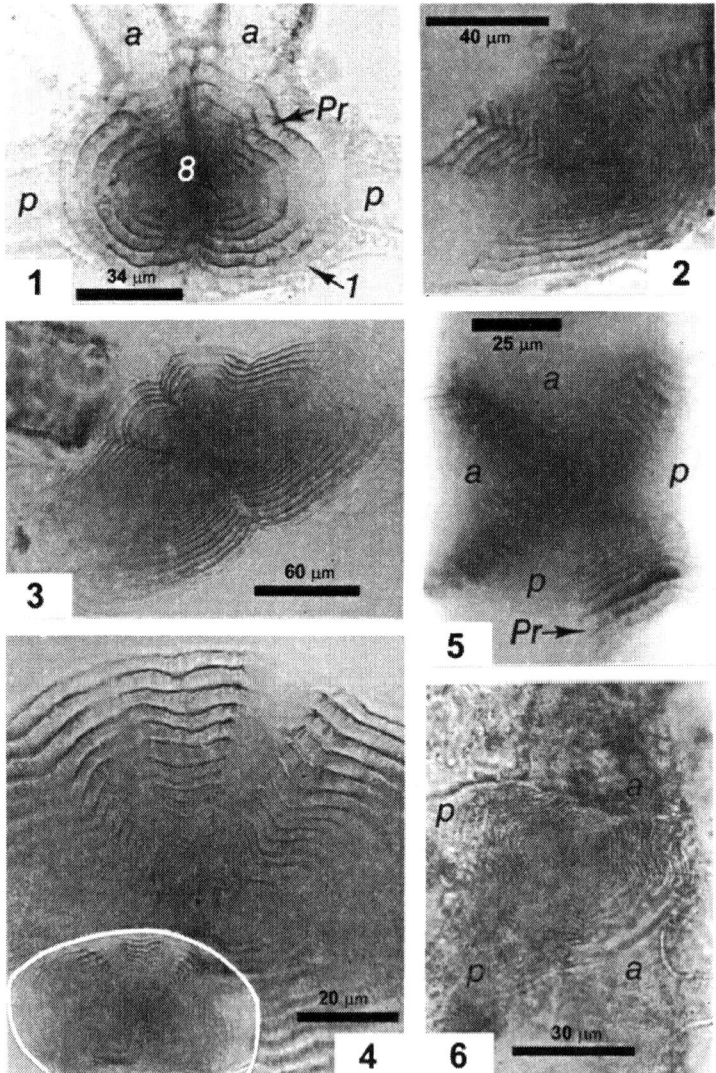

Plate 4: Figure 1—Ventral view of a Palmen body from an immature Heptageniidae sp. larvae. There are at least 8 "rings" (Palmen body units) but the center of the Palmen body cannot be clearly seen. Figure 2— Ventral view of a Palmen body from a mature female *Maccaffertium exiguum* larvae. Most of the Palmen body units (apparent rings) can be seen easily; however, the inner units are still difficult to see with assurance. The total number of Palmen body units is 16+. Figure 3—Ventral view of a Palmen body from a Heptageniidae sp. larvae. Most of the Palmen body units are readily visible in the light microscope (although not in this photograph due

to depth of field limitations). The total number of Palmen body units is approximately 17 to 18. Figure 4—Ventral view of a Palmen body from a Heptageniidae sp. larvae (inset entire Palmen body). The "rings" (Palmen body units) of only the two anterior "cones" are quite distinct. The total number of Palmen body units is approximately 15. Figure 5—Ventral view of the Palmen body from a last instar female *Hexagenia limbata* larvae. Total number of Palmen body units is approximately 20 to 21 (not all visible in photograph). Figure 6—Dorsal view of the Palmen body from a mature male *Callibaetis pretiosus* larvae. The overlapping edges of Palmen body units ("rings") are small but distinct and can be counted to the center (but not in this photograph due to depth of field limitations). Total number of Palmen body units is 19 to 20.

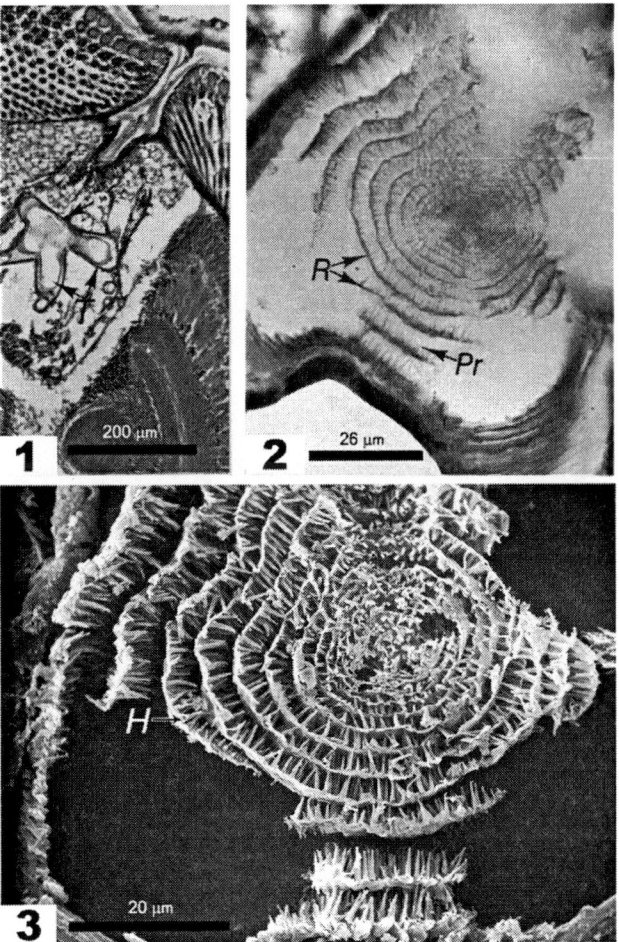

Plate 5: Figures are from 8- m thick paraffin sections from a final instar larvae of *Baetisca rogersi*. Figure 1—Compound light micrograph. Figure 2—Compound light micrograph. Figure 3—Scanning electron micrograph of the section seen in Fig. 2.

Palmen Body Structure as Viewed in the Scanning Electron Microscope. Plate 5 shows the Palmen body from a mature larva of *Baetisca rogersi.* Figs. 1 and 2 show a compound light microscope view of a paraffin section showing the four attached tracheae, false rings and false ring projections (Palmen body unit projections). Fig. 3 is an enlargement of the view seen in Fig. 2 but now viewed in the scanning electron microscope. This view indicates about 11 or 12 Palmen body units, which here appear as 11 "rings" and a central mass (Pescador 1974 determined 12 larval instars by morphological examination). The rings here though show that previous descriptions of Palmen body rings were in error in that rings described in previous papers (for the latest account see Ruffieux et al. 1996) are really pointing to the empty space between Palmen body units, and the fine lines or striations on the rings are really Palmen body projections that span the interunit space. These projections are hollow (Fig. 3). Other views of hollow projections in this and other species are seen in the following figures: Plate 6: Fig. 6, Plate 7: Figs. 5 and 6, Plate 8: Fig. 6, Plate 9: Figs. 3 and 4.

Plate 6 shows views of Palmen bodies of *Hexagenia limbata*. The strongly three dimensional nature of the Palmen body is clearly seen in scanning electron micrographs 1–5. The dorsal surface shows a strong chain like circular mesh, which is composed of ridges of cuticle secreted by the former overlying epidermal cells of the same size and shape (Figs. 1, 4, 6). It can clearly be seen where the four tracheae once attached. The Palmen body gives the illusion of being made up of four cone like structures due to the circular lip of the tracheal attachment and the progressively smaller circular lips of inner, earlier formed Palmen body units (Figs. 1–5). Looking through a "cone view" (Figs. 2, 3, 5) shows not rings (which really do not exist) but some of the earlier formed Palmen body units, each one with inward hollow projections. Unfortunately, it is not possible to see all of the Palmen body units from a cone view due to the contorted shape of the units, great depth of the "cones" and the presence of the obscuring projections. Fig. 6 shows the latest formed Palmen body unit from an external view. The holes indicate the beginning of the hollow projections.

The Palmen body units appear to be contorted where adjacent tracheae attach. Here the cuticle arches or deflects towards the center of the unit (Plate 6: Figs. 1 and 4). This can be seen in light microscope preparations (Plate 4: Figs. 1–3, Plate 7: Figs. 1 and 2) and in the paraffin (deparaffinized) thick sections viewed in the scanning electron microscope (Plate 7: Fig. 4).

The Palmen body unit's shape is also affected by the different sizes of the attaching tracheae in at least some species. In *Hexagenia* the posterior tracheae are larger and therefore the Palmen body posterior cones are larger. Like *Hexagenia*, most Palmen bodies are squarish in overall shape. Some Heptageniidae species mature Palmen bodies, however, are much more flattened dorsoventrally and elongated in the transverse direction (Plate 4: Fig. 3, Plate 8: Figs. 1 and 2).

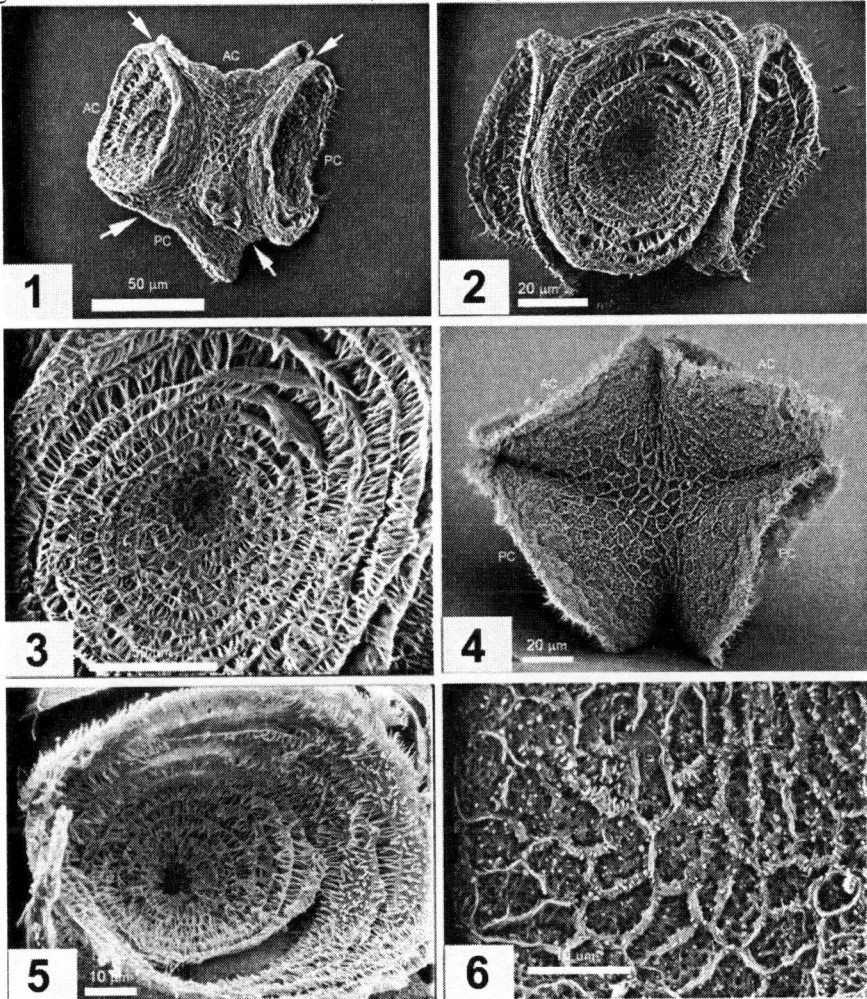

Plate 6: Scanning electron micrographs of the Palmen bodies from two *Hexagenia limbata* female subimagos. Figures 1–3 are from one female. Figures 4–6 are from the other. Arrows in Fig. 1 point to where adjacent "cones" abut (see text).

Many Palmen bodies may be arched dorsally as the Palmen body units are deflected down to meet the upward directed tracheae. This is certainly the case in *Palingenia longicauda* where the large Palmen body can be clearly seen to be arched up dorsally (Plate 3: Fig. 4). This is difficult to see in other Palmen bodies due to their small size. While the edges of most Palmen body units are straight, in the two

Siphlonuridae Palmen bodies examined (Plate 2: Fig. 6, and see Ruffieux et al. 1996) the edges are wavy.

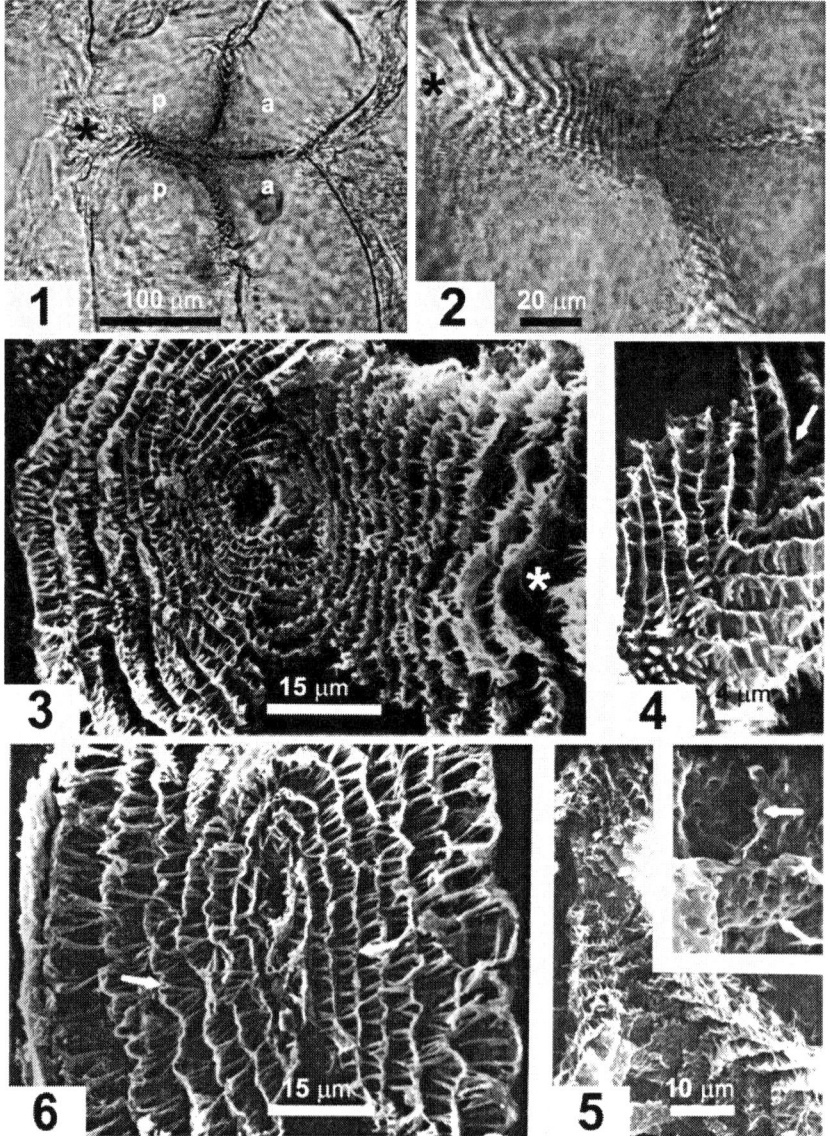

Plate 7: Figures 1 and 2—Light micrographs of the same *Hexagenia* sp. Palmen body at two focuses. * is a region of the Palmen body that is also shown below in Fig. 3. Figures 3, 4, and 5—are 8- m thick paraffin sections from three *Hexagenia limbata* adults (Figs. 3 and 4 are subimagos, Fig. 5 is an imago). Fig. 4 shows the region of the Palmen body where adjacent anterior tracheae once attached. Arrow shows the inward deflection of the cuticle. Fig. 5 shows the hollowness of the projections (arrows).

Figure 6 is a 5- m thick section of a Palmen body from a mature female *Baetisca rogersi* larvae. Arrows point to cuts in projections showing that they are hollow.

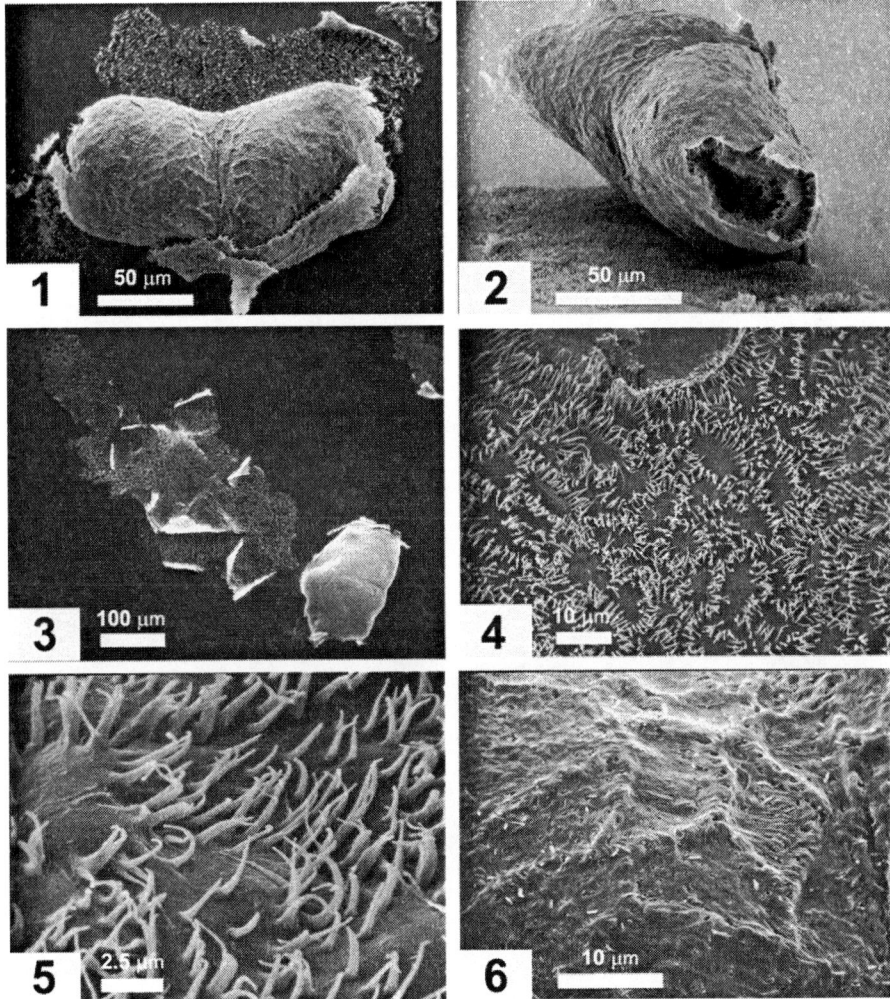

Plate 8: Figures 1–6 are scanning electron micrographs of *Maccaffertium carlsoni imagos*. Palmens in Figs. 1–5 have been rolled on double-sided sticky tape to partially dissect the Palmen body to reveal inner Palmen body units and projections (projections enlarged in Figs. 4–5). Fig. 6 shows the external surface of the Palmen body showing the epidermal cell cuticular ridges and the beginning point of the hollow projections.

In Plate 8 of scanning electron microscope views of Heptageniidae Palmen bodies, you can clearly see the multiunit nature of the Palmen body. In those figures the Palmen bodies were partially dissected by rolling the body on double-sided sticky tape. Thus, you can see previous Palmen bodies dissected off and stuck to the tape (Figs. 1–3). The dissected off Palmen body units also clearly reveal the internal arrangement of the projections. In these Heptageniidae, the projections occur along the edge of the circular cuticular imprint of the former epidermal cells (Figs. 1–5). In Figs. 1 and 2, and especially 6, you can see the external cuticular ridge that is the cuticular imprint of the overlying epidermal cells. Holes on the ridges mark the hollow invagination point of the projections. In *Hexagenia* the projections are not limited to the periphery of the epidermal cuticular imprint, and appear not to be located on the ridges (Plate 6: Fig. 6).

Scanning electron micrographs of *Dolania americana* Palmen bodies are shown in Plate 9. Multiple Palmen body units are clearly evident in Figs. 1 and 2. Epidermal cuticular ridges and the external starting point of the hollow projections are seen in Figs. 3 and 4.

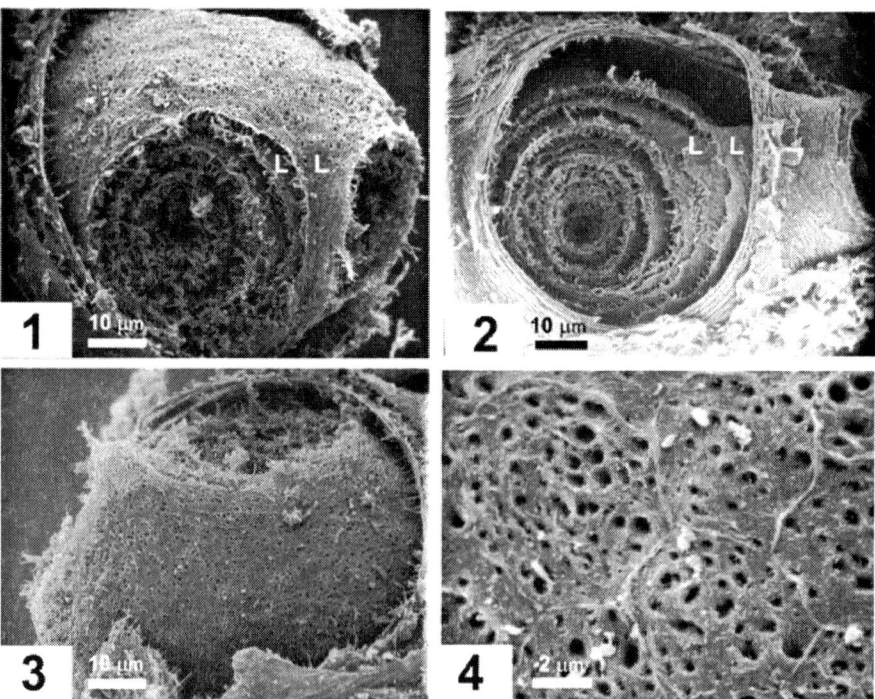

Plate 9: Figures 1–4 *Dolania americana* female subimago Palmen body (same specimen). Figs. 1 and 2 show the multiunit structure of the Palmen, where each subsequent Palmen body unit is a slightly larger facsimile of the one before. Fig. 4 is a highly magnified view of the surface seen in Fig. 3. The dark circular holes are the hollow interior of the projections. The outline of the epidermal cells that secreted the

Palmen body cuticle is indicated by the large circular raised ridges of cuticle that enclose an area with many hollow projections.

Palmen body ultrastructure. Light micrographs and TEM micrographs in Plate 10 illustrate for the first time that Palmen body cuticle is modified tracheal cuticle. In Figs. 1–4 you can see the change from tracheael taenidia to Palmen body unit and projection cuticle.

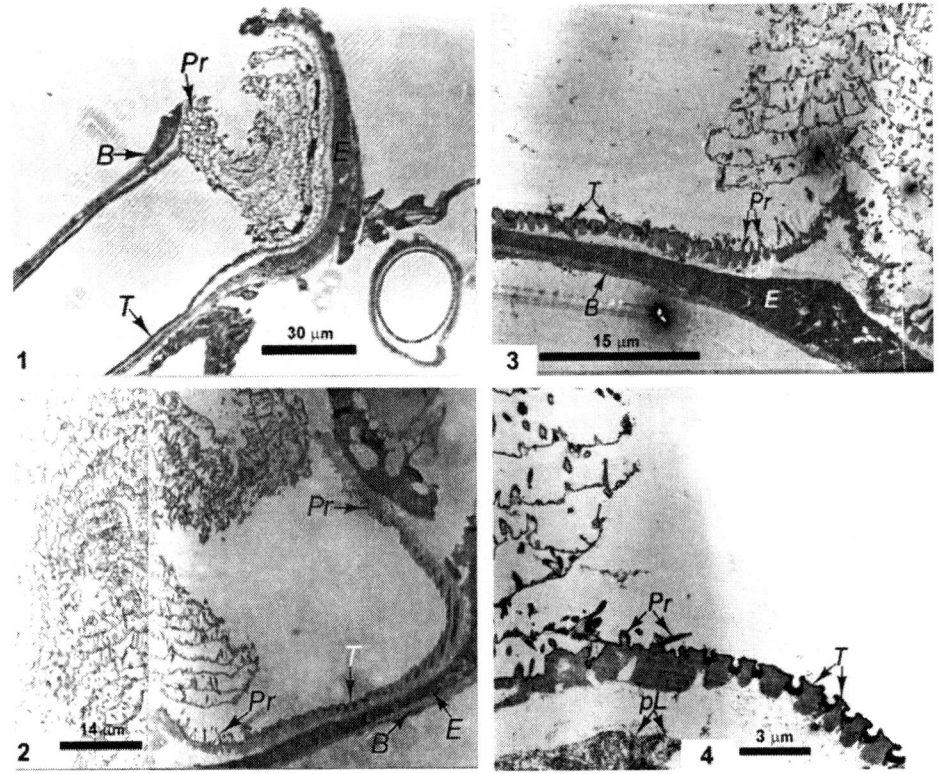

Plate 10: Figures 1–4—Thick (Figs. 1 and 2) and ultrathin (Figs. 3 and 4) epoxy sections through the Palmen body and attached tracheae of a mature female *Dolania americana* larvae, showing the change from tracheal cuticle to Palmen body cuticle. The thick sections are viewed in a compound light microscope, while the ultrathin sections were imaged in TEM.

Contrasting the TEM micrograph plates of tracheae (Plate 11) and Palmen body (Plate 12) shows that ultrastructurally that tracheae and Palmen body cuticle are similar. However, Palmen body epicuticle appears to lack the inner epicuticle and granular layer of tracheal cuticle.

The different three-dimensional molding of these similar ultrastructural components molds taenidial cuticle into Palmen body cuticle. By analogy with the Palmen body, contact mallet ultrastructure would be similar.

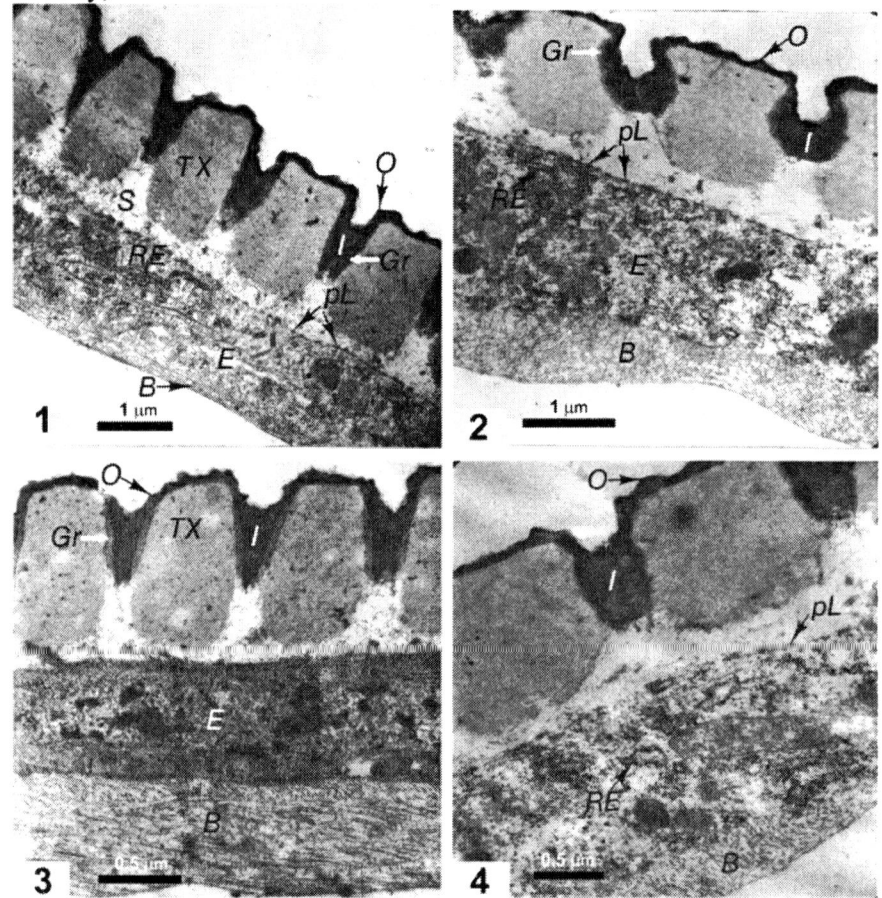

Plate 11: Figures 1–4—TEM micrographs of the Palmen body tracheae from a mature female *Dolania americana* larvae very near emergence to the subimago, showing typical tracheal ultrastructure. Apolysis has begun in this specimen. Compare this Plate with Plate 12 of Palmen body cuticle.

Palmen Body and Contact Mallet Function. The fact that all mayflies appear to have Palmen bodies (and certainly many with contact mallets as well) seems to argue for a possible function for Palmen bodies and contact mallets. They may be stronger tracheal junctions than normal tracheal junctions, but broken contact mallets are observed occasionally (Plate 1: Fig. 2), and even in the mayfly most tracheal junctions involve normal tracheal cuticle. A strong argument against a significant function of Palmen bodies and contact mallets is the lack of these structures in other insects. Palmén (1877) found no Palmen body like structures in members of the Insect orders Odonata, Plecoptera, Trichoptera, Megaloptera, Diptera, Hymenoptera, Lepidoptera and Coleoptera, and other arthropods. Plate 13 shows head tracheal cuticle in Odonata that in mayflies would show the presence of a Palmen body. Several specimens of the orders Trichoptera, Plecoptera, Megaloptera and Thysanura were briefly examined in this study and no Palmen like structures were observed.

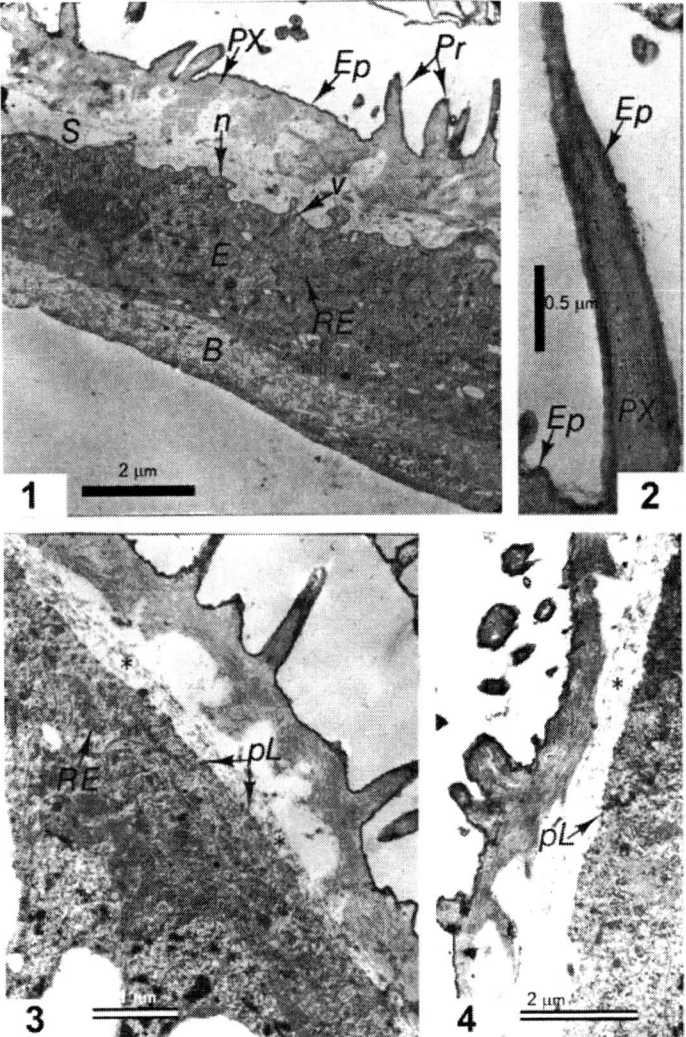

Plate 12: Figures 1–4—TEM micrographs of Palmen body cuticle from the same specimen in Plate 11. Compare with the tracheal cuticle shown in Plate 11. Palmen

body cuticle apparently lacks the inner epicuticle and granular layer of the tracheae. The * indicates a dense staining region that might be molting fluid.

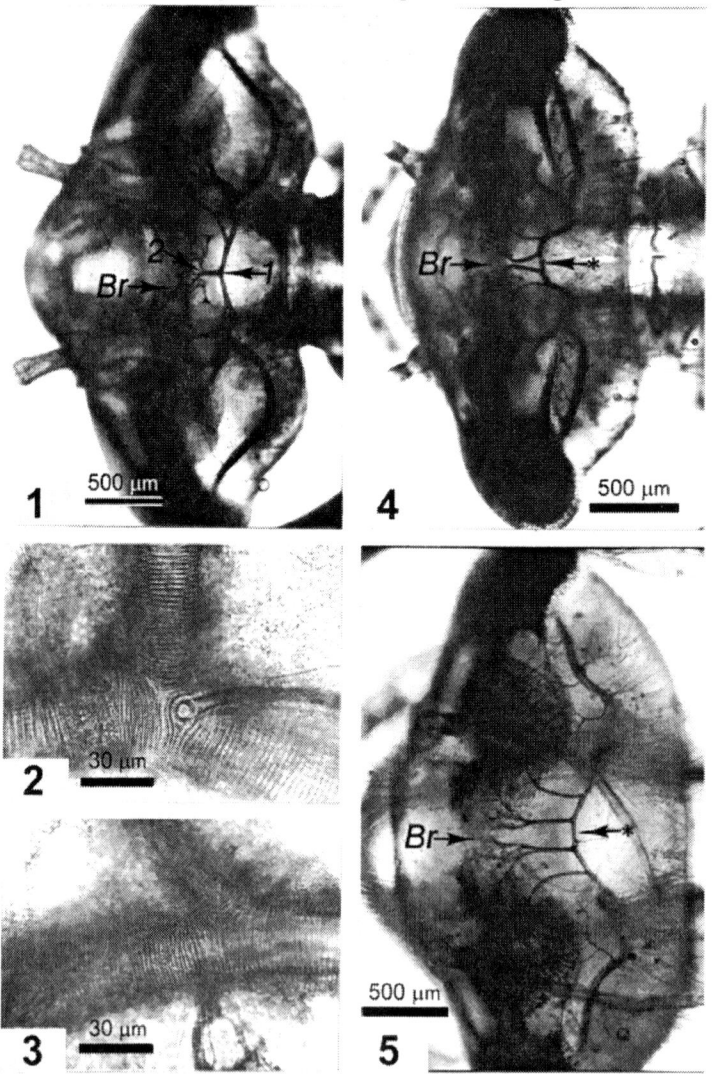

Plate 13: Figure 1—Dorsal view of the head of a mature *Enallagma* (Zygoptera, Coenagrionidae) sp. larvae cleared in 100% glycerin. A non-Palmen body-like tracheal anastomosis (arrow 1) occurs in a location that in mayflies is occupied by the Palmen body. Arrows 1 and 2 refer to areas enlarged in Figs. 2 and 3 respectively. Figure 2—Enlargement of area 1 of Fig. 1. Figure 3—Enlargement of area 2 of Fig. 1. Figure 4—Dorsal view of head of an immature *Pachydiplax longipennis* (Anisoptera, Libellulidae) larvae cleared in 100% glycerin. A non-Palmen body-like anastomosis (starred arrow) occurs in a location that in mayflies is

occupied by the Palmen body. Figure 5—Dorsal view of head of an immature *Celithemis* sp. (Libellulidae) larvae cleared in 100% glycerin. Explanation as for Fig. 4.

Palmén (1877) reports to have found Palmen body-like structures in almost all segments of *Geophilus*, a centipede (Chilopoda, Geophilomorpha). In the present study two live centipedes of the order Scolopendromorpha were examined and no Palmen body-like structures were found, including at the many tracheal commissures that occurred just above the cuticle of the abdominal sternites. Also examined were five live millipedes (Diplopoda) with similar results. However, it was so difficult to dissect the very hard external cuticle of the millipedes that Palmen body-like structures could have easily been destroyed without detection.

Landa (1948) could find no obvious function of Palmen body like structures and stated that perhaps they had some function in molting by providing precise nodes for breaking of tracheal connections so that molting of the tracheae could occur. However, other tracheal junctions in mayflies and in other insects molt quite well without these Palmen body like structures.

Gross (1903) and Brodskiy (1973) proposed, without providing any evidence, an orienting function of the Palmen body during flight.

Wodsedalek (1912) proposed and tested for an orienting-statocyst-like function in larvae of *Stenacron interpunctatum*, which normally inhabit the undersides of rocks. In the first experiment recently dead larvae with and without the Palmen body were dropped in a column of water and larvae with the Palmen body always fell on their dorsal surface while larvae without the Palmen body were equally as likely to fall on either surface. Wodsedalek concluded "..., a self directing process, that is, by the presence of the organ the nymph is swerved into position-a matter of physical equilibrium." Wodsedalek performed two other experiments (experiments 2 and 3) on live larvae, one group (experimental) had the Palmen body removed while the other group (control) had the head capsule similarly dissected but the Palmen left intact. In experiment 2, conducted in a darkened room to avoid phototactic responses, the experimental larvae remained on the tops of the rocks much longer (hours-weeks) than the control larvae after having been dropped in a container of water containing rocks or after the rocks were flipped over. The third experiment was similarly conducted except that rocks were suspended thereby negating certain thigmotactic responses (i.e., body wedging). The experimental group remained evenly scattered over the entire rock surface while most of the control group remained on the lower surface.

I repeated Wodsedalek's experiments using *Maccaffertium exiguum* larvae. This species is in the same family (Heptageniidae) as *Stenacron interpunctatum*, is about the same size, and displays similar larval behavior. In the water-column dead larvae (freshly killed with hot water) dropping experiment, 14 larvae (some several trials with and followed by several trials without the Palmen, while some other larvae were used in trials with the Palmen body and other larvae were used in trials without the Palmen body) were dropped with or without the Palmen body into an aquarium with water to a depth of 122 mm. There was no significant difference between larvae with or without the Palmen body; out of 59 drops per larval group 39 or 66% in each group fell on their dorsal surface.

In experiment two, three groups of larvae were reared in separate small aquariums with a single rock and larvae were acclimated for 12 hours. The three groups were 36 Palmen body extracted larvae (group E), 33 head dissected but Palmen body intact larvae (group D) and 80 normal larvae (Group N, these were on average smaller larvae than groups E and N since larger larvae were needed for dissection logistics). They were then checked for rock location (top, bottom or sides) eleven times (usually at night) on four days and total survival of groups E and D were noted on the eighth day. Mortality was high for groups E (6 of 36 survived to the eighth day) and D (15 of 33 survived to the eighth day). Larvae were very difficult to see on the rocks and this in itself involved more handling and possible increased mortality, and some nymphs detached from the rocks when the rock was briefly removed from the water for counting of larvae. Despite these problems, more than 80% of the specimens from all groups were found on the bottom of the rocks.

In experiment three, the same three aquariums and the remaining surviving larvae were used, but the same flattish rocks were suspended by rope approximately 8 to 10 cm off the bottom of the aquarium. Six individuals of group E, 15 of group D, and approximately 80 individuals of group N were available for the start and duration of this experiment. Larvae were checked for rock location five to seven times (seven times for group D, and five times each for groups E and N) over three days during both light and dark hours. In all three groups larvae were found on the bottom of the rock greater than 75% of the time.

These observations cast doubt on the validity of Wodsedalek's experiments and strongly indicate that the Palmen body does not have an effect on orientation.

It is extremely doubtful that the Palmen body has any role in orientation for several morphological reasons. First, no nervous connection to these bodies is known (Wodsedalek 1912, Landa 1948, this study). Second, the Palmen body is surrounded in muscle tissue and is not as loosely bound and free to move about as Wodsedalek (1912) stated "It is the writer's opinion that the chitinous organ being so loosely supported by the four tracheal tubes exerts a pressure on the surrounding tissues, whereby the disturbing stimulus reaches the central nervous system." Also the distribution of Palmen body-like structures, i.e., contact mallets, throughout the body appears, as Landa (1948) stated, to argue against an orienting function. Palmen body-like structures may even be a slight detriment to the mayfly since although air can pass through these structures, it must be impeded. However, most probably Palmen bodies and contact mallets are innocuous structures with no vital function.

The Future of the Palmen Body in Instar Determination of Mayflies. The Palmen body and rearing are the only accurate instar determination methods in mayflies (Fink 1980, Fink 1982, Fink 1984). The potential of the Palmen body is immense in that it records the number of molts for every individual of any age. Specimens can be analyzed for the number of molts and this information can then be correlated with developmental stages and with environmental parameters, such as temperature, pollution and relative ecosystem quality. Unfortunately, the very small size of the Palmen body and the multiunit structure hampered further by the presence of projections makes detection/handling and compound light microscopic observation very difficult.

Despite these difficulties, the Palmen body can be used in a very limited way to determine the approximate total number of instars for a species or population. This can be done by simple whole mounts and has worked very well for larger mayfly species in the present study (unpublished data, see the caption for Plate 4).

To unleash the full power of the Palmen body, accurate and fast instar determination of every specimen is necessary. Sectioning of Palmen bodies using TEM epoxy resins is certainly capable of accurate counts (Ruffieux et al. 1996), but it can be argued whether it is a fast method, or even practical for small species/small (young) specimens, or for investigators who lack necessary equipment (e.g., TEM microtomes) and/or skilled labor. In the study cited above, two to three days (including waiting time) were required to process only ten Palmen bodies and this was for a relatively large Palmen for a relatively large species, *Siphlonurus aestivalis* (Eaton).

Confocal laser scanning microscopy is a powerful technique that allows noninvasive optical sectioning of small structures and 3-D reconstruction with appropriate software. Since only focused light is captured, the image is greatly improved over conventional compound light microscopy. While most confocal light microscopy has been done on cells, recently several investigators working on small structures in insects, like genitalia, have used confocal microscopy where images often surpass SEM (see comprehensive paper and references in Klaus et al. 2003). Confocal microscopy could be a powerful tool for instar determination in mayflies if specimen processing and microscope time and analysis can be relatively rapid and fairly foolproof. Large mayfly individuals will require dissecting Palmen bodies out, while very small individuals might not require dissection, and observation of the Palmen body might be possible through the head. Secondly, access to these very expensive confocal light microscopes/computers needs to be routinely available and at an affordable price.

Acknowledgments

I would like to deeply thank Janice G. Peters and the late William L. Peters for their unwavering support over the many years I have known them.

Literature Cited

Benech, V. 1972. Le polyvoltinisme chez *Baetis rhodani* Pictet (Insecta, Ephemeroptera) dans un ruisseau a truites des Pyrénées-Atlantiques, le Lissuraga. Annales d'Hydrobiologie **3**:141–171.

Brodskiy, A. K. 1973. The swarming behavior of mayflies (Ephemeroptera). Entomological Review **52**:33–39.

Degrange, C. 1959. Nombre de mues et organe de Palmén de *Cloeon simile* Etn. (Ephéméroptères). Comptes Rendus des Séances de l'Académie des Sciences **249**:2118–2119.

Fink, T. J. 1980. A comparison of mayfly (Ephemeroptera) instar determination methods. Pages 367–380 *in* J. F. Flannagan, and K. E. Marshall, editors. Advances in Ephemeroptera Biology. Plenum Press, New York.

Fink, T. J. 1982. Why the simple frequency and Janetschek methods are unreliable for determining instars of Mayflies: an analysis of the number of nymphal instars of *Stenonema modestum* (Banks) (Ephemeroptera: Heptageniidae) in Virginia, USA. Aquatic Insects **4**:67–71.

Fink, T. J. 1984. Errors in instar determination of mayflies (Ephemeroptera) and stoneflies (Plecoptera) using the simple frequency, Janetschek, Cassie and Dyar's law method. Freshwater Biology **14**:347–365.

Gross, J. 1903. Ueber das Palmén'sche organ der Ephemeriden. Zoologische Jahrbٳcher Abteilung Fٳr Anatomie und Ontogenie der Tiere **19**:91–105.

Hsu, Y. C. 1933. Some new morphological findings in Ephemeroptera. Proceedings Fifth International Congress of Entomology Paris **2**:361–368.

Hsu, Y. C. 1935. Chapter IV. Internal anatomy. Respiratory system. Pages 38–42 *in* J. G. Needham, J. R. Traver, and Y. C. Hsu, editors. The Biology of Mayflies With a Systematic Account of North American Species. Comstock Publishing Company, Ithaca, New York.

Jones, J. 1977. The ecology, life cycle, and seasonal distribution of *Neoephemera (s.s) youngi* Berner (Ephemeroptera: Neoephemeridae). M.S. Thesis, Florida State University, Tallahassee, Florida, USA.

Klaus, A. V., V. L. Kulasekera, and V. Schawaroch. 2003. Three-dimensional visualization of insect morphology using confocal laser scanning microscopy. Journal of Microscopy **212**:107–121.

Landa, V. 1948. Contributions to the anatomy of Ephemerids larvae. I. Topography and anatomy of tracheal system. Vestnik Ceskoslovenské Spolecnosti Zoologické **12**:25–82.

Landa, V., T. Soldán, and W. L. Peters. 1980. Comparative anatomy of larvae of the family Leptophlebiidae (Ephemeroptera) based on ventral nerve cord, alimentary canal, malpighian tubules, gonads and tracheal system. Acta Entomologica Bohemoslovaca **77**:169–195.

Landa, V., T. Soldán, and G. F. Edmunds. 1982. Comparative anatomy of larvae of the family Ephemerellidae (Ephemeroptera). Acta Entomologica Bohemoslovaca **79**:241–253.

McLean, J. A. 1970. Studies on the larva of *Oniscigaster wakefieldi* (Ephemeroptera: Siphlonuridae) in Waitakere Stream, Auckland. New Zealand Journal Marine Freshwater Research **4**:36–45.

Palmén, J. A. 1877. Zur morphologic des tracheensystems. Wilhelm Engelmann, Leipzig.

Pescador, M. L., and W. L. Peters. 1974. The life history and ecology of *Baetisca rogersi* Berner (Ephemeroptera: Baetiscidae). Bulletin of the Florida State Museum, Biological Sciences, University of Florida, Gainesville, Florida **17**:151–209 (plus 3 figures prior to page 151).

Rawlinson, R. 1939. Studies on the life-history and breeding of *Ecdyonurus venosus* (Ephemeroptera). Proceedings Zoological Society London, B, **109**:377–450.

Ruffieux, L., M. Sartori, and G. L'Eplattenier. 1996. Palmen body: a reliable structure to estimate the number of instars in *Siphlonurus aestivalis* (Eaton) (Ephemeroptera: Siphlonuridae). International Journal of Insect Morphology and Embryology **25**:341–344.

Taylor, R. L., and A.G. Richards. 1963. The subimaginal cuticle of the mayfly *Callibaetis* sp. (Ephemeroptera). Annals Entomological Society of America **56**:418–426.

Whitten, J. M. 1972. Comparative anatomy of the tracheal system. Annual Review of Entomology **17**:373–402.

Wodsedalek, J. E. 1912. Palmen's organ and its function in nymphs of the Ephemeridae *Heptagenia interpunctata* (Say) and *Ecdyurus maculipennis* (Walsh). Biological Bulletin Woods Hole **22**:253–273.

Appendix A

Plate Abbreviations

a	anterior Palmen body "cone" (where the anterior tracheae attach)
AC	anterior Palmen body "cone" (where the anterior tracheae attach)
A	lack of a contact mallet where you would expect one
B	basement membrane
Br	brain
C	compound eye
Cm	contact mallet
Cs	coronal suture
Ct	illusion of a coating due to visualization of the projections from underlying Palmen body units
E	epidermal cells
Ep	epicuticle
F	gut
G	abdominal gills
Gr	granular cuticular layer
H	shows hollow Palmen body projection
I	inner epicuticle
L	Palmen body units (appearing like layers)
LT	lateral tracheal trunk
m	microvillate apical plasma membrane
M	median ocellus
Mt	mesothorax
n	new outer epicuticle
p	posterior Palmen body "cone" (where the posterior tracheae attach)
P	Palmen body
PC	posterior Palmen body "cone" (where the posterior tracheae attach)
PH	posterior margin of the head
pL	plaques of new outer epicuticle
Pr	Palmen body unit projections
Pt	prothorax
PX	Palmen body exocuticle
O	outer epicuticle
R	"rings" (the edges of Palmen body units)
Re	rough endoplasmic reticulum
S	apolysis space
T	tracheae (in Fig. 10 refers to taenidia)
TX	taenidial exocuticle

SYSTEMATIC REVIEW OF THE WINGLESS STONEFLIES, SCOPURIDAE

Y. H. Jin[1], T. Kishimoto[2] and Y. J. Bae[1]

[1] *Department of Biology, Seoul Women's University, Seoul 139-774, Korea*
[2] *Biological Laboratory, Tsukuba Kokusai University, Ibaraki 300-0051, Japan*

Abstract

The monophyletic family Scopuridae is unique among stoneflies due to their entirely wingless adult and nymphal stages as well as whorled filamentous gills on the terminal abdominal segment.   The family is geographically limited to the northeastern part of Palaearctic Asia and includes four species from the Korean Peninsular (*Scopura laminata* Uchida, *S. gaya* Jin and Bae, *S. scorea* Jin and Bae and *S. jiri* Jin and Bae) and four species from the Japanese Islands (*S. longa* Uéno, *S. montana* Maruyama, *S. bihamulata* Uchida and *S. quattuorhamulata* Uchida) in the monotypic genus *Scopura* Uéno.   Based on a comprehensive review of material and bibliographic sources of the family, we provide type information, diagnoses, distribution, taxonomic remarks, keys to all known stages and species and discussion on phylogeny and biogeography.   Two clades, "Korean-*Scopura*" and "Japanese-*Scopura*," in the Scopuridae are hypothesized and supported by the synapomorphic characters of the posteromedian process on the eighth abdominal sternum of female adult and the elongated process on the basal cerci of male adult, respectively.

Key words: Plecoptera; Scopuridae; *Scopura*; wingless stonefly; systematics; Korea; Japan.

Introduction

The rarely known stonefly family Scopuridae is unique among stoneflies due to their wingless adult and nymphal stages.   Although some brachypterous or reduced winged stoneflies have been reported from the Capniidae (Zenger and Baumann 2004, Zwick 2000) and the Austroperlidae (Theischinger 1996), all members of the Scopuridae entirely lack wings in the nymphal and adult stages. The nymph can also be characterized by the whorled filamentous gills on the terminal abdominal segment.

The Scopuridae is limited to the northeastern part of Palaearctic Asia, i.e., the Korean Peninsula and the Japanese Islands (Central and Northern Honshu and Hokkaido).   The habitat is also limited to headwater streams of high mountains, where the water is cold and shallow and the substrate is composed of large stones on a sandy bottom with abundant fallen leaves.   The nymphs are slow-growers showing a four-year life span (Ida 1994) and are surface scrapers utilizing leaves such as maples and oaks from the riparian forest.

395

Uéno (1929) described the first *Scopura* species, *Scopura longa*, from Japan and members of the Japanese *Scopura* have been comprehensively studied by Uchida and Maruyama (1987) with descriptions of three more species: *S. bihamulata* Uchida, *S. montana* Maruyama and *S. quattuorhamulata* Uchida. Members of the Korean *Scopura* have been studied by Uchida and Maruyama (1987) and Jin and Bae (2005a, 2005b) and four species are known: *S. laminata* Uchida, *S. gaya* Jin and Bae, *S. scorea* Jin and Bae and *S. jiri* Jin and Bae.

In this paper, we comparatively review all known stages and species of the family not only to provide differential diagnoses and identification keys but also to discuss further on the biogeography and phylogeny.

Materials and Methods

In order to complete species taxonomy, reference material (types and nontypes) of all known species of Scopuridae were comprehensively examined (see Jin and Bae 2005a, 2005b). Most holotype specimens are housed in the Aquatic Insect Collection of Seoul Women's University (SWU-AIC) in Seoul and National Science Museum (NSMT) in Tokyo. Characters and terminology follow Uchida and Maruyama (1987) and Jin and Bae (2005a).

Taxonomic Account

Family Scopuridae Uéno

Scopuridae: Uéno, 1935, p. 41 [*Type genus*: *Scopura* Uéno, 1929, p. 124; monotypy]; Uéno, 1938a, p. 154.

Diagnosis: Male adult has an interior process at the base of cerci (Figs. 1E, 1F, 2D, 2E, 4E–8E, 4F–8F) (one species lacks the process: Figs. 3E and 3F). Male epiproct has an internal sac, which is membranous, eversible and armed with teeth and spinules (Fig. 6D). Nymph has gill tuft between the abdominal segment 9 and 10 (Fig. 3J).

Remarks: Zwick (2000) redefined the concept of the family using phylogenetic characters and presented interfamilial relationships of the Scopuridae and related families.

Genus *Scopura* Uéno

Scopura Uéno, 1929, p. 124 [*Type species*: *Scopura longa* Uéno, 1929, p. 124–130, plates 13 & 14; original designation].

Diagnosis: Male adult thoracic nota have lateral expansions (Figs. 1A–8A). Posterior part of abdominal tergum 9 is elevated and setose. Epiproct has a pair of distinct lateral projections (Figs. 1B–8B, 1C–8C). Penis has a pair of longitudinal sclerites (Figs. 1G, 2F, 3G–8G). Female adult body is larger and paler than male. Membrane between the abdominal sternum 8 and 9 has setae. Ovary is loop-shaped. Nymphal thoracic nota have lateral expansions, but the lateral expansions are rounder than those in adult. Each membranous part of epiproct and paraprocts has single filamentous gill, but last instar of male nymph lacks this single filamentous gill on epiproct. Female nymph has a notch on the abdominal sternum 8.

Remarks: Kawai (1974) and Uchida and Maruyama (1987) redefined the concepts of the adult and nymph of *Scopura*.

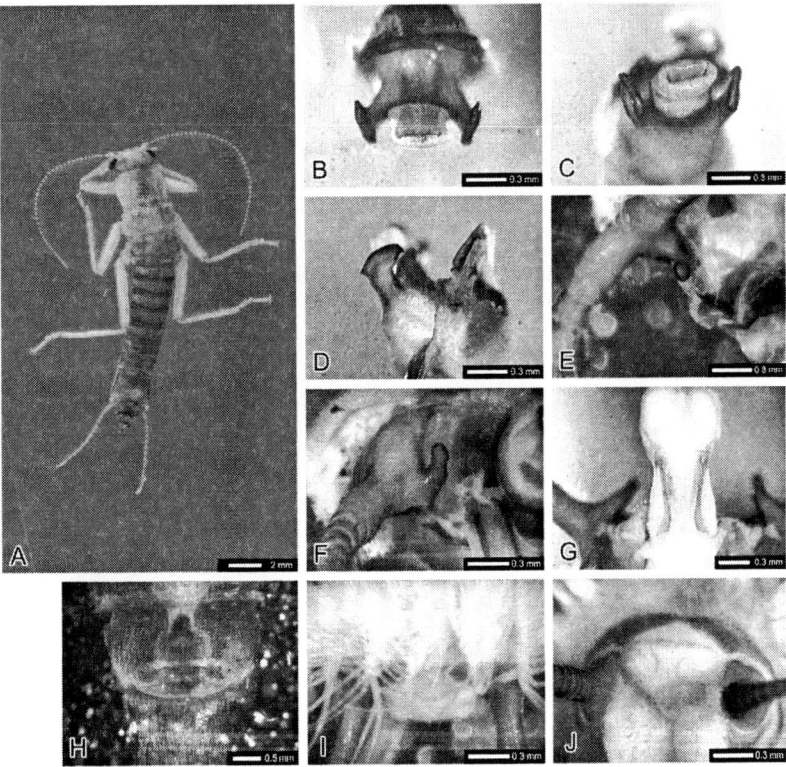

Figure 1. *Scopura bihamulata*, Male adult: A, habitus; B, epiproct, dorsal; C, epiproct, posterior; D, epiproct, lateral; E, interior process of cerci, dorsal; F, interior process of cerci, posterior; G, penis, dissected. Female adult: H, posteromedian margin of sternum 8. Mature male nymph: I, epiproct, dorsal; J, epiproct, posterior.

Scopura bihamulata Uchida

(Fig. 1)

Scopura bihamulata: Uchida and Maruyama, 1987, p. 704–705 [*Holotype*: Male adult, Japan, Hokkaido, Otaru-shi, NSMT].

Diagnosis: Male adult 22 mm; female adult 22 mm. Male adult epiproct has a pair of disk-shaped lateral projections, but lacks submedian projections (Figs. 1B–D). Interior process of cerci is greatly elongated and curved posterodorsally (Figs. 1E, 1F). Female adult posteromedian margin of sternum 8 is not concaved or produced, but has a pair of ventral sclerotized humps (Fig. 1H). Male nymph epiproct has a pair of large dorsolateral swellings that are produced dorsally; the borderline between the epiproct and swellings is present (Figs. 1I and 1J).

Distribution: Southwestern Hokkaido in Japan.

Remarks: Both *S. bihamulata* and *S. quattuorhamulata* occur in Hokkaido, but their distributions are not overlapping. *S. bihamulata* is distributed in the southwestern part of Hokkaido, while *S. quattuorhamulata* is distributed in central Hokkaido (Uchida and Maruyama 1987).

Scopura gaya Jin and Bae

(Fig. 2)

Scopura gaya: Jin and Bae, 2005a, p. 23–26 [*Holotype*: Male adult, South Korea, Gyeongsangnam-do, Hapcheon-gun, Gayasan (Mt.), SWU-AIC].

Diagnosis: Male adult 20 mm; female adult 25 mm; nymph 19–25 mm. Male adult epiproct has two pairs of projections (Figs. 2B and 2C); submedian projections are rudimentary and almost fused; lateral projections are elongated and hooked apically. Interior process of cerci is small and ball-shaped (Figs. 2D and 2E). Female adult vagina is elongated as in Fig. 2G. Posteromedian margin of sternum 8 is developed into a pair of long submedian projections (Fig. 2H). Male nymph epiproct has a pair of mid-sized dorsolateral swellings, a dorsal concavity and a dorsal transverse sclerite; lateral swellings are as high as the concavity; the borderline between the epiproct and swellings is absent (Figs. 2I and 2J).

Distribution: Southern Korean Peninsula.

Remarks: *S. gaya* is distributed in the intermediate area between *S. jiri* and *S. scorea* in southern Korea. Only the type locality is known, but they may occur in some distance from the type locality.

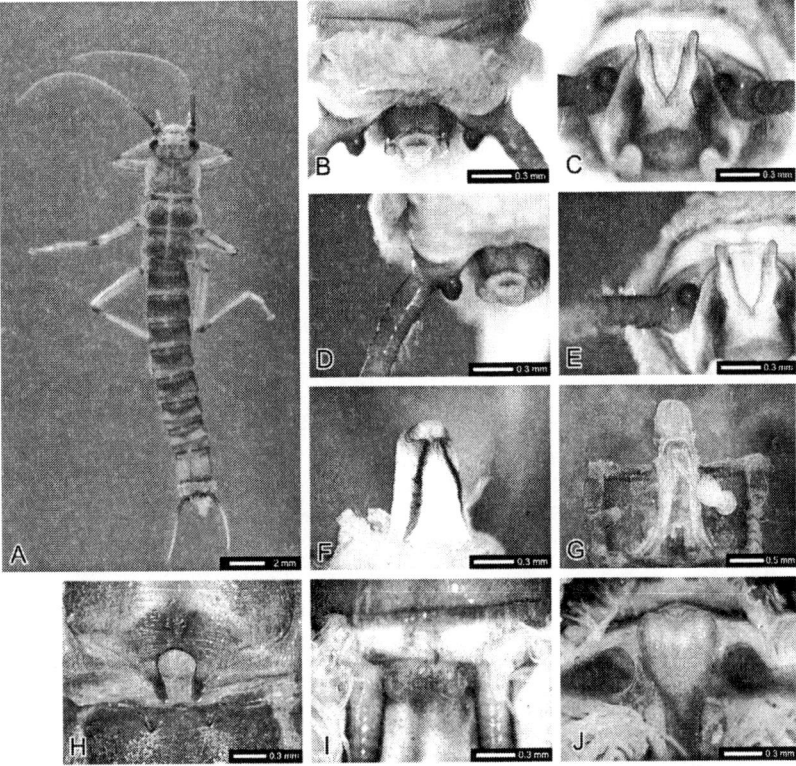

Figure 2. *Scopura gaya*, Male adult: A, habitus; B, epiproct, dorsal; C, epiproct, posterior; D, interior process of cerci, dorsal; E, interior process of cerci, posterior; F, penis, dissected. Female adult: G, vagina, dissected; H, posteromedian margin of sternum 8. Mature male nymph: I, epiproct, dorsal; J, epiproct, posterior.

Scopura jiri Jin and Bae

(Fig. 3)

Scopura jiri: Jin and Bae, 2005b, p. 89–92 [*Holotype*: Male adult, South Korea, Jeollanam-do, Gurye-gun, Jirisan (Mt.), SWU-AIC].

Diagnosis: Male adult 18–20 mm; nymph 17–23 mm. Male adult epiproct has two pairs of projections (Figs. 3B, 3C, 3D) as in *S. quattuorhamulata*, but the submedian projections are more widely separated and the lateral projections are more elongated than those of *S. quattuorhamulata* (Figs. 7B, 7C, 7D). Interior process of cerci is absent (Figs. 3E and 3F). Penis has a round median sclerite as well as lateral sclerites (Fig. 3G). Female adult is unknown. Male nymph epiproct has a pair of lateral swellings and a small median swelling; lateral swellings are located at basal 1/3 of epiproct (Figs. 3H, 3I, 3J).

Distribution: Southern Korean Peninsula.

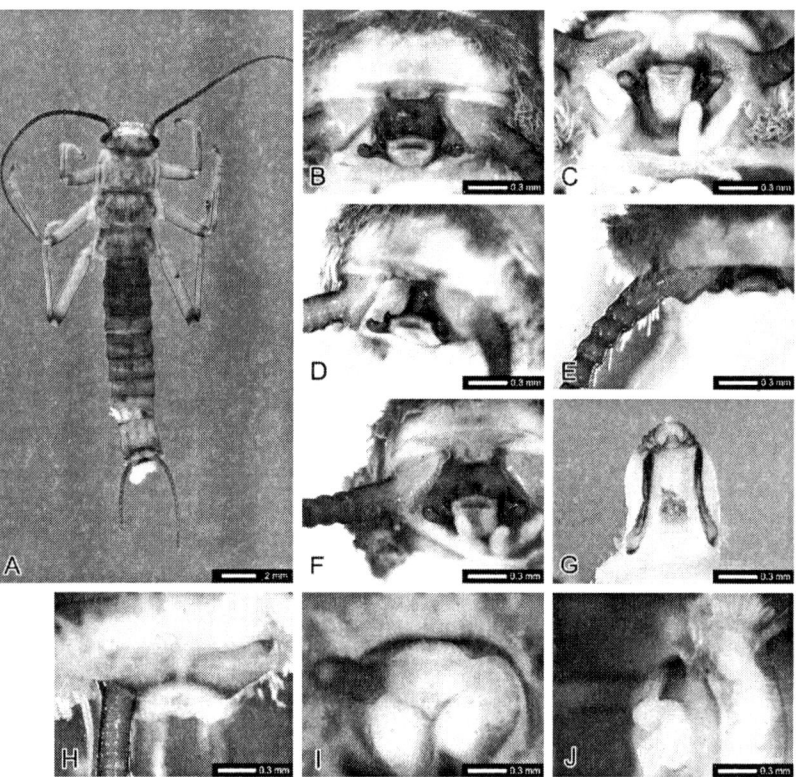

Figure 3. *Scopura jiri*, Male adult: A, habitus; B, epiproct, dorsal; C, epiproct, posterior; D, epiproct, dorsolateral; E, base of cerci, dorsal; F, base of cerci, posterior; G, penis, dissected. Mature male nymph: H, epiproct, dorsal; I, epiproct, posterior; J epiproct, lateral.

Scopura laminata Uchida

(Fig. 4)

Scopura longa: Yoon and Aw, 1985: 116 (misidentification).
Scopura laminata Uchida: Uchida and Maruyama, 1987, p. 707 [*Holotype*: Male nymph, South Korea, Gangwon-do, Odaesan (Mt.), NSMT]; Kim et al., 1998, p.

43 (checklist); Jin and Bae, 2005a, p. 26–30 (male and female adult's description).

Diagnosis: Male adult 16–24 mm; female adult 22–25 mm; nymph 16–24 mm. Male adult epiproct has two pairs of projections (Figs. 4B, 4C, 4D); submedian projections are small and moderately separated and lateral projections are relatively well developed and elongated. Interior process of cerci is small (Figs. 4E and 4F). Female adult posteromedian margin of sternum 8 is developed and has rudimentary submedian projections (Fig. 4H).   Male nymph epiproct has a pair of lateral swellings; lateral swellings are located lower than epiproct and directed laterally (Figs. 4I and 4J).

Distribution: Central Korean Peninsula.

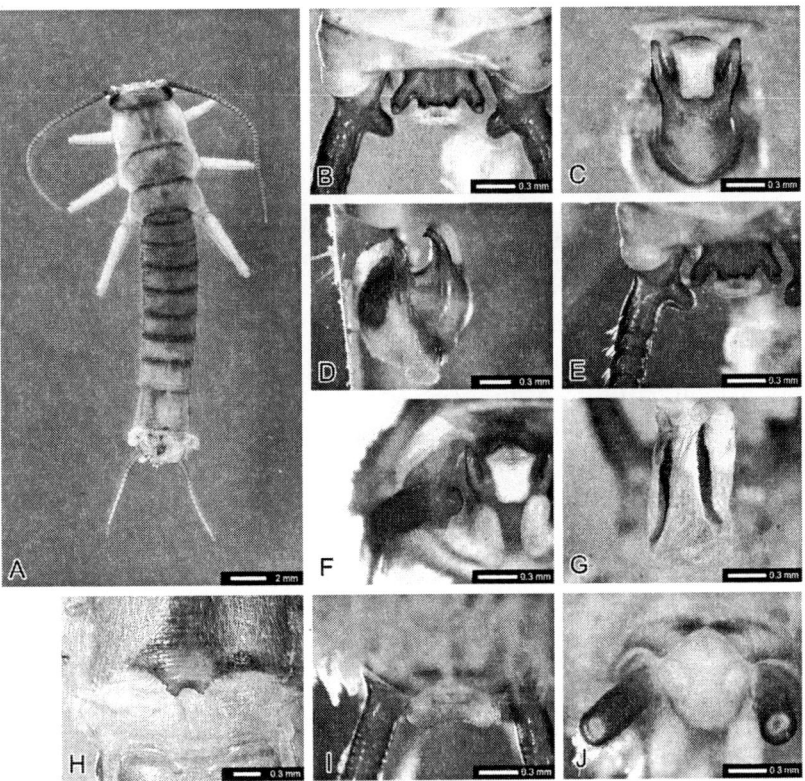

Figure 4. *Scopura laminata*, Male adult: A, habitus; B, epiproct, dorsal; C, epiproct, posterior; D, epiproct, lateral; E, interior process of cerci, dorsal; F, interior process of cerci, posterior; G, penis, dissected. Female adult: H, posteromedian margin of sternum 8. Mature male nymph: I, epiproct, dorsal; J, epiproct, posterior.

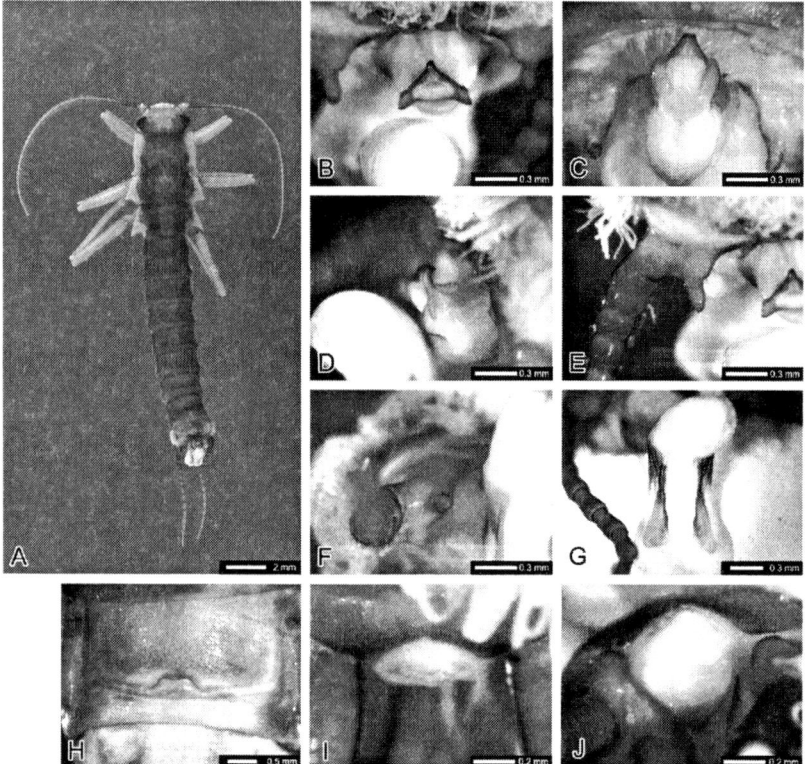

Figure 5. *Scopura longa*, Male adult: A, habitus; B, epiproct, dorsal; C, epiproct, posterior; D, epiproct, dorsolateral; E, interior process of cerci, dorsal; F, interior process of cerci, posterior; G, penis, dissected. Female adult: H, posteromedian margin of sternum 8. Mature male nymph: I, epiproct, dorsal; J, epiproct, posterior.

Scopura longa Uéno

(Fig. 5)

Scopura longa: Uéno,1929, p. 125 [*Holotype*: Male nymph, Japan, Northern Honshu, Akita-ken, Wainai (= Oide City), in T. Kawai's collection of Nara Women's University]; Uchida and Maruyama (1987), p. 702 (male and female adult's description).

Scopura prolifera Kawai,1974, p 275 [Holotype: Male adult, Japan, Sado Island, in T. Kawai's collection of Nara Women's University] (synonymized by Uchida and Maruyama 1987).

Diagnosis: Male adult 18–25 mm; female adult 23–40 mm. Male adult epiproct has a pairs of lateral lobe-like projections and one median projection (Figs. 5B, 5C, 5D). Interior process of cerci is elongated and directed posteriorly (Figs. 5E and 5F). Female adult posteromedian margin of sternum 8 is moderately concaved and M-shaped (Fig. 5H). Male nymph epiproct has a sclerotized dorsal knob (Figs. 5I and 5J).

Distribution: Northern Honshu and Sado Island in Japan.

Remarks: Uchida and Maruyama (1987) recognized the adult of *S. longa* from the type locality, which was confused with *S. montana* Maruyama (see below) and refined the concept of *S. longa* Uéno using adult characters. *S. prolifera* Kawai was synonymized with *S. longa* Uéno and Kawai's (1974) *S. longa* was corrected as *S. montana* (Uchida and Maruyama, 1987).

Scopura montana Maruyama

(Fig. 6)

Scopura longa: Kawai, 1974, p. 280 (misidentification)
Scopura montana: Uchida and Maruyama (1987), p. 701 [*Holotype*: Male adult, Japan, Tokyo, Hinohara-mura, Mitô-san, NSMT].

Diagnosis: Male adult 13–25 mm; female adult 25–35 mm. Male adult epiproct has a pair of lateral projections, but lacks submedian projections; lateral projections are horn-shaped (Figs. 6B, 6C, 6D). Interior process of cerci is narrow, elongated and hooked dorsally (Figs. 6E, 6F). Female adult posteromedian margin of sternum 8 is slightly concaved (Fig. 6H). Male nymph epiproct has a pair of sclerotized dorsolateral knobs (Figs. 6I, 6J).

Distribution: Central Honshu in Japan.

Remarks: *S. montana* is a common species of *Scopura* in central Honshu, which is geographically separated from the northern congener *S. longa* (Fig. 9).

Scopura quattuorhamulata Uchida

(Fig. 7)

Scopura quattuorhamulata: Uchida and Maruyama, 1987, p 706 [*Holotype*: Male adult, Japan, Hokkaido, Niikappu-chô, NSMT].

Diagnosis: Male adult 18–25 mm; female adult 22–30 mm. Male adult epiproct has two pairs of projections (Figs. 7B, 7C, 7D); submedian projections are widely

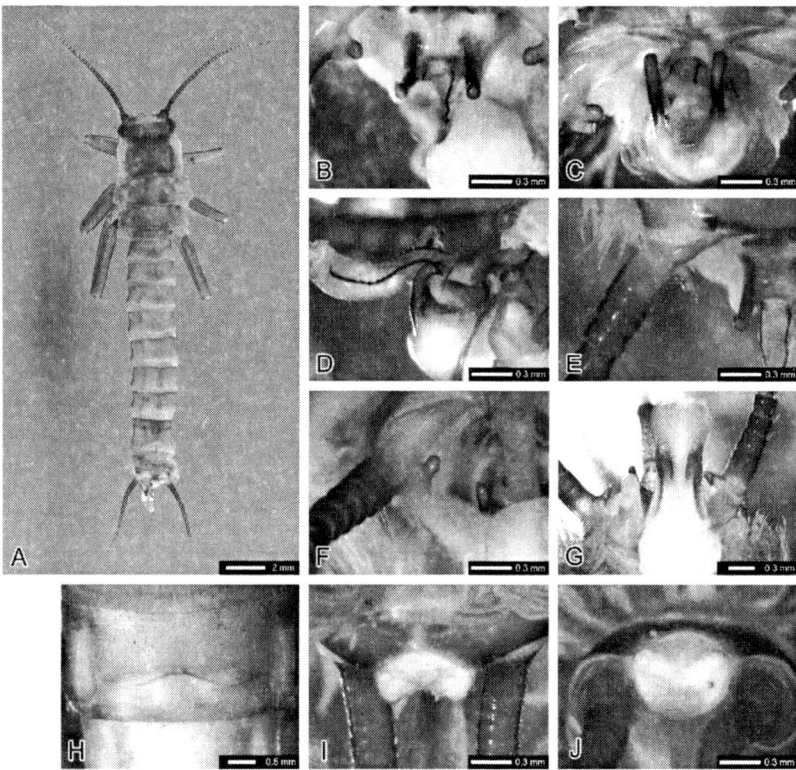

Figure 6. *Scopura montana*, Male adult: A, habitus; B, epiproct, dorsal; C, epiproct, posterior; D, epiproct, dorsolateral; E, interior process of cerci, dorsal; F, interior process of cerci, posterior; G, penis, dissected. Female adult: H, posteromedian margin of sternum 8. Mature male nymph: I, epiproct, dorsal; J, epiproct, posterior.

separated; lateral projections are relatively small and hooked anteriorly. The shape of the epiproct of *S. quattuorhamulata* is generally similar to that of *S. jiri*, but distinguished by the size and location of the projections. Interior process of cerci is elongated and hooked dorsally (Figs. 7E and 7F). Female adult posteromedian margin of sternum 8 is concaved as in Fig. 7H. Male nymph epiproct has no swelling or knob (Figs. 7I and 7J).

Distribution: Central Hokkaido in Japan.

Remarks: *S. quattuorhamulata* is the only species in Japan that has two pairs of projections on the male epiproct.

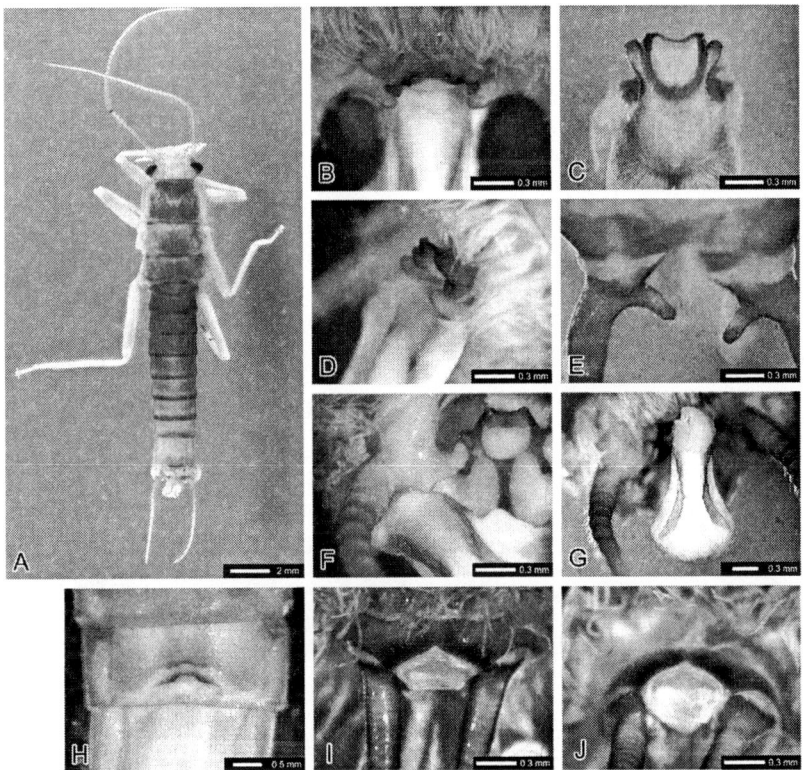

Figure 7. *Scopura quattuorhamulata*, Male adult: A, habitus; B, epiproct, dorsal; C, epiproct, posterior; D, epiproct, dorsolateral; E, interior process of cerci, dorsal; F, interior process of cerci, posterior; G, penis, dissected. Female adult: H, posteromedian margin of sternum 8. Mature male nymph: I, epiproct, dorsal; J, epiproct, posterior.

Scopura scorea Jin and Bae

(Fig. 8)

Scopura scorea: Jin and Bae, 2005a, p. 30–33 [*Holotype*: Male adult, South Korea, Gangwon-do, Wonju, Chiaksan (Mt.), SWU-AIC].

Diagnosis: Male adult 17–20 mm; female adult 24–28 mm. Male adult epiproct has two pairs of projections (Figs. 8B, 8C, 8D); submedian projections are moderately separated; lateral projections are small and round. Interior process of cerci is small (Figs. 8E and 8F). Female adult vagina is elongated (Fig. 8H). Posteromedian margin of sternum 8 is produced and has a pair of well developed submedian projections (Fig. 8I). Male nymph epiproct has a pair of lateral swellings; lateral swellings are located at the middle of epiproct (Figs. 8J, 8K, 8L). *S. scorea* is similar to *S. jiri* in nymphal epiproct, but they can be distinguished by the vertical position of the swelling: the swilling of *S. scorea* is higher than that of *S. jiri*.

Distribution: Central Korean Peninsula.

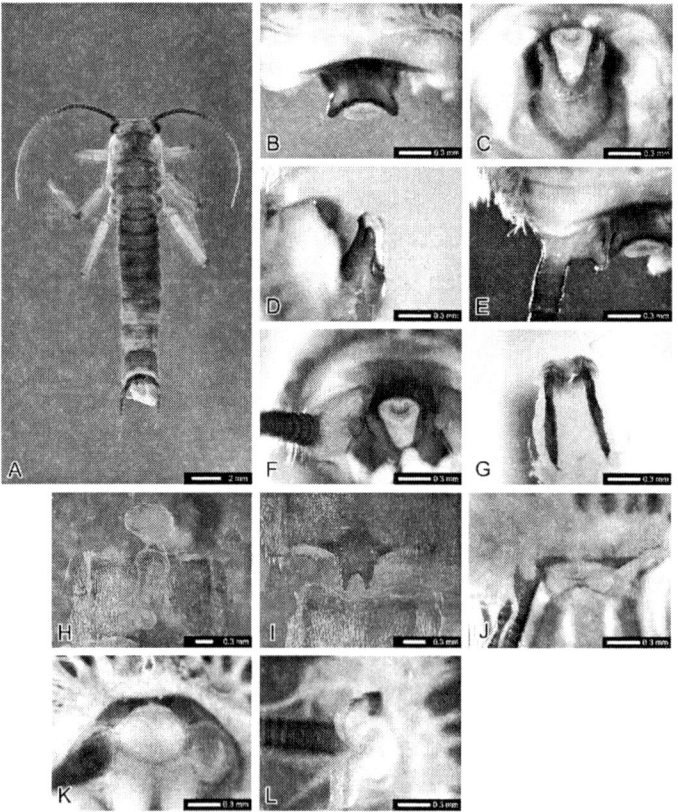

Figure 8. *Scopura scorea*, Male adult: A, habitus; B, epiproct, dorsal; C, epiproct, posterior; D, epiproct, lateral; E, interior process of cerci, dorsal; F, interior process of cerci, posterior; G, penis, dissected. Female adult: H, vagina dissected; I, posteromedian margin of sternum 8. Mature male nymph: J, epiproct, dorsal; K, epiproct, posterior; L, epiproct lateral.

Keys to Species

Male adult

1. Interior process at the base of cerci absent (Figs. 3E and 3F). Penis with a round median sclerite as well as a pair of lateral longitudinal sclerites (Fig. 3G)..*Scopura jiri*
   Interior process at the base of cerci present (Figs. 1E, 1F, 2D, 2E). Penis with only a pair of lateral longitudinal sclerites (Figs. 1G, 2F, 4G–8G)..........2

2. Epiproct with paired submedian projections and lateral projections (Figs. 2B, 2C, 4B, 4C, 4D, 7B, 7C, 7D, 8B, 8C, 8D)...3
   Epiproct with only paired lateral projections or lobes (Figs. 1B, 1C, 1D, 5B, 5C, 5D, 6B, 6C, 6D)..6

3. Epiproct lateral projections not elongated (Figs. 7B, 7C, 7D); distributed in Hokkaido, Japan .... *S. quattuorhamulata*
   Epiproct lateral projections elongated (Figs. 2B, 2C, 4B, 4C, 4D, 8B, 8C, 8D); distributed in Korea 4

4. Epiproct submedian projections rudimentary and almost fused (Figs. 2B and 2C) .... *S. gaya*
   Epiproct submedian projections distinct and clearly separated (Figs. 4B, 4C, 4D, 8B, 8C, 8D) 5

5. Epiproct lateral projections hooked and as high as apex of epiproct (Figs. 4B, 4C, 4D)...*S. laminata*
   Epiproct lateral projections round apically and lower than apex of epiproct (Figs. 8B, 8C, 8D) .... *S. scorea*

6. Epiproct with paired lateral lobes and a median hook (Figs. 5B, 5C, 5D) ....*S. longa*
   Epiproct with paired lateral projections (Figs. 1B, 1C, 1D, 6B, 6C, 6D).........7

7. Epiproct lateral projections fan-shaped (Figs. 1B, 1C, 1D) ....*S. bihamulata*
   Epiproct lateral projections cylindrical and horn-shaped (Figs. 6B, 6C, 6D) ....*S. montana*

Female adult

1. Posteromedian margin of abdominal sternum 8 developed into paired submedian projections (Figs. 2H, 4H, 8I) 2
   Posteromedian margin of abdominal sternum 8 not produced (Figs. 1H, 5H, 6H, 7H) 4

2. Submedian projections elongated and deeply concave (Ω-shaped) (Fig. 2H) .... *S. gaya*
   Submedian projections rudimentary or moderately concave (Figs. 4H and 8I) 3

3. Submedian projections rudimentary (Fig. 4H) ....*S. laminata*

Submedian projections moderately concave
(Fig. 8I) .... *S. scorea*

4.   Posteromedian margin of abdominal sternum 8 not concave and with paired
ventral humps (Fig. 1H)...*S. bihamulata*
Posteromedian margin of abdominal sternum 8 concave
(Figs. 5H, 6H, 7H) 5

5.   Posteromedian margin of abdominal sternum 8 weakly concave
(Fig. 6H) ....*S. montana*
Posteromedian margin of abdominal sternum 8 deeply concave
(Figs. 5H and 7H)...6

6.   Posteromedian margin of abdominal sternum 8 abruptly concave (basally M-
shaped) (Fig. 5H) .... *S. longa*
Posteromedian margin of abdominal sternum 8 gradually concave
(Fig. 7H) .... *S. quattuorhamulata*

Male nymph
1.   Epiproct without knobs or swellings (Figs. 7I and 7J) ..... *S. quattuorhamulata*
Epiproct with knob(s) or swelling(s) (Figs. 1I, 1J, 2I, 2J, 3H, 3I, 3J, 4I, 4J,
5I, 5J, 6I, 6J, 8J, 8K, 8L) 2

2.   Epiproct with a dorsomedian knob (Figs. 5I and 5J)*S. longa*
Epiproct with paired knobs or swellings (Figs. 1I, 1J, 2I, 2J, 3H, 3I, 3J, 4I,
4J, 6I, 6J, 8J, 8K, 8L)..3

3.   Epiproct with paired small knobs dorsolaterally (Figs. 6I and 6J) ..*S. montana*
Epiproct with paired swellings laterally (Figs. 1I, 1J, 2I, 2J, 3H, 3I, 3J, 4I, 4J,
8J, 8K, 8L)...4

4.   Epiproct with paired large swellings; borderline between epiproct and
swellings present (Figs. 1I and 1J)..*S. bihamulata*
Epiproct with paired moderate swellings; borderline between epiproct and
swellings absent (Figs. 2I, 2J, 3H, 3I, 3J, 4I, 4J, 8J, 8K, 8L)............................5

5.   Lateral swellings as high as concavity (Fig. 2J)...................................... *S. gaya*
Lateral swellings lower than concavity (Figs. 3I, 3J, 4J, 8K, 8L) 6

6.   Lateral swellings expanded laterally beyond epiproct
(Fig. 4I and 4J) ....*S. laminata*
Lateral swellings not expanded laterally beyond epiproct
(Figs. 3H, 3I, 8J, 8K) 7

7.   Lateral swellings located at lower 1/3 of epiproct (Figs. 3I and 3J)......... *S. jiri*
Lateral swellings located at middle of epiproct (Figs. 8K and 8L)..... *S. scorea*

Discussion

Up to date, eight species of the Scopuridae are known to be distributed in the Korean
Peninsula and the Honshu and Hokkaido Islands in Japan between 35 and 45 degrees
in latitude (Fig. 9), although some parts of mountain areas in North Korea, Northeast

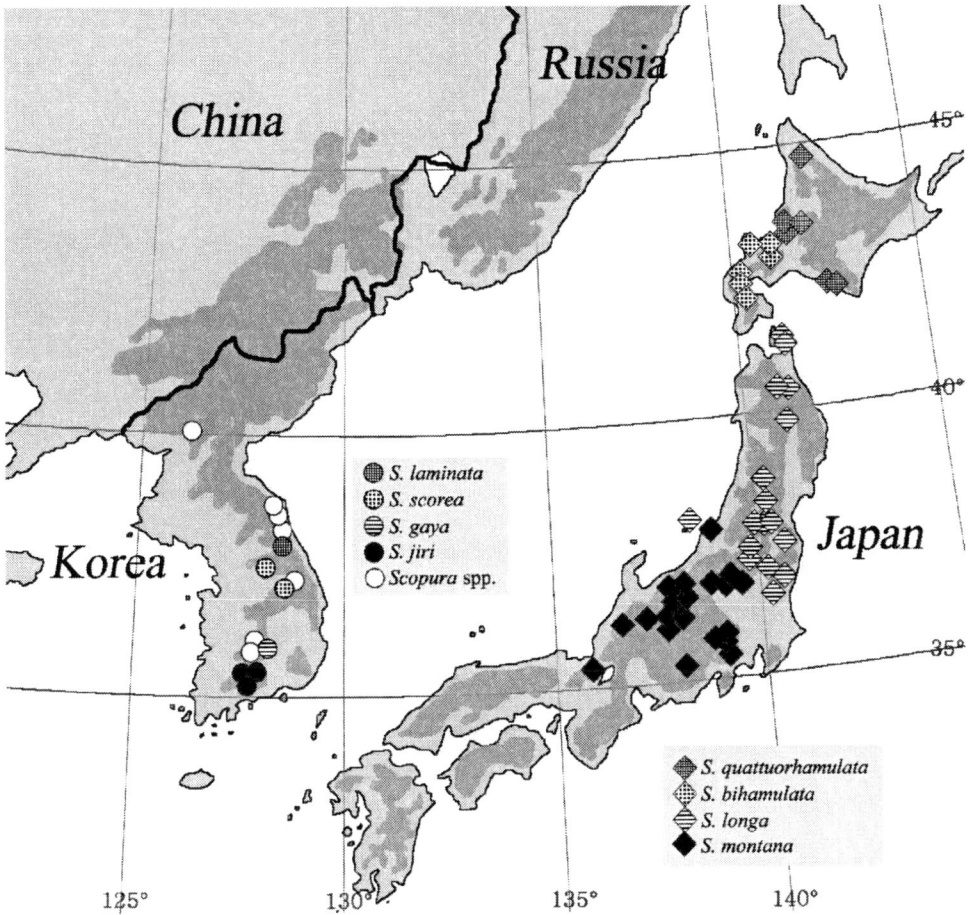

Figure 9. Distribution of *Scopura* spp.

China, and Russian Far East have not been well investigated. The only known record of Scopuridae from North Korea is the nymphs of *S. longa* from a locality (Uéno 1938b), but this may be a different species.

As shown in Fig. 9, distributions of the species in Japan (*S. montana, S. longa, s. bihamulata* and *S. quattuorhamulata*) and in Korea (*S. jiri, S. gaya, S. scorea* and *S. laminata*) are relatively well delineated by latitude. Their latitudinal and altitudinal distributions are also correlated. The lower latitude species such as *S. montana* and *S. longa* show a wider range of altitudinal distribution, while the higher latitude species such as *S. bihamulata* and *S. quattuorhamulata* show a lower altitudinal distribution (Fig. 10A, 10B).

Zwick (2000) presented interfamilial relationships of the Scopuridae being the family as the sister group of Nemouroidea and gave four synapomorphic characters

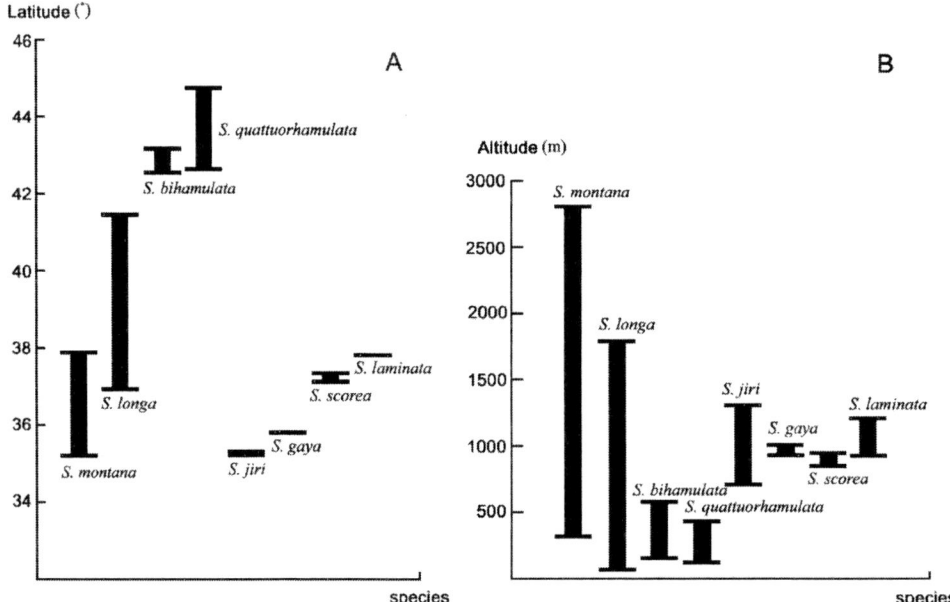

Figure 10. Latitudinal and altitudinal distributions of *Scopura* spp. A, latitudinal distribution; B, altitudinal distribution.

supporting a monophyly of the family: 1) restricted distribution in Japan and Korea, 2) nymphal gills rising from the intersegmental membrane between abdominal segment 9 and 10, 3) male epiproct possessing an eversible membranous tube armed with sclerites and 4) male possessing a prong on the basal cercus.

Based on comprehensive examinations of the adult and nymphal characters, we hypothesize two species groups or clades, the Korean-*Scopura* and the Japanese-*Scopura*, within the family (Fig. 11), which can be supported by the synapomorphic characters of the posteromedian projections on the female abdominal sternum 8 (Figs. 2H, 4H, 8I) and the elongated process at the base of male cerci (Figs. 1E, 1F, 5E, 5F, 6E, 6F, 7E, 7F), respectively. Except for genital characters, however, general morphology of the adults and nymphs are quite similar among the species of the family. A relatively recent glacial event in Northeast Asia, which could cause a major extinction of the family (Cox and Moore 1985), may explain the phenomena of a lower degree of species diversity and speciation, a north-south distribution pattern and the two geographic clades of the family presented above.

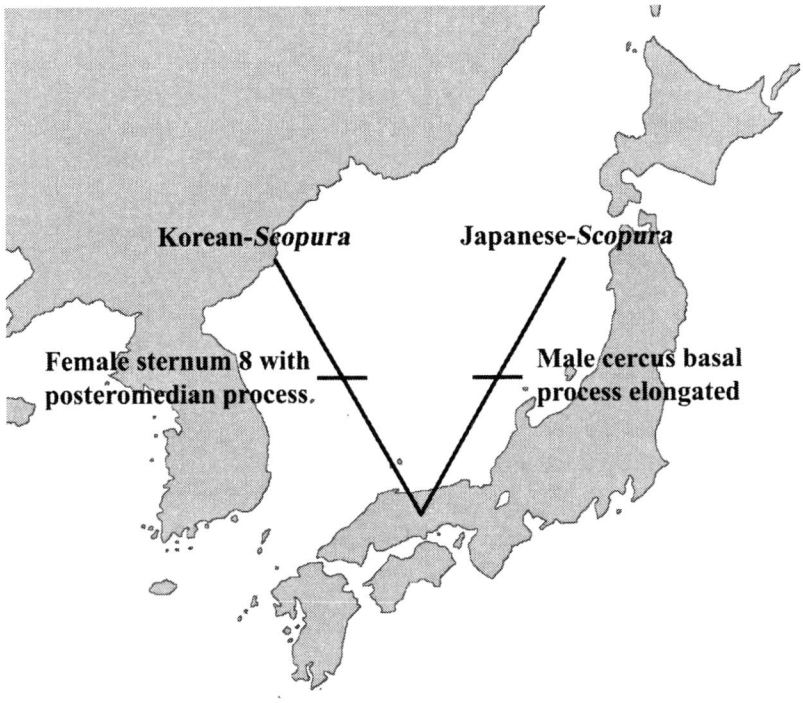

Figure 11. Phylogeny within the Scopuridae.

Acknowledgments

We are grateful to Dr. S. Uchida of Aichi Institute of Technology in Toyota, Japan, for providing useful type information and bibliographic data, Dr. M. Tomokuni of Natural Science Museum in Tokyo, Japan, for the loan of the holotype material of Japanese scopurids, Dr. K. Masunaga of Lake Biwa Museum in Kusatsu, Japan, for the loan of paratype and nontype material of Japanese scopurids and Dr. P. Zwick of Limnologische Fluss-Station des Max-Planck-Instituts fuer Limnologie, Schlitz, Germany, for useful comments and suggestions. This work was supported by the research project "Ecotophia 21" from the Ministry of Environment of Korea in 2005.

Literature Cited

Cox, C. B., and P. D. Moore. 1985. Biogeography. An Ecological and Evolutionary Approach, 4th edition. Blackwell Scientific Publications, Oxford, UK.

Ida, M. 1994. The life cycle of *Scopura montana* (Plecoptera: Scopuridae). Japanese Journal of Limnology **55**:23–25.

Jin, Y. H., and Y. J. Bae. 2005a. The wingless stonefly family Scopuridae (Plecoptera) in Korea. Aquatic Insects **27**(1):21–34.

Jin, Y. H., and Y. J. Bae. 2005b. A new wingless stonefly, *Scopura jiri* (Plecoptera: Scopuridae), from Korea. Entomological News **116**(2):89–92.

Kawai, T. 1974. The second species of the genus *Scopura* (Plecoptera, Scopuridae). Bulletin of Natural Science Museum, Tokyo **17**(4):275–281.

Kim, J. S., S. A. Ham, and Y. J. Bae. 1998. Checklist of South Korean Plecoptera (Insecta). Entomological Research Bulletin, Seoul **24**:43–48.

Theischinger, G.1996. Plecoptera. *in* CSIRO (ed.) The Insects of Australia. Melbourne University Press, Victoria.

Uchida, S., and H. Maruyama. 1987. What is *Scopura longa* Uéno, 1929 (Insecta, Plecoptera)? A revision of the genus. Zoological Science **4**:699–709.

Uéno, M. 1929. Studies on the stoneflies of Japan. Memoirs of the College of Science, Kyoto Imperial University, Series B **4**(2):97–155, pl. 24.

Uéno, M. 1935. Plecoptera. Pages 30–51 *in* M. Uéno, and D. Miyaji, editors. Aquatic Animals of the Azusa-Gawa River System and Kamikochi. Iwanami.

Uéno, M. 1938a. Scopuridae an aberrant family of the order Plecoptera. Insecta Matsumurana **2**:154–159.

Uéno, M. 1938b. Recorded localities of *Scopura longa* Uéno (Plecoptera – Scopuridae). Mushi **11**:201–203.

Yoon, I. B., and S. J. Aw. 1985. A taxonomic study on the stonefly (Plecoptera) nymphs of Korea (I) – Suborder Holognatha and Systellognatha. Entomological Research Bulletin, Seoul **11**:111–139.

Zenger, J. T., and R. W. Baumann. 2004. The holarctic winter stonefly genus *Isocapnia*, with an emphasis on the North American Fauna. Monographs of the Western North American Naturalist **2**:65–95.

Zwick, P. 2000. Phylogenetic system and zoogeography of the Plecoptera. Annual Review of Entomology **45**:709–746.